PRINCIPLES OF
FLUID MECHANICS

Andreas N. Alexandrou

Department of Mechanical Engineering

Worcester Polytechnic Institute

Prentice Hall
Upper Saddle River, New Jersey 07458

Library of Congress Cataloging-in-Publication Data

Alexandrou, Andreas
Principles of Fluid Mechanics
    Andreas Alexandrou — 1st ed.
       p. cm.
    Includes bibliographical references and index.
    ISBN 0-13-801762-X
    1. Fluid dynamics. 2. Mechanical Engineering. I. Title.
QA76.9.C643573 2000   00–7450
004.2′2–dc21   CIP

Acquistions Editor: Laura Curless
Vice-President of Editorial, ECS: Marcia Horton
Executive Managing Editor: Vince O'Brien
Managing Editor: David A. George
Marketing Manager: Holly Stark
Vice-President of Production and Manufacturing: David W. Riccardi
Art Director: Heather Scott

 © 2001 by Prentice-Hall, Inc.
Upper Saddle River, New Jersey 07458

The author and publisher of this book have used their best efforts in preparing this book. These efforts include the development, research, and testing of the theories and programs to determine their effectiveness. The author and publisher make no warranty of any kind, expressed or implied, with regard to these programs or the documentation contained in this book. The author and publisher shall not be liable in any event for incidental or consequential damages in connection with, or arising out of, the furnishing, performance, or use of these programs.

Printed in the United States of America

10 9 8 7 6 5 4 3 2 1

ISBN 0-13-801762-X

Prentice-Hall International (UK) Limited, *London*
Prentice-Hall of Australia Pty. Limited, *Sydney*
Prentice-Hall Canada Inc., *Toronto*
Prentice-Hall Hispanoamericana, S.A., *Mexico*
Prentice-Hall of India Private Limited, *New Delhi*
Prentice-Hall of Japan, Inc., *Tokyo*
Pearson Education Asia Pte. Ltd., *Singapore*
Editora Prentice-Hall do Brasil, Ltda., *Rio de Janeiro*

To
My parents, Electra and Neophytos Alexandrou
and
Lisa and Nadia

Preface

This textbook is an introduction to fluid dynamics. The first nine chapters form the basis for the first sophomore level course in fluid dynamics. In addition, Chapters 4, 5, 6, 8, 9, 10, 11, 12, and 13 form the basis for a second course in fluid dynamics. The main prerequisite for the book is a basic knowledge of calculus and physics.

Physical phenomena, and by extension fluid flow, are governed by the same basic laws. Therefore, particular topics can be deduced from the more general framework provided by the laws of nature. For this reason, whenever possible the material in this textbook is presented from a general, deductive viewpoint. This approach is also consistent with the needs of modern engineering analysis and design, and is achieved without sacrificing the quality or quantity of the discussion of "classical" fluid flow phenomena.

Traditionally, students consider fluid dynamics to be a difficult topic because of its mathematical nature and the apparent complexity of its concepts. Often, whether in class or textbooks, the material is presented as a collection of seemingly unrelated concepts, thus making it more difficult for students to fully comprehend the material. Here, the material is organized in a manner that avoids this confusion; similar themes are grouped and discussed together.

Some of the unique features of this book include (a) the point of view of the presentation, (b) the thematic organization of the material, and (c) the introduction of ideas from computational and experimental fluid dynamics. The material and concepts are demonstrated and reinforced through examples and problems for each section and chapter.

Since most concepts in fluid dynamics are quite mathematical, one of the appendixes reviews the basic mathematics required for the study of fluid dynamics. In my experience, this chapter prepares students for the more mathematically rigorous parts of the material.

The book is divided into three parts:

Theory

The material in this part is divided into the following categories: (a) introduction, (b) conservation laws (c) fluid kinematics, and (d) fluid dynamics (for finite and differential control volumes). Since the material is presented from a general point of view, in which the universality of the laws of nature is stressed, *Hydrostatics* is not discussed in a separate chapter but is presented as a special case of the momentum equation in Chapter 2.

Irrespective of the driving forces or dynamic conditions that induce flow, the motion and deformation of fluid particles are characterized by simple kinematic principles. Therefore, in Chapter 4, fluid flow is defined using purely kinematic arguments without reference to the dynamic effects.

Following the kinematics of fluid flow, Chapter 5 introduces and emphasizes the governing laws and constitutive relations. The concept of the constitutive behavior of fluids and therefore of the viscosity of fluids follows naturally from the discussion. Concepts such as laminar flow, turbulent flow, and non-Newtonian fluids are also presented within this framework. An entirely new section on boundary conditions completes the theoretical description of fluid systems.

Analysis

Following the theoretical and mathematical description of fluid systems, this part deals with the analysis of fluid problems. This step is also consistent with engineering design methodology. In Chapters 6, 7, and 8 and in Chapters 12 and 13, various solution procedures such as dimensionless analysis, analytic, experimental, and numerical solutions are introduced and applied to fundamental problems. Traditional concepts such as internal and external flows are presented in a separate chapter under a common theme dealing with the use of combined analytical and experimental methods in fluid dynamics.

Special Topics

Finally, special topics such as ideal, inviscid flow, compressible flow, and dynamics of rotating fluids are reserved for separate chapters. This avoids unnecessary confusion about the relation of these topics to the fundamental theory of fluid dynamics. Instructors can select, at will, the topics to cover in their course.

I would like to express my thanks to my colleagues Nikos Gatsonis, David Olinger, Hamid Johari, Jim Hermanson, and David Watt, who have helped me in the preparation of this manuscript by providing not only advice and encouragement, but also original problems and pictures from their work. Particular mention is reserved for Mark Richman, whose unfailing intuition and keen mathematical insight have helped me throughout the writing of the manuscript. I would also like to thank Nadeem Majaj and Professor Marios Soteriou for providing problems, and Lisa Majaj for her editing prior to submission of the manuscript.

ANA
WPI

Contents

Introduction and Basic Definitions

You cannot step twice in the same river.[1]
—Heraclitus

1.1 Introduction

The goal of this book is to introduce the reader to the basic principles that govern fluid flow phenomena and behavior. This field of study is commonly known as *fluid dynamics* or *fluid mechanics*. At the outset, therefore, it is necessary to define fluid flow, to identify what we mean by a fluid phenomenon, and to establish the importance of studying such phenomena.

In ancient times, Greek philosophers established the principle "$\pi\acute{\alpha}\nu\tau\alpha$ $\rho\epsilon\iota$," or "everything flows." Indeed, one can easily argue that everything is in relative motion and that $\sigma\tau\acute{\alpha}\sigma\iota\varsigma$ (immobility) is a special case of motion. However, in more practical terms, a *fluid* is defined as a material that cannot resist externally applied lateral (shear) forces, but instead *deforms continuously* under the influence of such forces. This property can be demonstrated easily by considering two hands immersed in water (or in air) held parallel with palms facing each other, moving in opposite directions (Figure 1.1(a)). As the hands move, the water (or air) between the hands deforms continuously without stopping. *Fluid flow* is defined as the effect of the fluid's continuous deformation under the action of externally applied forces.

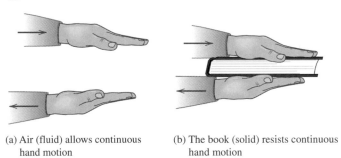

(a) Air (fluid) allows continuous hand motion

(b) The book (solid) resists continuous hand motion

FIGURE 1.1 (a) Fluids cannot resist externally applied shear stresses. (b) Solids can resist externally applied shear forces.

Another property that defines a fluid is the fact that fluids do not have a predetermined shape but, rather, assume the shape of their container. For this reason, a liquid can be easily transferred from one container to another. These fluid properties are in contrast to the behavior of "solids" that have predefined shapes (i.e., they cannot be easily "poured" into shapes other than their original shape) and that deform by a finite amount (not continuously)

---

[1] *The Concise Columbia Encyclopedia*, (J. Levey and A. Greenhall, eds.), 1983.

under the influence of externally applied forces. For instance, by placing a solid object such as a book between two hands and by repeating the same experiment mentioned previously, the hands will initially move but will eventually stop when the book reaches a final deformed state (Figure 1.1(b)).

These characteristic fluid properties are a reflection of the intrinsic microstructure of fluids, in which molecules, although densely packed, are held together by relatively weak intermolecular forces. These weak intermolecular forces then account for the fact that fluids are structurally more "flexible" than solids.

Because fluids deform easily and assume the shape of their container, during fluid flow one expects numerous special phenomena to develop. Indeed, fluid particles under the action of externally applied forces can flow through channels and around objects of different dimensions and shapes. In such flows, fluid particles can either follow the contour of an object (i.e., by remaining attached to the surface), or *separate* by moving away from it. Such separated flows yield zones of *circulation* where the particles move in a closed contour. Under certain conditions the flow can be either *laminar* or *turbulent*, depending on whether the fluid particles move in an orderly or in a random, chaotic manner. In addition, objects in contact with fluid experience forces that depend upon local flow conditions (such as pressure, fluid stresses, etc.) along the common interface (Figure 1.2).

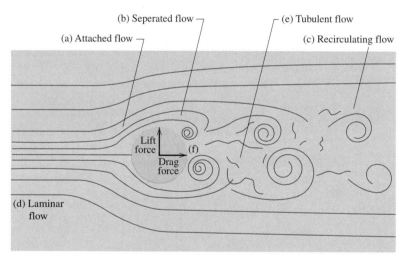

FIGURE 1.2 Sample fluid phenomena: (a) attached flow; (b) flow separation; (c) flow circulation; (d) laminar flow; (e) turbulent flow; (f) flow-induced forces.

The above are just a handful of fluid-related phenomena. However, the apparent simplicity of these phenomena does not reveal the degree of their complexity and their extreme importance for most aspects of life. Currently, there is hardly any technology that is not directly or indirectly related to fluid flow. For instance, consider the operation of a car: fuel flow in the engine, the lubrication of moving parts, the ventilation of the passenger cabin, the cooling of the engine block, the resistance force on the car due to airflow past it—all these demonstrate applications of fluid flow. Other applications include airplane flight, power generation, cooling of electronic devices, flood-control dams, and the operation of measuring devices. Many materials and products are manufactured and processed while in molten (fluid) condition. Knowledge of how fluids flow and the relation to the quality of the final products has been partially responsible for the remarkable advancements in the production of novel materials and of products with superior material or functional qualities.

In nature, many phenomena can be explained or predicted using concepts from fluid dynamics, including ocean waves, weather patterns and river flow. Furthermore, life itself is sustained or created by phenomena directly related to fluid flows and their properties. Examples include blood flow, with the movement of nutrients and biological products to and from vital organs, and breathing and other phenomena occurring in the lungs. Irregularities in the flow of biological fluids can be the cause of many health problems e.g., cardiac problems caused by the restriction of blood in arterial stenoses.

Many fluid flow phenomena in practical applications cannot be studied separately from other classical fields of study, such as solid mechanics, heat transfer, and thermodynamics. Because of the multidisciplinary nature of engineering problems, their solution requires also synergy from other fields of specialization typically considered unrelated to engineering. For example, engineering solutions to environmental problems require collaboration from all traditional fields of engineering as well as from chemists, biologists, economists, and sociologists. However, most fields of science are governed by a common set of conservation laws. Therefore, these laws provide a "natural" framework with which to develop fundamental concepts of fluid flow. For these reasons, and whenever possible, the material in this textbook is presented from a general viewpoint by stressing the conservation laws.

1.2 Fluid Dynamics and Engineering Analysis and Design

Engineering has been generally considered an "applied" discipline where the primary objective has been to use or develop theories and technologies to meet societal needs.

As rapid developments in engineering technologies in this century demonstrate, engineering analysis and design have been quite successful in meeting society's technological needs. Fluid dynamics is an integral part of this technological base and is central to its further development. Figure 1.3 shows a typical procedure used for engineering analysis and design.

As shown in the figure, one usually starts with an initial need or design (a). This may be to control environmental pollution, to design an optimum shape of an airfoil, or to study blood flow through an arterial stenosis. The procedure then calls for a formal, preferably mathematical, description of the physics of the problem (b). This description may take the form of a simple algebraic relation or of a more complicated system of

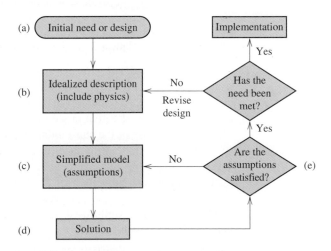

FIGURE 1.3 Typical procedure followed in engineering analysis and design.

differential equations or other mathematical formalisms. This mathematical description is commonly known as a *mathematical model* or simply a *model*. Ideally, one would ultimately like to obtain an exact solution to the posed problem. Unfortunately (as we will indicate in many fluid problems) this can rarely be accomplished. Therefore, we are forced to simplify the problem by making certain assumptions (c). This is an important step since our ability to simplify the problem, while maintaining the credibility of the description, may determine whether an accurate solution can be obtained. Ultimately, the ability to invoke reasonable assumptions may be the single most important attribute of a successful practicing engineer. The solution of the simplified model (d) and the evaluation of the results are the next steps in the procedure. The validity of the assumptions can often be checked by using the solution itself. For instance, if an effect is assumed to be small and, hence, removed from the simplified model, the solution can be used to check this assumption (e). The entire analysis may be incorporated in an "optimization" loop where an "optimum" solution is obtained by comparing the results to a set of objective criteria.

1.2.1   Solution Methods

The solution to the simplified model may be obtained by one, or a combination of the following methods:

1. *Analytic solutions using mathematical methods*. These solutions are usually the most accurate and, hence, the most attractive alternative. Analytic solutions are easily manipulated to yield optimum results, which, in general, can be achieved quickly and inexpensively. Unfortunately, however, very few problems of practical importance can be handled analytically.

2. *Experimental solutions*. Often, fluid problems cannot be formulated mathematically because of lack of knowledge of the exact behavior of fluid systems. For example, in turbulent flow, where the fluid particles move in a chaotic fashion, current knowledge is not complete enough to allow the generalization of this behavior using mathematical tools. An alternative approach is to reproduce and study the problem under controlled conditions in the laboratory. Experimentation is so pivotal to the study of fluid flow that experimentation and fluid dynamics have become closely interlinked. However, experimental solutions tend to be expensive and time-consuming. As a result of the size of real physical systems in relation to available experimental facilities, the design of experiments and the interpretation of the results are often very complicated. Furthermore, optimization of the fluid systems can also be problematic.

3. *Discrete computational solutions*. Many analytically intractable problems are solved using discrete solutions. These solutions are obtained by approximating the mathematical problem and then solving it using the computer. This relatively recent approach has become extremely effective in solving problems of practical importance. This is particularly true because of the substantial theoretical developments of discretization theories (finite elements, finite difference, finite volume, boundary elements, spectral methods), and the remarkable improvement in the computing power of computers. Figure 1.4 shows a numerical simulation of the flow past an entire spacecraft—a clear demonstration of the power and importance of numerical computations. However, discrete solutions presume a well-posed mathematical model which, as we will show later, may not always be possible because of complex fluid behavior.

4. *Solutions based on heuristic principles*. This is the least reliable method of analysis since it is based on phenomenological observations of physical systems. Obviously,

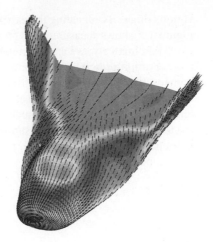

FIGURE 1.4 Numerical simulation of the flow past a spacecraft. (Courtesy of Amtec Engineering Co.)

this knowledge is personal and cannot be easily documented or transferred to other individuals. However, experience is important during the formulation of the problem as well as in the development of the model when simplifying assumptions are required.

Since there is not a single method that can solve all problems, fluid systems are often studied using any combination of the above solution methods.

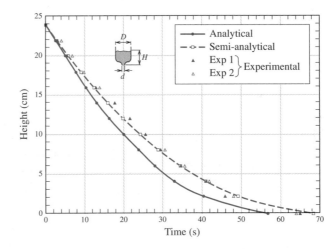

FIGURE 1.5 Water level in a cylindrical container with diameter $D = 15.24$ cm as a function of time while water is being discharged through a small circular hole with diameter $d = 0.9525$ cm at the bottom of a cylindrical container, using an experimental, an analytic, and a semi-analytic solution strategy. (Contributed by David J. Olinger.)

To demonstrate the need for a combined solution strategy, consider the simple flow problem shown in Figure 1.5. The objective in this problem is to determine how fast a cylindrical container with diameter D and filled with water drains through a circular hole with diameter d at the bottom of the cylinder.

The solution can be obtained experimentally by using a stopwatch to record either the volume of water in the cylinder or the volume of water collected in a catch bucket, at various times. By repeating the experiment, the measurements can be made quite consistent. Figure 1.5 shows measurements from two runs of this simple experiment.

Alternatively, we can formulate the problem mathematically using the laws of nature (conservation principles). These are general principles that describe the way natural phenomena behave. As we will discuss extensively in later chapters, these principles deal with the conservation of physical quantities such as mass, momentum, and energy. Unfortunately, an exact solution to such a "complete" mathematical model is not possible. The model can be simplified by assuming that the flow in the container and through the hole is frictionless and, consequently, that energy losses are negligible. Under these conditions, the draining time t_d is given by

$$t_d = \sqrt{\frac{2H}{g}\left(\frac{D^4}{d^4} - 1\right)},$$

where H is the original water level, and g the acceleration due to gravity. However, the solution to this simplified model, as shown in Figure 1.5, does not predict the measured flow rate with sufficient accuracy. This shows that the assumptions used were an oversimplification of the actual physics of the problem. This is a clear illustration that the accuracy of the solution depends on the ability of the analyst to invoke proper assumptions.

The simple analytic solution can be made more realistic by including the effects of friction and energy loss on the mass flow rate $\dot{m}$ kg/s by using an experimentally determined coefficient, $C = 0.82$, that corrects for these effects. C accounts for the difference between the ideal $\dot{m}_i$ and the actual flow rate $\dot{m}_a$ as $C = \dot{m}_i/\dot{m}_a$. Coefficients such as C are determined experimentally for a number of common flow geometries and are available in the literature (fluid mechanics handbooks, journal papers). In this case, the draining time is given as

$$t_d = \frac{1}{C}\sqrt{\frac{2H}{g}\left(\frac{D^4}{d^4} - 1\right)}.$$

As shown in Figure 1.5, the modified theory reproduces the experimental results quite accurately.

The use of experimental coefficients to represent the bulk behavior of complicated flows is commonplace in fluid dynamics. In this simple experiment, by "lumping" all unknown losses into an *a priori* determined experimental "loss coefficient," the theoretical prediction is identical to the experimental result.

Note, however, that in many problems, lumped parameters are not sufficient and more detailed analysis is required. In such cases we must return to "complete" mathematical models and seek solutions using more sophisticated methods of analysis such as numerical methods.

1.3 Control Volume Analysis: Open System vs. Closed System

In order to study a phenomenon, we need to unambiguously define the object of our study; in other words, we must identify the physical system being studied. To achieve this we must isolate the physical system from its surroundings and then develop the mathematical model that accurately describes it. The isolated view of the physical system is called a *control volume*, very much like the control groups used other fields of science e.g., medicine and psychology. This approach is also similar to the use of free-body diagrams in solid mechanics.

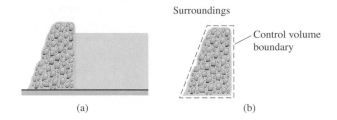

FIGURE 1.6   Hydraulic dam: (a) simplified physical view; (b) isolated view using a control volume.

For example, to determine the size and the required strength of a hydraulic dam, it is convenient to isolate the structure from the surrounding water. The schematic in Figure 1.6(a) shows a simplified view of a hydraulic dam. Figure 1.6(b) shows the isolated view of the structure, which is highlighted by defining its *boundary* with fictitious lines. Anything outside the defining boundary is defined as the *surroundings*. The same procedure is followed when the objective is to study the elastic response of a soccer ball when it is kicked, as shown in Figure 1.7(a); a fictitious boundary isolates the ball from the shoe of the player kicking the ball, as shown in Figure 1.7(b).

Similarly, if the focus of study is the dynamics of the human heart, we must isolate the heart from the surrounding tissue and muscle as well as from all the arteries and veins that emanate from or arrive at the heart, as shown in Figure 1.8(a). The fictitious line again identifies the boundary of the heart (Figure 1.8(b)).

Each control volume, then, is characterized by a boundary that at first glance provides a visual representation of the object of study. The boundary clearly identifies the *volume*, as well as the amount of *mass* enclosed within this boundary. The volume is a measure of the entire physical space the system occupies and is measured in units of $meters^3$. Mass is the measure of the total amount of substance enclosed by the boundary and is measured in units of *kilograms*. The amount of mass m of a certain substance enclosed in a volume $\mathcal{V}$, defined by the ratio $\dfrac{m}{\mathcal{V}}$, is known as the density ρ with units of kg/m^3, which is a *property* (to be defined formally later) of the particular substance.

The defining boundary may but does not necessarily *deform and may be a function of time*. This is obviously the case with the boundaries of the heart and of the soccer ball, both of which deform during the development of the phenomena under study. When the volume does not change, the control volume is defined as a *fixed volume*.

A closer look at the above examples reveals another important distinction: no mass crosses the boundary of the dam and the ball, but mass crosses the boundary of the heart. Consequently, when no mass crosses the boundary, the control volume is known

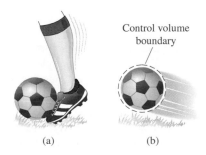

FIGURE 1.7   Soccer ball: (a) simplified physical view; (b) isolated view using a control volume.

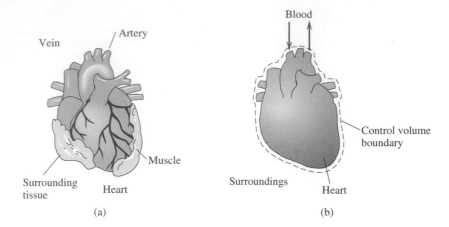

FIGURE 1.8   Human heart: (a) simplified physical view; (b) isolated view using a control volume.

as a *closed system* or *system of fixed mass*. In contrast, if mass crosses the boundary, the control volume is known as an *open system*. In this text, we use the term closed system to denote a system of fixed mass and the term open system to denote a control volume in which mass crosses the boundary. When the discussion applies to both, we use the term control volume.

1.4   The Continuum Assumption, Differential Analysis

When the size of the control volume is finite i.e., its dimensions are similar to the physical size of the problem under consideration, it is defined as a *finite control volume*. In most applications, finite control volume descriptions ignore the fact that properties can vary not only with time but also with location *within the material*. As a consequence, finite control volume analyses provide information only about the average (or bulk) behavior of the physical system. Hence, phenomena with characteristic dimensions much smaller than the size of the volume cannot be studied. For example, a finite control volume analysis of flow through a pipe with a varying cross-sectional area (where the entire pipe is selected as the control volume) yields results on bulk quantities such as the volumetric flow rate and the overall pressure drop across the pipe. The analysis cannot predict possible flow reversal and formation of recirculation zones. In practice, however, entire designs or operations depend on such detailed information. The *continuum assumption* and the use of *differential calculus* described subsequently make possible such detailed study of fluid flows.

Before proceeding any further, we need to address some key questions: (a) Does the size of the control volume impose any restrictions on the fidelity of the analysis? (b) What is the theoretical basis for the continuum assumption?

The following example from sports can help set the proper framework within which to view these issues. Consider a densely packed stadium during a football game. The game is usually broadcast for television audience using cameras mounted on a blimp flying high above the stadium. From the normal flying elevation few details can be distinguished with the naked eye and, therefore, very little can be said about the game or the spectators. Indeed, from that level the spectators appear as a large continuous colorful mass in constant motion, and the field as a green area with dark spots representing the players.

With the help of the camera, and by incrementally zooming in, more details of the game can be observed and broadcast. At some intermediate zoom level, while the spectators appear still as a continuous mass in bulk motion, the "wave" performed occasionally by the spectators appears very "real," closely resembling actual ocean waves. At this level, the details of the game are still not clear, and zooming closer may be necessary in order to observe the movement of the players and, thus, the development of the game. Of course, at the appropriate zoom level, the game can be covered with sufficient accuracy. However, by zooming in to the extent that only individual players are shown on the TV screen, a single camera is not sufficient to cover the entire game. At this level of detail, additional cameras (and hence TV sets) are required to fully broadcast and view the game. Of course, this is not practical and, therefore, it is never actually used.

If we consider the football game as analogous to fluid flow (i.e., consider each player to be a fluid molecule), the above illustrates that the same flow (the football game) appears to be different depending on the size of the control volume (the zoom level). A very detailed analysis of the flow (following a single player), which is equivalent to following individual fluid particles becomes impractical to use. A more formal discussion of the same concepts is presented below, using a simple classical thought experiment.

Suppose we have an experimental probe that can sample air and measure the mass m of the collected air. Selecting the probe as a control volume with a given volume $\mathcal{V}$, the density of the air is calculated as $\rho = \dfrac{m}{\mathcal{V}}$. If the volume of the collecting probe is reduced and the experiment is repeated for the same conditions (pressure, temperature), the measured density is expected, within some experimental error, to be the same as the one calculated with a larger probe. The experimental error may be due to inaccuracies in measuring the volume of the probe or the mass of the air.

The measured density from this thought experiment can be plotted against the volume of the probe as shown in Figure 1.9. As the size of the probe is reduced further, it will inevitably become comparable to the average distance air molecules will travel before they collide i.e., the mean free path of the molecules. Because of the small size of the probe, the number of molecules collected will be small and far less than those collected by a bigger probe. Furthermore, because of the molecules' random motion, their number will vary significantly between experimental tries; the total number collected depends on the number of molecules that happen to be at the vicinity of the probe during the experiment. The calculated density under these conditions will not be a constant number, but will vary erratically, as shown in Figure 1.9.

Obviously, the measured air density depends on the size of the control volume only when the dimension of the volume is of the same order of magnitude as the mean free path of the molecules. The same conclusion then is valid for all physical systems and

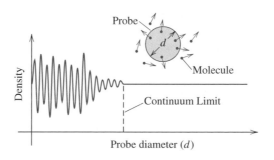

FIGURE 1.9 Density as a function of the diameter of the probe.

most measured physical quantities. At this point, a natural question is obvious: Does this limitation affect the way we study engineering? The answer is no, because most measured quantities such as density, temperature, or velocity are based on a locally averaged molecular behavior and not on that of individual molecules. Moreover, measuring instruments are not capable of following individual molecules; instead, they measure the average behavior or state of the molecules in the neighborhood of the sensing parts of the instruments.

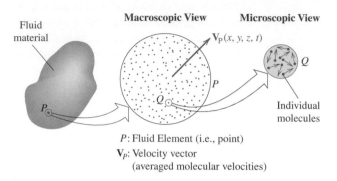

FIGURE 1.10   Different perspectives for flow analysis. The microscopic view is based on individual molecular behavior, and the macroscopic on locally averaged molecular behavior.

The above discussion points to two possible alternative methods of analysis, which are discussed on the following paragraphs.

(a) The *microscopic* approach, in which the material is considered as a collection of individual molecules. The overall phenomenon is the net effect of these molecular motions and inter-molecular collisions and dynamics (Figure 1.10). To analyze the physical phenomena using the microscopic approach, one must identify and follow each molecule by tracing its position, velocity, etc. Because typical systems (even the simplest ones) are made up of billions of molecules, this task is impossible to achieve. However, this approach is appropriate in applications where the mean free path of the molecules is large compared to the size of the system. Such cases occur in rarefied gas dynamics.

(b) The *macroscopic* approach, in which the analysis is based on a local average of the behavior of the molecules. The average is obtained by assuming that properties are continuous throughout the extent of the matter. This approach can be derived directly from the microscopic picture described earlier, where individual molecules move in random fashion and collide with each other; each molecule can be characterized by its instantaneous location through the position vector $\mathbf{r}$, and velocity $\mathbf{u} = \dfrac{d\mathbf{r}}{dt}$. The motion of the molecules then can be expressed as the combination of an average (or drift) velocity and the instantaneous velocity deviation from this mean velocity. Locally, then, at a point P, we can define an average velocity $\mathbf{V}$ as the average among the molecules' average velocity (Figure 1.10). To compensate for the assumption that the medium is continuous, each mathematical point is also assigned a density, which, upon integration, accounts for the total mass of the material m_T

$$\int \rho \, d\mathcal{V} = m_T.$$

According to the macroscopic approach, we assume that at a mathematical point with zero dimensions, not only the velocity but other quantities such as the temperature and pressure are in reality the local average of the behavior of the molecules around this point. This assumption is known as the *continuum assumption*. The continuum assumption allows the use of differential calculus and other related mathematical tools in the analysis of distributed physical systems. This approach is known as the *differential approach*. In this introductory text, fundamental fluid dynamics concepts are presented using both the finite control volume and differential approaches.

1.5 Fluid Properties

The study of fluid dynamics is closely linked to the physical properties of fluids such as density, viscosity, and pressure. As an introduction to some of the issues in the mechanics of fluids, a preliminary discussion of a few such properties follows.

Density

The density of a fluid ρ in units of kg/m^3 is defined as the mass per unit volume of the fluid

$$\rho = \frac{dm}{d\mathcal{V}},$$

where m is the mass and $\mathcal{V}$ the volume. In general, ρ can also be a function of the local pressure and temperature. While in gases the dependence on pressure and temperature is significant, in liquids the density is a weak function of pressure but depends on temperature. For instance, the density of water at $20°C$ is about $998.2\ kg/m^3$ and at $80°C$ is $971.8\ kg/m^3$. Therefore, for constant temperature flows (i.e., isothermal flows), the density for most liquids is constant, in which case they are defined as *incompressible*. For incompressible fluids, a fixed mass of fluid occupies a fixed volume. The term *gas* usually signifies a fluid that is *compressible*. Appendix A shows the variation of the density of air as a function of the elevation above sea level. Unless otherwise stated, liquids in this textbook will be treated as incompressible and isothermal (constant temperature), with properties taken at $20°C$.

Pressure

Pressure is defined as the local normal force F_n, per unit area A, with units of $kg\ m/sec^2 = N/m^2$, otherwise known as Pascal (Pa). This pressure is the result of molecular collisions with solid surfaces. Because the numerical value of 1 Pascal is small, pressure is usually expressed in kiloPascal, kPa. In formal mathematical form, the pressure force is expressed as

$$P = \frac{dF_n}{dA},$$

where F_n is the force normal to surface A. Often we need to calculate the total force **F** on a surface due to the local pressure $P(x, y, z)$ — that is,

$$\mathbf{F} = \int_A d\mathbf{F},$$

where

$$d\mathbf{F} = -\mathbf{n}\,P(x, y, z)\,dA.$$

n is the *outward pointing unit normal vector* i.e., it is a unit vector normal to the surface and it is directed *away* from the surface as shown in Figure 1.11. The normal vector **n** is included to make sure that the resulting force acts normal to the surface. The (−) sign is included to ensure that the force acts in a compressive direction. The total force is determined by integration over the entire surface as

$$\mathbf{F} = \int_A d\mathbf{F} = \int_A -\mathbf{n}P(x, y, z)\,dA.$$

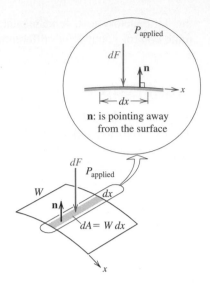

FIGURE 1.11   Force on a surface due to an applied pressure.

EXAMPLE 1.1

Problem Statement   Consider a flat surface with 2.0 m length and 0.5 m width, as shown in Figure 1.12. If the ambient air pressure P_a is 150 kPa, find the direction and magnitude of the total pressure force acting on the surface.

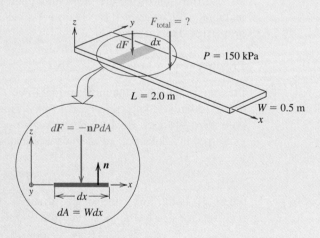

FIGURE 1.12   Schematic for Example 1.1.

Governing Equations   The force due to the ambient pressure on an infinitesimal area dA as shown in Figure 1.12 is

$$d\mathbf{F} = -\mathbf{n}P_a \, dA.$$

Assumptions   Assume that the pressure along the surface is constant.

SOLUTION   Consistent with the coordinate system shown in the figure, the *outward pointing normal vector* is $\mathbf{k}$ and $dA = W \, dx$, where W is the width of the surface.

Therefore,

$$d\mathbf{F} = -\mathbf{k}\, P_a\, W\, dx.$$

The total force $\mathbf{F}$ is obtained by integration as

$$\mathbf{F} = \int dF = -\mathbf{k} \int_0^L P_a\, W\, dx = -P_a W L\, \mathbf{k}.$$

As expected then, the force is pushing against the surface i.e., in the $-\mathbf{k}$ direction. The magnitude of the force is

$$|\mathbf{F}| = \underbrace{P_a}_{Pressure}\ \underbrace{W L}_{area} = 150\ kPa \times 0.5\ m \times 2.0\ m = 150\ kN.$$

In this case, since the pressure is constant, the same result could be obtained easily by using instead

$$P_a = \frac{F}{A} \Rightarrow F = P_a A$$

and by noting that the force is in the compressive direction.

Of the pressure forces, the most widely known is the one due to the *atmospheric* pressure i.e., the force due to the surrounding air pressure. The local atmospheric pressure is variable and is a function of the local elevation and weather conditions. However, in practical applications and for design purposes, a numerical value of 101325 Pa is considered as a *standard atmosphere*, P_{atm}. Appendix A shows the variation of the atmospheric pressure as a function of the elevation above sea level. Unless otherwise stated, the local atmospheric pressure will be assumed to be equal to the standard atmospheric pressure.

Normally, a given pressure P is expressed relative to the local atmospheric pressure P_{a_l}, as shown in Figure 1.13, using the difference between the actual pressure and the local atmospheric pressure, $P - P_{a_l}$. If the difference is a positive number, this difference is called *gage pressure*. Therefore,

$$P_{gage} = P - P_{a_l}, \quad \text{if} \quad P_{gage} > 0.$$

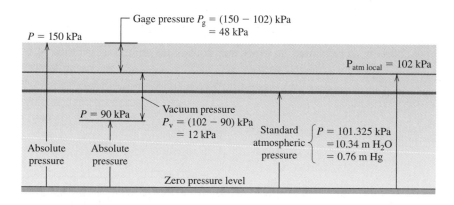

FIGURE 1.13 Various ways to represent pressure.

If the difference is a negative number, the absolute value of the difference is called *vacuum pressure*. Hence,

$$P_{vacuum} = |P - P_{a_l}|, \quad \text{if} \quad P_{gage} < 0.$$

Often the pressure is expressed in terms of an equivalent height h of a liquid column with density ρ, and cross-sectional area A. By equating the pressure force to the weight of the column

$$PA = \rho Ahg \Rightarrow P = \rho gh,$$

the equivalent height h is then

$$h = \frac{P}{\rho g}.$$

For instance, one standard atmosphere is equivalent to a height of 10.34 m of a water column. It is also equivalent to 0.76 m of a mercury column.

In hydraulic applications, it is customary to define the weight per unit volume $\gamma = \rho g$ as a new quantity known as the *specific weight* of the fluid. The ratio γ/γ_w at a given temperature where γ_w is the specific weight of water is known as the *specific gravity*, *SG*.

Viscosity

Another important fluid property is *viscosity*. At this point, viscosity is defined simply as a property related to *internal fluid friction*. It is formally defined and discussed more extensively in Chapter 5.

The presence of friction can be demonstrated by the simple experiment shown in Figure 1.14, where the gap between two long plates is filled with fluid. In order to move one plate with velocity U relative to the other as shown in the figure, a finite force F in the direction of motion is required. It can be shown experimentally that this force is proportional to the velocity gradient du/dy at the surface of the moving plate, i.e.,

$$\frac{F}{A} \propto \left.\frac{du}{dy}\right|_{y=H} \Rightarrow \frac{F}{A} = \mu \left.\frac{du}{dy}\right|_{y=H}$$

where A is the surface of the upper plate in contact with the liquid. As we will discuss in Chapter 5 the term F/A in this unidirectional flow (e.g., flow in one direction) is the viscous shear stress denoted by τ.

The proportionality constant μ is the *absolute viscosity*, otherwise known as the *dynamic viscosity*. This is a characteristic property of the fluid. By definition, viscosity is measured in units of $Kg/m\ sec$. The ratio $\nu = \dfrac{\mu}{\rho}$ in units of m^2/s is known as the *kinematic viscosity*.

As is demonstrated in Chapter 4, for this flow the term $\dfrac{du}{dy}$ is a measure of the rate at which the fluid is deforming as a result of a lateral (shear) force. Therefore, the

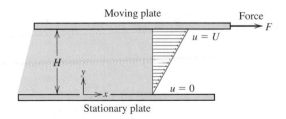

FIGURE 1.14   Simple experiment showing the effects of fluid friction.

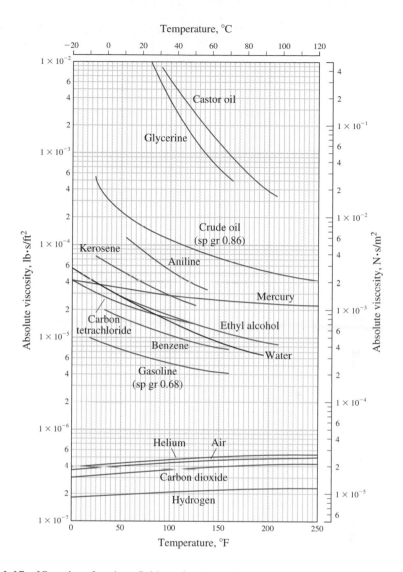

FIGURE 1.15 Viscosity of various fluids as function of temperature. (Taken from Steeter et al., *Fluid Mechanics*, 9th Ed., McGraw-Hill, by permission.)

viscosity is a fluid property that determines the force required to establish a given rate of deformation.

For most fluids, viscosity is a function of temperature; in general, the viscosity for liquids is decreasing with increasing temperature (Figure 1.15). Moreover, the viscosity of many fluids can be a function of the rate deformation of the fluid (velocity gradient). Fluids with μ that *is not a function* of deformation, (e.g., water or air), are known as *Newtonian fluids*, and fluids whose viscosity depends on the velocity gradient are known as *Non-Newtonian fluids* (Figure 1.16). Non-Newtonian fluids are considered *shear thinning*, or *pseudoplastic*, when their viscosity is decreasing with increasing velocity gradient. Molten plastics and paint exhibit such behavior. For instance, the faster a paint brush is moved, the easier the paint flows, because its viscosity decreases with increasing shear. Other fluids show *shear thickening* (otherwise known as *dilatant* behavior) when the viscosity increases with increasing velocity gradient. This behavior is usually observed in mixtures of liquid and solid particles.

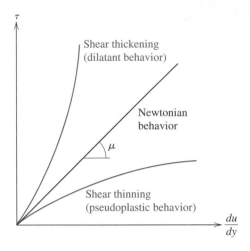

FIGURE 1.16   Different types of fluid behavior.

EXAMPLE 1.2

Problem Statement   Consider the same arrangement as in Figure 1.14, where the gap between two plates is filled with water at $25°\,C$. The gap height H is $0.2\;m$. The upper plate with dimensions of $2\;m$ by $3\;m$ is moving to the right with velocity $U = 5.0\;m/s$. Find the total friction force acting on the moving plate.

Governing Equations   Since water is a Newtonian fluid, we have

$$\frac{F}{A} = \mu \frac{\partial u}{\partial y}$$

where, according to Table A in Appendix A, the viscosity of water at $25°\,C$ is $\mu = 0.894 \times 10^{-3}\;N \cdot s/m^2$. A is the area of the plate in contact with the water—that is, $A = 2.0\;m \times 3.0\;m$

Assumptions   Assume that the velocity distribution across the gap height is linear.

SOLUTION   Since the velocity across the height of the gap is linear, $u = ay + b$ where a and b are appropriate constants and y is the distance from the lower (stationary) plate. According to the boundary conditions, at $y = 0$, $u = 0$ and at $y = H$, $u = U$. Therefore, the constants are $a = \dfrac{U}{H}$ and $b = 0$. Consequently, the velocity distribution is given by $u = \dfrac{U}{H}y$. Therefore, $\dfrac{\partial u}{\partial y} = \dfrac{U}{H}$

The total friction force, acting along the surface of the plate is then

$$F = \mu A \frac{U}{H}.$$

This general expression shows that the force is proportional to the viscosity (i.e., the higher the viscosity, the higher is the developed friction force). The force is also inversely proportional to the gap height (i.e., the force increases with decreasing gap height.)

Substituting the numerical values, the magnitude of the force is

$$F = 0.894 \times 10^{-3} \ N \cdot s/m^2 (2.0 \ m \times 3.0 \ m) \frac{5.0 \ m/s}{0.2 \ m} = 0.134 \ N.$$

The effect of this force is to slow down the motion of the plate. By reaction, then, the same force is responsible for the induced fluid motion.

Surface Tension

Surface tension is the surface force that develops at the interface between two immiscible liquids or between liquid and gas or at the interface between a liquid and a solid surface. Because of surface tension, small water droplets, gas bubbles, drops of mercury, and the like tend to maintain spherical shapes (Figure 1.17).

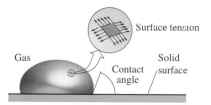

FIGURE 1.17   Schematic of surface tension physics.

The presence of surface tension and its dymanics are due to complex interactions at the molecular level along interfaces. Away from interfaces, molecules are surrounded by like molecules on all sides. Therefore, intermolecular force interactions result in a zero net force. However, molecules at interfaces interact with molecules of the same fluid on only one side. Consequently, such molecules experience a net force that puts the interface under tension. The ultimate magnitude and direction of this tension force is determined not only by what happens on either side of the interface, but by the way molecules of the two fluids interact with each other. Therefore, surface tension is a property specific to the "participating" fluids. For instance, surface tension between water and air is different from that between water and another gas. Surface tension forces are also sensitive to the physical and chemical condition of the solid surface in contact, such as its roughness, cleanliness, or temperature.

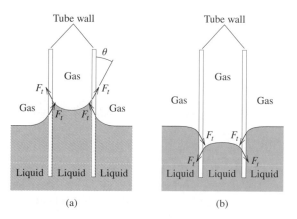

FIGURE 1.18   Schematic of surface tension physics: (a) wetting fluid; (b) nonwetting fluid.

A phenomenon related to surface tension is the *contact angle*, the angle the interface makes with solid surfaces, as shown in Figure 1.17. Under static conditions, the contact angle is also a property of the liquid-liquid or liquid-gas system.

Because of surface tension some liquids can adhere to solid surfaces. For example, liquid in a small tube rises as shown in Figure 1.18(a). In such fluid-surface systems, the fluid is said to *wet* the surface. Other fluids can be *nonwetting*; for instance, because of cohesion, surface tension can depress the liquid meniscus in a small tube, as shown in Figure 1.18(b). Mercury, for instance, exhibits nonwetting behavior as ball-like mercury droplets form on solid surfaces.

A convenient way to visualize surface tension is to consider a force F_t needed to stretch an interface made up of molecules (that have the tendency to attract each other) into a very thin layer with thickness comparable to the size of the molecules (Figure 1.19). By definition, F_t lies in the plane of the interface and acts in a direction tangent to it (Figure 1.19). The magnitude of F_t is modeled using the concept of *surface energy* per unit area σ measured in units of N/m, where σ represents the amount of energy per unit interface area needed to overcome the molecular attraction and to keep the film stretched. In terms of σ, then we have

$$\sigma = \frac{d\mathcal{E}}{dA},$$

where $\mathcal{E}$ is the energy and A the area of the interface.

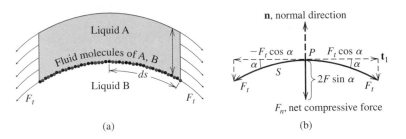

FIGURE 1.19   (a) Determination of surface tension; (b) the net effect of surface tension on a curved surface.

Consider now a part of the interface stretched by a force F_t a length dx as shown in Figure 1.19(a). By definition, $d\mathcal{E} = F_t\,ds$ and $dA = L\,ds$, where L is the width of the interface and corresponds to the length along which tension is acting. Therefore,

$$\sigma = \frac{d\mathcal{E}}{dA} = \frac{F_t\,ds}{L\,ds} = \frac{F_t}{L}.$$

In vector form, then, the force due to surface tension **F** is

$$\mathbf{F} = \sigma L \mathbf{t},$$

where **t** is a unit vector along the tangent direction to the interface. If the interface forms a circular ring (for instance, half of a spherical droplet) with radius r, surface tension acts along the perimeter of the ring, and $L = 2\pi r$. Therefore,

$$\mathbf{F} = (\sigma 2\pi r)\mathbf{t}.$$

However, as shown in Figure 1.19(b), because of the symmetry along a curved interface S, the tangential forces on either side of any point P along S cancel each other. Consequently, the net tension force is in the normal direction **n**. Therefore, the net effect of surface tension along a curved fluid-fluid surface is to subject a *concave surface* to a normal *compressive* *force*, and a *convex surface* to a normal *tensile force*. By using elementary calculus, it can be shown that the surface tension force can be expressed also as

$$\mathbf{F} = -\int_{A_s} \sigma \kappa \mathbf{n} \, dA_s,$$

where σ is the surface energy per unit area; κ is the local principal curvature, considered positive for a concave surface; A_s is the area; and $\mathbf{n}$ is the usual outward-pointing unit normal vector. If the interface is part of a circular arc, then $\kappa = 1/r$, where r is the radius of the arc. In this case, since κ is constant, the force due to surface tension is

$$\mathbf{F} = -\frac{\sigma}{r} A_s \mathbf{n},$$

where A_s is the area of the interface. If the interface is part of a sphere $\kappa = 2/r$ (due to the fact that the curvature for each of two mutually perpendicular directions is $1/r$). The above results show that the effect of surface tension is more important in interfaces with a small radius of curvature.

EXAMPLE 1.3

Problem Statement As a result of surface tension, liquid in a narrow tube rises as shown in Figure 1.20. For the conditions shown in the figure, find the the average height H as a function of the surface tension σ, contact θ, liquid density ρ, and tube radius R. If the liquid in the tube is water, $\theta = 60°$, and $r = 1 \, mm$, find the average height of the water in the tube.

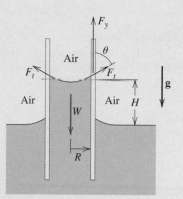

FIGURE 1.20 Schematic for Example 1.3.

Governing Equations For the column of water to be in static equilibrium as shown in the figure, the resultant force $\mathbf{F}$ must be zero — that is,

$$\sum \mathbf{F} = \mathbf{0}.$$

Since the intersection of the liquid's free surface and the tube is a circle with the same radius as the tube R, the tangential tensile force due to surface tension is given by

$$F_t = \sigma L = \sigma 2\pi R,$$

where σ is the surface energy.

Assumptions Assume that the fluid is in static equilibrium.

SOLUTION   If the water column is selected as the control volume, the forces in the vertical direction are those due to the weight W and the component of the surface tension force F_t in the vertical direction (y) $F_{t_y} = F_t \cos \theta$. For static equilibrium, then,

$$\sum F_y = F_t \cos \theta - W = 0.$$

The weight of the water column is

$$W = \rho_w \mathcal{V} g = \rho_w \pi R^2 H g$$

and the surface tension force is

$$F_t = \sigma 2\pi R.$$

The average column height, then, is

$$H = \frac{\sigma 2\pi R}{\rho_w \pi R^2 g} \cos \theta = \frac{2\sigma}{\rho g R} \cos \theta.$$

Substituting the numerical values,

$$H = \frac{2 \times 0.074 \ N/m}{998 \ Kg/m^3 \times 9.81 \ m/s^2 \times 0.001 \ m} \cos 60 = 7.56 \times 10^{-3} \ m.$$

EXAMPLE 1.4

Problem Statement   Consider a spherical air bubble immersed in water (Figure 1.21). The radius of the bubble is $r = 0.005 \ m$, and the water is exerting a pressure of $110 \ kPa$ on the bubble. Find the pressure of the air inside the bubble.

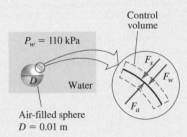

FIGURE 1.21   Schematic for Example 1.4.

Governing Equations   For the bubble to be in static equilibrium the resultant force on the bubble must be zero—that is,

$$\sum \mathbf{F} = \mathbf{0}.$$

$\mathbf{F}$ represents the forces on acting on the control volume.

Assumptions   Assume that the pressure along the surface of the bubble is constant.

SOLUTION   By selecting a small section of the bubble surface dA, as our control volume (Figure 1.21), static equilibrium is expressed by considering all forces normal to the control volume as

$$\sum \mathbf{F} = F_a - F_w - F_s = 0,$$

where F_a is the force exerted on the control volume due to the air pressure inside the bubble. Similarly, F_w is the force due to the water pressure, and F_s is the force due to surface tension. Note that the pressure forces are normal to the boundary and act in a compressive manner (i.e., in the opposite direction as the outward pointing unit normal vector). Using the definition of pressure

$$F_a = P_a \, dA,$$

and

$$F_w = P_w \, dA.$$

The force due to surface tension is given as $F_s = \dfrac{\sigma}{R} \, dA$, where R is the principal radius of curvature of the bubble. Since the bubble is spherical,

$$\frac{1}{R} = \frac{1}{r_1} + \frac{1}{r_2} = \frac{2}{r},$$

where r_1 and r_2 are the radii of curvature along two mutual perpendicular directions. Here, for the combination of air in water at $20° C$, $\sigma = 0.074 \, N/m$.

By substitution and by simplifying, the terms we get

$$P_a = P_w + \frac{2\sigma}{r},$$

which shows that the pressure inside the bubble is higher than the pressure of the surrounding water. Note also that this extra pressure is inversely proportional to the radius of the bubble. This implies that as the bubble size increases, the importance of surface tension decreases. Therefore, in the extreme limit when the surface is flat, surface tension plays no role.

Note that the same final results could be obtained by considering as control volume half the bubble. The balance of forces in the vertical direction then gives

$$P_a \pi r^2 = P_w \pi r^2 + \sigma 2\pi r \Rightarrow P_a = P_w + \frac{2\sigma}{r}.$$

The actual numerical value of the pressure inside the bubble is then

$$P_a = 110 \, kPa + \frac{2 \times 0.074 \times 10^{-3} \, kN/m}{0.005 \, m} = 110.0296 \, kPa.$$

1.6   Dimensions and Systems of Units

Physical parameters are distinguished by their *dimensions*. For instance, parameters representing distance are expressed in dimensions of length, symbolically denoted by L. Similarly, mass is expressed by M, time by T, and temperature by Θ. By convention, we will use $[q]$ to represent the dimensions of a variable q. For example, if D is a length term such as the diameter of a pipe or cylinder, $[D]$ indicates the dimensions of D. Symbolically then, $[D] = L$, reads: "D has dimension of length."

Any equation describing physical phenomena must be dimensionally consistent, as we cannot add or subtract dissimilar quantities. Physical parameters are also quantities that need to be measured and, therefore, require a method with which to quantify them. This quantification is achieved by comparing the "size" of the parameters with "standard values." These standard values, known as *units*, are quantities with fixed magnitude that have been adopted internationally as base quantities. Each dimension is also assigned an appropriate unit. For instance, one *meter* is a length scale whose magnitude is fixed *a priori* by international treaties. A distance of 2.5 meters is thus two-and-a-half times that fixed length. However, since units have arbitrary magnitude, length terms can alternatively be measured using various "standard values," such as *feet*, *inches*, and the like. The variables can be converted from one unit system to another using *conversion factors*. For instance,

$$1 \; meter = 0.3048 \; feet$$

Similarly, mass is measured in units of *kilograms*, *slugs*, or *pound mass*; time is measured in *seconds*, *minutes*, and *hours*; and temperature in degrees *Kelvin* or in degrees *Rankine*.

A small number of dimensions, defined as *primary dimensions*, can represent the dimensions of all other variables. In fluid dynamics, this is achieved by using dimensions of length (L), time (T), mass (M), and temperature (Θ). By assigning a unit to each primary dimension, the flow variables can be fully described. For instance, according to the definition, density has dimensions of M/L^3. Using appropriate units, density can be expressed in Kg/m^3.

The choice of these primary dimensions, though, is arbitrary. Currently, using Newton's Second Law, $\mathbf{F} \equiv m\mathbf{a}$, as the basis to ensure consistency of dimensions, we have various different *systems of units*. Four such systems relevant to fluid dynamics are described in the following paragraphs.

1. *International (SI) system* (M, L, T, Θ). In the SI system of units, the primary dimensions are those of mass (M), length (L), time (T), and temperature (Θ). Consequently, force is represented as a secondary dimension, defined by Newton's Second Law as $[\mathbf{F}] = ML/T^2$. In this system, mass is expressed in *kilograms* (Kg), length in *meters* (m), time in *seconds* (sec, or s), and temperature in degrees *Kelvin* (K). The unit of force known as the *Newton* (N) is defined as

$$1 \; N \equiv 1 \; Kg \cdot m/s^2$$

2. *Absolute metric system* (M, L, T, Θ). In the absolute metric system of units, the primary dimensions are those of mass (M), length (L), time (T), and temperature (Θ). Again, the force is represented as a secondary dimension, defined by Newton's Second Law as $[\mathbf{F}] = ML/T^2$. In this system, mass is expressed in *grams* (gr), length in *centimeters* (cm), time in *seconds* (sec or s), and temperature in degrees *Kelvin* (K). The unit of force known as the *dyne* ($dyne$) is defined as

$$1 \; dyne \equiv 1 \; gr \cdot cm/s^2$$

3. *British gravitational system of units* (F, L, T, Θ). In the British gravitational system of units, the primary dimensions are those of force (F), length (L), time (T), and temperature (Θ). In this system, mass m is a secondary dimension determined by Newton's Second law as $[m] = FT^2/L$. The unit of force in this system is defined as the *pound* (lbf). Length is measured in *feet* (ft), time in *seconds* (sec or s), and temperature in degrees *Rankine* (R). The unit of mass *slug* ($slug$) is defined accordingly as

$$1 \; slug \equiv 1 \; lbf \cdot sec^2/ft$$

TABLE 1.1 Conversion factors for principal parameters.

Conversion Factors	SI $\Rightarrow$ BGS	BGS $\Rightarrow$ SI
Length (L)	$1\ m = 3.281\ ft$	$1\ ft = 0.3048\ m$
Area (L^2)	$1\ m^2 = 10.76\ ft^2$	$1\ ft^2 = 0.0929\ m^2$
Volume (L^3)	$1\ m^3 = 35.32\ ft^3$	$1\ ft^3 = 0.028317\ m^3$
Force (ML/T^2)	$1\ N = 0.224809\ lbf$	$1\ lbf = 4.448222\ N$
Pressure (M/LT^2)	$1\ kPa = 0.145038\ psi$	$1\ psi = 6.894757\ kPa$
Energy (ML^2/T^2)	$1\ J = 0.7376\ lbf\ ft$	$1\ lbf\ ft = 1.3558\ J$
Temperature (Θ)	$T_K = 5/9\ T_R$	$T_R = 9/5\ T_K$

4. *English Engineering System of Units* (F, M, L, T, Θ). In the English engineering system of units, the primary dimensions are those of force (F), mass (M), length (L), time (T), and temperature (Θ). Notice that in this system both the force and mass are primary dimensions. For this to be possible and not violate the consistency of dimensions when applied to Newton's second law, we must introduce a proportionality constant as

$$\mathbf{F} = \frac{m\mathbf{a}}{g_c}.$$

The units of force are expressed in *pounds of force (lbf)*, the mass is expressed in *pounds of mass (lbm)*, length is expressed in *feet (ft)*, time is expressed in *seconds (sec, or s)*, and temperature is expressed in degrees *Rankine (R)*. One *lbf* is defined as the force required to accelerate 1 *lbm* to the standard gravitational acceleration of $32.2\ ft/sec^2$. Therefore,

$$1\ lbf = \frac{1\ lbm \times 32.2\ ft/sec^2}{g_c},$$

which gives

$$g_c = 32.2\ ft\ lbm/lbf \cdot sec^2.$$

Therefore g_c has both dimensions and units. By comparing the British Gravitational with the English Engineering system of unit we get the conversion factor

$$1\ slug \equiv 32.2\ lbm.$$

Conversion factors between the SI and the British gravitational systems of units given in Table 1.1. Note that the pressure is given in terms of units that are used in practice, rather than those defined by the system of units. This textbook uses the SI system of units.

REFERENCES

G.A. TOKATY, *A History and Philosophy of Fluid Mechanics*, New York, Dover, 1994.

P.H. WRIGHT, *Introduction to Engineering*, New York, Wiley, 1989.

R. MAYNE and S. MARGOLIS, *Introduction to Engineering*, New York, McGraw-Hill, 1982.

E.A. AVALONE, and T. BAUMEISTER, *Marks' Standard Handbook for Mechanical Engineers*, 9th ed., New York, McGraw-Hill, 1987.

R.B. BIRD, W.E. STEWART, and E.N. LIGHTFOOT, *Transport Phenomena*, New York, Wiley, 1960.

R.B. BIRD, R.C. ARMSTRONG and O. HASSAGER, *Dynamics of Polymeric Liquids*, 2nd ed., New York, Wiley-Interscience, 1987.

N.B. VARGAFTIK, *Tables on the Thermophysical Properties of Liquids and Gases*, 2nd ed., Washington D.C., Hemisphere Publishing Corp., 1975.

Y.S. TOULOUKIAN, S.C. SAXENA, and P. HESTERMANS, "Thermophysical Properties of Matter," The TPRC Data Series, Vol. **11** — Viscosity, New York, Plenum Press, 1975.

Handbook of Chemistry and Physics, 62nd ed., Cleveland, Chemical Rubber Publishing Co., 1981–1982.

PROBLEMS

1. For each of the following physical problems, identify what system you would use for analysis, sketch the system boundary and show any mass crossing the boundary:

   • A sprinter during a race.

   • A jet engine mounted on an airplane.

   • Heating of a cup of coffee.

2. For each of the following physical problems, identify what system you would use for analysis, sketch the system boundary and show any mass crossing the boundary:

   • A rocket during flight.

   • A propeller-powered boat.

   • A stamping process where an amount of solid is stamped into a final product.

3. For each of the following physical problems, identify what system you would use for analysis, sketch the system boundary and show any mass crossing the boundary:

   • A centrifugal pump

   • A boiler

   • A sprinkler system

4. For each of the following physical problems, identify what system you would use for analysis, sketch the system boundary and show any mass crossing the boundary:

   • A sailboat

   • Human lung

   • Vacuum cleaner

5. For each of the following physical problems, identify what system you would use for analysis, sketch the system boundary and show any mass crossing the boundary:

   • A power plant

   • An injection molding machine

   • A heat exchanger

6. For each of the following physical problems, identify what system you would use for analysis, sketch the system boundary and show any mass crossing the boundary:

   • A gas turbine

   • A windmill

   • A glider

7. The Avogadro's number defines the number of molecules per mole. This number is 6.023×10^{23} *molecules/mole.*

Evaluate the number of molecules in one liter of: (a) air, (b) water.

8. The density of water is $998 \ kg/m^3$. How much volume 10.5 kg of water will occupy?

9. A 2 m^3 container is filled with 2.7 kg of air. What is the density of air at these conditions?

10. Two different drugs, considered as incompressible liquids, with densities $1050 \ kg.m^3$ and $960 \ kg/m^3$ are mixed in a 0.5 m^3 vessel. If the volume of the denser drug prior to mixing was 60% of that of the less dense drug, determine the density of the mixture and the mass of each drug.

11. A mixture of two incompressible fluids has density of $920 \ kg/m^3$, which is 80% of the density of one of the fluids. Find the density of each fluid when the ratio of the volume each fluid occupies is 0.75.

12. A mixture of two incompressible fluids has density of $920 \ kg/m^3$, which is 75% of the density of one of the fluids. Find the density of each fluid by itself occupies the same volume.

13. Find the amount of fluid with density of $1000 \ kg/m^3$ required to mix to 0.15 kg of fluid with density of $1160 \ kg/m^3$ to form a mixture with density $1080 \ kg/m^3$.

14. Express an absolute pressure of 372 kPa in terms of: (a) gage pressure using a standard atmosphere, (b) an equivalent column of water and (c) an equivalent column of mercury.

15. Express an absolute pressure of 235 kPa in: (a) gage pressure with respect to a standard atmosphere, (b) an equivalent height of water and (c) equivalent height of mercury.

16. Find the absolute pressure that corresponds to a vacuum equivalent to a column height of 2.3 m of water.

17. Express a gage pressure of 130 kPa into an absolute pressure in: (a) kPa, (b) equivalent column of water, and (c) equivalent column of mercury.

18. What is the absolute pressure inside a tank when the gage pressure is measured to be the equivalent to a 1.2 m column of water?

19. What is the absolute pressure inside a tank when the vacuum pressure is measured to be the equivalent to a 2.5 m column of water.

20. Find the gage/vacuum pressure expressed in a equivalent height of water of: (a) 18.2 m, (b) 8.4 m, (c) 22 m, (d) 4.6 m.

21. A constant force given by $\mathbf{F} = -120 \ kN\mathbf{i} - 140 \ kN\mathbf{j}$ is acting on a flat surface aligned with the x-direction as shown in Figure 1.22. If the area of the surface is 4.3 m^2, find the pressure on the surface.

22. Find the resulting pressure produced by a constant, vertical force of 180 kN acting on a inclined flat surface with total area 6.2 m^2. The surface makes an angle of 30° degrees with respect to the horizontal direction.

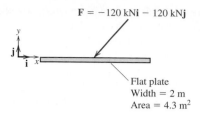

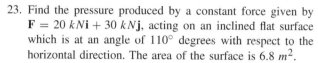

FIGURE 1.22   Schematic for Problem 21.

23. Find the pressure produced by a constant force given by $\mathbf{F} = 20\,kN\mathbf{i} + 30\,kN\mathbf{j}$, acting on an inclined flat surface which is at an angle of $110°$ degrees with respect to the horizontal direction. The area of the surface is $6.8\,m^2$.

24. Find the total force on a flat square surface with area of $3.3\,m^2$ due to an externally applied constant pressure of $145\,kPa$. Use an appropriate schematic and coordinate system to indicate the force as a vector. Where is the point of application of the resultant force?

25. Calculate the net force vector on a $2.3\,m \times 3.8\,m$ flat surface due to a constant pressure that it is equivalent to a $22.5\,m$ column of water. Where is the point of application of the resultant force? Sketch the problem using an appropriate coordinate system, and show on the same schematic the pressure distribution and the resulting force vector.

26. Find the total force due to an externally applied pressure on a flat square surface with $0.8\,m$ side. The applied pressure is constant in one direction and varies linearly from $80\,kPa$ to $165\,kPa$ in the other direction. Where is the point of application of the resultant force? Sketch the problem using an appropriate coordinate system, and show on the same schematic the pressure distribution and the resulting force vector.

27. Find the net force vector on a flat $2 \times 3\,m$ plate due to the applied pressure shown in Figure 1.23. Where is the point of application of the resultant force?

28. Find the net force vector on the curved surface due to the constant applied pressure shown in Figure 1.24. Where is the point of application of the resultant force?

29. Find the net force vector on the curved surface due to an applied pressure which varies according to $P = 14 - y$, as

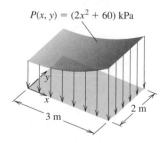

FIGURE 1.23   Schematic for Problem 27.

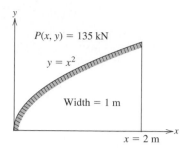

FIGURE 1.24   Schematic for Problem 28.

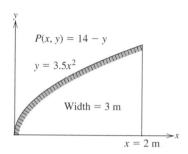

FIGURE 1.25   Schematic for Problem 29.

shown in Figure 1.25. Where is the point of application of the resultant force?

30. The pressure distribution on a rectangular wing with length $L = 4\,m$ and width $W = 1\,m$ is given as $P = 101.3 + 80\sin(\pi x/L)\,kPa$, where x is the direction along the length of the wing with $x = 0$ is the forward edge. Sketch the pressure distribution and then calculate the net force acting on it.

31. The pressure load on a rectangular platform with length $L = 8\,m$ and width $W = 1.5\,m$ is given as $P = 200(1 + \cos(\pi x/L))\,kPa$, where x is the direction along the length of the platform with $x = 0$ is the forward edge. Sketch the pressure distribution and then calculate the net force acting on it.

32. What is the pressure in kPa $16.5\,m$ below the ocean's surface?

33. At what depth below the ocean's surface the pressure is equivalent to $1.3\,m$ of mercury? How does this compare with the atmospheric pressure?

34. At what depth below the ocean's surface the pressure is equivalent to 2.3 standard atmospheres?

35. Consider a $15\,cm$-diameter cylinder fitted with a $12.8\,kg$ piston, filled with air. Heat is transferred showly to the cylinder causing it to move. Find the pressure inside the cylinder during the process.

36. Design a $10\,cm$ cylinder fitted with a piston so that, when the gas inside the cylinder is heated, the pressure increases linearly from $120\,kPa$ to $140\,kPa$ while the piston moves

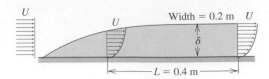

FIGURE 1.26   Schematic for Problem 41.

8.15 *cm*. For the design consider that the piston is attached to a linear spring (force F is proportional to the displacement x of the spring $F = kx$) with unknown stiffness k. Consider an outside pressure of 101 *kPa*.

37. In a pressure cooker the pressure regulator is 10 *kg* and the hole where the connector is connected to is 0.4 *cm*. At what pressure will the cooker start releasing steam?

38. Design a pressure cooker that will operate at a constant pressure of 140 *kPa*. Unknown parameters are the mass of the pressure regulator and the diameter of the hole. Justify your selection of the design parameters.

39. The viscosity of water at 20° *C* is 1.002×10^{-3} *Ns/m²*, and the density 998.2 *kg/m³*. Calculate the kinematic viscosity.

40. The kinematic viscosity of air at 20° *C* is 1.51×10^{-5} *m²/s*, and the density 1.204 *kg/m³*. What is the dynamic viscosity?

41. The velocity near a solid surface can be approximated by $u = U \sin\left(\frac{\pi}{2}\frac{y}{\delta}\right)$, where δ is the thickness of the viscous layer as shown in Figure 1.26. In reality the thickness of this layer increases in the direction of flow. Here, estimate the total viscous force due to water flowing past a flat plate with velocity $U = 20$ *m/s* by assuming constant thickness $\delta = 1$ *mm*, over a distance $L = 0.4$ *m* along a plate which is 0.2 *m* wide.

42. Consider flow of water between two flat plates each 2.3 *m* long and $2H = 0.5$ *m* apart. If the velocity of the liquid is $u = 1.5\left(1 - \frac{y^2}{H^2}\right)$ as shown in Figure 1.27, find the magnitude of the force and its direction of action on the surface of the plate.

43. Find and sketch the distribution of the internal fluid friction force along the height of the channel in the previous problem. What is the friction force along the centerline of the channel? What is the physical explanation of the result?

44. Find the power required to maintain constant velocity of a square plate relative to a fixed solid surface, as a function of the thickness of a liquid film between the plate and the solid surface (note: power = force × velocity).

45. Show that the expression for the power described in Problem 44 is dimensionally consistent (Hint: express both sides of the expression in terms of the fundamental dimensions).

46. Find the power required to maintain constant velocity of 1.2 *m/s* of a square plate with a side length of 80 *cm* relative to a fixed solid surface, when the thickness of a water film between the plate and the solid surface is 1 *cm* (note: power = force × velocity).

47. Find and plot the power required to maintain constant velocity of a square plate with a side length of 80 *cm* relative to a fixed solid surface as a function of the velocity of the plate for a water thickness between the plate and the solid surface of 1 *cm* (note: power = force × velocity).

48. The velocity profile for steady, fully developed flow down an inclined plane is given by $u = \frac{\rho g}{\mu} \sin\alpha \left(Hy - \frac{y^2}{2}\right)$, ρ is the density, g the gravitational acceleration, μ the viscosity of the fluid and H the depth of the flow (Figure 1.28). Determine the total force an inclined surface 10 *m* long and 14 *m* wide. Assume water with $H = 0.8$ and $\alpha = 28°$.

49. Find and sketch the distribution of the internal fluid friction force along the thickness of the fluid in the previous problem. What is the physical explanation of the result?

50. Consider two concentric cylinders with diameters d_1 and d_2 ($d_1 < d_2$) with $d_2 - d_1$ being very small and with the same height H. Find the power required to rotate the inner cylinder at a constant rotation (revolutions per minute) when the gap between the cylinders is filled with liquid. What is the power required to rotate the outer cylinder?

51. Consider two concentric cylinders with diameters 12 *cm* and 12.5 *cm* and the same height of 40 *cm*. Find the power required to rotate the inner cylinder at a constant rate of 130 revolutions per minute when the gap between the cylinders is filled with glycerin.

52. Consider the same concentric cylinder-cylinder arrangement as in the above two problems, and find the number of revolutions per minute as a function of the gap size for a fixed amount of power.

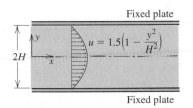

FIGURE 1.27   Schematic for Problem 42.

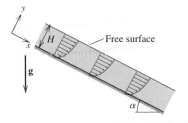

FIGURE 1.28   Schematic for Problem 48.

53. A 12 *cm*-diameter shaft rotates at 2000 *rev/min* inside a 12.4 *cm* bearing. The gap between the shaft and the bearing is filled with oil with viscosity of 0.021 *kg/ms*. What is the power required to maintain the given rotation. Assume linear velocity variation across the gap.

54. The schematic in Figure 1.29 shows a viscous brake where, due to the action of viscous friction, a rotating disk can be slowed down. Assuming that the velocity of the liquid between the fixed plate and the rotating disk is linear, evaluate the total net torque experienced by the disk when the disk is rotating at a constant rotational speed.

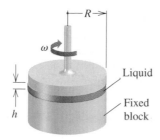

FIGURE 1.29   Schematic for Problem 54.

55. Show that the expression for the total viscous torque obtained for the viscous break described in Problem 54 is dimensionally consistent (Hint: express both sides of the expression in terms of the fundamental dimensions).

56. Using the result from the above problem, calculate the total torque developed on the viscous break. The disk with radius of 25 *cm* rotates at 3800 *rev/min*. The 0.15 *cm* gap between the disk and the fixed plate is filled with oil with $\mu = 8 \times 10^{-2}$ *N s/m²*.

57. Consider a modified version of the viscous break described in Problem 54, where now the action of the fluid is confined in an annular ring defined by an inner and outer radii of r_i and r_o respectively, calculate the total force.

58. Using the data given in Problem 56, calculate the total viscous torque developed in a modified viscous break where the action of the liquid is confined into an annular ring defined by $r_i = 0.2\ r$ and $r_o = r$, where r is the radius of the rotating disk.

59. A 4.2 *m²* plate slides relative to a solid surface due to a thin film of water between the plate and the surface. If the velocity of the plate is 1.6 *m/s* and the thickness of the film is 1.4 *cm*, find the total force required to sustain the motion of the plate. Assume that the velocity of the water within the film varies linearly across the film's thickness. What is the force if the film thickness is (a) increased by 10 percent, (b) decreased by 10 percent.

60. Find the height water in a 0.25 *mm*-diameter glass tube will rise due to capillary action.

61. A 2 *mm*-diameter tube is inserted in a water reservoir. At what height will the water rise inside the tube?

62. Find the pressure difference across a soap bubble with surface tension σ and diameter D. Compare the result with the pressure difference across a spherical air bubble in water (Hint: consider all fluid-fluid interphases).

63. What is the diameter of a water droplet exposed in air when the gage pressure inside the droplet is 0.8 *kPa*?

64. Calculate the pressure difference across a cylindrical air pocket with radius of 1 *mm*, ($\sigma = 0.074 \times 10^{-3}$ *kN/m*).

65. Convert 100 *ft²* in units of *m²*.

66. Convert 2.4 *m²* into *ft²*.

67. Convert 56 *Psi* in: (a) kPa, (b) equivalent height of water in meters, (c) equivalent height of mercury in *ft*.

68. Convert a standard atmosphere (101.3 *kPa*) in: (a) Psi, (b) equivalent height of mercury in *ft*.

69. Using the definition of surface tension, express σ in terms of fundamental dimensions of L, M, T, Θ.

70. Express the absolute viscosity μ in terms of fundamental dimensions of L, M, T, Θ. Repeat the same for the kinematic viscosity ν.

71. The streamfunction ψ is defined as $u = \dfrac{\partial \psi}{\partial y}$, and the velocity potential as $u = \dfrac{\partial \phi}{\partial x}$ where u is the velocity and x and y the coordinate directions. What are the dimensions of ψ and ϕ?

72. Extensional strain defined as $\dot{\epsilon}_{xx} = \dfrac{\partial u}{\partial x}$, gives the rate at which fluid particles extend in the axial direction. Find the dimensions of $\dot{\epsilon}_{xx}$.

73. The rate of rotation of fluid particles are expressed by the vorticity vector ω. For instance, it can be shown that the rate of rotation around the z-axis is given by $\omega_z = \dfrac{\partial v}{\partial x} - \dfrac{\partial u}{\partial y}$. What are the dimensions of ω_z?

74. Rotation in the flow is often expressed using the concept of circulation Γ, defined as $\Gamma = \oint \mathbf{u} \cdot \mathbf{t}\, ds$ where the integral is defined over a closed loop and $\mathbf{u}$ is the velocity vector, $\mathbf{t}$ is the unit tangent vector along the loop and s the arc-length along the integration path. What are the dimensions of Γ?

75. The maximum rotational speed ω at which a circular container with radius R, filled with $\mathcal{V}$ volume of liquid, can be rotated without exposing the bottom of the container is given by $\omega^2 = 4\mathcal{V}g/(\pi R)$, where g is the gravitational acceleration. Show that the given expression is dimensionally consistent.

76. The pressure variation $P = P_0 + \rho g(z - z_0) + 0.5\rho\omega^2 r^2$, in a partially filled container with liquid rotating with angular velocity ω is a function of the distance r from the center of rotation, the local depth z from a free surface exposed to a pressure P_0, the density of the liquid ρ, and the gravitational acceleration g. Show that the expression is dimensionally consistent.

77. When fluid is flowing past a solid surface we typically assume that the fluid and solid in contact move with the

same velocity. Therefore, when the solid surface is stationary, we consider the fluid velocity there u to be zero. However, for some fluids this assumption is not quite valid as experimental evidence indicates that the fluid along a solid surface can slip with a velocity assumed to be proportional to the local wall shear τ_w according to $u = \lambda \tau_w$. What are the proper units for λ?

78. The Weber number We is defined as $we \equiv \rho L V^2 / \sigma$ where ρ is the density of the fluid, L and V are respectively characteristic length and velocity terms and σ surface tension. Show that We has no dimensions.

79. An important parameter in fluid mechanics is the Reynolds number Re defined as $Re \equiv \rho V L / \mu$, where ρ and μ are respectively the density and viscosity of the fluid, and V and L are characteristic velocity and length terms. So that Re is non-dimensional, i.e., it has no dimensions.

80. Flow past a solid surface develops a thin layer of fluid of thickness δ within which shear stresses are large. The thickness of the layer grows according to $\delta^2 \sim \mu x / (\rho V)$ where μ and ρ are respectively the viscosity and density of the fluid, V is the velocity away from the surface and x is the distance in the main flow direction. Show that the expression is dimensionally consistent.

81. The volumetric flow rate $\dot{Q}$ in a circular pipe with diameter D and length L due to a pressure difference ΔP across the pipe is given by $\dot{Q} = \pi (\Delta) P D^4 / (128 \mu L)$. Show that the expression is dimensionally consistent.

82. A power-law fluid is one whose viscosity changes with the rate of deformation. For a simple one dimensional flow the viscosity is expressed by $\tau = K \left(\dfrac{du}{dy} \right)^{n-1} \dfrac{du}{dy}$, where n is the power law index, (for $n = 1$ we recover the usual Newtonian fluid behavior) u is the fluid velocity, y is the direction normal to the direction of u, and K the consistency index. What are the proper units of K?

Conservation Laws for Closed Systems

This chapter discusses the fundamental principles that govern fluid flow behavior. We stress the universality of these principles and show that all physical phenomena, not just fluid phenomena, are governed by the same set of general principles known as the laws of nature. The performance of a steel structure, for instance, is governed by the same principles that govern the operation of a gas turbine. This is true despite the fact that the two situations appear to be fundamentally different from each other.

The governing laws, expressed as a list of universal conservation principles, represent our current understanding of how nature behaves. These laws are analogous to the rules according to which any game is played. In basketball, for example, although individual games are very distinct from each other, they are controlled by a set of rules that are enforced by the referees. In the same manner, all natural phenomena, although distinct from each other, follow a set of universal principles that are activated instantaneously as the phenomena develop.

The initial acceleration and eventual flight of vehicles into space, for instance, is in part governed by a principle that relates the rate of change of linear momentum to the applied resultant force. Or, to take another example, the water in a kettle set on the stove becomes hot, while a cup of coffee left on the kitchen counter gets cold. These simple thermal phenomena are also governed by a universal principle, which states that the net energy gained or lost by an object is respectively equal to the total heat transfer to or from it.

2.1 General Conservation Law for a Closed System

In addition to their universality, the conservation laws are expressed in a simple mathematical form that is remarkably similar for all of them. For the closed system shown in Figure 2.1, while the mass m of the system moves and deforms, the conservation of a generic quantity $\mathcal{A}$ specific to the system is described by the general expression

$$\frac{D\mathcal{A}}{Dt} \equiv B. \tag{2.1}$$

The derivative in the form $D(\cdot)/Dt$ denotes rate of change with respect to time (t) while following the closed system. Therefore, all quantities in Equation (2.1) are *expressed relative to the closed system*. This means that they are expressed relative to an observer moving with the closed system. The significance of this derivative is explained in greater detail in Chapter 4. The right-hand side B is a quantity specific to the particular conservation law.

---

[1]G.A. Tokaty, *A History and Philosophy of Fluid Mechanics*.

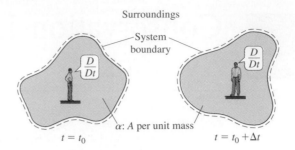

FIGURE 2.1 A deforming, fixed fluid mass defined by a closed system, at two instances t_0 and $t_0 + \Delta t$, where Δt is a short time interval.

By defining

$$\alpha = \frac{d\mathcal{A}}{dm},$$

where α is the density of property $\mathcal{A}$ (i.e., $\mathcal{A}$ per unit mass) the general form of the conservation laws is expressed as

$$\frac{D}{Dt} \int_m \alpha \, dm \equiv B. \tag{2.2}$$

Finally, by using the definition of density $dm = \rho d\mathcal{V}$, the preceding equation is converted into a more convenient form as an integral over the volume $\mathcal{V}$ that m occupies according to

$$\frac{D}{Dt} \int_{\mathcal{V}} \alpha \rho d\mathcal{V} \equiv B. \tag{2.3}$$

The ($\equiv$) symbol is used to denote the fact that the conservation laws cannot be derived mathematically, but are accepted to be true by definition. This is because the laws are *phenomenological* and, hence, their validity is established by repeated observations of natural phenomena. For example, by experience we know that an amount of water in a glass, in the absence of any leaks, evaporation, or other losses, remains the same when it is transferred to another container. Unlike the rules of sports games, however, challenges to the conservation laws are remarkable and rare events in the history of science and have profound implications for humankind. One challenge to the applicability limits of the laws is Einstein's theory of relativity, which has completely changed our notion about the relation of space and time. However, as long as the velocity is less than the speed of light (as is the case for all practical problems), the classical form of the conservation laws is still valid.

Equation (2.1) is also a *general expression that is valid irrespective of the nature of the medium to which it is applied*. The universal laws can be made specific to a certain material by using material constants such as density, viscosity and conductivity. The material response to physical changes — for instance, the extent a gas will compress due to the action of a piston, is expressed through *constitutive relations*. These are empirical relations established experimentally. The well-known ideal gas relation $P\mathcal{V} = mRT$ is a typical constitutive relation which, through the experimentally determined gas constant (R), indicates a unique relation between the gas variables—pressure (P), volume ($\mathcal{V}$), mass (m), and temperature (T)—as the gas undergoes change. Unlike the conservation laws, constitutive relations can be challenged and, hence, changed. For instance, the behavior of non-Newtonian fluids is constantly being updated as new experimental data are obtained and new theories are developed.

To elaborate further on the issue of the constitutive behavior of materials: the difference in the way materials respond to the same external factors may be considered analogous to the way different basketball teams play basketball. As water flows easier than honey when subjected to the same lateral force because of the lower viscosity of the water (viscosity being an intrinsic material property), as a result of the higher skill level of the players under the same basic rules a professional basketball team plays a much better game than an elementary school team (the skill level being an intrinsic property of the team).

Often, because of their broad acceptance, some constitutive relations are also defined as "laws," such as Fourier's law of conduction and Newton's law of viscosity. Throughout this book we will make clear the distinction between governing equations and constitutive relations.

2.2 Conservation of Mass

The phenomenological observation that in physical systems the total mass is conserved is expressed by the conservation of mass law, or as otherwise known in fluid dynamics, the continuity equation. For a closed system, using the general framework expressed by Equation (2.3), $\mathcal{A} = m$, $\alpha = 1$, and $B = 0$, the conservation of mass is expressed as

$$\frac{D}{Dt} \int_m dm = \left[\frac{Dm}{Dt} \right]_{system} \equiv 0. \tag{2.4}$$

The conservation of mass equation is a scalar equation and, hence, is independent of the choice of the coordinate system. Consistent with the definition of a closed system, upon integration Equation (2.4) gives

$$m = \text{constant.}$$

In closed systems then, the conservation of mass is satisfied automatically.

2.3 Conservation of Linear Momentum

Aristotle,[2] in the third century B.C., observed that

> a smaller and lighter weight will be given more movement if the force acting on it is the same...

Indeed, the observation that the rate at which a quantity defined as linear momentum $m\mathbf{V}$ is changing with time is equal to the resultant force $\sum \mathbf{F}$ applied to a closed system with mass m and velocity $\mathbf{V}$, as shown in Figure 2.2, is expressed by the conservation of linear momentum principle or simply, the momentum equation. This principle is also known as *Newton's second law* as a tribute to Isaac Newton who articulated the principle in mathematic form.

Following the general expression given by Equation (2.3), $\mathcal{A} = m\mathbf{V}$, $\alpha = \mathbf{V}$, and $B = \sum \mathbf{F}$. Accordingly, the conservation of linear momentum is expressed as

$$\underbrace{\frac{D}{Dt} \int_m \mathbf{V} dm}_{Kinematics} \equiv \underbrace{\sum \mathbf{F}}_{Dynamics} , \tag{2.5}$$

---

[2] Aristotle, *De caelo*. Taken from Sir Thomas Heath, *A History of Greek Mathematics*

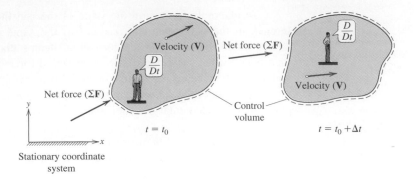

FIGURE 2.2 Conservation of momentum for a deforming, fixed fluid mass defined by a closed system, at two instances t_0 and $t_0 + \Delta t$, where Δt is a short time interval.

or, in terms of the volume $\mathcal{V}$ that encloses the mass of the system,

$$\underbrace{\frac{D}{Dt} \int_{\mathcal{V}} \rho \mathbf{V} d\mathcal{V}}_{Kinematics} \equiv \underbrace{\sum \mathbf{F}}_{Dynamics} \ . \qquad (2.6)$$

Since the velocity and forces are vector quantities, the momentum equation is a vector equation and must be expressed relative to an appropriate coordinate system. In the form given either by Equation (2.5) or by Equation (2.6), the momentum equation is valid for an *inertial coordinate system* (i.e., a coordinate system that *is either stationary or moving with constant velocity*).

This law states that the rate of change in the system's linear momentum is balanced by the resultant of the forces acting on it. Consequently, a mass of fluid originally at rest is set in motion (accelerates) only when there is a net resultant force applied to it. The mass, then, will move in the direction of the applied resultant force.

Alternatively, the momentum equation relates fluid motion (i.e., kinematics) to the applied forces (i.e., dynamics). The term *kinematics* derives from the Greek word for *motion*. Therefore, fluid motion and kinematics are synonymous terms. Also, since the term *dynamics* derives from the Greek word for *force*, fluid forces and fluid dynamics are equivalent terms.

To apply the momentum equation, we must define and understand each term that appears in Equations (2.5) and (2.6).

Mass and Volume

By definition, the mass m within the closed system is fixed. The mass and the volume it occupies, $\mathcal{V}$, can move and deform as a function of time. At no time, however, does mass cross the system boundary.

Velocity Vector

$\mathbf{V}$ is the velocity vector of the fluid within the closed system expressed relative to the coordinate system as shown in Figure 2.2. In principle, the velocity can vary both with time and with location within the mass of the system — that is, $\mathbf{V} = \mathbf{V}(x, y, z, t)$.

Force Vector

The momentum equation defines formally the concept of force by stating that

$$Force = Mass \times Acceleration$$

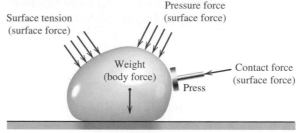

FIGURE 2.3 Classification of the different types of forces.

Following this definition, the force is measured in units of $kg\ m/sec^2$, which is commonly known as a Newton (N). A Newton, then, is the force needed to accelerate a mass of 1 kg at a rate of 1 meter per second per second. Depending on their nature, forces may be classified either as *body forces* or as *surface forces* (Figure 2.3).

Body forces.   These are forces that act throughout the body. The most commonly known body force is the weight representing the influence of gravity. The magnitude of the weight **W** is given as

$$\mathbf{W} = m\mathbf{g},$$

where m is the mass and $\mathbf{g} = 9.81\ m/s^2$ (standard value) is the local acceleration vector due to gravity; the weight vector always points in the same direction as the local gravity vector.

Other body forces may include electromagnetic forces. These are forces that develop in electrically conducting fluids as a result of the combination of fluid flow and applied electric or magnetic fields. The study of such flows is known as electro-magneto-hydrodynamics or magneto-Hydrodynamics (MHD). Given the introductory nature of this text, these MHD forces are not considered further.

Surface forces.   These are external forces that act along part of or the entire boundary of the closed system. Typical surface forces are

- *The pressure forces.* These are surface forces acting in a compressive direction due to the local pressure. By definition, the total force on a surface A_s due to pressure P is expressed as

$$\mathbf{F} = -\int_{A_s} P\,\mathbf{n}\,dA_s,$$

  where **n** is the *outward pointing unit normal vector*. Again ($-$) and **n** are included to make sure that the force acts in a compressive direction perpendicular to the surface.

- *The contact forces.* These are surface forces that develop because of contact with neighboring solid or fluid surfaces such as the tangential viscous forces due to fluid friction along a surface. Contact forces are also those applied externally—for example, by a mechanical press. Overall, contact forces can act in any direction. The total force on a arbitrary surface A_s due to a surface contact force per unit area $\mathbf{f}_s$ is

$$\mathbf{F}_{contact} = \int_{A_s} \mathbf{f}_s\,dA_s.$$

- *Surface tension.* As discussed in Chapter 1, while surface tension lies in the plane of the interface, its net effect is a normal surface force (tensile or compressive) that develops at the interface between two immiscible liquids or at the interface between liquid and gas. By definition, the force due to surface tension is expressed as

$$\mathbf{F}_{surface\ tension} = -\int_{A_s} \frac{\sigma}{R_c}\,\mathbf{n}\,dA_s,$$

where σ is the surface energy per unit area, R_c is the local of radius curvature, and A_s is the surface area. The radius of curvature is considered positive when the surface is concave. In this case, the force acts in a compressive direction.

EXAMPLE 2.1

Problem Statement   Find the initial velocity V_o and angle α with which a baseball player should throw the ball to second base 50 m away (Figure 2.4).

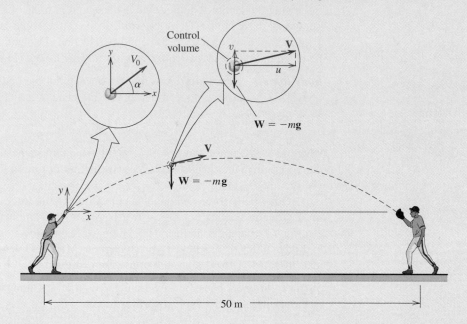

FIGURE 2.4   Schematic of Example 2.1.

Governing Equations   The schematic in Figure 2.4 shows the expected path of the ball. By selecting the system boundary as enveloping the ball, we have a closed system for which the conservation of momentum with respect to a fixed coordinate system as shown in Figure 2.4 is given as

$$\frac{Dm\mathbf{V}}{Dt} \equiv \sum \mathbf{F}.$$

Basic Assumptions   From experience, we know that the motion of the ball is retarded by air resistance. For simplicity, however, such resistance is neglected in this problem. The pressure around the ball is also assumed to be constant.

Initial Conditions   With respect to the chosen coordinate system, and according to the description of the problem, at $t = 0$, $u = V_o \cos\alpha$, and $v = V_o \sin\alpha$, where u and v are the components of the velocity vector, respectively, along the x- and y-directions. Initially, the ball is assumed to be at the origin of the selected coordinate system ($x = y = 0$).

SOLUTION   The momentum equation in terms of individual components is expressed as

$$\frac{Dmu}{Dt} \equiv \sum F_x, \quad \frac{Dmv}{Dt} \equiv \sum F_y.$$

The only relevant force here is the body force due to the weight **W** that acts in the negative y-direction — that is,

$$\mathbf{W} = -mg\,\mathbf{j}.$$

Since pressure is assumed constant around the ball, there is no net pressure force. In the absence of other forces, the conservation of momentum in the x-direction simplifies to

$$\frac{Dmu}{Dt} \equiv 0,$$

or, since m is constant

$$\frac{Du}{Dt} = 0,$$

which upon integration gives

$$u = C_1.$$

The integration constant C_1 is determined from the initial conditions: since at $t = 0$, $u = V_o \cos\alpha$, then $C_1 = V_o \cos\alpha$. Hence,

$$u = V_o \cos\alpha.$$

Using the definition

$$u = \frac{Dx}{Dt},$$

the distance traveled by the ball in the horizontal direction is obtained by integration as

$$x = \int_0^t u\, Dt = \int_0^t V_o \cos\alpha \, Dx = (V_o \cos\alpha)\, t. \tag{2.7}$$

Similarly, the momentum equation in the y-direction is

$$\frac{Dmv}{Dt} \equiv -mg, \quad \text{or} \quad \frac{Dv}{Dt} \equiv -g,$$

which, upon integration, yields

$$v = -gt + C_2.$$

Since at $t = 0$, $v = V_o \sin\alpha$, then $C_2 = V_o \sin\alpha$. Therefore,

$$v = V_o \sin\alpha - gt. \tag{2.8}$$

This shows that the ball reaches the maximum height when $v = 0$, after

$$t_{y_{max}} = \frac{V_o}{g} \sin\alpha \ \ seconds.$$

Using the definition

$$\frac{Dy}{Dt} = v,$$

and by substituting v using Equation (2.8) the distance traveled in the vertical direction as a function of time is found by integration as

$$y = (V_o \sin \alpha) t - \frac{1}{2} g t^2. \tag{2.9}$$

By eliminating time t from Equations (2.7) and (2.9), the trajectory of the ball (path followed by the ball) is

$$y = V_o \sin \alpha \frac{x}{V_o \cos \alpha} - \frac{1}{2} g \left(\frac{x}{V_o \cos \alpha} \right)^2.$$

By simplifying

$$y = x \left(\tan \alpha - \frac{g}{2 V_o^2 \cos^2 \alpha} x \right). \tag{2.10}$$

This is a parabolic trajectory symmetric with respect to a vertical line passing through the point of maximum height. The ball is at ground level ($y = 0$) at the moment of release ($x = 0$), and also at the end of its trajectory ($x = x_{max}$) when

$$\tan \alpha - \frac{g}{2 V_o^2 \cos^2 \alpha} x_{max} = 0,$$

or

$$x_{max} = \frac{2 V_o^2}{g} \cos \alpha \, \sin \alpha = \frac{V_o^2}{g} \sin 2\alpha.$$

This is the maximum distance the ball will reach in the horizontal direction.

According to Equation (2.7), from the moment of its release, the ball reaches its maximum range after

$$t = \frac{x_{max}}{V_o \cos \alpha t} = \frac{2 V_o}{g} \sin \alpha \; seconds.$$

This is twice the time it takes the ball to reach the maximum height, which is expected as a result of the symmetry of the ball's trajectory.

In this problem, second base is 50 m away or $x_{max} = 50$ m. Therefore,

$$x_{max} = \frac{V_o^2}{g} \sin 2\alpha = 50 \; m,$$

which gives

$$V_o^2 \sin 2\alpha = 50 \; m \; g = 50 \; m \times 9.81 \; m/s^2 = 490.5 \; m^2/s^2.$$

Trajectory problems such as this one do not have a unique solution: for any velocity V_o there are two possible solutions for the throwing angle, one corresponding to a lower and another to a higher trajectory. This is due to the fact that the solution to

$$2\alpha = \sin^{-1} g \frac{x_{max}}{V_o^2},$$

has two solutions α_1 and α_2, with $0 \leq 2\alpha_1 \leq 90$, and $90 \leq 2\alpha_2 \leq 180$. Therefore, since α_1 and α_2 are in the range $0 \leq \alpha_1, \alpha_2 \leq 90$, both are solutions to the original problem. For the conditions given here, $\alpha_1 = 25.85°$ and $\alpha_2 = 64.15°$. By assuming an initial velocity $V_o = 25 \ m/s$, Figure 2.5 shows the two possible solutions to the original question.

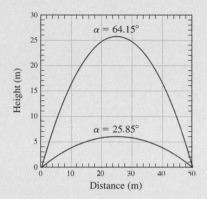

FIGURE 2.5 Solution of Example 2.3 showing the trajectory of the ball thrown at two different initial angles—$\alpha_1 = 25.85°$ and $\alpha_2 = 64.15°$.

Since the higher trajectory corresponds to a longer time of flight, in order to minimize time, the player will obviously select the lower trajectory to throw the ball.

This example demonstrates the application of the conservation of linear momentum law to a very simple problem. Despite its simplicity, however, it clearly demonstrates that *simplifying assumptions* yield an exact solution that can be easily manipulated. In a more realistic problem, however, the validity of assumptions such as no air resistance must be established, for instance, by *estimating* the effect of this force on the final solution. The problem also highlights the *design* dilemma of selecting the solution that provides an optimum answer to the posed problem. These issues as well as other similar modeling and analysis issues are central to the study of fluid mechanics for more realistic everyday applications.

EXAMPLE 2.2

Problem Statement Aircraft flight and dynamics are determined by the interplay between the weight of the aircraft, the lift force (i.e., the force normal to the direction of motion), and the drag force (i.e., the resistance to motion acting to inhibit flight). These forces are customarily expressed in terms of overall lift (C_L) and drag (C_D) coefficients according to $F_{L,D} = C_{L,D} 1/2 \rho V^2 A$, where ρ is the density of the surrounding air, V is the velocity of the aircraft, and A is the platform (i.e., the projected area responsible for lift). Gliders are light-weight aircraft designed for maximum lift and minimum drag so that they could remain aloft for a long time. Find the distance a glider with $C_L = 1.1$ and $C_D = 0.06$ will travel from a level-flight elevation of $200 \ m$ before it will reach the ground.

Governing Equations The schematic in Figure 2.6 shows the expected path of the glider. By selecting the system boundary as enveloping the glider, we have a closed system for which the conservation of momentum with respect to a coordinate system fixed to the glider (as shown in Figure 2.6) is given as

$$\frac{Dm\mathbf{V}}{Dt} \equiv \sum \mathbf{F}.$$

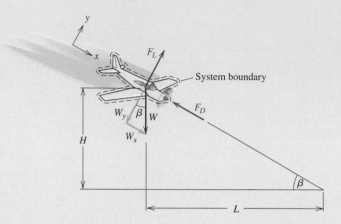

FIGURE 2.6   Schematic of Example 2.2.

Basic Assumptions   We are assuming that the glider is descending with constant velocity. The pressure around the glider is assumed to be constant.

SOLUTION   The momentum equation in terms of individual components is expressed as

$$\frac{Dmu}{Dt} \equiv \sum F_x, \quad \frac{Dmv}{Dt} \equiv \sum F_y,$$

Since the velocity is constant, the acceleration is zero — that is,

$$\frac{Dmu}{Dt} = \frac{Dmv}{Dt} = 0.$$

The forces acting on the glider are the body force due to the weight $\mathbf{W}$ that acts in the direction of gravity, the drag force $\mathbf{F}_D$ that is acting to oppose motion, and the lift force $\mathbf{F}_L$ that acts to keep the glider flying. According to the selected coordinate system

$$\mathbf{F}_D = -F_D\mathbf{i}, \quad \mathbf{F}_L = F_L\mathbf{j},$$

and

$$\mathbf{W} = W_x\mathbf{i} + W_y\mathbf{j} = W\left(\sin\beta\mathbf{i} - \cos\beta\mathbf{j}\right),$$

where β is the angle the flight path makes with respect to the horizontal direction and W is the weight. Since pressure is assumed constant around the glider, there is no net pressure force. By substitution, the two components of the momentum equation reduce to

y-direction:

$$W_y + F_L = 0 \Rightarrow -W\cos\beta + F_L = 0.$$

Therefore,

$$F_L = W\cos\beta. \tag{2.11}$$

x-direction:

$$W_x - F_D = 0 \Rightarrow W \sin \beta - F_D = 0.$$

Therefore,

$$F_D = W \sin \beta. \tag{2.12}$$

By dividing Equation (2.11) with (2.12) and by expressing F_L and F_D in terms of the lift C_L and drag C_D coefficients, we get

$$G_r \equiv \frac{F_L}{F_D} = \frac{1/2\rho V^2 A C_L}{1/2\rho V^2 A C_D} = \frac{C_L}{C_D} = \frac{cos\beta}{\sin\beta} = \frac{1}{\tan\beta} = \frac{L}{H}.$$

The lift-to-drag ratio G_r is known as the "glide ratio." From the geometry of the glider's trajectory, it is obvious that G_r is equal to the ratio of the horizontal distance traveled L to altitude lost H. In general, gliders tend to be low-performance, unpowered aircraft and sailplanes are more expensive high-performance aircraft whose lift-to-drag ratio is rather large, often in excess of 40. Therefore, sailplanes are designed to travel significant distances before reaching the ground. In this example we have

$$G_r \equiv \frac{C_L}{C_D} = \frac{1}{\tan\beta} = \frac{L}{H} = \frac{1.1}{0.06} = 18.33.$$

Therefore, for an altitude of $H = 200 \ m$, the glider will travel $L = 18.33 \times 200 \ m = 3666.7 \ m$ before reaching the ground!

2.3.1   Hydrostatics

Many practical problems are associated with the hydrostatic forces in fully or partially submerged bodies when the fluid is not moving (i.e., $\mathbf{V} = \mathbf{0}$). For example, consider a scuba diver underwater as shown in Figure 2.7. Since human physiology limits the ability of divers to dive to great depths, it is important to calculate the pressure force exerted by the water as a function of the depth. This is the same problem as finding the maximum depth to which a submarine can descend without collapsing under the pressure of the external fluid forces. Similarly, the design of hydraulic dams depends on the effects of these forces.

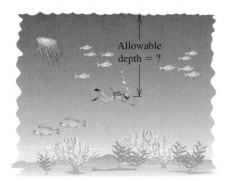

FIGURE 2.7   Because of the hydrostatic pressure, a diver experiences a net force that ultimately detrmines how deep the diver can descend.

Problem Statement   The general problem here is to determine the force experienced by a body at a certain depth resulting from the weight of the stationary fluid above it. More appropriately, though, we will determine the normal force per unit exposed area (i.e., the pressure P at a point below the free surface).

Governing Equations   By selecting as a system a column of liquid as shown in Figure 2.8, we have a closed system for which the conservation of momentum with respect to the coordinate system shown in the figure is

$$\frac{D}{Dt}\int_{\mathcal{V}}\rho\mathbf{V}d\mathcal{V}\equiv\sum\mathbf{F},\tag{2.13}$$

where $\mathcal{V}$ is the volume of the selected liquid column.

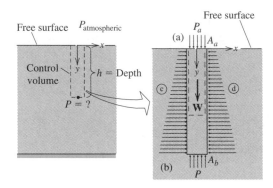

FIGURE 2.8   Balance of forces on a column of liquid below the free surface.

Basic Assumptions   In this problem, the fluid is assumed to be at rest ($\mathbf{V}=\mathbf{0}$).

SOLUTION   For static conditions ($\mathbf{V}=\mathbf{0}$), the momentum equation reduces to

$$\sum\mathbf{F}=\mathbf{0},\tag{2.14}$$

which is the well-known, almost axiomatic statement that *"sum of forces is equal to zero."* This statement is the foundation of many engineering subfields, such as statics and stress analysis in solids. The dynamics of fluids at rest is called *hydrostatics* or simply *fluid statics*.

   Note that if instead of a stationary system we have a fixed mass moving with a constant velocity $\mathbf{V}=\mathbf{V}_c$, Equation (2.13) reduces to the same result $\left(\sum\mathbf{F}=\mathbf{0}\right)$. Therefore, the derivation that follows as well as the final results are general and apply not only to the case of static fluid but also to the more general case in which the *entire mass* is moving with the same constant velocity.

   By considering both the body and surface forces acting on the closed system, we have

$$\sum\mathbf{F}=\sum\left(\mathbf{F}_{body}+\mathbf{F}_{surface}\right)=\mathbf{0}.$$

The only relevant body force is the weight of the column. Likewise, the only surface forces are those due to the pressure acting along each section of the system boundary. Hence,

$$\sum \mathbf{F} = \mathbf{F}_{weight} + \sum \mathbf{F}_{pressure} = \mathbf{0}.$$

For the general case in which the density ρ is a function of the depth, the weight in vector form is

$$\mathbf{F}_{weight} = mg\,\mathbf{j} = \mathbf{j}g \int_{\mathcal{V}} \rho d\mathcal{V} = \mathbf{j}g A_s \int_0^h \rho dy,$$

where $d\mathcal{V} = A_s dy$, with A_s being the cross-sectional area of the liquid column. For constant density, the weight is given by

$$\mathbf{F}_{weight} = mg\,\mathbf{j} = \rho \mathcal{V} g\,\mathbf{j}.$$

The force along each segment of the system boundary is evaluated by integrating the pressure according to the definition

$$\mathbf{F} = \int_{A_s} -P\mathbf{n}\,dA_s,$$

where $\mathbf{n}$ is the outward pointing unit normal vector as defined previously.

For the bottom boundary, using the coordinate system shown in Figure 2.8, we can write

$$\mathbf{F}_b = -\int_{A_b} P\mathbf{j}\,dA_b = -\mathbf{j}\int_{A_b} P\,dA_b.$$

Since the pressure along this boundary is constant, the integral simplifies to

$$\mathbf{F}_b = -PA_b\mathbf{j}.$$

The same is true for the top boundary, where the pressure is equal to the local external atmospheric pressure P_a. Hence,

$$\mathbf{F}_a = P_a A_a \mathbf{j}.$$

The momentum equation along j direction reduces to

$$P_a A_a - P A_b + g A_s \int_0^h \rho dy = 0.$$

Recognizing that $A_a = A_b = A_s$ and that $\mathcal{V}/A_s = h$, where h is strictly the depth of the liquid column (i.e., the distance from the *free surface*), we get

$$P(h) = P_a + g \int_0^h \rho dy = 0. \tag{2.15}$$

The result expressed by Equation (2.15) is not surprising. It indicates that the pressure at point below a free surface is equal to the ambient pressure at the free surface, plus the weight of the liquid column above the point per unit exposed area.

For constant density Equation (2.15) reduces to

$$P(h) = P_a + g\rho h = 0, \tag{2.16}$$

which shows that the pressure P at any point in the fluid is only a function of the depth h, defined as the vertical distance from the free surface.

Using Equation (2.15), the pressure forces along the vertical boundaries are evaluated as

$$\mathbf{F}_c = \mathbf{i}\int_{A_c} (P_a + \rho g h)\,dA_c, \quad \mathbf{F}_d = -\mathbf{i}\int_{A_d} (P_a + \rho g h)\,dA_d.$$

However, since the pressure variation along each vertical boundary is identical and $A_c = A_d$, the forces on the two vertical boundaries cancel out. Therefore, the net resultant force in the horizontal direction is zero,

$$\mathbf{F}_c + \mathbf{F}_d = \mathbf{0}.$$

This is not surprising, because according to the conservation of momentum, any finite resultant force in this direction would have otherwise induced motion by accelerating the fluid.

EXAMPLE 2.3

Problem Statement   A *manometer* is a simple device used to measure the pressure difference between two points connected by a continuous line of fluid. Consider the manometer tube shown in Figure 2.9, filled with three different incompressible fluids. The closed end of the tube is connected to a reservoir with constant pressure. For this configuration, find the pressure in the reservoir in (a) kPa; (b) as gage pressure in kPa; (c) the equivalent height of water; (d) the equivalent height of mercury.

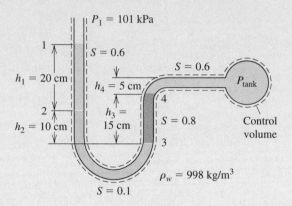

FIGURE 2.9   Pressure calculation using a manometer.

Governing Equations   By considering the entire manometer as a closed system as shown in Figure 2.9, the pressure variation within the tube is expressed by

$$P(h) = P_{0_k} + \rho_k g h_k,$$

where subscript k refers to conditions of the kth fluid. P_{0_k} is a reference pressure, and h_k is the depth from the elevation at which P_{0_k} is taken.

Basic Assumptions   The fluids in the manimeter are assumed stationary ($\mathbf{V} = \mathbf{0}$).

SOLUTION   Using the definition of the hydrostatic pressure, the solution to manometer-type problems can be facilitated by noting that

(a) *the pressure increases in the direction of increasing depth, and*

(b) *two liquid interfaces at the same elevation experience the same pressure, as long as they are connected by a continuous line of fluid particles.*

Returning to the example problem and according to the above observations, the pressure at free surface (1) is related to the pressure at (2) by

$$P_1 + \rho_1 g h_1 = P_2,$$

and the pressure at (2) is related to that at (3) by

$$P_2 + \rho_2 g h_2 = P_3.$$

Similarly,

$$P_3 - \rho_3 g h_3 = P_4,$$

and

$$P_4 - \rho_4 g h_4 = P_{tank}.$$

The negative sign in the last two expressions is included to denote the fact that by moving from interface (3) to interface (4) and from interface (4) to the tank, the depth decreases.

By combining the preceding expressions we get

$$P_1 + \rho_1 g h_1 + \rho_2 g h_2 - \rho_3 g h_3 - \rho_4 g h_4 = P_{tank}.$$

This expression can be generalized for N fluid interfaces between an initial pressure P_1 and a final pressure P_f as

$$P_f = P_1 \pm \sum_{k=2}^{N} \rho_k g h_k = P_1 \pm \rho_{water} g \sum_{k=2}^{N} SG_k h_k, \tag{2.17}$$

where SG is the specific gravity of each liquid. The positive sign $(+)$ is used whenever by going from interface $(k-1)$ to interface k, we move in the direction of increasing depth. The negative sign $(-)$ is used whenever we move in the direction of decreasing depth. The starting interface P_1, is selected to correspond to a known pressure. The reader can verify this general expression using the preceding example.

By substituting the numerical values

$$P_{tank} = 101 \ kPa + 998 \ kg/m^3 \times 9.81 \ m/s^2/1000 \times (0.6 \times 0.2 \ m +$$

$$1.0 \times 0.1 \ m - 0.8 \times 0.15 \ m - 0.6 \times 0.05 \ m) = 101.68 \ kPa.$$

The corresponding gage pressure is $P_g = P_{tank} - 101.3 \ kPa = 0.38 \ kPa$. The total pressure as an equivalent height of a water column h_w is evaluated using the definition

$$P_{tank} = \rho_w g h_w \Rightarrow h_w = \frac{P_{tank}}{\rho_w g} = \frac{101.68 \ kPa \times 1000}{998 \ m^3/kg \times 9.81 \ m/s^2} = 10.39 \ m.$$

Similarly, the same pressure is expressed in terms of an equivalent height of mercury with specific gravity $SG = 13.6$ as

$$h_m = \frac{P_{tank}}{\rho_m g} = \frac{P_{tank}}{SG \rho_w g} = \frac{101.68 \ kPa \times 1000}{13.6 \times 998 \ kg/m^3 \times 9.81 \ m/s^2} = 0.76 \ m.$$

EXAMPLE 2.4

Problem Statement   Find the force on the piston required to maintain the mercury level as shown in Figure 2.10. The mass of the piston is 2 *kg*.

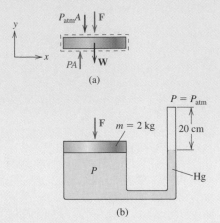

(a)

(b)

FIGURE 2.10   Schematic of Example 2.4.

Governing Equations   By considering the piston as a closed system, for static equilibrium we must have

$$\sum \mathbf{F} = \mathbf{0}.$$

The pressure variation within the tube as a function of the depth h is expressed by

$$P(h) = P_a + \rho g h,$$

where P_a is the pressure at zero depth $h = 0$. In this problem, P_a is the local atmospheric pressure.

Basic Assumptions   The pressure variation is obtained by assuming hydrostatic conditions ($\mathbf{V} = \mathbf{0}$).

SOLUTION   By considering diagram (a) and the selected coordinate system, the forces on the piston are the body force due to the weight of the piston $\mathbf{W} = -m_p g\,\mathbf{j}$ and the surface forces due to the applied force $\mathbf{F}$, the local atmospheric pressure $\mathbf{F}_a = -P_a A\mathbf{j}$, and the force due to the pressure from the liquid $\mathbf{F}_p = PA\mathbf{j}$.

For static equilibrium, we must have

$$\mathbf{W} + \mathbf{F}_a + \mathbf{F}_p + \mathbf{F} = \mathbf{0} \Rightarrow -m_p g\,\mathbf{j} - P_a A\mathbf{j} + PA\mathbf{j} + \mathbf{F} = \mathbf{0},$$

or

$$\mathbf{F} = m_p g\,\mathbf{j} + P_a A\mathbf{j} - PA\mathbf{j}.$$

Using the "manometer approach," the pressure P due to the liquid is related to the local atmospheric pressure by

$$P = P_a + \rho_{Hg}\, gh = P_a + SG_{Hg}\,\rho_w gh.$$

By combining the preceding relations

$$\mathbf{F} = m_p g\,\mathbf{j} + P_a A\mathbf{j} - \left(P_a + SG_{Hg}\rho_w gh\right) A\mathbf{j}.$$

Hence,

$$\mathbf{F} = m_p g\,\mathbf{j} - SG_{Hg}\rho_w gh A\mathbf{j}.$$

In terms of numerical values

$$\mathbf{F} = 2.0\ kg \times 9.81\ m/s^2\mathbf{j} - 13.55 \times 998\ kg/m^3 \times 9.81\ m/s^2 \times 0.2\ m \times \frac{\pi \times 0.2^2 m^2}{4},$$

$$\mathbf{F} = -813.9\ N\mathbf{j}.$$

The negative sign shows that the force is acting downwards.

As a result of the hydrostatic pressure, fully or partially submerged bodies are subjected to forces and moments. Therefore, dams, sluice gates, and the like need to be designed appropriately to withstand the effects of such forces and moments. Problems and discussion on such bodies are provided in Section 2.7.

2.3.2 The Conservation of Momentum for a Non-Inertial Coordinate System

The conservation of momentum as introduced previously implies that all the velocities and their derivatives in Equation (2.6) are expressed relative to an *inertial coordinate system*.

However, when the coordinate system is *non-inertial* and it is accelerating relative to a fixed coordinate system, the momentum equation must be modified according to

$$\underbrace{\frac{d}{dt}\int_{\mathcal{V}} \rho\mathbf{V}d\mathcal{V}}_{\text{Relative to the moving coord. system}} \equiv \sum\mathbf{F} - \underbrace{\int_{\mathcal{V}} \rho\mathbf{a}_{xyz}}_{\text{Relative to a fixed coord. system}} d\mathcal{V}, \qquad (2.18)$$

where $\mathbf{a}_{xyz}$ is the acceleration of the coordinate system relative to an inertial (i.e., fixed) coordinate system. Again, $\sum\mathbf{F}$ is the resultant of the applied forces, and $\mathcal{V}$ the volume occupied by the fixed mass. The velocity vector $\mathbf{V}$ and its time derivative are expressed relative to the moving coordinate system. This discussion applies, for instance, to the case in which the coordinate system is fixed on an accelerating car. Therefore, the term d/dt (and not D/Dt) is used here to denote the time derivative.

EXAMPLE 2.5

Problem Statement Reconsider the problem as stated in Example 2.1 by using a Cartesian coordinate system that is attached to and moving with the ball.

In this case, the coordinate system is *non-inertial* since the coordinate system is accelerating relative to a coordinate system attached to the ground (i.e., a fixed coordinate system). Obviously, then, the acceleration of the coordinate system is the same as the acceleration of the ball,

$$\mathbf{a}_{xyz} = \frac{du}{dt}\mathbf{i} + \frac{dv}{dt}\mathbf{j}.$$

The momentum equation then takes the form

$$\frac{d}{dt}\int_{\mathcal{V}} \rho\mathbf{V}d\mathcal{V} \equiv \sum\mathbf{F} - \int_{\mathcal{V}} \rho\left(\frac{du}{dt}\mathbf{i} + \frac{dv}{dt}\mathbf{j}\right)d\mathcal{V}.$$

However, since the coordinate system is attached to the ball, the acceleration of the ball relative to the moving coordinate system is zero and $\frac{d}{dt}\int_{\mathcal{V}} \rho\mathbf{V}d\mathcal{V} = \mathbf{0}$. The momentum equation then simplifies to

$$m \left(\frac{du}{dt}\mathbf{i} + \frac{dv}{dt}\mathbf{j} \right) \equiv \sum \mathbf{F},$$

which is identical to the equation used in Example 2.1. Therefore, irrespective of the choice of the coordinate system, the result is the same (as it should be, in as much as nature does not follow particular coordinate systems).

2.4   Conservation of Energy

The statement for the conservation of energy E — known also as the *first law of thermodynamics* or simply as the *energy equation* — for a closed system is also obtained from the general expression established by Equation (2.3) by using $\mathcal{A} = E$, $\alpha = \epsilon = dE/dm$, and $B =$ Rate of Net Energy Exchange as

$$\frac{D}{Dt} \int_m \epsilon \, dm \equiv \sum Rate \; of \; Net \; Energy \; Exchange. \qquad (2.19)$$

In terms of the volume ($\mathcal{V}$) that the mass of the closed system (m) occupies, Equation (2.19) becomes

$$\frac{D}{Dt} \int_{\mathcal{V}} \rho \epsilon \, d\mathcal{V} \equiv \sum Rate \; of \; Net \; Energy \; Exchange, \qquad (2.20)$$

where ϵ is the energy per unit mass of the matter that composes the system.

This rather intuitive statement expresses the fact that the rate of increase or decrease of energy inside a closed system is equal to the net rate of exchange/transfer of energy to and from the system. Since all terms in the energy equation are scalar quantities, the energy equation is a scalar equation independent of the coordinate system.

For the time being, although the concept of energy has not been formally introduced, we can express the energy per unit mass of the system ϵ as

$$\epsilon = \frac{1}{2}|\mathbf{V}|^2 + u.$$

The first term in the preceding expression is the kinetic energy per unit mass (note that often $|\mathbf{V}|^2$ is denoted by $\mathbf{V}^2$ or simply by V^2). The kinetic energy is a function of the magnitude of the macroscopic velocity $\mathbf{V}$ (i.e., the local mean of the molecular-level motions) of the fluid. The second term, u, is the internal energy of the system and corresponds to the energy associated with the fluctuations around the local mean velocity. These fluctuations are proportional to the local temperature.

In thermodynamics, it is customary to use the symbol u for the internal energy, v for specific volume, h for enthalpy, and s for specific entropy (all to be defined subsequently). However, these symbols in fluid dynamics are used for other variables such as the components of the velocity vector, the elevation and the arc length. In order to keep with tradition and to avoid possible conflict with the established practice in thermodynamics, the traditional symbols used in both fluids and thermodynamics are used here as well. However, depending on the context, each variable is clearly defined and explained to avoid confusion.

In many fluid problems, energy is exchanged by the transfer of heat across the system boundary and by work done as a result of the interaction of the boundary with

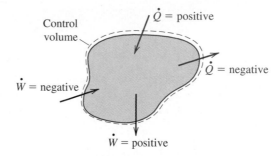

FIGURE 2.11   Sign convention consistent with Equation (2.21).

its surroundings. During heating, the total energy of the system increases, and when the system does work on its surroundings, the net energy of the system decreases. To correctly represent the physics of such energy exchanges, the conservation of energy is given as

$$\frac{D}{Dt} \int_m \epsilon \, dm \equiv \dot{\mathcal{Q}} - \dot{\mathcal{W}}. \tag{2.21}$$

In Equation (2.21), $\dot{\mathcal{Q}}$ is the net rate of heat transfer *from the system to the surroundings*, and $\dot{\mathcal{W}}$ is the net rate of work that *the system does on the surroundings* (both considered as positive quantities). The sign convention, as shown in Figure 2.11, then allows for all possible directions of energy transfer. The dot notation in $\dot{\mathcal{Q}}$ and $\dot{\mathcal{W}}$ indicates rate of change with respect to time,

$$\dot{\mathcal{Q}} \equiv \frac{\delta \mathcal{Q}}{\delta t}, \quad \text{and} \quad \dot{\mathcal{W}} \equiv \frac{\delta \mathcal{W}}{\delta t},$$

where the symbol δ is used to represent the derivative.

Mechanical Work, Energy, Power

The concept of energy can be introduced formally by using the definition of mechanical *work* ($\mathcal{W}_{mech}$)

$$\mathcal{W}_{mech} = \int \mathbf{F} \cdot d\mathbf{r}, \tag{2.22}$$

where $\mathbf{F}$ is the force vector and $\mathbf{r}$ the displacement vector. The vector product in the expression indicates that work is a scalar quantity. By definition, work is measured in units of $N \cdot m$, which is otherwise known as a Joule (J). One Joule, then, is the total work associated with a force of one Newton applied over a distance of one meter.

If the force and position vectors are collinear (acting in the same direction) the expression can be simplified to

$$W = FL,$$

where L is the total distance along which the force with magnitude F is applied.

Physically, work can be interpreted as the "effort" needed to perform a task that may involve the application of a force F over a distance L or an equivalent situation that could be presented by a weight F over a height H. An obvious example is the case of an electric motor producing work for a certain application; irrespective of the task performed by the motor, we could imagine the same motor fitted with proper attachments (i.e., gear system, pulleys) to lift a body with weight F to a height L.

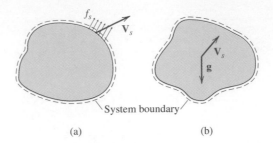

FIGURE 2.12   Work done by surface (a) and body (b) forces acting on the system.

Energy is defined as a quantity that has the "potential" of producing work. This work-producing potential may be associated with the physical condition of the system, such as its chemistry, motion, relative location, or temperature, in which cases it is called, respectively, chemical, kinetic, potential, or thermal energy.

The rate at which energy E is spent or produced is defined as power $\mathcal{P}$,

$$\mathcal{P} = \frac{dE}{dt} \equiv \dot{E},$$

which is measured in *Joules per second* otherwise known as a *Watt* (W).

Energy can also change forms such as in electric power generation where chemical energy stored in oil is changed after combustion into thermal energy and then converted into kinetic energy by the turbine. Finally, this energy is transformed into electrical energy by a generator.

The rate of work $\dot{W}$ in the energy equation accounts for the rate at which the applied forces (surface and body forces) do work on the system. Therefore, the total rate of work $\dot{W}$ includes the rate of work done by the body $\dot{W}_b$ and surface forces $\dot{W}_s$ as

$$\dot{W} = \dot{W}_s + \dot{W}_b,$$

The rate of work due to the surface forces per unit surface area $\mathbf{f}_s$ is

$$\dot{W}_s = \int_{A_s} (\mathbf{f}_s \cdot \mathbf{V_s}) \, dA_s,$$

where $\mathbf{V_s}$ is the velocity of the point of application of the force at the surface element dA_s (Figure 2.12a).

Similarly, the rate of work due to the body force is

$$\dot{W}_b = \int_{\mathcal{V}} \rho \, (\mathbf{g} \cdot \mathbf{V_s}) \, d\mathcal{V},$$

where $\mathbf{V}_s$ is the velocity of the mass enclosed by the system's volume (Figure 2.12b). The energy due to the body force is known as *potential energy* (PE), defined as

$$PE = mgz,$$

where z is the elevation from an arbitrary reference plane normal to the direction of gravity.

The rate at which potential energy changes with respect to time over the system's volume is expressed as

$$\dot{PE} = \frac{D}{Dt} \int_{\mathcal{V}} \rho g z \, d\mathcal{V}.$$

Traditionally, the potential energy per unit mass is combined with the internal and kinetic energy terms to form the total energy of the entire system as

$$\epsilon = \frac{1}{2}V^2 + gz + u.$$

The final form of the energy equation then is

$$\frac{D}{Dt}\int_m \epsilon \, dm = \frac{D}{Dt}\int_{\mathcal{V}} \rho \underbrace{\left(\frac{1}{2}V^2 + gz + u\right)}_{\epsilon} d\mathcal{V} \equiv \dot{Q} - \dot{\mathcal{W}}_s, \qquad (2.23)$$

where $\dot{\mathcal{W}}_s$ is strictly the rate of work of the surface forces acting at the boundary of the system.

To include other forms of energy such as chemical or nuclear, ϵ must be modified accordingly as

$$\epsilon = \frac{1}{2}V^2 + gz + u + \epsilon_{chem} + \epsilon_{nuclear}\ldots + other\ forms\ of\ energy.$$

Thermal Energy, Heat Transfer

Thermal energy is the form of energy that is associated with the temperature of the physical system, whose transfer is commonly referred to as *heat transfer*. Heat is transferred by one or a combination of three distinct modes: conduction, convection, and radiation. In the absence of heat transfer, the process is called *adiabatic*. A well-insulated surface, for instance, can be assumed to be adiabatic.

Conduction is the mode of heat transfer in both fluid and solid media due to molecular-level diffusion. Using *Fourier's law of conduction* (which in reality is a constitutive equation and not a law), the rate of heat transfer $\dot{Q}_{cond}$ is assumed to be proportional to the conducting area A_s and to the temperature gradient $\dfrac{\partial T}{\partial n}$ normal to that area

$$\dot{Q}_{cond} = -A_s k \frac{\partial T}{\partial n}.$$

The proportionality constant k is a material constant known as the thermal conductivity. Depending on the value of k, different materials conduct heat at different rates. The magnitude of k is determined experimentally for different materials. The negative sign is included to indicate the fact that heat is conducted in the direction of decreasing temperature (i.e., from higher to lower temperature).

Convection is similar to heat conduction with the exception that it takes place only in the presence of fluid flow. Using *Newton's law of cooling* (which again is not a law but a constitutive equation), the rate of heat transfer by convection is assumed to be proportional to the temperature difference ΔT between the exchanging bodies—for instance, between a liquid and a solid.

$$\dot{Q}_{conv} = A_s h_c \Delta T,$$

where the proportionality constant h_c is the convection heat transfer coefficient determined experimentally. Convection heat transfer is traditionally distinguished as *forced convection* and *free or natural convection*, depending respectively, on whether the flow is forced by mechanical or other means or is simply allowed to flow freely without external forcing—for instance, due to density variations throughout the fluid. A fan in a room produces cooling by forced convection, while water in a kettle heats up by free convection.

Radiation heat transfer occurs because all objects emit discrete packets of energy called photons. These photons impinge on other objects, become absorbed, and transmit their energy to the object. The energy transfer can take place in the absence of any medium (either solid or liquid). Heat transferred from the sun, for instance, takes place by radiation. The rate of heat transfer by radiation between two objects at absolute temperatures T_1 and T_2 is given by

$$\dot{Q}_R = e\sigma\, F_{12} A_s \left(T_1^4 - T_2^4 \right),$$

where σ is the Stefan-Boltzmann constant. The numerical value of the constant is $5.6 \times 10^{-8}\ Watts/m^2\ K^4$. The quantity e is a dimensionless number between 0 and 1, which compares the photon emission of the objects in question with that of an ideal emitter (a black body). The factor F_{12} is a view factor that quantifies the geometrical relationship between objects as radiation leaving one object may be intercepted by the body itself, and only a fraction of it arrives at the other surface.

EXAMPLE 2.6

Problem Statement   Show that the total energy E_T during the trajectory of the ball in Example 2.1 is conserved.

Governing Equations   The total energy within the closed system is the sum of the internal energy (U), the kinetic energy (KE), and the potential energy (PE). Therefore,

$$E_T = U + KE + PE.$$

Basic Assumptions   We will assume that the ball is in thermal equilibrium with its surroundings, which implies that there is no heat transfer between the ball and its surroundings.

SOLUTION   For isothermal conditions, the change in the internal energy is zero, $\Delta U = 0$. Using the results of Example 2.1:

$$u = V_o \cos\alpha, \quad x = (V_o \cos\alpha)\, t,$$

$$v = V_o \sin\alpha - gt, \quad y = (V_o \sin\alpha)\, t - g\frac{t^2}{2}.$$

The kinetic energy is

$$KE = m\frac{u^2 + v^2}{2} = \frac{m}{2}\left[(V_o \cos\alpha)^2 + (V_o \sin\alpha - gt)^2 \right],$$

$$KE = m\left[\frac{V_o^2}{2} - gt V_o \sin\alpha + \frac{g^2 t^2}{2} \right],$$

and the potential energy, with respect to the ground is

$$PE = mgy = mg\left[(V_o \sin\alpha)\, t - g\frac{t^2}{2} \right].$$

By adding the two forms of energy we get

$$E_T = m\frac{V_o^2}{2}.$$

Since both the mass m and velocity V_o are constant, the total energy E_T is conserved.

EXAMPLE 2.7

Problem Statement *Work done during a polytropic process.* A cylinder filled with gas is fitted with a piston. Find the work done by the piston during a special process where, while the piston is moving, the pressure and volume are related by $PV^n = C$, where C and n are constants. This particular process is known as a polytropic process.

Governing Equations By definition, the work done is defined as

$$W = \int_s \mathbf{F} \cdot d\mathbf{r},$$

where the integral is performed along the path of the piston.

Basic Assumptions: Assume that the process is closely controlled so that the pressure and volume are related as described.

SOLUTION By multiplying and dividing by the cross-sectional area A_s, the expression for work can be expressed as

$$dW = \underbrace{\frac{\mathbf{F}}{A_s}}_{pressure} \cdot \underbrace{A_s d\mathbf{r}}_{volume} = P dV.$$

Using $P = C/V^n$,

$$W = \int_{V_1}^{V_2} P dV = \int_{V_1}^{V_2} \frac{C}{V^n} dV,$$

then,

$$W = C \int_{V_1}^{V_2} \frac{dV}{V^n} = C \frac{\left(V_2^{1-n} - V_1^{1-n}\right)}{1 - n}.$$

Since $C = P_1 V_1^n = P_2 V_2^n$, we get

$$W = \frac{(P_2 V_2 - P_1 V_1)}{1 - n}.$$

The preceding expression is valid for all n except $n = 1$.

EXAMPLE 2.8

Problem Statement Consider the gas within a piston-cylinder arrangement as shown in Figure 2.13. During a part of the power generation cycle, $550 \; kJ/kg$ of heat is added while the piston is still moving. If during the process the net increase of the energy is $1300 \; kJ/kg$, find the mechanical work associated with the movement of the piston.

Governing Equations By selecting the gas within the cylinder as a closed system, the conservation of energy requires

$$\dot{E} = \dot{Q} - \dot{W},$$

where E is the total energy within the closed system.

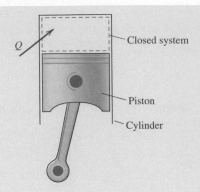

FIGURE 2.13 Piston-cylinder arrangement.

Basic Assumptions   Assume that the gas in the piston is at thermal equilibrium with the surroundings.

SOLUTION   Integrating the energy equation between the original and final thermodynamic states and dividing by the total mass, we have

$$\Delta\epsilon = \Delta q - \Delta w = q_{1\to2} - w_{1\to2},$$

where $\epsilon = E/m$, $q = Q/m$, and $w = W/m$. $q_{1\to2}$ shows the total amount of heat transfer as the system changes from the original to the final thermodynamic state. Instead of $q_{1\to2}$, we could have written $q_2 - q_1$. However, this is not quite correct, since heat transfer is an integral quantity that must be integrated along the path of the process. Therefore, heat transfer between two states (1) and (2) is denoted by $q_{1\to2}$. The same also holds for the work term.

Following the statement of the problem, $550\ kJ/kg$ of heat is added. Therefore, $\Delta q = q_{1\to2}$ is a positive quantity. Similarly, since the net energy of the system increased by $1300\ kJ/kg$, $\Delta\epsilon = \epsilon_2 - \epsilon_1$ is also a positive quantity. Substituting the numerical values into the conservation of energy yields

$$1300\ kJ/kg = 550\ kJ/kg - w_{1\to2}.$$

Therefore,

$$w_{1\to2} = 550\ kJ/kg - 1300\ kJ/kg = -750\ kJ/kg.$$

According to the sign convention, the negative sign shows that the piston is doing work on the gas (i.e., it is compressing the gas).

2.5   Second Law of Thermodynamics

The second law of thermodynamics stipulates that any physically realizable process must satisfy the condition

$$\left[\frac{D\mathcal{S}}{Dt}\right]_{isolated\ closed\ system} \geq 0, \tag{2.24}$$

where $\mathcal{S}$ is the entropy defined using the heat transfer Q and the absolute temperature T as

$$dS \equiv \frac{\delta Q}{T}.$$

The above statement is true only when the closed system is thermally *isolated*. This means that the system *does not thermally interact with its surroundings*. A well-insulated body, for instance, can be such an isolated system.

This law provides the means by which to identify processes that cannot physically exist. For instance, the second law of thermodynamics can evaluate the validity of the statement that "a cup of warm coffee, say at $35°C$, left undisturbed for certain time in a constant temperature room $(25°C)$ will heat up to $45°C$." The first law of thermodynamics for this problem predicts that as long as heat is transferred at a sufficient rate, the coffee will heat up! In other words, it fails to show that, in the absence of external energy input, it is physically impossible to have heat flow from a body of a lower temperature lower to one of a higher temperature.

The second law of thermodynamics, however, as shown subsequently, proves that the process is not possible. Since heat is transferred between the cup $\mathcal{V}_c$ and the room $\mathcal{V}_r$ these two systems are thermally interacting. However, it is reasonable to assume that the combined system $\mathcal{V}_c + \mathcal{V}_r$ is a thermally isolated system (Figure 2.14).

The net entropy change for the combined control volume then is

$$\Delta \mathcal{S}_{isolated\ control\ volume} = \Delta \mathcal{S}_c + \Delta \mathcal{S}_r,$$

where $\Delta \mathcal{S}_c$ and $\Delta \mathcal{S}_r$ are, respectively, the entropy change for the cup and the room.

In order for the coffee to heat up, a total heat (ΔQ) must be transferred from the room to the cup. While heat is being transferred to the coffee, its temperature T_c changes from $308\ K$ to $318\ K$. For the purpose of calculating the entropy change, we use a constant average temperatue of $313\ K$. According to the sign convention, then, we have

$$\Delta \mathcal{S}_c = +\frac{\Delta Q}{T_c} = +\frac{\Delta Q}{313}.$$

Since the same amount of heat must come from the surroundings, the entropy change for the surroundings is

$$\Delta \mathcal{S}_r = -\frac{\Delta Q}{T_r} = -\frac{\Delta Q}{298},$$

where T_r is the absolute room temperature $(T_r = 298\ K)$. The negative sign for ΔQ reflects the fact that heat *leaves* the surroundings.

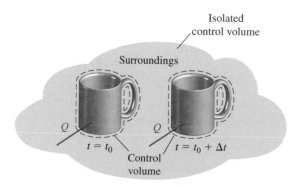

FIGURE 2.14   Schematic of a cup with thermal interaction with its surroundings.

The net entropy change for the combined system then is

$$\Delta S_{isolated \ control \ volume} = \Delta Q \left(\frac{1}{313} - \frac{1}{298} \right) < 0.$$

Obviously, since the expression is negative, it violates the second law of thermodynamics. Formally, then, we conclude that this process cannot physically take place. Under the same conditions, the reader can use the same reasoning to show that it is possible for the cup to cool from $35°C$ to a room temperature of $25°C$.

2.6   Review of Basic Thermodynamic Principles[3]

In Section 2.4, the symbols D, d, and δ were used to represent differential quantities. Consistent with the traditional mathematical notation, the symbol d denotes an exact differential, where as δ denotes an inexact differential. This distinction is not coincidental but intentional, in order to help introduce some important thermodynamic concepts.

The exact differential dA expresses the change of A, as $dA = A_2 - A_1$, where subscripts 1, 2 indicate the two end conditions. In other words, the difference is independent of the actual path that the quantity follows in changing from (1) to (2). Physical quantities that obey this rule are known as *properties*. Formally, a property is defined as an observable, macroscopic physical quantity whose change depends on the end conditions, not the actual path followed. Typical properties are temperature, pressure, density, and internal energy. Properties that are independent of the mass, such as pressure, temperature, and density are called *intensive* properties. *Extensive* properties are those properties (e.g., internal energy, mass, total volume) whose values vary directly with the mass. Extensive properties can be made into intensive properties by dividing by the mass. For instance, the total volume $\mathcal{V}$ divided by the mass m yields the specific volume $v = \mathcal{V}/m$, which is an intensive property. Traditionally, upper-case letters are used to signify extensive properties, and lower-case letters signify intensive properties. For instance, the specific enthalpy h is denoted by $h = H/m$.

A chemically homogeneous and invariable substance is known as a *pure substance*. The *thermodynamic state* of a pure substance is a condition of the system characterized by a number of properties. In order to uniquely define a state, we need two independent intensive properties. In the two-phase region, while the substance is *changing phase*—for example, from liquid to gas—both the pressure and temperature remain constant. Therefore, in this region the pressure and temperature are not independent properties. Once the thermodynamic state is uniquely defined, intensive properties at that state have the same numerical value. For example, water vapor exposed to atmospheric pressure boils at a temperature of $100°C$. The *thermodynamic path* followed by a system changing from an initial thermodynamic state to a final state is, therefore, the result of all the thermodynamic states that the system goes through to reach the final state (Figure 2.15).

A system under *thermodynamic equilibrium* is one in which there is no tendency for change in the thermodynamic state anywhere inside a control volume (open or closed systems). In this case, the thermodynamic state of a substance has meaning for the entire control volume. *Thermal*, *mechanical*, and *chemical* equilibrium states are defined, similarly, as situations in which there is no tendency for changes in the temperature, pressure, and chemical composition, respectively.

Unlike properties, heat transfer and work done are quantities that depend on the path followed by the system; therefore, they are evaluated by proper integration along that

---

[3]This section can be skipped without loss of continuity.

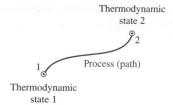

FIGURE 2.15 Schematic of a thermodynamic process.

path. For this reason, their change is expressed mathematically as inexact differentials. Consequently, heat transfer and work are not properties. For example, the work done by a force $\mathbf{F}$ acting on a system boundary is given as

$$\mathcal{W} = \int_{path} \mathbf{F} \cdot d\mathbf{r},$$

where $\mathbf{r}$ is the displacement vector. Therefore, the final result depends on $\mathbf{r}$ and, hence, on the path followed by the boundary.

Using the above thermodynamic concepts and definitions, we can define some other useful properties. The specific heat coefficient under constant volume C_v is defined as

$$\left. \frac{\partial u}{\partial T} \right|_{constant\ volume} \equiv C_v,$$

where u is the specific internal energy and T is the absolute temperature. Similarly, the specific heat under constant pressure C_p is defined as

$$\left. \frac{\partial h}{\partial T} \right|_{constant\ pressure} \equiv C_p,$$

where h is the specific enthalpy defined as $h = u + Pv$ (note the use of u, v, and h to denote the internal energy, specific volume, and enthalpy, respectively). Traditionally, the ratio $\gamma = C_p/C_v$ is called the *specific heat ratio*.

For incompressible fluids and solids, $d(pv) \approx 0$. Therefore, $dh \approx du$ and $C_v \approx C_p = C$, where C is again a constant specific to the fluid or solid.

2.6.1 Dynamics of Ideal Gases

During changes in their thermodynamic state, gases such as helium and argon as well as other gases at relatively low pressure (or density) behave in a manner that can be expressed by the simple constitutive relation

$$PV = mRT, \tag{2.25}$$

where P and T are the absolute pressure and temperature. V is the volume and m is the mass of the gas. R is a constant specific to the gas under consideration, and it is related to the universal gas constant $\bar{R} = 8.3144\ kNm/kmol\ K$ through $R = \bar{R}/M$, where M is the molecular weight of the gas.

Equation (2.25) is the well-known "ideal gas relation," which is valid only under special conditions. Gases that follow this constitutive relation are known as ideal gases. The constitutive relations of many pure substances are often given in tabular form. The properties of water and steam, for instance, are given by the well-known steam tables.

In 1843, Joule demonstrated experimentally that the enthalpy and internal energy for ideal gases are only functions of the temperature — that is, $u = u(T)$ and $h = h(T)$. Therefore,

$$\frac{\partial u}{\partial T} = \frac{du}{dT} = C_v,$$

and

$$\frac{\partial h}{\partial T} = \frac{dh}{dT} = C_p.$$

By definition,

$$h = u + Pv.$$

then,

$$dh = du + d(Pv).$$

Using the ideal gas relation,

$$dh = du + RdT.$$

Furthermore, using the definition of the specific heats

$$\underbrace{C_p \, dT}_{dh} = \underbrace{C_v \, dT}_{du} + R \, dT,$$

which yields

$$C_p - C_v = R.$$

The above relation can be manipulated further by first dividing it by C_p and then by C_v. After rearrangement, we get the following useful relations in terms of γ:

$$\frac{C_p}{R} = \frac{\gamma}{\gamma - 1} \quad \text{and} \quad \frac{C_v}{R} = \frac{1}{\gamma - 1}.$$

Isentropic Flow

Consider a closed system with changing thermodynamic state. If we assume that the mechanical work is due to a simple boundary movement, the first law of thermodynamics can be expressed as

$$dU = \delta Q - P d\mathcal{V},$$

where Q is the heat transfer, U is the internal energy, P is the pressure, and $\mathcal{V}$ is the volume. Using the definition of entropy $\delta Q = T d\mathcal{S}$,

$$dU = T d\mathcal{S} - P d\mathcal{V}.$$

However,

$$P d\mathcal{V} = d(P\mathcal{V}) - \mathcal{V} dP.$$

Therefore, $dU = T d\mathcal{S} - d(P\mathcal{V}) + \mathcal{V} dP$, which yields

$$d(U + P\mathcal{V}) = T d\mathcal{S} + \mathcal{V} dP.$$

Using the definition of enthalpy H and by dividing by the mass m, we get

$$T ds = dh - \frac{dP}{\rho}.$$

Note the use of the symbol s as the entropy per unit mass. Thus, the entropy change for an ideal gas can be expressed as

$$ds = \frac{dh}{T} - \frac{dP}{T\rho} = \frac{dh}{T} - R\frac{dP}{P}.$$

By integration,

$$s_2 - s_1 = \int_1^2 C_p(T)\frac{dT}{T} - R\ln\frac{P_2}{P_1}.$$

Additionally, if we assume that C_p is constant, we have

$$s_2 - s_1 = C_p \ln\frac{T_2}{T_1} - R\ln\frac{P_2}{P_1}.$$

For the special process in which the entropy remains constant, the flow is said to be isentropic and $s_2 - s_1 = 0$. Therefore,

$$C_p \ln\frac{T_2}{T_1} = R\ln\frac{P_2}{P_1},$$

which gives

$$\ln\frac{T_2}{T_1} = \frac{R}{C_p}\ln\frac{P_2}{P_1}.$$

In terms of the specific heat ratio γ,

$$\frac{T_2}{T_1} = \frac{\gamma - 1}{\gamma}\ln\frac{P_2}{P_1} \rightarrow \frac{T_2}{T_1} = \left(\frac{P_2}{P_1}\right)^{\frac{\gamma-1}{\gamma}}.$$

Using the ideal gas relation, the preceding expression can be manipulated to yield

$$P_1 v_1^\gamma = P_2 v_2^\gamma,$$

which shows that isentropic flow is a special case of a polytropic process $P_1 v_1^n = P_2 v_2^n$, when $n = \gamma$.

EXAMPLE 2.9

Problem Statement *Cooling by quenching.* A $10\ kg$ copper block at $200°$ is quenched in $100\ kg$ of water (Figure 2.16). If the initial temperature of the water is $25°$, find the final temperature of the water and the copper block ($C_{copper} = 0.386\ kJ/kgK$, $C_{water} = 4.184\ kJ/kgK$).

Governing Equations By selecting the entire steel block, we have a closed system for which the energy equation is expressed as

$$\dot{E} = \dot{Q} - \dot{W}.$$

Basic Assumptions Assume that sufficient time is allowed so that at the final state the steel block reaches thermal equilibrium.

SOLUTION Using the assumption of final thermal equilibrium by integrating in time

$$E_2 - E_1 = Q_{1\rightarrow2} - W_{1\rightarrow2},$$

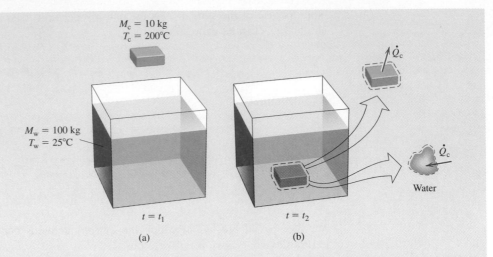

$M_c = 10\ \text{kg}$
$T_c = 200°\text{C}$

$M_w = 100\ \text{kg}$
$T_w = 25°\text{C}$

$\dot{Q}_c$

$\dot{Q}_c$

Water

$t = t_1$ $t = t_2$

(a)   (b)

FIGURE 2.16   Schematic of Example 2.9.

where, as established earlier, $E = U + KE + PE$. $\mathcal{Q}_{1\to2}$ shows the total amount of heat transfer as the system goes from the original to the final thermodynamic state. Again, since $\mathcal{Q}_{1\to2}$, and $\mathcal{W}_{1\to2}$ are integral quantities, they are not thermodynamic properties.

Since the block is not moving, $KE = 0$, and as long as the height from an arbitrary reference level remains constant, $\Delta PE = 0$. A check around the boundary of the control volume shows that there is no work associated with this problem (i.e., $\mathcal{W}_{1\to2} = 0$). However, the block *loses* heat $\mathcal{Q}_{1\to2}$ by a combination of conduction and convection. Therefore, the energy equation simplifies to

$$U_2 - U_1|_{block} = -\,\mathcal{Q}_{1\to2}|_{block}\,. \qquad (2.26)$$

The negative sign was included to indicate that heat is being lost (i.e., heat leaves the control volume). Using identical arguments, the energy equation for the water bath simplifies to

$$U_2 - U_1|_{water} = \mathcal{Q}_{1\to2}|_{water}\,, \qquad (2.27)$$

where the positive sign was included to indicate that the water receives heat. Physically this is obvious; the heat lost by the block is gained by the water—that is, $\mathcal{Q}_{1\to2}|_{block} = \mathcal{Q}_{1\to2}|_{water}$. By adding Equations (2.26) and (2.27),

$$U_2 - U_1|_{block} + U_2 - U_1|_{water} = 0,$$

which, in terms of specific quantities, is expressed as

$$m_{block}\,(u_2 - u_1)_{block} + m_{water}\,(u_2 - u_1)_{water} = 0.$$

Using the definition of C, the internal energy change for a solid is expressed in terms of the temperature $u_2 - u_1 = C\,(T_2 - T_1)$. Therefore,

$$(mC)_{block}\,(T_2 - T_1)_{block} + (mC)_{water}\,(T_2 - T_1)_{water} = 0,$$

$$\Rightarrow T_2 = \frac{(mCT_1)_{block} + (mCT_1)_{water}}{(mC)_{block} + (mC)_{water}}.$$

In terms of numerical values,

$$T_2 = \frac{(10\ kg \times 0.386\ kJ/kgK \times 473\ K)_{block} + (100\ kg \times 4.184\ kJ/kgK \times 298\ K)_{water}}{(10\ kg \times 0.386\ kJ/kgK)_{block} + (100\ kg \times 4.184\ kJ/kgK)_{water}},$$

$$\Rightarrow T_2 = 299.6\ K.$$

During cooling, the block loses heat equal to

$$Q_{1 \to 2} = mC\,(T_2 - T_1) = 10\ kg \times 0.386\ kJ/kgK \times (299.6\ K - 473\ K) = -669.3\ kJ.$$

Alternatively, the above result could be obtained by considering the block and the water as a single system. For this, we can write

$$\Delta E = 0 \Rightarrow E_2 = E_1, \tag{2.28}$$

where $E = U_{block} + U_{water}$. The result obtained using this approach is (as it should be) identical to the one obtained previously.

EXAMPLE 2.10

Problem Statement   A $10\ kg$ object is falling from a height of $10\ m$ (Figure 2.17). Using energy concepts, derive an expression for the final velocity attained before the object hits the ground.

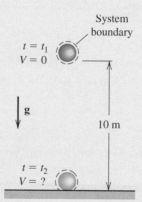

FIGURE 2.17   Schematic of Example 2.10.

Governing Equations   Selecting the solid object, we have, again, a closed system for which the conservation of energy is

$$\dot{E} = \dot{Q} - \dot{W}.$$

Basic Assumptions   Assume that the body is in thermal equilibrium with its surroundings (i.e., there is no heat transfer between the body and its surroundings).

SOLUTION   In the absence of heat transfer and work, the energy equation upon integration between the original and final states reduces to

$$\Delta E = 0 \Rightarrow E_1 = E_2.$$

The total energy is made up of the internal U, kinetic KE, and potential PE energies — that is, $E = U + (KE) + (PE)$. During falling, then,

$$\Delta U + \Delta (KE) + \Delta (PE) = 0,$$

or

$$U_2 - U_1 + (KE)_2 - (KE)_1 + (PE)_2 - (PE)_1 = 0.$$

For isothermal conditions, $U_1 = U_2$. Prior to release, the velocity of the object $V = 0$; therefore, $(KE)_1 = 0$. The initial potential energy, with respect to the ground as the *arbitrary* reference frame, is $(PE)_1 = mgH$, where H is the maximum height. At the moment the object hits the ground, $(PE)_2 = 0$ and $(KE)_2 = \frac{1}{2}mV_2^2$. At this point, the velocity is maximum. Therefore,

$$\underbrace{(U_2 - U_1)}_{\Delta U = 0} + \underbrace{(mgH - 0)}_{\Delta PE} + \underbrace{\left(0 - \frac{1}{2}mV^2\right)}_{\Delta KE} = 0.$$

The final velocity is then

$$V = \sqrt{2gH} = \sqrt{2 \times 9.81 \ m/s^2 \times 10 \ m} = 14.01 \ m/s.$$

The same result could also be obtained by considering the momentum equation

$$\frac{DV}{Dt} = +g.$$

The positive sign is consistent with a coordinate system pointing downwards. The velocity V and distance traveled y are, respectively,

$$V = gt \quad \text{and} \quad y = g\frac{t^2}{2}.$$

The object arrives at the ground after $t = \sqrt{2H/g}$ seconds. At that moment, according to the velocity expression, the actual velocity is $V = \sqrt{2gH}$, which is identical to the result obtained previously.

EXAMPLE 2.11

Problem Statement   A cylinder of 20 cm diameter, filled with air, is fitted with a 20 kg piston, as shown in Figure 2.18. Find the amount of heat transfer needed to bring the piston to the stops as shown in the figure. Assume that the properties of the air inside the cylinder can be described using the ideal gas constitutive relation $PV = mRT$.

Governing Equations   Considering the gas as the control volume as shown in Figure 2.18, the conservation of energy is given as

$$\dot{E} = \dot{Q} - \dot{W}.$$

Basic Assumptions   It is assumed that the process happens slowly so that acceleration of the system is negligible. Moreover, sufficient time is allowed so that the system reaches

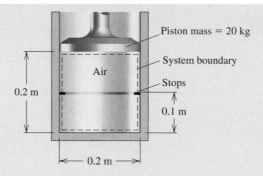

FIGURE 2.18   Schematic of Example 2.11.

thermal equilibrium. It is also assumed that changes in the kinetic and potential energies are small.

SOLUTION   Using the assumptions, the energy equation simplifies to

$$\Delta E = E_2 - E_1 = U_2 - U_1 = \mathcal{Q}_{1\to2} - \mathcal{W}_{1\to2}.$$

Since the contents of the gas are under a constant force determined by the weight of the piston and the ambient atmospheric pressure, the process is isobaric (constant pressure).

$$P_{gas} = P_a + \frac{mg}{Area}.$$

In this example,

$$P_{gas} = 101.3\ kPa + \frac{20\ kg \times 9.81\ m/s^2}{\pi \times 0.1^2 m^2 \times 1000} = 107.55\ kPa.$$

Using the definition $\mathcal{Q}_{1\to2} = mC_v\,(T_2 - T_1)$ and $\mathcal{W}_{1\to2} = P_{gas}\,(\mathcal{V}_2 - \mathcal{V}_1)$, the heat transfer needed is then

$$\mathcal{Q}_{1\to2} = mC_v\,(T_2 - T_1) + P_{gas}\,(\mathcal{V}_2 - \mathcal{V}_1).$$

According to the ideal gas relation, $\dfrac{P\mathcal{V}}{R} = mT$. Therefore,

$$\mathcal{Q}_{1\to2} = C_v \frac{P_g}{R}\,(\mathcal{V}_2 - \mathcal{V}_1) + P_{gas}\,(\mathcal{V}_2 - \mathcal{V}_1),$$

which simplifies to

$$\mathcal{Q}_{1\to2} = P_{gas}\left(\frac{C_v + R}{R}\right)(\mathcal{V}_2 - \mathcal{V}_1).$$

For air, $C_v = 0.7165\ kJ/kg$ and $R = 0.287\ kJ/kg$, and for this problem, $\mathcal{V}_2 - \mathcal{V}_1 = \pi r^2\,(h_2 - h_1)$. Upon substitution,

$$\mathcal{Q}_{1\to2} = 101.3\ kPa \times \pi \times 0.1^2\ m^2 \times (0.1\ m - 0.2\ m) \times \left(\frac{0.7165 + 0.287}{0.287}\right),$$

$$\mathcal{Q}_{1\to2} = -1.11\ kJ/kg.$$

EXAMPLE 2.12

Problem Statement Consider the cooling of the block discussed in Example 2.12. Find the total entropy change for (a) the block; (b) the water; and (c) the combined block and water combination.

Governing Equations According to the definition, the entropy change for either solids or liquids (since $dP \approx 0$) is given by

$$T \, d\mathcal{S} = \delta Q = mC \, dT \Rightarrow \mathcal{S}_2 - \mathcal{S}_1 = mC \ln\left(\frac{T_2}{T_1}\right).$$

Assumptions Assume that sufficient time is allowed so that at the final state, the control closed system reaches thermal equilibrium.

SOLUTION Using the definition, the entropy change for the copper is

$$\Delta \mathcal{S}_{copper} = 10 \; kg \times 0.386 \; kJ/kgK \times \ln\left(\frac{299.6 \; K}{473 \; K}\right) = -1.763 \; kJ/K,$$

and for the water,

$$\Delta \mathcal{S}_{water} = 100 \; kg \times 4.184 \; kJ/kgK \times \ln\left(\frac{299.6 \; K}{298 \; K}\right) = 2.240 \; kJ/K.$$

The total entropy change is, then,

$$\Delta \mathcal{S}_{total} = \Delta \mathcal{S}_{copper} + \Delta \mathcal{S}_{water} = 0.477 \; kJ/K.$$

EXAMPLE 2.13

Problem Statement Air originally at 408 kPa and 610 K expands to 150 kPa and 500 K. Find the entropy change for this process. If instead the process is isothermal, for the same pressure conditions find the entropy change and the net heat transfer per unit mass during the process.

Governing Equations By choosing the entire air content as the control volume, the second law of thermodynamics per unit mass is given as

$$ds = C_p \ln\left(\frac{T_2}{T_1}\right) - R \ln\left(\frac{P_2}{P_1}\right).$$

Assumptions Assume that sufficient time is allowed so that at the final state the closed system reaches thermal equilibrium.

SOLUTION The second law of thermodynamics can be readily used to evaluate the entropy change as

$$ds = 1.0035 \; kJ/kgK \times \ln\left(\frac{500 \; K}{610 \; K}\right) - 0.287 \; kJ/kgK \times \ln\left(\frac{150 \; kPa}{408 \; kPa}\right),$$

$$ds = 0.0876 \; kJ/kgK.$$

For an isothermal process, the entropy change is

$$ds = -R \ln\left(\frac{P_2}{P_1}\right) = -0.287 \; kJ/kgK \times \ln\left(\frac{150}{408}\right) = 0.287 \; kJ/kgK.$$

The total heat transferred dq per unit mass is

$$dq = T\,ds = 610\ K \times 0.287\ kJ/kgK = 175.1\ kJ/kg.$$

2.7 Hydrostatic Forces on Submerged Bodies[4]

The pressure distribution under static conditions (or when the entire fluid is moving with constant velocity) given by Equation (2.16),

$$P(h) = P_a + \rho g h,$$

is a fundamental result that can be readily used to calculate the forces and moments on fully or partially submerged bodies.

Consider the schematic of an arbitrary two-dimensional body as shown in Figure 2.19. As a result of the hydrostatic pressure on a small part of the surface of the body, dA experiences a force $d\mathbf{F}$ according to

$$d\mathbf{F} = -P\,\mathbf{n}\,dA,$$

where $\mathbf{n}$ is the *outward pointing unit normal vector.*

The net force then on the entire surface is obtained by integration

$$\mathbf{F}_{net} = -\int_A P\,\mathbf{n}\,dA.$$

For the surface shown in the figure, the outward unit normal vector is $\mathbf{n} = \dfrac{dy}{ds}\mathbf{i} - \dfrac{dx}{ds}\mathbf{j}$ and $dA = W\,ds$, where s is the distance along the boundary measured in the counterclockwise direction (arc length) and W is the width of the surface (note the use of s here as the arc length). The net force $\mathbf{F}_{net}$ then is

$$\mathbf{F}_{net} = -\int_0^{s_{max}} P\left(\frac{dy}{ds}\mathbf{i} - \frac{dx}{ds}\mathbf{j}\right) W\,ds,$$

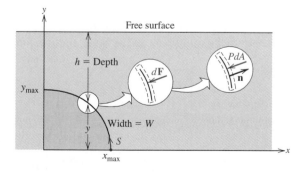

FIGURE 2.19   Forces on submerged bodies.

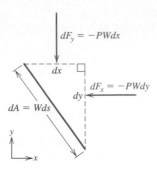

FIGURE 2.20   Alternative calculation of the forces on submerged bodies.

where the integration now is over the entire bounding surface of the body. By substitution and by multiplying through by ds,

$$\mathbf{F}_{net} = -\int_A P\,(dy\,\mathbf{i} - dx\,\mathbf{j})\,W.$$

In terms of the individual components $\mathbf{F} = F_x\mathbf{i} + F_y\mathbf{j}$,

$$F_x = -\int_0^{y_{max}} PW dy, \tag{2.29}$$

and

$$F_y = \int_{x_{max}}^{0} PW dx = -\int_0^{x_{max}} PW dx. \tag{2.30}$$

Note that the same result could have been obtained by using the projections dx and dy of the elementary area dA normal to the y-axis and x-axis as shown in Figure 2.20 and by integrating each force component separately.

Substituting now the expression for the hydrostatic pressure $P = P(h)$,

$$F_y = -\int_0^{x_{max}} (P_a + \rho g h)\,W dx = -P_a W x_{max} - \rho g \int_0^{x_{max}} h\,W dx.$$

However, the term

$$\int_0^{x_{max}} h\,W dx = \mathcal{V},$$

is the volume of the liquid $\mathcal{V}$ above the body. Therefore, $\mathcal{W} = \rho \mathcal{V} g$ is the weight of this liquid. Consequently, the vertical force F_y on a submerged body is the sum of the weight of the liquid above the body and the force due to the local atmospheric pressure P_a.

$$F_y = -\left(P_a \underbrace{W x_{max}}_{projected\ area} + \underbrace{\mathcal{W}}_{Liquid\ weight} \right).$$

The force in the horizontal direction is calculated using

$$F_x = -\int_0^{y_{max}} PW dy - -\int_0^{y_{max}} (P_a + \rho g h)\,W dy.$$

According to the schematic, the depth is $h = H - y$. Hence,

$$F_x = -\int_0^{y_{max}} \left[P_a + \rho g\,(H - y) \right] W dy.$$

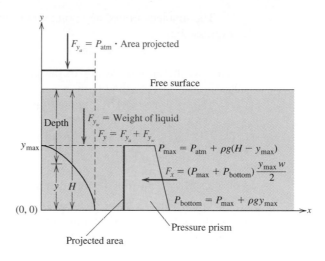

FIGURE 2.21 Alternative calculation of the forces on submerged bodies.

The preceding expression can be rearranged by adding and subtracting a constant pressure $\rho g y_{max}$ as

$$F_x = - \int_0^{y_{max}} \left[P_a + \rho g \left(H - y_{max} \right) + \rho g \left(y_{max} - y \right) \right] W \, dy.$$

In this form, the term $P_a + \rho g \left(H - y_{max} \right) = P_{y_{max}}$, where $P_{y_{max}}$ is the pressure at the point of maximum y. Upon integration, then,

$$F_x = - \left[P_{y_{max}} y_{max} W + \int_0^{y_{max}} \rho g \left(y_{max} - y \right) W \, dy \right],$$

or

$$F_x = - \left[P_{y_{max}} y_{max} W + \rho g \frac{y_{max}^2}{2} W \right].$$

This result is illustrated schematically in Figure 2.21: the magnitude of the net force in the x-direction is equal to the total volume of the *pressure prism* formed by the pressure distribution along the projected surface as shown in the figure.

In conclusion, then, the hydrostatic forces can be evaluated either by using the "direct," more mathematical solution procedure that led to Equations (2.29) and (2.30) or by using the observations that

(a) the vertical component of the force is equal to the weight of the liquid above the body (and whenever appropriate, the force due to the atmospheric pressure),

(b) the horizontal component of the force can be found by evaluating the volume of the prism formed by the pressure distribution on the projected area of the body normal to the horizontal direction.

The hydrostatic forces acting on submerged bodies are also responsible for generating a moment $\mathbf{M}_H$. In addition to $\mathbf{M}_H$, moments are also produced by all other forces applied to the system. For static equilibrium, we need not only $\sum \mathbf{F} = \mathbf{0}$; we also need to ensure that the sum of all moments with respect to any point is zero $\sum \mathbf{M} = \mathbf{0}$. Subsequently, we develop the methodology to evaluate the moment $\mathbf{M}_H$ produced by hydrostatic forces.

The moment around any point due to the force $d\mathbf{F}$ on a small infinitesimal area dA is expressed as

$$d\mathbf{M} = \mathbf{r} \times d\mathbf{F} = -P\,(\mathbf{r} \times \mathbf{n})\,dA,$$

where $\mathbf{r}$ is the position vector of the point of application of the force. If the moment is taken with respect to the origin, $\mathbf{r} = x\,\mathbf{i} + y\,\mathbf{j}$.

The total moment $\mathbf{M}_H$ around the origin then is obtained by integration:

$$\mathbf{M}_H = \int_A \mathbf{r} \times d\mathbf{F} = \int_A -P\,(\mathbf{r} \times \mathbf{n})\,dA.$$

For the surface shown in the figure, the outward unit normal vector is $\mathbf{n} = \dfrac{dy}{ds}\,\mathbf{i} - \dfrac{dx}{ds}\,\mathbf{j}$ and $dA = W\,ds$, where s is the distance along the boundary measured in the counterclockwise direction (arc length). By multiplying through by ds and by substituting the various terms, the moment due to the hydrostatic forces is

$$\mathbf{M}_H = -\int_A P\,(x\,\mathbf{i} + y\,\mathbf{j}) \times (dy\,\mathbf{i} - dx\,\mathbf{j})\,W.$$

Alternatively, the net moment can be evaluated using the moments produced by each force acting on the projected areas $W\,dx$ and $W\,dy$, as shown in Figure 2.22

$$\mathbf{M}_H = \mathbf{k} \int_0^{y_{max}} P\,y\,W\,dy - \mathbf{k} \int_0^{x_{max}} P\,x\,W\,dx.$$

The coordinates (x_R, y_R) of the point of application of the resultant hydrostatic force $\mathbf{F}_R$ are found by requiring that

$$(x_R\,\mathbf{i} + y_R\,\mathbf{j}) \times \mathbf{F}_R = \mathbf{M}_H.$$

Using Figure 2.22, the point of application of the resultant force can be found by considering the pressure prism and the volume of fluid above the submerged body and by noting that

> *The point of application of the weight of the liquid above the submerged body, and the point of application of the equivalent forces for each pressure prism is at the <u>centroid</u> of these volumes.*

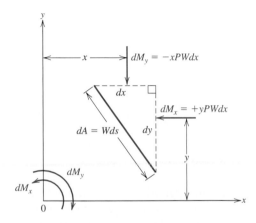

FIGURE 2.22   Alternative calculation of the moments on submerged bodies.

This alternative solution procedure is convenient for problems with simple geometry but becomes less practical for more complicated shapes. In such cases, the direct (more mathematical) approach may be more appropriate.

EXAMPLE 2.14

Problem Statement *Force on a vertical gate.* Consider the submerged gate as shown in Figure 2.23. Find the total resultant force and its point of application on the gate. What is the minimum force required to keep the gate closed?

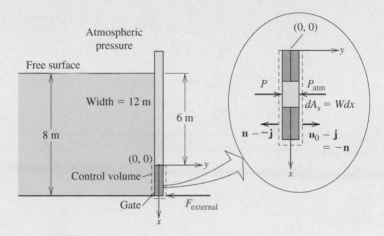

FIGURE 2.23 Hydrostatic force on a vertical surface.

Governing Equations By considering the gate as the closed system and by using a Cartesian coordinate system with the origin at the hinge of the gate as shown in Figure 2.23, the total force $\mathbf{F}_R$ on the gate is

$$\mathbf{F}_R = \int_{A_s} d\mathbf{F}.$$

$d\mathbf{F}$ is the force due to the pressure difference across an infinitesimal gate area dA_s that is,

$$d\mathbf{F} = -(P - P_a)\,\mathbf{n}\,dA_s,$$

and $\mathbf{n}$ is the *outward pointing unit normal vector* (Figure 2.23).

For static equilibrium, we also require that the sum of all moments around any point be zero,

$$\sum \mathbf{M} = \mathbf{0}.$$

Basic Assumptions Assume hydrostatic conditions—that is, the fluid is at rest ($\mathbf{V} = \mathbf{0}$).

SOLUTION Under hydrostatic conditions, the pressure along the gate is expressed by

$$P(h) = P_a + \rho g h,$$

where h is the depth from the free surface and P_a is the local atmospheric pressure. With respect to the chosen coordinate system, the depth from the free surface is $h = 6 + x$.

From the geometry of the problem, $\mathbf{n} = -\mathbf{j}$, and $dA_s = Wdx$, with W being the width of the gate (into the plane of the paper). Therefore, the force on dA_s is

$$d\mathbf{F} = -\rho g \underbrace{(x+6)}_{depth} \underbrace{(-\mathbf{j})}_{normal \ vector} \underbrace{Wdx}_{area}.$$

The total resultant force on the gate is obtained by integration

$$\mathbf{F}_R = \int_{x=0}^{x=2} \rho g \,(x+6)\, Wdx\,\mathbf{j} = \rho g W \left[6x + \frac{x^2}{2} \right]_{x=0}^{x=2} \mathbf{j} = 14\rho g W\,\mathbf{j},$$

$$\mathbf{F_R} = 14 \times 998 \ kg/m^3 \times 9.81 \ m/s^2 \times 12 \ m/1000\,\mathbf{j} = 1644.8 \ kN\,\mathbf{j}.$$

Note that the same result could have been obtained by using the volume of the pressure prisms as shown in Figure 2.24. The magnitude of the net force in the horizontal direction is the difference between the volume of the prism on the left-hand side of the gate and the pressure prism on the right-hand side,

$$\mathbf{F_R} = (\text{Volume})_{ABCD} - (\text{Volume})_{ADEF}\,.$$

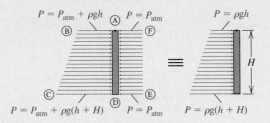

FIGURE 2.24   Hydrostatic forces in terms of pressure prisms.

The force is acting in the direction of the larger of the two volumes.
If, for instance, Volume$_{ABCD}$ > Volume$_{ADEF}$, the resultant force acts in the positive $\mathbf{i}^{th}$ direction. Using the schematic and defining H as the height of the gate, these volumes are evaluated as

$$\text{Volume}_{ABCD} = \frac{1}{2} \left[\left(\underbrace{P_a + \rho g h}_{P_{top}} \right) + \left(\underbrace{P_a + \rho g\,(h + H)}_{P_{bottom}} \right) \right] \underbrace{HW}_{Area},$$

or

$$\text{Volume}_{ABCD} = \left[P_a + \rho g \left(h + \frac{H}{2} \right) \right] HW,$$

and

$$\text{Volume}_{ADEF} = \underbrace{P_a}_{Pressure} \underbrace{HW}_{Area}.$$

Therefore, the net force is the difference of the two volumes

$$\mathbf{F}_R = \left(P_a + \rho g \left(h + \frac{H}{2} \right) \right) HW - P_a HW,$$

which finally gives,

$$\mathbf{F}_R = \rho g \left(h + \frac{H}{2} \right) H W = 14 \rho g W.$$

Since the result is a positive number, the force is acting in the positive $\mathbf{i}^{\text{th}}$ direction. This result is identical to the one obtained above from direct integration.

The position vector of the resultant force $\mathbf{r_R} = x_R \mathbf{i}$ is determined by taking the moment around the origin and requiring that the moment produced by the resultant force $\mathbf{M}_R$ be equal to the moment due to the distributed hydrostatic pressure $\mathbf{M}_H$. Therefore,

$$\mathbf{M}_R = \mathbf{r}_R \times \mathbf{F}_R = (x_R \mathbf{i}) \times (14 \rho g W \, \mathbf{j}) = x_R (14 \rho g W) \, \mathbf{k},$$

and

$$\mathbf{M}_{II} = \rho g W \int_{x=0}^{x=2} (x\mathbf{i}) \times (x+6) \, \mathbf{j} dx = \rho g W \int_{x=0}^{x=2} \left(x^2 + 6x \right) dx \, \mathbf{k},$$

$$\mathbf{M}_H = \rho g W \left[\frac{x^3}{3} + 3x^2 \right]_{x=0}^{x=2} \mathbf{k} = \frac{44}{3} \rho g W.$$

Using the condition

$$\mathbf{M}_R = \mathbf{M}_H,$$

the position of the resultant force is

$$x_R = \frac{44}{3} \frac{\rho g W}{14 \rho g W} = 1.05 \ m.$$

Again, by considering the equivalent pressure prism as shown in Figure 2.25 and by defining the height of the gate as H, the point of application of the resultant force is found by using the equivalent forces $\mathbf{F}_1$ and $\mathbf{F}_2$ shown in the schematic. The magnitude of $\mathbf{F}_1$ is

$$\mathbf{F}_1 = \rho g h \, H W.$$

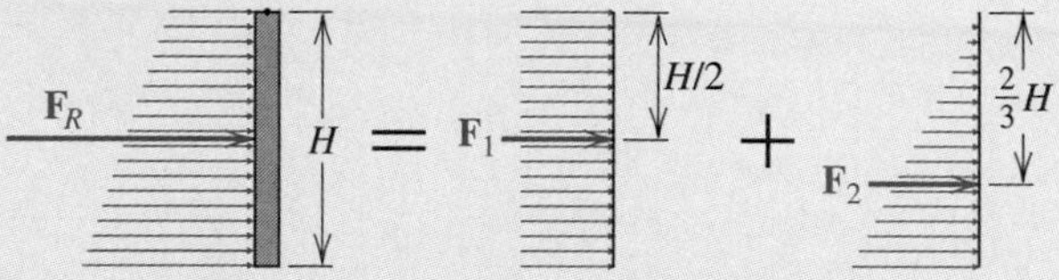

FIGURE 2.25   Point of application of the resultant force.

The point of application of $\mathbf{F}_1$ is at the centroid of the rectangular pressure distribution — that is, at the center of the gate $L_1 = \frac{H}{2}$. Similarly, the magnitude of $\mathbf{F}_2$ is

$$\mathbf{F}_2 = \frac{1}{2} \rho g \, H W,$$

and its point of application is at the centroid of the triangular pressure distribution (i.e., at a distance $L_2 = \frac{2}{3} H$ from the pivot point). By definition, the moment of the resultant force around the pivot point is equal to the moment of the equivalent forces $\mathbf{F}_1$ and $\mathbf{F}_2$,

$$L\,\mathbf{F}_R = L_1\,\mathbf{F}_1 + L_2\,\mathbf{F}_2 = \left(\frac{H}{2}\right)(\rho g h\, HW) + \left(\frac{2}{3}H\right)\left(\frac{1}{2}\rho g HW\right).$$

Substituting $\mathbf{F}_R = 14\rho g W$, $h = 6\ m$, and $H = 2\ m$, the point of application L of the resultant force is

$$L = \left(12 + \frac{8}{3}\right)\frac{1}{14}\ m = 1.05\ m,$$

which is the same result obtained from direct integration.

The gate remains closed as long as an external force $\mathbf{F}_e$ applied at the other side of the gate produces a moment $\mathbf{M}_{\mathbf{F}_e}$ such that $\sum \mathbf{M} = \mathbf{0}$, or

$$\mathbf{M}_R = \mathbf{M}_{\mathbf{F}_e}.$$

$\mathbf{F}_e$ is minimum when the moment arm is maximum (i.e., when the force is applied at the edge of the gate which is the furthest point from the pivot). Therefore,

$$\mathbf{M}_R = |\mathbf{F}|_e\, L_{max} = x_R|\mathbf{F}_R|,$$

$$|\mathbf{F}|_e = \frac{x_R|\mathbf{F}_R|}{L} = \frac{1.05\ m \times 1644.8\ kN}{2\ m} = 863.5\ kN.$$

EXAMPLE 2.15

Problem Statement *Force on an inclined gate.* Consider the submerged gate as shown in Figure 2.26. Find the total force and the point of application of the resultant force on the gate. What is the minimum force required to keep the gate closed?

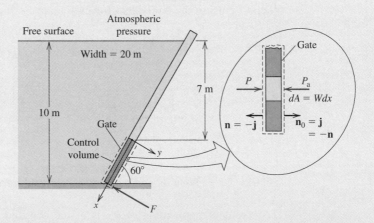

FIGURE 2.26 Hydrostatic force on an inclined surface.

Governing Equations By considering the gate as the closed system with respect to the Cartesian coordinate system with the origin at the hinge as shown in Figure 2.26, the total force on the gate is

$$\mathbf{F}_R = \int_{A_s} d\mathbf{F}.$$

$d\mathbf{F}$ is the force due to the pressure difference $(P - P_a)$ across an infinitesimal gate area dA_s — that is,

$$dF = -(P - P_a)\, \mathbf{n}\, dA_s,$$

and $\mathbf{n}$ is the *outward pointing unit normal vector* (Figure 2.26).

For static equilibrium, it is required that the sum of all moments around any point be zero

$$\sum \mathbf{M} = \mathbf{0}.$$

Basic Assumptions   Assume hydrostatic conditions ($\mathbf{V} = \mathbf{0}$).

SOLUTION   Using hydrostatic conditions, the local pressure is

$$P(h) = P_a + \rho g h,$$

where h is the depth and P_a is the local atmospheric pressure. From the geometry of the problem $\mathbf{n} = -\mathbf{j}$ and $dA_s = W dx$, where W is the width of the gate. With respect to the selected coordinate system, $h = 7 + x \sin 60$. The force on dA_s, then, is

$$dF = -\rho g\, \underbrace{(7 + x \sin 60)}_{depth}\ \underbrace{(-\mathbf{j})}_{normal\ vector}\ \underbrace{W dx}_{area}.$$

Consequently, the total resultant force on the gate is obtained by integration:

$$\mathbf{F}_R = \int_{x=0}^{x=3/\sin 60 = 3.464} \rho g\, (7 + x \sin 60)\, W dx\, \mathbf{j}$$

$$= \rho g W \left[7x + \frac{x^2}{2} \sin 60 \right]_{x=0}^{x=3.464} \mathbf{j} = 998\ kg/m^3 \times 9.81\ m/s^2 \times 20\ m$$

$$\times \left[7 \times 3.464 + \frac{(3.464)^2}{2} \sin 60 \right] \mathbf{j} \Rightarrow \mathbf{F}_R = 5765.3\ kN\, \mathbf{j}.$$

Using the alternative solution procedure as shown in Figure 2.27, the force in the vertical direction F_y is equal to the weight of the liquid above the gate (the effects of the atmospheric pressure cancel out).

Therefore,

$$F_y = -\rho \mathcal{V} g\, \mathbf{j}.$$

The volume of the liquid $\mathcal{V}$ is

$$\mathcal{V} = \left(\frac{h_1 + H}{2} \right) L_x W,$$

where $L_x = h_2 / \tan 60$ is the projected length of the gate in the horizontal direction. A negative sign is included to indicate the fact that the force vector is acting downward. Therefore,

$$F_y = -\rho W g \left(\frac{h_1 + H}{2} \right) \frac{h_2}{\tan 60}\, \mathbf{j}.$$

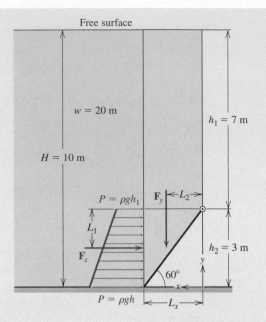

FIGURE 2.27   Hydrostatic force on an inclined surface using pressure prisms.

In terms of numerical values,

$$F_y = -998 \ kg/m^3 \times 9.81 \ m/s^2 \left(\frac{7 \ m + 10 \ m}{2}\right) \times \frac{3 \ m}{\tan 60} \times 20 \ m \, \mathbf{j} = -2882.8 \ kN \, \mathbf{j}.$$

The force in the horizontal direction is equal to the volume of the pressure prism as shown in Figure 2.27,

$$F_x = \left(\frac{\rho g h_1 + \rho g H}{2}\right) h_2 W \, \mathbf{i}.$$

In terms of numerical values,

$$F_x = 998 \ kg/m^3 \times 9.81 \ m/s^2 \times \left(\frac{7 \ m + 10 \ m}{2}\right) \times 3 \ m \times 20 \ m = 4993.1 \ kN \, \mathbf{i}.$$

The resultant force,

$$F_R = \sqrt{F_x^2 + F_y^2} = \sqrt{2882.8^2 + 4993.1^2} = 5765.6 \ kN,$$

which (except for a small round-off error) is identical to the force obtained from direct integration. Note also that

$$\tan^{-1}\left(\frac{F_x}{F_y}\right) = 60°,$$

showing that the resultant force is, indeed, acting in a direction normal to the gate.

The position vector of the resultant force $\mathbf{r_R} = x_R \mathbf{i}$ is evaluated by taking the moment around the origin and requiring that the moment produced by the resultant force $\mathbf{M}_R$ be equal to the moment of the distributed hydrostatic pressure force $\mathbf{M}_H$.

Hence,

$$\mathbf{M}_H = \rho g W \int_{x=0}^{x=3.464} x\mathbf{i} \times (7 + x \sin 60)\,\mathbf{j}dx = \rho g W \int_{x=0}^{x=3.464} \left(7x + x^2 \sin 60\right) dx\mathbf{k},$$

$$\mathbf{M}_H = \rho g W \left[\frac{7}{2}x^2 + \frac{x^3}{3} \sin 60\right]_{x=0}^{x=3.464} \mathbf{k} = 10572.9\ kNm\,\mathbf{k},$$

and

$$\mathbf{M}_R = \mathbf{r_R} \times \mathbf{F}_R = (x_R\mathbf{i}) \times (5765.3\ kN\mathbf{j}) = x_R|\mathbf{F}_R|\,\mathbf{k}.$$

Applying

$$\mathbf{M}_R = \mathbf{M}_H,$$

we get

$$x_R|\mathbf{F}_R| = 10572.9\ kNm.$$

Therefore,

$$x_R = \frac{10572.9\ kNm}{5765.3\ kN} = 1.834\ m.$$

The distance L_R of the point of application of the resultant force from the hinge of the gate (pivot point) can be found also by considering that F_x and F_y are acting at the centroid of their respective volumes and requiring that

$$L_R F_R = F_x L_1 + F_y L_2,$$

where L_1 and L_2 are, respectively, the distance of F_x and F_y from the hinge as shown in the figure (left as an exercise for the reader).

EXAMPLE 2.16

Problem Statement *Force on an curved surface.* Consider the dam with a curved wall as shown in Figure 2.28. By considering the information provided with the figure, find the total resultant force and the point of application of the force on the dam.

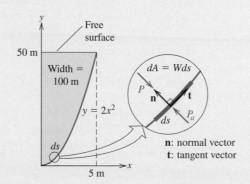

FIGURE 2.28 Hydrostatic force on a curved surface.

Governing Equations By considering the dam as the closed system, according to the selected Cartesian coordinate system the total force is

$$\mathbf{F} = \int_{A_s} d\mathbf{F}.$$

$d\mathbf{F}$ is the force due to the pressure difference $(P - P_a)$ across an infinitesimal gate area dA_s

$$d\mathbf{F} = -(P - P_a)\,\mathbf{n}\,dA_s,$$

and $\mathbf{n}$ is the *outward pointing unit normal vector* (Figure 2.28).

For static equilibrium, it is required that sum of all moments around any point be zero,

$$\sum \mathbf{M} = \mathbf{0}.$$

Basic Assumptions   Assume hydrostatic conditions ($\mathbf{V} = \mathbf{0}$).

SOLUTION   Using the assumption of hydrostatic conditions, the local pressure is

$$P(h) = P_a + \rho g h,$$

where h is the depth, which, according to the coordinate system, is $h = y_{max} - y$, and P_a is the local atmospheric pressure.

The vector $\mathbf{n}$ can be constructed from the tangent vector $\mathbf{t} = \dfrac{dx}{ds}\mathbf{i} + \dfrac{dy}{ds}\mathbf{j}$, and the definitions $\mathbf{t} \cdot \mathbf{n} = 0$ and $\mathbf{t} \times \mathbf{n} = \mathbf{k}$, which yield

$$n_y\frac{dx}{ds} - n_x\frac{dy}{ds} = 1 \quad \text{and} \quad n_x\frac{dx}{ds} + n_y\frac{dy}{ds} = 0.$$

When solved simultaneously for the components of the $\mathbf{n}$, n_x, and n_y we get $\mathbf{n} = -\dfrac{dy}{ds}\mathbf{i} + \dfrac{dx}{ds}\mathbf{j}$.

The infinitesimal area $dA_s = W ds$, where W is the width of the gate. The force $d\mathbf{F}$ is then

$$d\mathbf{F} = -\rho g \underbrace{(y_{max} - y)}_{depth} \underbrace{\left(-\frac{dy}{ds}\mathbf{i} + \frac{dx}{ds}\mathbf{j}\right)}_{normal\ vector} \underbrace{W ds}_{area}.$$

Upon simplification,

$$d\mathbf{F} = \rho g\,(y_{max} - y)\,(dy\,\mathbf{i} - dx\,\mathbf{j})\,W.$$

Alternatively, the problem could be formulated by considering the projections of the area dA, dx and dy, normal to the y- and x-directions respectively. Using these projections, we can readily write

$$F_x = \int_0^{y_{max}} P W\,dy,$$

which, as the schematic shows, is pointing in the positive $\mathbf{i}$ direction. Also,

$$F_y = -\int_0^{x_{max}} P W\,dx,$$

which is acting in the negative $(-\mathbf{j})$ direction.

By integrating separately for the components of the force in each direction we get

$$F_x = \int_0^{y_{max}} dF_x = \int_0^{y_{max}} \rho g \, (y_{max} - y) \, W dy = \rho g W \left[y_{max} y - \frac{y^2}{2} \right]_0^{y_{max}}$$

$$= \rho g W \frac{y_{max}^2}{2} = 998 \; kg/m^3 \times 9.81 \; m/s^2 \times 100 \; m \times 50^2 \, m^2/(2 \times 1000)$$

$$\Rightarrow F_x = 1.22 \times 10^6 \; kN,$$

and

$$F_y = \int_0^{x_{max}} dF_y = - \int_0^{x_{max}} \rho g \left(y_{max} - 2x^2 \right) W dx$$

$$= F_y = -\rho g W \left[y_{max} x - 2 \frac{x^3}{3} \right]_0^{x_{max}}$$

$$= -998 \; kg/m^3 \times 9.81 \; m/s^2 \times 100 \; m \times \left[50 \; m \times 5 \; m - 2 \times \frac{125}{3} \, m^2 \right]$$

$$\Rightarrow F_y = -0.163 \times 10^6 \; kN.$$

The position vector $\mathbf{r}_R = x_R \mathbf{i} + y_R \mathbf{j}$ of the point of application can be found using the moments about the origin by requiring that the resultant force produce the same moment $\mathbf{M}_R$ as the distributed hydrostatic pressure force $\mathbf{M}_H$. Therefore,

$$\mathbf{M}_R = (x_R \mathbf{i} + y_R \mathbf{j}) \times \left(F_x \mathbf{i} + F_y \mathbf{j} \right) = x_R F_y - y_R F_x;$$

by substituting the numerical values

$$\mathbf{M}_R = x_R \left(-0.163 \times 10^6 \; kN \right) - y_R \left(1.22 \times 10^6 \; kN \right);$$

and

$$\mathbf{M}_H = \int (x \mathbf{i} + y \mathbf{j}) \times \left(dF_x \mathbf{i} + dF_y \mathbf{j} \right) = \int \left(x \, dF_y - y \, dF_x \right),$$

or

$$\mathbf{M}_H = -\rho g W \left[\int_0^{x_{max}} x \left(y_{max} - 2x^2 \right) dx + \int_0^{y_{max}} y \left(y_{max} - y \right) dy \right]$$

$$= -\rho g W \left[y_{max} \frac{x_{max}^2}{2} - 2 \frac{x_{max}^4}{4} + y_{max} \frac{y_{max}^2}{2} - \frac{y_{max}^3}{3} \right].$$

By substituting the numerical values

$$\mathbf{M}_H = -998 \; kg/m^3 \times 9.81 \; m/s^2 \times 100 \; m$$

$$\times \left[50 \; m \times \frac{5^2}{2} \, m^2 - 2 \frac{5^4}{4} \, m^3 + 50 \; m \times \frac{50^2}{2} \, m^2 - \frac{50^3}{3} \, m^3 \right]$$

$$\mathbf{M}_H = -20.7 \times 10^6 \; kNm.$$

Using the condition $\mathbf{M}_R = \mathbf{M}_H$, the force is applied along the line

$$0.163 \, x_R + 1.22 \, y_R = 20.7,$$

which intersects the dam boundary at

$$0.163\, x_R + 1.22 \left(2x_R^2\right) = 20.7 \Rightarrow x_R = 2.88 \; m.$$

Therefore, since $y_R = 2\, x_R^2$, the point of application is at

$$y_R = 16.59 \; m.$$

For most curved surfaces, there is no real advantage in using the less mathematical solution procedure shown in previous problems (except possibly for calculating the horizontal component of the force) since integration may not be avoided.

EXAMPLE 2.17

Problem Statement   *Archimedes's Principle.*[5] Show that any partially or fully submerged body in static equilibrium displaces fluid equal to its weight. What are some of the implications of this observation?

Governing Equations   By considering the closed system around a body with arbitrary shape immersed in a fluid bath, as shown in Figure 2.29, for static equilibrium the conservation of momentum reduces to

$$\sum \mathbf{F} = \mathbf{0}.$$

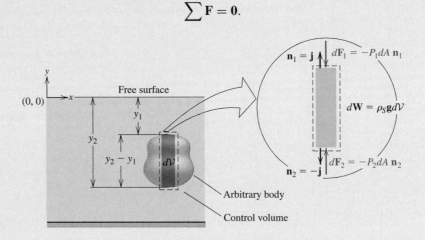

FIGURE 2.29   Schematic of a fully submerged body with arbitrary shape.

Basic Assumptions   The above expression is due to the assumption of hydrostatic conditions ($\mathbf{V} = \mathbf{0}$).

SOLUTION   Consider first the fully submerged case as shown in Figure 2.29. For a narrow strip of solid with volume dV, extending from the bottom to the top surface of the body as shown in the Figure 2.29, the balance of forces required for static equilibrium is expressed as

$$\underbrace{d\mathbf{W}}_{body\ force} + \underbrace{d\mathbf{F}_1 + d\mathbf{F}_2}_{surface\ forces} = \mathbf{0},$$

[5] Demonstrated by Archimedes, 287–212 B.C.

where $d\mathbf{W}$ is the weight of the strip and $d\mathbf{F}_1$, $d\mathbf{F}_2$ are, respectively, the surface forces at the top and bottom boundaries due to the hydrostatic pressure $P(y) = P_a + \rho g y$. The local atmospheric pressure and depth are respectively denoted by P_a and y. According to the figure, then,

$$d\mathbf{W} = -\rho_s g \, d\mathcal{V} \, \mathbf{j},$$

$$d\mathbf{F}_1 = -P \, \mathbf{n}_1 \, dA = -(P_a + \rho_f \, g \, y_1)(\mathbf{j}) \, dA,$$

$$d\mathbf{F}_2 = -P \, \mathbf{n}_2 \, dA = -(P_a + \rho_f \, g \, y_2)(-\mathbf{j}) \, dA,$$

where subscripts s and f refer to the solid and fluid properties, respectively, and dA to the cross-sectional area of the strips. By combining the terms, the balance of forces in the vertical direction gives

$$-\rho_s g \, d\mathcal{V} \, \mathbf{j} - \rho_f g y_1 \, dA \, \mathbf{j} + \rho_f g y_2 \, dA \, \mathbf{j} = \mathbf{0},$$

or,

$$\rho_s g \, d\mathcal{V} = \rho_f g \, (y_2 - y_1) \, dA = \rho_f g y \, dA.$$

However, $y \, dA$ is equal to the volume of the closed system $d\mathcal{V}$. Therefore,

$$\rho_s g \, d\mathcal{V} = \rho_f g \, d\mathcal{V}.$$

Finally, by integrating over the entire volume of the closed system,

$$\int_{\mathcal{V}_b} \rho_s g \, d\mathcal{V} = \int_{\mathcal{V}_f} \rho_f g \, d\mathcal{V},$$

we get

$$\rho_s g \mathcal{V}_b = \rho_f g \mathcal{V}_f \Rightarrow W_b = W_f,$$

demonstrating that the weight of the displaced volume of fluid is indeed equal to the weight of the body. The vertical force W_f that the body experiences is known as the *buoyant force*. It can be also shown (using $\sum \mathbf{M} = \mathbf{0}$) that the *buoyant force acts through the centroid of the displaced volume of fluid*, known as the *center of buoyancy*.

Now, by considering the partially submerged case as shown in Figure 2.30, the balance of forces reduces to

$$\rho_s g \, d\mathcal{V} = \rho_f \, g \, y_2 \, dA,$$

where $y_2 \, dA$ is the volume of the displaced fluid, and it is equal to the volume of the submerged part of the body $d\mathcal{V}_f$. By integration then, we get the same result as above:

$$\rho_s g \mathcal{V}_b = \rho_f g \mathcal{V}_f \Rightarrow W_b = W_{fs},$$

where W_{fs} is the weight of displaced liquid.

The above results show that

1. A body remains fully submerged at a certain depth only when the density of the body is equal to the density of the fluid. The body then is known to be *neutrally buoyant*.

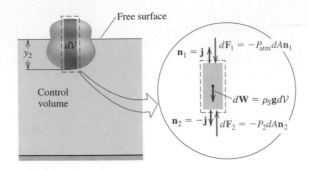

FIGURE 2.30   Schematic of a partially submerged body with arbitrary shape.

2. A body lighter than the fluid ($\rho_s < \rho_f$) will float at the free surface. In this case, the weight of the displaced fluid is equal to the weight of the body. It also follows naturally that a body with $\rho_s > \rho_f$ will sink to the bottom.

3. A body lighter than the fluid $\rho_s < \rho_f$ can be forced to remain submerged by a force **F**, equal to the difference between the weight of the displaced fluid and the weight of the body — that is, $\mathbf{F}_b = (\rho_f \mathbf{g} \mathcal{V}_b - \mathbf{W}_b)$.

EXAMPLE 2.18

Problem Statement   A spherical buoy with 1.0 m diameter floats on the surface of a lake. The buoy is designed so that while it floats freely, only one-third of its volume remains under water. If the buoy is secured in place by a cable, what is the maximum tension that the cable must be designed for?

Governing Equations   Following the results established previously, a partially submerged body displaces liquid with weight W_w equal to the weight of the body W_b. Therefore,

$$\underbrace{\rho_w \, g \, \mathcal{V}_{sb}}_{\text{displaced water weight}} = \underbrace{\rho_b \, g \, \mathcal{V}_b}_{\text{buoy weight}},$$

where ρ_w and ρ_b are, respectively, the density of the water and the density of the buoy. $\mathcal{V}_b$ and $\mathcal{V}_{sb}$ are, respectively, the volume of the buoy and the volume of the displaced water.

A body lighter than the fluid is forced to remain submerged by a force **F** equal to the difference between the weight of the displaced fluid and the weight of the body

$$\mathbf{F}_c = \underbrace{\rho_w g \mathcal{V}_b}_{\text{displaced water weight}} - \underbrace{\mathbf{W}_b}_{\text{buoy weight}}.$$

Basic Assumptions   The fluid is assumed to be at rest.

SOLUTION   When the buoy is floating freely at the free surface

$$\rho_w g \mathcal{V}_{sb} = \rho_b g \mathcal{V}_b.$$

However, since one-third of the body is submerged, $3 \mathcal{V}_{sb} = \mathcal{V}_b$, the preceding expression gives $\rho_w = 3 \rho_b$.

The cable will experience its maximum tension when the level of the lake rises to make the buoy fully submerged. In this case, the force on the cable F_c, is equal to

$$\mathbf{F}_c = \rho_w \mathbf{g} \mathcal{V}_b - \mathbf{W}_b = \rho_w \mathbf{g} \mathcal{V}_b - \rho_b \mathbf{g} \mathcal{V}_b = \rho_w \mathbf{g} \mathcal{V}_b - \frac{1}{3}\rho_w \mathbf{g} \mathcal{V}_b = \frac{2}{3}\rho_w \mathbf{g} \mathcal{V}_b.$$

Substituting the numerical values, the magnitude of the force on the cable is

$$|\mathbf{F_c}| = \frac{2}{3} \times 998 \ kg/m^3 \times 9.81 \ m/s^2 \times \frac{4}{3} \times \pi \times 0.5^3 \ m^3/1000 \ N/kN = 3.42 \ kN.$$

REFERENCES

G.A. Tokaty, *A History and Philosophy of Fluid Mechanics*, New York: Dover, 1994.

E.A. Avalone, and T. Baumeister, *Marks' Standard Handbook for Mechanical Engineers*, 9th ed., New York, McGraw-Hill, 1987.

I. Newton, *Principia* (by S. Pepys, London, 1886), Vol. 1, *The Motion of Bodies*; translated by F. Cajori, Berkeley University of California Press, 1966.

Archimedes, *On Floating Bodies*, Books I and II (in T. L.Heath, *The Works of Archimedes*) Cambridge, England, Cambridge University Press, 1897.

G.K. Batchelor, F.R.S., *An Introduction to Fluid Dynamics*, Cambridge, England, Cambridge University Press, 1970.

R.B. Bird, W.E. Steward, and E.N. Lightfoot, *Transport Phenomena*, New York, Wiley, 1960.

R.L. Panton, *Incompressible Flow*, New York, Wiley, 1984.

PROBLEMS

1. A droplet of mercury is resting on an inclined surface as shown in Figure 2.31. By selecting an appropriate control volume and coordinate system, identify the relevant forces on the droplet. Express the conservation of linear momentum for the droplet. Indicate clearly all assumptions.

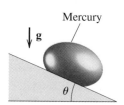

FIGURE 2.31   Schematic for Problem 1.

2. A rectangular block is sliding down an inclined plate, with angle θ, with respect to the horizontal and friction coefficient f. Develop a mathematical model that describes the velocity of the block. Indicate clearly all assumptions.

3. During a heavy downpour, a car brakes suddenly and skids. By selecting an appropriate control volume, coordinate system, and conservation laws, model the car in order to calculate the time it will take to come to rest. Please state clearly your assumptions.

4. In ice skating, because of the motion of the skates, snow melts and forms a thin layer between the skates and the snow. Use appropriate control volume and conservation laws to develop a mathematical model that describes the motion of the skater.

5. Model the motion (velocity, acceleration) of a cyclist on a bicycle. What are the important factors affecting the motion? Suggest ways to optimize the performance.

6. Consider the parachutist shown in Figure 2.32. Select an appropriate control volume and coordinate system and identify the relevant forces. Develop the appropriate expression for conservation of linear momentum for the selected control volume.

7. Using the appropriate conservation laws, develop a mathematical model to calculate the force on the beam supporting the aircraft model in the wind tunnel shown in Figure 2.33.

FIGURE 2.32   Schematic for Problem 6.

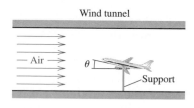

FIGURE 2.33   Schematic for Problem 7.

Press
Die
Charge

FIGURE 2.34   Schematic for Problem 8.

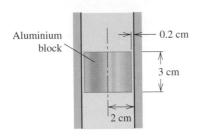

Aluminium
block
0.2 cm
3 cm
2 cm

FIGURE 2.35   Schematic for Problem 9.

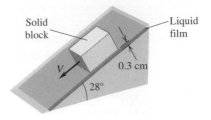

Solid
block
Liquid
film
0.3 cm
V
28°

FIGURE 2.36   Schematic for Problem 11.

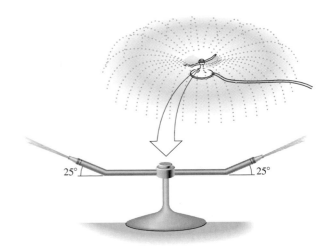

25°   25°

FIGURE 2.37   Schematic for Problem 18.

8. In compression molding, a fixed amount of material (charge) is compressed into shape by a press. Using an appropriate control volume, describe the deformation of the simple charge-press-die arrangement shown in Figure 2.34 by applying the appropriate conservation laws.

9. Find the steady, terminal velocity of the aluminum slug shown in Figure 2.35, where the cylindrical slug (radius = 5 cm) under the influence of its weight drops in a cylinder filled with water. The gap between the slug and the cylinder is 0.2 cm. Assume that the fluid velocity within the gap varies linearly.

10. Repeat Problem 9, now assuming that the fluid is glycerin.

11. A 250 N block is sliding down a wet inclined surface with 28° with respect to the horizontal as shown in Figure 2.36. The gap between the block and the surface is 0.3 cm, and the contact surface is 0.35 m^2. Assume that the liquid between the block and the surface is water and that the velocity within the gap is linear. If the block starts from rest, find the terminal velocity and the time it takes to attain it.

12. A car with total mass M and initial velocity V suddenly brakes. Develop the appropriate model for the time it takes for the car to come to rest and the distance it will travel. Use the model to determine the distance traveled for a car with initial velocity of 70 km/hr and total mass of 1200 kg. Consider a friction coefficient of $f = 0.53$.

13. A 1000 kg car with initial velocity of 35 $miles/hr$ brakes suddenly and skids in a heavy downpour. Find the time and distance the car will travel before it comes to a complete stop as a function of the thickness of the water between the tires and the road surface. Assume that the width of the tires is 20 cm and the contact line is 15 cm long. Please state clearly your assumptions.

14. A glider is dropped from a height of 50 m and travels a distance of 300 m before it reaches the ground. What is the glider's drag-to-lift ratio?

15. A fountain is designed to operate by releasing water in the vertical direction with an initial velocity of 5 m/s. What is the maximum height the water will reach?

16. Compare the time of flight of two projectiles that reach the same maximum range, having the same initial velocity but a different initial angle.

17. Design a fountain that delivers water at a vertical height of 3.5 m.

18. The sprinkler in Figure 2.37 sprays jets of water with velocity of 12 m/s at an angle of 25° degrees with respect to the horizontal. Find (a) the maximum height the fluid will reach; (b) the maximum distance the fluid will reach; (c) the time it takes to reach maximum height.

19. An outfielder throws the ball to home base 60 m away at an initial angle of 30°. What is the initial velocity when (a) the air resistance is neglected and (b) resistance is included only in the x-direction where the force is proportional to the velocity in the x-direction. Consider the proportionality constant to be 0.3 kg/s and the mass of the ball to be 0.2 kg.

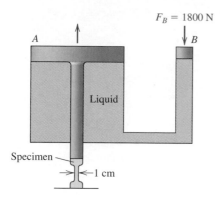

FIGURE 2.38   Schematic for Problem 22.

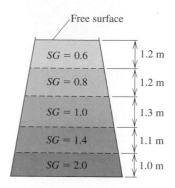

FIGURE 2.40   Schematic for Problem 24.

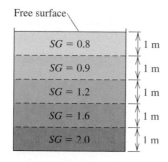

FIGURE 2.39   Schematic for Problem 23.

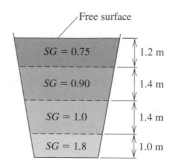

FIGURE 2.41   Schematic for Problem 25.

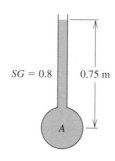

FIGURE 2.42   Schematic for Problem 26.

20. Assume that a ball traveling at a given velocity experiences resistance force that it is proportional to the velocity of the ball. If in 1 *sec* the velocity of a 0.12 *kg* ball drops from an initial velocity of 10 *m/sec* by 40%, find and plot the resistance force as a function of time. Is there anything peculiar about the final result?

21. Assume that the resistance force on a projectile traveling with velocity u is of the form $\alpha + \beta u$, where α and β are constants. Find the velocity of the projectile as a function of time, given that the initial velocity $t = 0$ is u_o.

22. Figure 2.38 is a schematic of a hydraulic testing machine. The machine is designed to produce 1800 N at point B and 300 MPa on the specimen (based on a specimen diameter of 1 *cm*). What is the area ratio of sections A and B?.

23. Find and plot the pressure as a function of the depth within the tank shown in Figure 2.39.

24. Find and plot the pressure as a function of the depth within the tank shown in Figure 2.40.

25. Find and plot the pressure as a function of the depth within the tank shown in Figure 2.41.

26. Find the pressure at point A in the tube shown in Figure 2.42. The height of the liquid in the tube is 0.75 *m*, and $SG = 0.8$.

27. Find the height of the liquid ($SG = 1.2$) in the tube shown in Figure 2.43.

28. Find the pressure difference between the two endpoints of the manometer arrangement shown in Figure 2.44.

29. Find the pressure difference between the two endpoints of the manometer arrangement shown in Figure 2.45.

30. Find the pressure inside the tank shown in Figure 2.46.

31. A manometer device can be used to measure the pressure drop between two points in the flow. Consider the pipe-manometer arrangement shown in Figure 2.47 with water flowing in the pipe. For the conditions shown, find the pressure difference between points A and B.

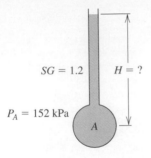

FIGURE 2.43   Schematic for Problem 27.

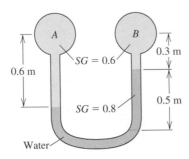

FIGURE 2.44   Schematic for Problem 28.

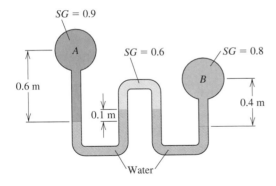

FIGURE 2.45   Schematic for Problem 29.

32. Consider the pipe-manometer arrangement shown in Figure 2.48. For the conditions shown, find the pressure at point A.

33. Find the specific gravity of the liquid at the bottom of the manometer shown in Figure 2.49.

34. Find the level of the liquid in the manometer arrangement shown in Figure 2.50.

35. Find the level of the liquid in the manometer arrangement shown in Figure 2.51.

36. Find the pressure at point A in the manometer shown in Figure 2.52.

37. A U-shaped manometer is filled with equal amounts of water and oil ($SG = 0.75$). Find the difference between the liquid levels at equilibrium conditions.

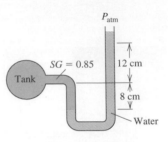

FIGURE 2.46   Schematic for Problem 30.

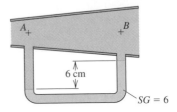

FIGURE 2.47   Schematic for Problem 31.

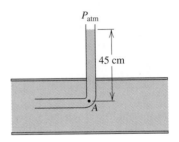

FIGURE 2.48   Schematic for Problem 32.

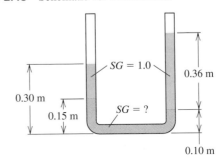

FIGURE 2.49   Schematic for Problem 33.

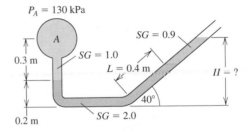

FIGURE 2.50   Schematic for Problem 34.

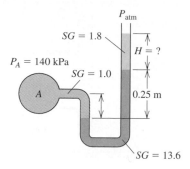

FIGURE 2.51 Schematic for Problem 35.

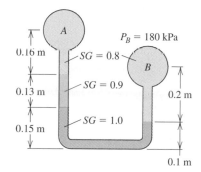

FIGURE 2.52 Schematic for Problem 36.

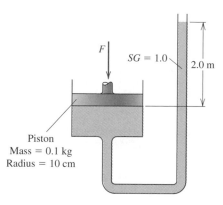

FIGURE 2.53 Schematic for Problem 39.

38. A U-shaped manometer is filled with equal amounts of two dissimilar liquids. If at equilibrium the difference in the liquid levels is 0.15 m, find the ratio of the specific gravities of the two liquids.

39. Calculate the force necessary to keep the level of water in the tube as shown in Figure 2.53.

40. Calculate the minimum force required to lift the circular tap with diameter of 0.8 m shown in Figure 2.54.

41. Determine the minimum force required to open the rectangular gate shown in Figure 2.55.

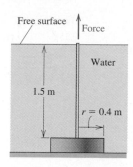

FIGURE 2.54 Schematic for Problem 40.

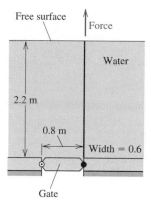

FIGURE 2.55 Schematic for Problem 41.

42. Find the minimum force required to open the rectangular gate shown in Figure 2.56.

43. Find the resultant force and its point of application acting on the dam shown in Figure 2.57.

44. Consider the schematic in Figure 2.58. If the width of the gate is 4 m find (a) the total hydrostatic force on the gate; (b) the point of application of the force; (c) the minimum force required to keep the gate closed.

45. Given the geometry and conditions shown in Figure 2.59, calculate (a) the maximum shear force; (b) the maximum bending moment experienced by the dam.

46. At what water level will the gate in Figure 2.60 open?

47. Find the location of the pivot point on the vertical gate shown in Figure 2.61 such that the gate remains closed.

48. Find the location of the hinge such that the gate remains closed under the conditions shown in Figure 2.62.

49. Find the total resultant force and its point of application on the dam shown in Figure 2.63.

50. Calculate the magnitude of the resultant hydrostatic force on the dam shown in Figure 2.64. Where is the point of application of the force?

51. Find the minimum force required to keep the gate shown in Figure 2.65 closed.

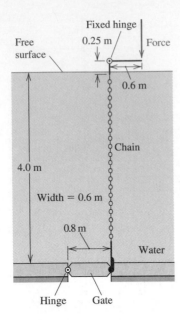

FIGURE 2.56   Schematic for Problem 42.

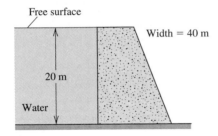

FIGURE 2.57   Schematic for Problem 43.

52. Determine the net horizontal force on the dam shown in Figure 2.66. Where is the point of application of the force?

53. Find the minimum force required to keep the plate in place as shown in Figure 2.67.

54. Find the location of the pivot point on the gate shown in Figure 2.68 so that the gate remains vertical.

55. Find the total resultant hydrostatic force and its point of application on the dam shown in Figure 2.69.

56. Calculate the net horizontal force on the curved dam shown in Figure 2.70.

57. Determine the minimum force required to keep the gate in place as shown in Figure 2.71.

58. Calculate the force on the support shown in Figure 2.72 in order to keep the gate closed.

59. Find the resultant hydrostatic force and its line of action on the curved dam shown in Figure 2.73.

60. Determine the magnitude of the hydrostatic force and its line of action on the curved dam shown in Figure 2.74.

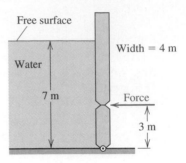

FIGURE 2.58   Schematic for Problem 44.

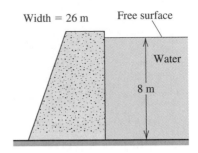

FIGURE 2.59   Schematic for Problem 45.

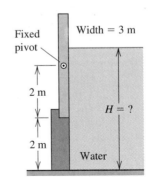

FIGURE 2.60   Schematic for Problem 46.

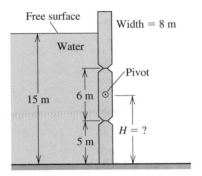

FIGURE 2.61   Schematic for Problem 47.

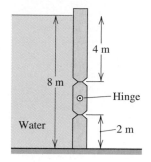

FIGURE 2.62   Schematic for Problem 48.

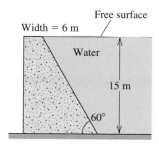

FIGURE 2.63   Schematic for Problem 49.

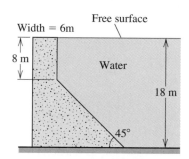

FIGURE 2.64   Schematic for Problem 50.

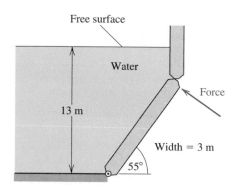

FIGURE 2.65   Schematic for Problem 51.

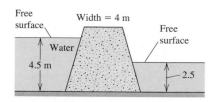

FIGURE 2.66   Schematic for Problem 52.

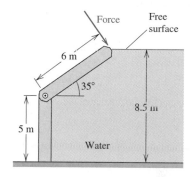

FIGURE 2.67   Schematic for Problem 53.

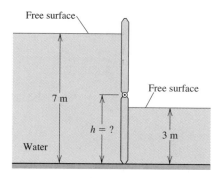

FIGURE 2.68   Schematic for Problem 54.

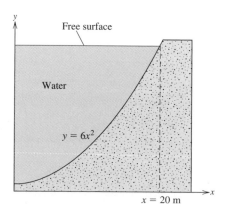

FIGURE 2.69   Schematic for Problem 55.

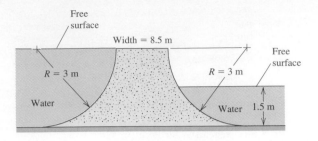

FIGURE 2.70   Schematic for Problem 56.

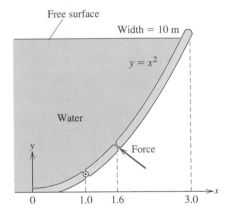

FIGURE 2.71   Schematic for Problem 57.

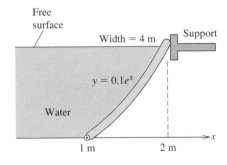

FIGURE 2.72   Schematic for Problem 58.

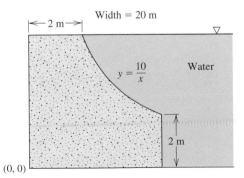

FIGURE 2.73   Schematic for Problem 59.

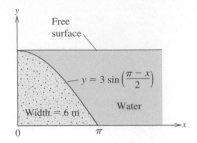

FIGURE 2.74   Schematic for Problem 60.

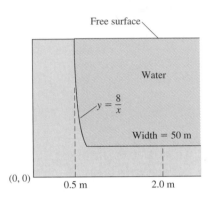

FIGURE 2.75   Schematic for Problem 61.

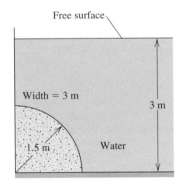

FIGURE 2.76   Schematic for Problem 62.

61. Calculate the resultant hydrostatic force and its line of action on the dam shown in Figure 2.75.

62. Find the total hydrostatic force and its line of action on the structure shown in Figure 2.76.

63. Find the minimum force required to keep the gate in place as shown in Figure 2.77.

64. Evaluate the net hydrostatic force and its line of action on the dam shown in Figure 2.78 as a function of the water height H, the radius of the dam R, and width W.

65. Determine the total hydrostatic force and its line of action on the structure shown in Figure 2.79.

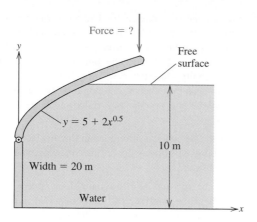

FIGURE 2.77   Schematic for Problem 63.

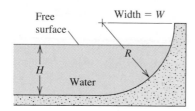

FIGURE 2.78   Schematic for Problem 64.

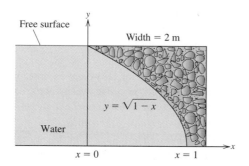

FIGURE 2.79   Schematic for Problem 65.

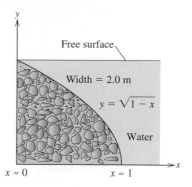

FIGURE 2.80   Schematic for Problem 66.

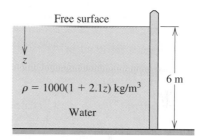

FIGURE 2.81   Schematic for Problem 69.

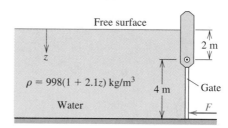

FIGURE 2.82   Schematic for Problem 70.

66. Evaluate the total hydrostatic force and its line of action on the structure shown in Figure 2.80.

67. The density of a fluid changes with the depth z from a free surface according to $\rho = \rho_o (1 + \alpha z)$, where ρ_o is the density at the free surface. Find the pressure distribution as a function of the depth.

68. The density of a fluid changes with the depth z from a free surface according to $\rho = \rho_o \left(1 + \alpha z^2\right)$, where ρ_o is the density at the free surface. Find the pressure distribution as a function of the depth.

69. Evaluate the total hydrostatic force and its line of action on the vertical wall shown in Figure 2.81.

70. Evaluate the minimum force required to keep the gate shown in Figure 2.82 closed.

71. Evaluate the total hydrostatic force and its line of action on the vertical wall shown in Problem 70 by assuming that the density of the liquid varies according to $\rho = 998 \left(1 + 1.2 z^2\right) \ kg/m^3$.

72. A solid object is found to weigh 2.3 N in air and 1.72 N when measured in water. What is the volume of the object?

73. The water line when the rectangular barge shown in Figure 2.83 is empty is 3 m below the deck level. Find the weight of the load when the water level is 1 m below the deck level.

74. What is the maximum load the barge described in Problem 73 can be loaded with before it sinks?

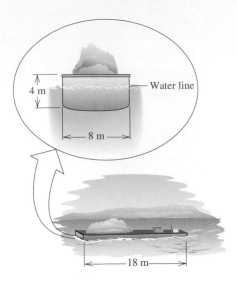

FIGURE 2.83 Schematic for Problem 73.

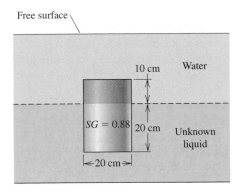

FIGURE 2.84 Schematic for Problem 76.

75. A solid block 1 m long, 0.5 m wide, and 0.3 m high floats half in water and half in oil, with specific gravity of 0.88. What is the specific gravity of the block?

76. Assume that the cylindrical block (specific gravity 0.88) shown in Figure 2.84 is in equilibrium. Find the specific gravity of the unknown fluid. Can the system be in equilibrium as shown? Where will the cylinder be when the system reaches equilibrium?

77. The combined density of a diver and his equipment is 950 kg/m^3. How much extra weight will the diver need to overcome the effects of buoyancy?

78. With increasing hydrostatic pressure, the overall density of a diver and his equipment increases with increasing pressure. Divers compensate for this by inflating a buoyancy-control device (i.e., increase their volume). If the density changes according to $\rho_d = 950 \ kg/m^3 \times e^{0.001\left(\frac{P-P_a}{P_a}\right)}$, find the additional volume the diver needs as a function of the depth in order to remain submerged there.

79. A *hydrometer* is a device used to measure the specific gravity SG of various liquids. The operation of the device is based on Archimedes's principle: the device is floated first in water and then in the liquid with the unknown specific gravity. In each case, the displaced volume is recorded. Using the theory, develop the expression for the unknown specific gravity as a function of the displaced volumes.

80. What is the percentage difference in the displaced volume when a buoyant body is floated first in water and then in mercury?

81. Find the specific gravity of a square block (side = 40 cm) that floats at the interface of two immiscible fluids, with specific gravities of 0.95 and 0.80. Consider that two-thirds of the block is in the heavier fluid.

82. What percentage of the volume of an iceberg floats above the surface of the water? Assume the density of the ice to be 916.3 kg/m^3 and that of the water to be 999.6 kg/m^3.

83. Approximate the shape of an iceberg by a cone 50 m height and 30 m in diameter. How much of the iceberg will show above the water? Assume the density of the ice to be 916.3 kg/m^3 and that of the water to be 999.6 kg/m^3.

84. A boat initially loaded with a large rock unloads the rock in the water. Will the water level rise or fall? Explain your answer.

85. An empty barge has a cross-sectional area of 2100 m^2 and the water line is 8 m^2 above its bottom. If the water line cannot be more than 9.2 m above the bottom, what is the maximum load it can carry? For the density of salt water use 995.6 kg/m^3.

86. A submarine adds 60% of its weight in order to remain submerged. What is the effective density of the submarine. Assume the density of water to be 999.6 kg/m^3.

87. A solid block with density ρ_s is floating at the interface between two immiscible liquids with density ρ_1 and ρ_2. Find the percentage of the solid's volume in each liquid.

88. Find the tension needed to keep a 4 m diameter balloon filled with helium at 110 kPa. The temperature of the gas and the surrounding air is 20°C. Assume standard atmospheric conditions.

89. Design a balloon that can lift a total weight of 42 kg to a maximum height of 800 m.

90. Consider the pilot of an aircraft during practice maneuvers. If the mass of the pilot is 68 kg, find the force on her seat when the plane is flying vertically with an acceleration of (a) 5 m/s^2 upward; (b) 5 m/s^2 downward; (b) 9.8 m/s^2 downward. What do you conclude from the results?

91. Reconsider the conditions of Problem 19. Find and plot the kinetic, potential, and total energy of the ball as a function of time.

92. The atmospheric pressure changes with altitude z according to $dp = -\rho g dz$. However, the density and temperature are not constant but change with z as well. It appears that at an altitude less than 11 km, the temperature changes linearly with altitude according to $dT = -\lambda dz$, where $\lambda = 6.5$ K/km (λ is the lapse rate). This part of the atmosphere is known as the troposphere. In the stratosphere 11 $km < z < 20.1$ km, the temperature remains constant. By assuming that the atmospheric air behaves like an ideal gas, calculate the pressure as a function of altitude and compare the results with those published in Appendix A.

93. In the case in which the atmospheric temperature decreases linearly with altitude, the final result is equivalent to a polytropic process $p/\rho^n = constant$, where p is pressure and ρ is density. The exponent n in this case is $n = g/(g - \lambda R)$, where λ is the lapse rate, R is the gas constant, and g is the gravitational acceleration. When n is equal to γ, the ratio of the specific heats λ is known as the adiabatic lapse rate.

94. Find the pressure at $3,000$ m above ground, when the pressure at ground level is 101 kPa and the temperature $20°C$, by assuming (a) isothermal conditions, (b) for constant lapse time ($dT/dz = -\lambda$), with $\lambda = 1/102$ K/m.

95. A balloon is said to be taut when it is completely filled with gas. In this state, the internal and external pressure and temperature are nearly equal. Show that the maximum height a taut ballon can reach may be found by measuring the tension needed to hold the balloon to the ground. Assume an adiabatic lapse rate as discussed in the preceding problems.

96. A balloon 2 m in diameter is filled with hydrogen at ground level and sealed. The weight of the balloon is 20 N. By assuming constant volume, find the maximum level the balloon will reach. By assuming that the temperature of the gas and outside air are equal, find the pressure difference between the inside and outside pressures. At the ground, assume 100 kPa pressure and $20°C$ temperature.

97. A balloon filled with hydrogen is designed to reach 4 $miles$ above earth. Find the volume of the hydrogen in the balloon at take-off and at the maximum height. Consider that the weight of the gondola and equipment (excluding the gas) is 9800 N. The tension required to keep the balloon at the ground is 10% of its weight. At the ground, assume 100 kPa pressure and $20°C$ temperature.

98. Consider the balloon shown in Figure 2.85 flying at a constant height of 100 m. If 5% of the total weight of the balloon is discarded, find the time it will take for the balloon to reach a height of 500 m. Explain all your assumptions (Hint: the balloon is now moving and density is a function of elevation). Consider isothermal atmospheric conditions.

99. Find the height of the atmosphere defined as the altitude where the pressure is equal to zero when atmospheric air

undergoes an isentropic process. According to the previous problem, $\gamma = g/(g - \lambda R)$. Use standard conditions at ground level (i.e., 101 kPa and $20°C$).

100. Calculate the energy stored in a spring compressed from an initial length x_1 to a final length x_2. Assume that the force on the spring is given by (a) $F = kx$; (b) $F = kx^2$, where k is a constant.

101. An aircraft flies at a speed of 475 km/h while climbing at an angle of $20°$ with respect to the horizontal. If the mass of the aircraft is 18,000 kg and the engines produce a combined thrust of 300 kN in the direction of flight, calculate the power required.

102. A 21.5 kg copper block initially at $82°C$ is dropped into an insulated tank that contains 146 L of water at $22°C$. Determine the final temperature and the total entropy change for this process.

103. A piston-cylinder device initially contains 0.5 m^3 of helium gas at 152 kPa and $20°C$. The gas is then compressed to 412 kPa and $140°C$ in a polytropic process ($PV^n = constant$). Determine (a) the entropy change for the gas; (b) the entropy change for the surroundings; (c) whether the process is reversible, irreversible, or possible. Assume the surroundings to be at $21°C$.

104. Under the influence of a force, a 10 kg block moves along an inclined surface. During the motion, the center of gravity of the block is raised 4 m, while the kinetic energy of the block decreases by 60 J. If the surface is at $30°$ with respect to the horizontal, find the average applied force.

105. Calculate the work required to accelerate a 176 kg motorcycle and a 70 kg driver from rest to 18 m/s. Indicate clearly your assumptions.

106. A closed system contains 53 gr of gas. How much heat is needed to produce 1957 $Joules$ of work while the energy of the system decreases by 123 $Joules$.

107. Find pressure p as a function of altitude z by assuming ideal gas relation and isothermal atmosphere (Hint: use $dP = -\rho g dz$, where ρ is density and g is gravity).

5% weight discarded

FIGURE 2.85   Schematic for Problem 98.

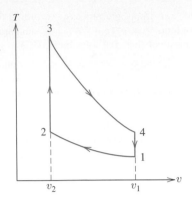

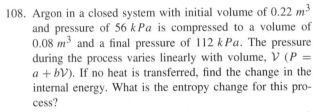

FIGURE 2.86   Schematic for Problem 110.

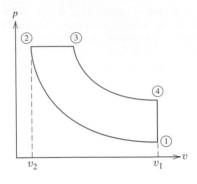

FIGURE 2.87   Schematic for Problem 112.

108. Argon in a closed system with initial volume of $0.22\ m^3$ and pressure of $56\ kPa$ is compressed to a volume of $0.08\ m^3$ and a final pressure of $112\ kPa$. The pressure during the process varies linearly with volume, $\mathcal{V}$ ($P = a + b\mathcal{V}$). If no heat is transferred, find the change in the internal energy. What is the entropy change for this process?

109. Helium in a closed system with initial volume of $0.28\ m^3$ and pressure of $80\ kPa$ is compressed to a volume of $0.15\ m^3$ and a final pressure of $200\ kPa$. The pressure P during the process varies according to $P = a + b\mathcal{V}^2$, where $\mathcal{V}$ is the volume. If no heat is transferred, find the change in the internal energy. What is the entropy change for this process?

110. Figure 2.86 shows the diagram of an ideal Otto cycle. This is an idealization of the operation of an automobile engine. According to the diagram, process $1 \rightarrow 2$ is an isentropic compression. During this part of the cycle, the piston compresses the mixture of air and fuel. Process $2 \rightarrow 3$ represents the ignition process, in which under constant volume, the mixture is ignited. Process $3 \rightarrow 4$ is the part of the cycle in which work is produced. This is idealized as an isentropic expansion. Finally, $4 \rightarrow 1$ represents heat rejection (exhaust) at constant volume. Consider, then an Otto cycle for which at the beginning of compression the pressure is $120\ kPa$ and temperature $18°C$. The compression ratio $r_v = v_1/v_2$, where v_1, v_2 are the specific volume at states 1 and 2, is 7.5, and $2,000\ kJ/kg$ of heat is added during ignition. Find the pressure and tempera-

ture at the end of each process in the cycle. Assume that the air-fuel mixture has the same properties as air and that it behaves as an ideal gas.

111. For the Otto cycle described in Problem 110 calculate the entropy change between the endpoints of each process.

112. Figure 2.87 shows the diagram of an ideal diesel cycle. This is an idealization of the operation of diesel engine. According to the diagram, process $1 \rightarrow 2$ is an isentropic compression. During this part of the cycle, the piston compresses the mixture of air and fuel. Process $2 \rightarrow 3$ represents the ignition process, in which under constant pressure, the mixture is ignited. Process $3 \rightarrow 4$ is the part of the cycle in which work is produced. This is idealized as an isentropic expansion. Finally, $4 \rightarrow 1$ represents heat rejection (exhaust) at constant volume. Consider, then a diesel cycle for which at the beginning of compression the pressure is $120\ kPa$ and temperature $18°C$. The compression ratio $r_v = v_1/v_2$, where v_1, v_2 are the specific volumes at states 1 and 2, is 18, and $2,200\ kJ/kg$ of heat is added during ignition. Find the pressure and temperature at the end of each process in the cycle. Assume that the air-fuel mixture has the same properties as air and that it behaves as an ideal gas.

113. For the diesel cycle described in Problem 112, calculate the entropy change between the endpoints of each process.

114. Air is compressed slowly in a polytropic process $p\mathcal{V}^1.5 = Constant$, where p is pressure and $\mathcal{V}$ is the volume, from an initial volume of $1\ m^3$ and pressure of $120\ kPa$ to a final volume of $0.6\ m^3$. Find (a) the final pressure; (b) the total work needed for the compression.

3 Conservation Laws for Open Systems

> *Methods and means cannot be separated from the ultimate aim.*[1]
> —Emma Goldman

The concept of a closed, fixed-mass system provides a useful framework of analysis for many engineering problems. However, whenever flowing fluids are involved, an obviously convenient choice of a control volume may not avoid having mass crossing its boundary. This happens, for instance, when the control volume envelops flow devices such as pumps, turbines, or compressors. Therefore, for such problems the closed-system approach is not a convenient one.

To analyze situations where mass crosses the boundary of the selected control volume, the concept of an open system is used. The relevant theory for open-system analysis follows naturally from the *Reynolds transport theorem*. This theorem is formulated using simple mathematical arguments and provides the means by which to convert the governing laws as articulated for closed systems into an equivalent form for open systems.

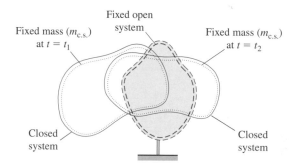

FIGURE 3.1 A deforming fixed mass of fluid at two different times (closed system) entering a fixed volume in space (open system).

3.1 Reynolds Transport Theorem

The differences between closed and open systems are seen in Figure 3.1. As the volume $\mathcal{V}_{cs}$ of the closed system moves and deforms, it always envelops the same, fixed amount of mass (m_{cs}). By contrast, the volume $\mathcal{V}_{os}$ of the open system, while moving and deforming, allows mass to cross its boundary. The differences between open and closed systems suggest that they are described by different mathematical forms. Obviously, though, physical problems and their solutions are independent of any mathematical formalism used to describe them. Therefore, regardless of the differences between open and closed systems, the two must be equivalent—when applied to the same problem, they must yield the same answer. The relationship between the two approaches is developed in this following text by using the schematic of the open and closed systems shown in Figure 3.2.

---

[1]Emma Goldman, *My Further Disillusionment (1924).*

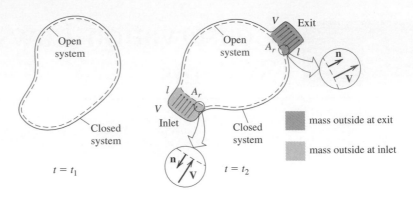

FIGURE 3.2 Schematic of the relative position of a closed system and an open system. At $t = t_1$ the two systems coincide with each other. At $t = t_2$ they have moved relative to each other.

At time $t = t_1$ the boundaries of both systems coincide as shown in the figure, and at an instant $t = t_1 + \Delta t = t_2$ later, the two move relative to each other. In general, mass could enter or leave the open system at various locations. For simplicity, we assume that this happens at only two locations: one where mass *enters*, and another where mass *leaves* the open system, designated respectively as *inlet* and *exit*.

Returning to the form of the conservation laws for a closed system, for a general quantity $\mathcal{A}$, with $\alpha = \dfrac{d\mathcal{A}}{dm}$, we seek to mathematically relate

$$\underbrace{\left.\frac{D\mathcal{A}}{Dt}\right|_{Closed\ System} \quad ? \quad \left.\frac{D\mathcal{A}}{Dt}\right|_{Open\ System}}_{How\ are\ they\ related?}$$

It is worth pointing out that the objective here is not to introduce a new set of conservation principles; rather, by using mathematics, we seek to relate quantities between open and closed systems.

The simplified schematic of the problem shows the situation at the two distinct times t_1 and t_2. For the closed system $\mathcal{V}_{cs}$ we can write

$$\left(\frac{\Delta\mathcal{A}}{\Delta t}\right)_{Closed\ System} = \left(\frac{\mathcal{A}_{at\ t_2} - \mathcal{A}_{at\ t_1}}{\Delta t}\right)_{Closed\ System}, \tag{3.1}$$

where $\Delta t = t_2 - t_1$. At $t = t_1$, we have

$$\underbrace{\mathcal{A}_{Closed\ System}}_{at\ t_1} = \underbrace{\mathcal{A}_{Open\ System} + \mathcal{A}_{Inlet} + \mathcal{A}_{Exit}}_{at\ t_1}.$$

Despite the fact that initially ($t = t_1$), there is no mass at both the inlet and exit and that, hence, $\mathcal{A}_{Inlet} = 0$ and $\mathcal{A}_{Exit} = 0$, they have been included in the mathematical derivation for convenience. At $t = t_2$, we have

$$\underbrace{\mathcal{A}_{Closed\ System}}_{at\ t_2} = \underbrace{\mathcal{A}_{Open\ System} + \mathcal{A}_{Inlet} + \mathcal{A}_{Exit}}_{at\ t_2},$$

where $\mathcal{A}_{Inlet}$ and $\mathcal{A}_{Exit}$ represent the amount of the property $\mathcal{A}$ outside the open system.

Substituting these two expressions into Equation (3.1) and collecting terms, we get

$$\left(\frac{\Delta \mathcal{A}}{\Delta t}\right)_{Closed\ System} = \left(\frac{\mathcal{A}_{t_2} - \mathcal{A}_{t_1}}{\Delta t}\right)_{Open\ System} + \left(\frac{\mathcal{A}_{t_2} - \mathcal{A}_{t_1}}{\Delta t}\right)_{Inlet} + \left(\frac{\mathcal{A}_{t_2} - \mathcal{A}_{t_1}}{\Delta t}\right)_{Exit}.$$

In the limit $\Delta t \to 0$,

$$\left(\frac{D\mathcal{A}}{Dt}\right)_{Closed\ System} = \left(\frac{d\mathcal{A}}{dt}\right)_{Open\ System} + \underbrace{\left(\frac{d\mathcal{A}}{dt}\right)_{Inlet} + \left(\frac{d\mathcal{A}}{dt}\right)_{Exit}}_{Boundary\ Flux\ Term}. \qquad (3.2)$$

The derivative D/Dt denotes the rate of change with respect to time as seen by an observer moving with the *same* velocity as that of the fixed mass. The derivative d/dt indicates the rate of change with respect to time as seen by an observer moving with a velocity *different* from the velocity of the enclosed mass. The distinction between the two derivatives is important and it is discussed in greater detail in Chapter 4. For the time being, the simple definition given previously is sufficient for the discussions in this chapter.

The surprisingly intuitive result expressed by Equation (3.2) shows that the rate of change of a property within a closed system differs from the rate of change within an open system by the mass inflow and ouflow at the system's boundary.

The amount of $\mathcal{A}$ at either the inlet or the exit can be expressed using local flow conditions and the mass m there by writing

$$\mathcal{A} = m\alpha = \alpha \underbrace{\rho l A_r}_{mass},$$

where the length l and the cross-sectional area A_r define the region just outside the open system. In this region, ρ is the density and α the specific value of property of $\mathcal{A}$.

The rate of flow of $\mathcal{A}$ (or the flux of $\mathcal{A}$) through the boundary, then, is expressed as

$$Flux\ of\ \mathcal{A}\ at\ the\ boundary \equiv \mathcal{F}_l = \left(\frac{d\mathcal{A}}{dt}\right)_{Inlet/Exit} = \frac{d(\rho \alpha l A_r)}{dt}.$$

For constant flow conditions (α, ρ, and A_r), the flux simplifies to

$$\mathcal{F}_l = \rho \alpha \frac{dl}{dt} A_r,$$

where $\frac{dl}{dt}$ is the rate at which material enters or leaves the boundary. This is identical to the magnitude of the fluid velocity V_n normal to the surface area A_r, expressed relative to the moving boundary. The flux then is

$$\mathcal{F}_l = \rho \alpha V_n A_r.$$

Recall that at all times the total mass for the closed system is constant. This constant mass is distributed between the open system and the regions defined as the inlet and exit. Therefore, the amounts of A at the inlet (A_i) and exit (A_e) are finite. Consequently, as mass continuously enters the open system, A_i decreases with time and $\frac{d\mathcal{A}_i}{dt} < 0$. The opposite is true at the exit, and $\frac{d\mathcal{A}_e}{dt} > 0$. To reflect these situations, the flux condition is more correctly expressed as

$$\mathcal{F}_l = \rho \alpha (\mathbf{V} \cdot \mathbf{n}) A_r,$$

where $\mathbf{V}$ is the velocity vector, and $\mathbf{n}$ the unit normal vector *pointing away from the surface* (Figure 3.3). This expression automatically takes care of the proper sign both at inlets and at exits.

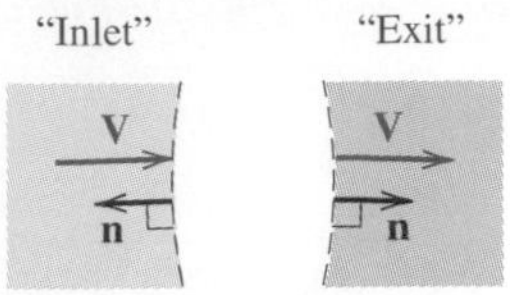

If conditions at locations where mass crosses the boundary are uniform (i.e., they are constant along A_r), the flux term simplifies to

$$\mathcal{F}_l = \pm \rho \alpha V_n A_r = \pm \dot{m}\alpha,$$

where $\dot{m}$ is the mass flow rate defined by $\dot{m} = \rho V_n A_r$ and V_n is the velocity normal to the boundary. The proper sign ($\pm$) is determined by the $(\mathbf{V} \cdot \mathbf{n})$ term; the $(-)$ sign is used at *inlets* and the $(+)$ sign at *exits*.

Equation (3.2) then is expressed as

$$\left(\frac{D\mathcal{A}}{Dt}\right)_{Closed\ System} = \left(\frac{d\mathcal{A}}{dt}\right)_{Open\ System} \pm \underbrace{\dot{m}\alpha}_{Boundary\ Term}. \tag{3.3}$$

When mass crosses the boundary of the open system at N locations with uniform flow conditions, Equation (3.3) simplifies to

$$\frac{D\mathcal{A}}{Dt}\bigg|_{Closed\ System} = \frac{d\mathcal{A}}{dt}\bigg|_{Open\ System} \pm \underbrace{\sum_{k=1}^{k=N} \dot{m}_k \alpha_k}_{Boundary\ Term}. \tag{3.4}$$

By replacing the summation in Equation (3.4) with an integral

$$\frac{D}{Dt}\int_{Closed\ System} \rho\alpha\,d\mathcal{V} = \frac{d}{dt}\int_{Open\ System} \alpha\rho\,d\mathcal{V} + \underbrace{\int_{A_r} \rho\alpha(\mathbf{V}\cdot\mathbf{n})\,dA_r}_{Boundary\ Term}. \tag{3.5}$$

Equation (3.5) is general, and applies to any number of locations where mass crosses the boundary with flow conditions that can vary along A_r. Equation (3.5) is known as the *Reynolds Transport Theorem*.

3.2   General Conservation Law for an Open System

Using the Reynolds transport theorem (Equation (3.5)), the general form of the conservation laws for an open system is expressed as

$$\underbrace{\frac{d}{dt}\int_{\mathcal{V}_{os}} \alpha\rho\,d\mathcal{V}}_{Rate\ of\ change\ of\ A\ within\ the\ open\ system} + \underbrace{\int_{A_r} \rho\alpha(\mathbf{V}\cdot\mathbf{n})\,dA_r}_{Net\ flux\ of\ A\ at\ boundary} \equiv B, \tag{3.6}$$

where $\mathcal{V}_{os}$ is the volume of the open system, A_r is the part of the system boundary where mass crosses, and $\alpha = d\mathcal{A}/dm$. The right-hand side B is a quantity specific to the particular conservation law. Again, all quantities are *expressed relative to the open system*. Equation (3.6) is also a *general expression that is valid irrespective of the nature of the medium it is applied to*.

In applying Equation (3.6), care must be exercised in evaluating the boundary term. As stressed earlier, this term is valid along the boundary of the open system at locations where mass crosses the boundary. At these locations, $\mathbf{V}$ is the fluid velocity vector and $\mathbf{n}$ is the unit *normal vector pointing away from the surface* (Figure 3.2).

"Inlet"   "Exit"

FIGURE 3.3 Evaluation of the term $(\mathbf{V} \cdot \mathbf{n})$ at the boundary, where $\mathbf{V}$ is the velocity vector and $\mathbf{n}$ the outward point unit normal vector to the boundary.

3.3 Conservation of Mass

Consistent with the general Equation (3.6), by using $\alpha = 1$ and $B = 0$, the conservation of mass for an open system is expressed as

$$\underbrace{\frac{d}{dt} \int_{\mathcal{V}_{os}} \rho d\mathcal{V}}_{\text{Rate of change of mass within the open system}} + \underbrace{\int_{A_r} \rho(\mathbf{V} \cdot \mathbf{n}) \, dA_r}_{\text{Net flux of mass across the boundary}} \equiv 0, \qquad (3.7)$$

where $\mathcal{V}_{os}$ is the volume of the open system and A_r is the area where mass crosses the boundary of the $\mathcal{V}_{os}$.

For steady flow

$$\frac{d}{dt} \int_{\mathcal{V}_{os}} \rho d\mathcal{V} = 0,$$

the conservation of mass simplifies to

$$\int_{A_r} \rho(\mathbf{V} \cdot \mathbf{n}) dA_r \equiv 0.$$

If mass crosses the boundary at N segments with uniform normal velocity V_k, cross-sectional area A_{r_k}, and constant density ρ, the expression simplifies further to

$$\sum_{k=1}^{k=N} (\pm) \rho V_k A_{r_k} = \sum_{k=1}^{k=N} (\pm) \dot{m}_k \equiv 0,$$

where $\dot{m}_k$ is the mass flow rate entering or leaving the system at each location k. Again, the proper sign ($\pm$) is obtained from $\mathbf{V} \cdot \mathbf{n}$); $(\mathbf{V} \cdot \mathbf{n}) < 0$ at an inlet and $(\mathbf{V} \cdot \mathbf{n}) > 0$ at an exit. Hence,

$$\sum_{k_e=1}^{k_e=N_e} \dot{m}_e - \sum_{k_i=1}^{k_i=N_i} \dot{m}_i \equiv 0.$$

N_i and N_e are the number of locations where mass respectively enters (subscript i) and leaves (subscript e) the open system.

In the special case of steady flow with mass crossing the boundary at two locations (i.e., one inlet and one exit), we have

$$\rho_i V_i A_{r_i} = \rho_e V_e A_{r_e} = \dot{m}. \qquad (3.8)$$

This implies that the mass flow rate in a channel at any cross section normal to the direction of flow is constant.

The integral,

$$\dot{Q} = \int_{A_r} (\mathbf{V} \cdot \mathbf{n}) \, dA,$$

with units of m^3/s, is the volumetric flow rate $\dot{Q}$. In case of uniform conditions across the area, the volumetric flow rate is given by $\dot{Q} = V_n A_r$, where V_n is the velocity normal to area A_r. The average velocity through A_r is evaluated as

$$\bar{V} = \frac{1}{A_r} \int_{A_r} (\mathbf{V} \cdot \mathbf{n}) \, dA = \frac{\dot{Q}}{A_r}.$$

Note that it is customary to always give both $\dot{Q}$ and $\dot{m}$ as positive quantities irrespective of the sign implied by $(\mathbf{V} \cdot \mathbf{n})$ in the preceding expressions. Therefore, we typically take the absolute value of the term $|(\mathbf{V} \cdot \mathbf{n})|$.

EXAMPLE 3.1

Problem Statement *Flow in a diverging channel.* Consider water flowing in the diverging channel shown in Figure 3.4. The velocity V_1 at the inlet with cross-sectional area 0.2 m^2 is 1 m/s. What are the mass and volumetric flow rates through the channel? If the cross-sectional area at the exit is 0.35 m^2, find the velocity at the exit of the channel.

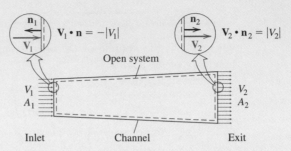

FIGURE 3.4   Flow in a diverging channel with uniform conditions at the inlet and exit.

Governing Equations   By selecting the section of the channel shown in Figure 3.4 as a control volume, we have an open system (mass crosses the boundary). The mass flow rate $\dot{m}$ through an area A_r is defined as

$$\dot{m} = \int_{A_r} \rho(\mathbf{V} \cdot \mathbf{n}) \, dA_r,$$

and the volumetric flow rate $\dot{Q}$ as

$$\dot{Q} = \int_{A_r} (\mathbf{V} \cdot \mathbf{n}) \, dA_r,$$

where ρ is the density of the fluid and $\mathbf{V} \cdot \mathbf{n} = V_n$ is the velocity normal to the area A_r.
    The conservation of mass for an open system is

$$\frac{d}{dt} \int_{\mathcal{V}_{os}} \rho d\mathcal{V} + \int_{A_r} \rho(\mathbf{V} \cdot \mathbf{n}) \, dA_r \equiv 0,$$

where $\mathcal{V}_{os}$ is the volume of the open system.

Basic Assumptions   It is assumed that the flow is incompressible (constant density ρ), steady, and uniform across the channel.

SOLUTION   By assuming uniform conditions at the inlet and exit, the mass flow rate in kg/s simplifies as

$$\dot{m} = \int_{A_r} \rho |(\mathbf{V} \cdot \mathbf{n})| dA_r = \rho V_n A_r,$$

and the volumetric flow rate in m^3/s as

$$\dot{Q} = \int_{A_r} |(\mathbf{V} \cdot \mathbf{n})| \, dA_r = V_n A_r.$$

Using the numerical values, the mass and volumetric flow rates are readily calculated as

$$\dot{m} = \rho V_{n_1} A_1 = 998 \ kg/m^3 \times 1 \ m/s \times 0.2 \ m^2 = 199.6 \ kg/s.$$

and

$$\dot{Q} = 1 \ m/s \times 0.2 \ m^2 = 0.2 \ m^3/s.$$

Similarly, since we have uniform conditions at the inlet and exit, conservation of mass reduces to

$$\rho V_{n_1} A_1 = \rho V_{n_2} A_2.$$

The velocity at the exit then is

$$V_{n_2} = \frac{\rho V_1 A_1}{\rho A_2} = \frac{1 \ m/s \times 0.2 \ m^2}{0.35 \ m^2} = 0.57 \ m/s.$$

This problem shows that the velocity of an incompressible fluid in a diverging channel decreases, whereas in a converging channel it increases.

EXAMPLE 3.2

Problem Statement *Ground water infiltration.* Consider the hydraulic dam shown in Figure 3.5, releasing water at a rate of $200 \ m^3/s$ into a dry river bed $50 \ m$ wide. The ground absorbs water at a rate of $0.01 \ m/s$ (velocity of the water entering the ground). Find the extent of the river that will be affected by the released water.

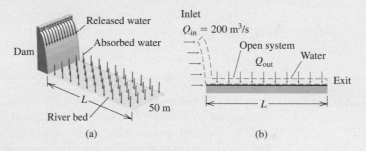

FIGURE 3.5 (a) Water released from a dam infiltrates the ground at a given velocity rate; (b) the selected open system considers both the released and the infiltrating water.

Governing Equations By considering the water as an open system, the conservation of mass is expressed using

$$\frac{d}{dt} \int_{\mathcal{V}_{os}} \rho \, d\mathcal{V} + \int_{A_r} \rho (\mathbf{V} \cdot \mathbf{n}) \, dA_r \equiv 0.$$

Basic Assumptions Assume steady state and incompressible flow, with uniform conditions at the locations the water enters and leaves the boundary of the selected open system.

SOLUTION In this problem, mass crosses the boundary of the open system at the locations shown in Figure 3.5 as inlet and exit. For steady flow $\frac{d}{dt} \int_{\mathcal{V}_{os}} \rho \, d\mathcal{V} = 0$, the continuity equation reduces to

$$\int_{A_r} (\mathbf{V} \cdot \mathbf{n}) \, dA_r = \int_{in} (\mathbf{V} \cdot \mathbf{n}) \, dA_i + \int_{out} (\mathbf{V} \cdot \mathbf{n}) \, dA_o = 0.$$

At the inlet we have $\mathbf{V} = V_i \mathbf{i}$ and $\mathbf{n} = -\mathbf{i}$, where V_i is the magnitude of the velocity of the released water at the dam. Therefore,

$$\int_{in} (\mathbf{V} \cdot \mathbf{n})\, dA_r = \int_{in} (V_i \mathbf{i}) \cdot (-\mathbf{i})\, dA_i = -V_i \int_{in} dA_i = -V_i A_i = -\dot{Q},$$

where $\dot{Q}$ is the volumetric rate of water release. At the boundary, where water leaves the open system and enters the ground, we have $\mathbf{V} = -V_e \mathbf{j}$ and $\mathbf{n} = -\mathbf{j}$, where V_e is the velocity of the water being absorbed by the ground. Therefore,

$$\int_{out} (\mathbf{V} \cdot \mathbf{n})\, dA_e = \int_{out} (-V_e \mathbf{j}) \cdot (-\mathbf{j})\, dA_e = V_e \int_{in} dA_e = V_e A_e,$$

where A_e is the affected area. From the geometry of the problem, $A_e = LW$, where W is the width of the river and L the length of this area.

By adding the terms at the inlet and exit, the conservation of mass yields

$$-\dot{Q} + V_e A_e = -\dot{Q} + V_e LW = 0.$$

Therefore,

$$L = \frac{\dot{Q}}{V_e W} = \frac{200\ m^3/s}{0.01\ m/s \times 50\ m} = 400\ m.$$

Since conditions at the inlet and exit are uniform, we could have alternatively used Equation (3.8) to easily arrive at the same result (left as an exercise for the reader).

EXAMPLE 3.3

Problem Statement   *Flow ejection due to boundary layer effects.* Consider the flow past a plate as shown in Figure 3.6. Because of friction, the velocity of the fluid at the plate is zero. Therefore, in the direction normal to the incoming flow direction (i.e., in the y-direction) the velocity changes from zero at the plate to the value of the freestream velocity (u_0) away from it. The layer across which the velocity varies is known as the *boundary layer*. The thickness of this layer (δ) increases with the distance x from the front end of the plate in the direction of flow — that is, $\delta = \delta(x)$.

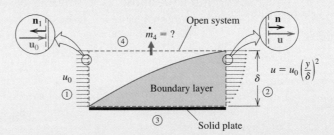

FIGURE 3.6   Due to the zero velocity at the plate, the velocity profile changes from a uniform profile at the inlet u_0 into a nearly parabolic profile $u = u_0(y/\delta)^2$ at the exit at a distance L away from the inlet. As a result, excess fluid is deflected away from the plate.

In this example, consider a uniform flow u_0 approaching a flat plate that changes into a parabolic profile $u = u_0 \left(\dfrac{y}{\delta}\right)^2$ at a distance L away from the front edge of the plate where $\delta = \delta(x = L)$. For the conditions shown in the figure, find the mass flow rate across each boundary of the selected open system.

Governing Equations The conservation of mass for the selected open system is given by

$$\frac{d}{dt} \int_{\mathcal{V}_{os}} \rho \, d\mathcal{V} + \int_{A_r} \rho(\mathbf{V} \cdot \mathbf{n}) \, dA_r = 0.$$

Basic Assumptions The flow is assumed to be steady, two-dimensional and incompressible.

SOLUTION By assuming steady flow $\frac{d}{dt} \int_{\mathcal{V}_{os}} \rho d\mathcal{V} = 0$ and by considering all four boundaries of the open system, the conservation of mass reduces to

$$\underbrace{\int_1 \rho(\mathbf{V} \cdot \mathbf{n}) W dS}_{I_1} + \underbrace{\int_2 \rho(\mathbf{V} \cdot \mathbf{n}) W dS}_{I_2} + \underbrace{\int_3 \rho(\mathbf{V} \cdot \mathbf{n}) W dS}_{I_3} + \underbrace{\int_4 \rho(\mathbf{V} \cdot \mathbf{n}) W \, dS}_{I_4} = 0.$$

At each boundary numbered as shown in Figure 3.6, we have the following conditions:

$$\mathbf{V}_1 = u_o \mathbf{i}, \quad \mathbf{n}_1 = -\mathbf{i}, \quad dS_1 = dy, \quad \mathbf{V}_2 = u_o \left(\frac{y}{\delta}\right)^2 \mathbf{i}, \quad \mathbf{n}_2 = \mathbf{i}, \quad dS_2 = dy,$$

$$\mathbf{V}_3 = 0, \quad \mathbf{n}_3 = \mathbf{j}, \quad dS_3 = dx, \quad \mathbf{V}_4 = ?, \quad \mathbf{n}_4 = -\mathbf{j}, \quad dS_4 = dx,$$

Each of the boundary integrals is evaluated as follows:

$$\int_{0_1}^{\delta} \rho(\mathbf{V} \cdot \mathbf{n}) W dy = \int_{0_1}^{\delta} \rho(u_o \mathbf{i}) \cdot (-\mathbf{i}) W dy = - \int_{0_1}^{\delta} \rho u_o W dy = -\rho u_o \delta W.$$

Similarly,

$$\int_{0_2}^{\delta} \rho(\mathbf{V} \cdot \mathbf{n}) W dy = \int_{0_2}^{\delta} \rho u_o \left(\frac{y}{\delta}\right)^2 W dy = \rho u_o \left[\frac{y^3}{3\delta^2}\right]_0^{\delta} W = \rho u_o \frac{\delta}{3} W.$$

Since the velocity along the wall is zero,

$$\int_3 \rho(\mathbf{V} \cdot \mathbf{n}) W dx = 0.$$

In order to conserve mass,

$$\underbrace{-\rho u_o \delta W}_{I_1} + \underbrace{\rho u_o \frac{\delta}{3} W}_{I_2} + \underbrace{0}_{I_3} + \underbrace{\dot{m}_4}_{I_4} = 0.$$

The mass flow rate at edge of the boundary layer then is

$$\dot{m}_4 = \rho u_o \delta W - \rho u_o \frac{\delta}{3} W \Rightarrow \dot{m}_4 = \frac{2}{3} \rho u_o \delta W.$$

Therefore, because of the fluid friction induced slowdown in the flow, two-thirds of the incoming mass flow rate is deflected away from the surface.

EXAMPLE 3.4

Problem Statement *Entrance flow.* Consider the entrance flow in the pipe as shown in Figure 3.7. Because of fluid-solid friction at the pipe wall, after a certain distance downstream of the inlet, the velocity develops into a parabolic profile. For the conditions shown, find the maximum velocity in the pipe.

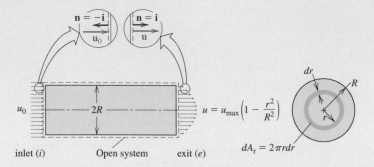

FIGURE 3.7 Uniform flow velocity entering a pipe changes into a parabolic flow. As result, the maximum velocity in the pipe increases and ultimately achieves a constant value.

Governing Equations The conservation of mass for the selected open system is given by

$$\frac{d}{dt}\int_{\mathcal{V}_{os}}\rho\,d\mathcal{V}+\int_{A_r}\rho(\mathbf{V}\cdot\mathbf{n})\,dA_r=0.$$

Basic Assumptions This two-dimensional flow is assumed to be steady and incompressible.

SOLUTION For steady flow $\frac{d}{dt}\int\rho\,d\mathcal{V}=0$, the conservation of mass simplifies to

$$\int_{A_r}\rho(\mathbf{V}\cdot\mathbf{n})\,dA_r=0.$$

The surface integrals vanish everywhere except at the inlet (subscript i) and exit (subscript e) of the pipe where mass crosses the boundary of the selected system. At the inlet, we have $\mathbf{V}_i = u_o\mathbf{i}$ and $\mathbf{n}_i = -\mathbf{i}$. At the exit, we have $\mathbf{V}_e = u_{max}\left(1-\frac{r^2}{R^2}\right)\mathbf{i}$ and $\mathbf{n}_e = \mathbf{i}$. For the cylindrical geometry, an infinitesimal cross-sectional area is expressed as $dA_r = 2\pi r\,dr$. Therefore,

$$\int_{inlet}\rho(\mathbf{V}\cdot\mathbf{n})dA_r+\int_{exit}\rho(\mathbf{V}\cdot\mathbf{n})dA_r=0\Rightarrow$$

$$\int_{inlet}\rho\underbrace{(u_o\mathbf{i})}_{\mathbf{V}_i}\cdot\underbrace{(-\mathbf{i})}_{\mathbf{n}_i}\underbrace{2\pi r\,dr}_{dA_i}+\int_{exit}\rho\underbrace{\left(u_{max}\left[1-\left(\frac{r}{R}\right)^2\right]\mathbf{i}\right)}_{\mathbf{V}_e}\cdot\underbrace{(\mathbf{i})}_{\mathbf{n}_e}\underbrace{2\pi r\,dr}_{dA_e}=0,$$

then,

$$-\rho u_o\pi\int_0^R 2r\,dr+\rho u_{max}2\pi\int_0^R\left[r-\frac{r^3}{R^2}\right]dr=0$$

$$-\rho u_o\pi R^2+2\pi\rho u_{max}\left[\frac{r^2}{2}-\frac{r^4}{4R^2}\right]_0^R=0$$

$$-\rho u_o \pi R^2 + 2\pi \rho u_{max}\left[\frac{R^2}{2} - \frac{R^4}{4R^2}\right] = -\rho u_o \pi R^2 + 2\pi \rho u_{max}\frac{R^2}{4} = 0$$

$$\Rightarrow u_{max} = 2u_o.$$

The velocity profile

$$u = u_0\left(1 - \frac{r^2}{R^2}\right)$$

is characteristic of a type of flow known as laminar pipe flow and is encountered well downstream from the inlet.

EXAMPLE 3.5

Problem Statement *Mixing.* Consider a medical device used to mix and deliver to a patient two drugs with roughly the same density 1050 kg/m^3 (Figure 3.8). For the conditions shown in the figure, find the maximum velocity after the two drug streams mix.

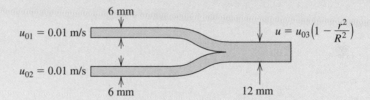

FIGURE 3.8 Mass conservation during mixing of two drug streams.

Governing Equations The conservation of mass for the selected open system is given by

$$\frac{d}{dt}\int_{\mathcal{V}_{os}} \rho\, d\mathcal{V} + \int_{A_r} \rho(\mathbf{V}\cdot\mathbf{n})\, dA_r = 0.$$

Basic Assumptions The flow is assumed to be steady and incompressible. Also, the velocity profile at the exit of the delivery tube is assumed to be parabolic of the form $u = u_0\left(1 - \frac{r^2}{R^2}\right)$.

SOLUTION For steady flow then $\dfrac{d}{dt}\displaystyle\int_{\mathcal{V}_{os}} \rho d\mathcal{V} = 0$, we have

$$\int_{A_r} \rho_i(\mathbf{V}_i\cdot\mathbf{n}_i)\, dA_r = 0,$$

where i denotes places where mass crosses the system boundary. At these locations we have the following conditions:

$$\mathbf{V}_1 = u_{o1}\mathbf{i} \quad \mathbf{n}_1 = -\mathbf{i}$$

$$\mathbf{V}_2 = u_{o2}\mathbf{i} \quad \mathbf{n}_2 = -\mathbf{i}$$

$$\mathbf{V}_3 = u_{o3}\left(1 - \frac{r^2}{R^2}\right)\mathbf{i}, \quad \mathbf{n}_3 = \mathbf{i}.$$

For the cylindrical geometry, an infinitesimal cross-sectional area is expressed as $dA_r = 2\pi r dr$. Therefore,

$$\int_1 \rho \mathbf{V}_1 \cdot \mathbf{n}_1 dA_1 + \int_2 \rho \mathbf{V}_2 \cdot \mathbf{n}_2 dA_2 + \int_3 \rho \mathbf{V}_3 \cdot \mathbf{n}_3 dA_3 = 0 \Rightarrow$$

$$\int_1 \rho \underbrace{(u_{o1}\mathbf{i})}_{\mathbf{V}_1} \cdot \underbrace{(-\mathbf{i})}_{\mathbf{n}_1} \underbrace{2\pi r dr}_{dA_1} + \int_2 \rho \underbrace{(u_{o2}\mathbf{i})}_{\mathbf{V}_2} \cdot \underbrace{(-\mathbf{i})}_{\mathbf{n}_2} \underbrace{2\pi r dr}_{dA_1} +$$

$$\int_3 \rho \underbrace{u_{o3} \left[1 - \left(\frac{r}{R} \right)^2 \right] (\mathbf{i})}_{\mathbf{V}_3} \cdot \underbrace{(\mathbf{i})}_{\mathbf{n}_3} \underbrace{2\pi r dr}_{dA_3} = 0,$$

by eliminating $2\rho\pi$ and by performing the scalar product between the velocity and the outward normal vectors, the preceding expression simplifies to

$$-u_{o1} \int_0^{R_1} r dr - u_{o2} \int_0^{R_2} r dr + u_{o3} \int_0^{R_3} \left[r - \frac{r^3}{R^2} \right] dr = 0$$

$$-u_{o1} \frac{R_1^2}{2} - u_{o2} \frac{R_2^2}{2} + u_{o3} \left[\frac{r^2}{2} - \frac{r^4}{4R^2} \right]_0^{R_3} = 0$$

$$-u_{o1} \frac{R_1^2}{2} - u_{o2} \frac{R_2^2}{2} + u_{o3} \frac{R_3^2}{4} = 0$$

$$u_{o3} = \frac{2u_{o1} R_1^2 + 2u_{o2} R_2^2}{R_3^2}.$$

By substituting the numerical values,

$$u_{o3} = \frac{2 \times 0.01 \ m/s \times 0.003^2 \ m^2 + 2 \times 0.01 \ m/s \times 0.003^2 \ m^2}{0.006^2 \ m^2} = 0.01 \ m/s.$$

EXAMPLE 3.6

Problem Statement *Reservoir filling*. Consider an initially empty reservoir with volume $2 \ m^3$ connected to a supply line through a valve as shown in Figure 3.9. Find the time it takes to fully fill the reservoir when the valve opens.

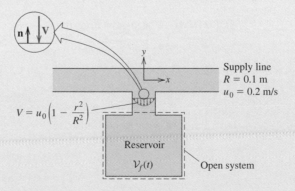

FIGURE 3.9 A tank with fixed volume is filled with water from a supply line.

Governing Equations The conservation of mass for the selected open system is given by

$$\frac{d}{dt}\int_{\mathcal{V}}\rho d\mathcal{V} + \int_{1}\rho(\mathbf{V}\cdot\mathbf{n})dA_r = 0.$$

In this problem, however, only part of the selected volume is filled with fluid. Therefore, at any instant t, the density ρ is defined only over the volume $\mathcal{V}_f(t)$ that the fluid occupies. Therefore, the unsteady term reduces to

$$\frac{d}{dt}\int_{\mathcal{V}}\rho d\mathcal{V} = \frac{d}{dt}\int_{\mathcal{V}_f}\rho d\mathcal{V}_f.$$

Basic Assumptions The flow at the supply line is assumed to be constant and incompressible.

SOLUTION At the inlet of the reservoir we have the following conditions: $\mathbf{V}_i = u_o\left(1 - \dfrac{r^2}{R^2}\right)(-\mathbf{j})$ and $\mathbf{n}_i = \mathbf{j}$. Substituting the appropriate terms in the conservation of mass

$$\frac{d}{dt}\int_{\mathcal{V}_f}\rho d\mathcal{V}_f + \int_{1}\rho\underbrace{u_o\left[1 - \frac{y^2}{R^2}\right](-\mathbf{j})}_{\mathbf{V}_i}\cdot\underbrace{(\mathbf{j})}_{\mathbf{n}_i}\underbrace{2\pi r dr}_{dA_i} = 0,$$

$$\rho\frac{d}{dt}\int_{\mathcal{V}_f}d\mathcal{V}_f - 2\pi\rho u_o\int_{1}\left[1 - \frac{r^2}{R^2}\right]r dr = 0,$$

then,

$$\frac{d\mathcal{V}_f}{dt} = 2\pi u_o\left[\frac{r^2}{2} - \frac{r^4}{4R^2}\right]_0^R$$

$$\frac{d\mathcal{V}_f}{dt} = \pi u_o\frac{R^2}{2} \Rightarrow d\mathcal{V}_f = \pi u_o\frac{R^2}{2}dt$$

By integration,

$$\int_0^{\mathcal{V}_t}d\mathcal{V}_f = \pi u_o\frac{R^2}{2}\int_0^{t_{fill}}dt.$$

The upper limit of the integration is set to $\mathcal{V}_f = \mathcal{V}_t$, where $\mathcal{V}_t$ is the volume of the tank (since the entire tank is finally filled with fluid). The final result then is

$$\mathcal{V}_t = \pi u_o\frac{R^2}{2}t_{fill}.$$

The reservoir will fill in

$$t_{fill} = \frac{2\mathcal{V}_t}{\pi u_o R^2} = \frac{2 \times 2\ m^3}{\pi \times 0.2\ m/s \times 0.1^2\ m^2} = 636.6\ seconds.$$

3.4   Conservation of Linear Momentum

If in Equation (3.6) we substitute $\alpha = \mathbf{V}$ and $B = \sum \mathbf{F}$, the conservation of linear momentum for an open system that is *stationary or moving with constant velocity (i.e., for an inertial coordinate system)*, emerges as

$$\underbrace{\frac{d}{dt} \int_{\mathcal{V}} \rho \mathbf{V} d\mathcal{V}}_{Rate\ of\ change\ of\ momentum} + \underbrace{\int_{A_r} \rho \mathbf{V}(\mathbf{V} \cdot \mathbf{n}) dA_r}_{Momentum\ flux\ at\ boundary} \equiv \underbrace{\sum \mathbf{F}}_{Net\ Force} . \tag{3.9}$$

Again, $\sum \mathbf{F}$ is a vector representing the resultant of the applied forces. The velocity vector $\mathbf{V}$ and its time derivative are expressed *relative to the selected coordinate system*.

If flow conditions at locations where mass crosses the boundary are uniform, the momentum equation simplifies to

$$\frac{d}{dt} \int_{\mathcal{V}} \rho \mathbf{V} d\mathcal{V} \pm \sum_{k=1}^{k=N} (\dot{m}\mathbf{V})_k \equiv \sum \mathbf{F}, \tag{3.10}$$

where N is the number of segments where mass crosses the boundary. Again, the proper sign is determined from the product $(\mathbf{V} \cdot \mathbf{n})$; the negative sign $(-)$ is associated with inlets, and the positive sign $(+)$ with exits.

In steady flows the equation simplifies to

$$\sum_{k_e=1}^{k=N_e} (\dot{m}_k \mathbf{V}_k)_{exit} - \sum_{k_i=1}^{k=N_i} (\dot{m}_k \mathbf{V}_k)_{inlet} \equiv \sum \mathbf{F}, \tag{3.11}$$

where subscripts (i) and (e) denote, respectively, inlets and exits. In cases of steady flow with a single inlet and a single exit we have

$$(\dot{m}\mathbf{V})_{exit} - (\dot{m}\mathbf{V})_{inlet} \equiv \sum \mathbf{F}.$$

But according to the continuity equation $\dot{m}_{in} = \dot{m}_{out} = \dot{m}$. Therefore,

$$\dot{m}(\mathbf{V}_{exit} - \mathbf{V}_{inlet}) \equiv \sum \mathbf{F}, \tag{3.12}$$

For the conditions stated above, Equation (3.12) is a useful alternative form of the momentum equation.

EXAMPLE 3.7

Problem Statement   *Force on a solid surface generated by an impinging jet of fluid.* During severe flooding, the manhole cover of the drainage system is lifted as shown in Figure 3.10. If the diameter D of the cover is 85 *cm* and the thickness 6 *cm*, find the volume of water per second out of the manhole. Consider a cast-iron cover whose density is 8200 kg/m^3.

Governing Equations   For the analysis, we have a choice for the selection of the control volume; we can either select the jet and the cover as a single control volume, or select the jet and the cover as separate control volumes. Irrespective of the choice of control volume, however, the analysis should yield the same answer. Selecting the cover and water as shown in Figure 3.10 as the control volume, the conservation of momentum

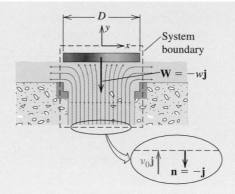

FIGURE 3.10   Force on a solid surface generated by a jet of fluid.

is expressed as

$$\frac{d}{dt}\int_{\mathcal{V}_{os}} \mathbf{V}\rho d\mathcal{V} + \int_{A_r} \rho\mathbf{V}(\mathbf{V}\cdot\mathbf{n})dA_r \equiv \sum\mathbf{F}.$$

Basic Assumptions   The flow of water is assumed to be steady and uniform at those locations where it crosses the boundary of the selected control volume.

SOLUTION   For steady flow $\frac{d}{dt}\int_{\mathcal{V}_{os}} \mathbf{V}\rho d\mathcal{V} = 0$, the momentum equation reduces to

$$\int_{A_r} \rho\mathbf{V}(\mathbf{V}\cdot\mathbf{n})dA_r \equiv \sum\mathbf{F}.$$

Because of symmetry, it is obvious that there is no net force in the horizontal direction. Hence, we need to consider only the y-component of the momentum equation

$$\int_{A_r} \rho v(\mathbf{V}\cdot\mathbf{n})dA_r \equiv \sum F_y.$$

The surface integral is evaluated at those locations where mass crosses the boundary and where v is not zero. This is true only at the location of the water jet (noted as inlet). The velocity vector at this location is $\mathbf{V} = v_o\,\mathbf{j}$, with v_o being the velocity of the water, and the outward unit normal vector is $\mathbf{n} = -\,\mathbf{j}$. The only force acting on the selected system is the weight of the cover $\mathbf{W} = -\rho_c g\mathcal{V}_c\,\mathbf{j}$, where ρ_c and $\mathcal{V}_c$ are, respectively, the density and volume of the cover and g is the gravitational acceleration.

Substituting the various terms into the momentum equation,

$$\int_{A_r} \rho(v_o\mathbf{j})(v_o\mathbf{j})\cdot(-\mathbf{j})dA_r = -\rho_c\mathcal{V}_c g\mathbf{j},$$

upon simplification and by using $\mathcal{V}_c = A_r H$, we get

$$-\rho v_o^2 A_r\mathbf{j} = -\rho_c\mathcal{V}_c g\mathbf{j} \Rightarrow v_o = \sqrt{g\frac{\rho_c}{\rho}\frac{A_r H}{A_r}} = \sqrt{g\frac{\rho_c}{\rho}H},$$

where H is the thickness of the cover. In terms of numerical values, the velocity of the water coming out of the manhole is

$$v_o = \sqrt{9.81 \ m/s^2 \times \frac{8200 \ kg/m^3}{998 \ kg/m^3} \times 0.06 \ m/s} = 2.2 \ m/s.$$

The volumetric flow rate $\dot{Q}$ is

$$\dot{Q} = v_o A_r = v_o \pi \frac{D^2}{4} = 2.2 \ m/s \times 3.14 \times \frac{0.85^2 \ m^2}{4} = 1.25 \ m^3/s.$$

EXAMPLE 3.8

Problem Statement *Force induced by flow in a 90° elbow.* Consider flow of water in a 90° elbow in a constant 0.06 m diameter pipe as shown in Figure 3.11. If the velocity u_o in the pipe is 0.15 m/s, find the required force to keep the elbow in place.

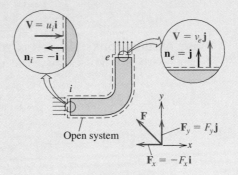

FIGURE 3.11   Force induced by flow in a 90° elbow.

Governing Equations By selecting a segment of the pipe as shown in the figure, we have an open system for which the conservation of momentum is expressed as

$$\frac{d}{dt} \int_{\mathcal{V}_{os}} \mathbf{V} \rho d\mathcal{V} + \int_{A_r} \rho \mathbf{V}(\mathbf{V} \cdot \mathbf{n}) dA_r \equiv \sum \mathbf{F},$$

and the conservation of mass as

$$\frac{d}{dt} \int_{\mathcal{V}_{os}} \rho d\mathcal{V} + \int_{A_r} \rho \mathbf{V} \cdot \mathbf{n} dA_r \equiv 0.$$

Basic Assumptions The flow is assumed to be steady, one-dimensional, incompressible, and with negligible gravity effects. We also neglect the pressure difference between the two ends of the pipe. Note that this is an oversimplification of the real flow, since a pressure difference across the pipe is necessary in order to induce and sustain the flow.

SOLUTION The surface integrals in the conservation laws are evaluated at the two locations where mass crosses the boundary denoted as *inlet* (i) and *exit* (e). At the *inlet*, $\mathbf{V}_i = u_i \mathbf{i}$ and $\mathbf{n}_i = -\mathbf{i}$. At the *exit*, $\mathbf{V}_e = v_e \mathbf{j}$ and $\mathbf{n}_e = \mathbf{j}$. The cross-sectional areas at both ends are equal; $A_i = A_e = A_r$.

After simplification, the conservation of mass for the open system shown in Figure 3.3 yields

$$\rho u_i A_r = \rho u_e A_r \Rightarrow u_i = v_e = u_o = 0.15 \ m/s.$$

The only relevant surface force acting on the system is the contact force $\mathbf{F}$ required to support the elbow. Assuming that the force acts in the direction shown in the figure, it is expressed as $\mathbf{F} = -F_x \ \mathbf{i} + F_y \ \mathbf{j}$. The momentum equation then becomes

$$\int_{inlet} \rho \mathbf{V}(\mathbf{V} \cdot \mathbf{n}) dA_r + \int_{exit} \rho \mathbf{V}(\mathbf{V} \cdot \mathbf{n}) dA_r = -F_x \mathbf{i} + F_y \mathbf{j}.$$

Substituting the appropriate expressions for the velocity and unit normal vectors, we get

$$\int_{inlet} \rho \mathbf{V}(\mathbf{V} \cdot \mathbf{n}) dA_r = \int_{inlet} \rho u_o \mathbf{i}(u_o \mathbf{i}) \cdot (-\mathbf{i}) dA_r = -\rho u_o^2 A_r \mathbf{i};$$

similarly,

$$\int_{exit} \rho \mathbf{V}(\mathbf{V} \cdot \mathbf{n}) dA_r = \int_{exit} \rho u_o \mathbf{j}(u_o \mathbf{j}) \cdot (\mathbf{j}) dA_r = \rho u_o^2 A_r \mathbf{j}.$$

The momentum equation then takes the form

$$-\rho u_o^2 A_r \mathbf{i} + \rho u_o^2 A_r \mathbf{j} = -F_x \mathbf{i} + F_y \mathbf{j}.$$

The same result could have been obtained by using Equation (3.12) and by setting $\mathbf{V}_1 = u_o \mathbf{i}$ and $\mathbf{V}_2 = u_o \mathbf{j}$.

By substituting the numerical values

$$F_x = \rho u_o^2 A_r = 998 \ kg/m^3 \times 0.15^2 \ m^2/s^2 \times \pi \times \frac{0.06^2}{4} \ m^2 = 0.0635 \ N.$$

and

$$F_y = \rho u_o^2 A_r = 998 \ kg/m^3 \times 0.15^2 \ m^2/s^2 \times \pi \times \frac{0.06^2}{4} \ m^2 = 0.0635 \ N.$$

Since F_x and F_y are positive quantities, the force directions are as assumed.

EXAMPLE 3.9

Problem Statement *Force induced by boundary layer effects.* Find the horizontal force induced along the plate due to the development of the boundary layer (Figure 3.6). Since the boundary layer develops as a result of fluid-solid friction, the sought-for force is the friction force at the plate surface that resists the flow.

Governing Equations For the selected open system, the conservation of momentum is expressed as

$$\frac{d}{dt} \int_{\mathcal{V}_{os}} \mathbf{V} \rho d\mathcal{V} + \int_{A_r} \rho \mathbf{V}(\mathbf{V} \cdot \mathbf{n}) dA_r \equiv \sum \mathbf{F},$$

and the conservation of mass as

$$\frac{d}{dt} \int_{\mathcal{V}_{os}} \rho d\mathcal{V} + \int_{A_r} \rho (\mathbf{V} \cdot \mathbf{n}) dA_r \equiv 0.$$

Basic Assumptions   The flow is assumed to be steady, two-dimensional, incompressible, and with negligible gravity effects.

SOLUTION   For steady flow $\dfrac{d}{dt}\displaystyle\int_{\mathcal{V}_{os}}\mathbf{V}\rho\,d\mathcal{V}=0$, and by considering the four boundaries of the open system, the momentum equation in the x-direction (since we are interested only in the force in this direction) reduces to

$$\int_1 \rho u_1(\mathbf{V}\cdot\mathbf{n})_1 W\,dS_1 + \int_2 \rho u_2(\mathbf{V}\cdot\mathbf{n})_2 W\,dS_2$$

$$+\int_3 \rho u_3(\mathbf{V}\cdot\mathbf{n})_3 W\,dS_3 + \int_4 \rho u_4(\mathbf{V}\cdot\mathbf{n})_4 W\,dS_4 = F_x.$$

Along the boundaries, we have the following conditions:

$$\mathbf{V}_1=u_o\mathbf{i}, \quad \mathbf{n}_1=-\mathbf{i}, \quad dS_1=dy, \quad \mathbf{V}_2=u_o\frac{y^2}{\delta^2}\mathbf{i}, \quad \mathbf{n}_2=\mathbf{i}, \quad dS_2=dy,$$

$$\mathbf{V}_3=0\mathbf{j}, \quad \mathbf{n}_3=-\mathbf{j}, \quad dS_3=dx, \quad \mathbf{V}_4=?, \quad \mathbf{n}_4=\mathbf{j}, \quad dS_4=dx,$$

By substitution and by considering the proper sign according to $(\mathbf{V}\cdot\mathbf{n})$, we get

$$F_x=-\int_{0_1}^{\delta}\rho u_o^2\,W\,dy+\int_{0_2}^{\delta}\rho u_o^2\left(\frac{y^2}{\delta^2}\right)^2 W\,dy+0+\int_{0_4}^{L}\rho u_o(\mathbf{V}\cdot\mathbf{n})W\,dx,$$

where L is the length of the plate. By integration,

$$F_x=-\rho u_o^2\delta W+\rho u_o^2\left[\frac{y^5}{5\delta^4}\right]_0^\delta W+\int_{0_4}^{L}\rho u_o(\mathbf{V}\cdot\mathbf{n})_4 W\,dx.$$

For mass conservation, we can make use of the result obtained in Example 3.3, where $\dot m_4=\dfrac{2}{3}\rho u_o\delta W$. The last term in the momentum equation simplifies to

$$\int_{0_4}^{L}\rho u_o(\mathbf{V}\cdot\mathbf{n})_4 W\,dx=u_o\underbrace{\int_{0_4}^{L}\rho(\mathbf{V}\cdot\mathbf{n})_4 W\,dx}_{\dot m}=u_o\dot m_4=\frac{2}{3}\rho u_o^2\delta W.$$

Finally,

$$F_x=-\rho u_o^2\delta W+\rho u_o^2\frac{1}{5}\delta W+\frac{2}{3}\rho u_o^2\delta W=-\frac{2}{15}\rho u_o^2\delta W.$$

This is the friction force the plate exerts on the fluid. By reaction, the force on the plate has the same magnitude but acts in the opposite direction

$$\Rightarrow F_{plate}=\frac{2}{15}\rho u_o^2\delta W.$$

3.4.1 Conservation of Linear Momentum for a Non-Inertial Coordinate System

The conservation of momentum relative to a *non-inertial* coordinate system (i.e., relative to a coordinate system that is accelerating with respect to a fixed coordinate system) is expressed as

$$\underbrace{\frac{d}{dt}\int_{\mathcal{V}}\rho\mathbf{V}d\mathcal{V} + \int_{A_r}\rho\mathbf{V}(\mathbf{V}\cdot\mathbf{n})dA_r}_{\textit{Relative to the accel. coord. system}} \equiv \sum\mathbf{F} - \underbrace{\int_{\mathcal{V}}\rho\mathbf{a}_{xyz}d\mathcal{V}}_{\textit{Relative to a fixed coord. system}} \qquad . \quad (3.13)$$

where $\mathbf{a}_{xyz}$ is the acceleration of the coordinate system relative to an inertial (i.e., fixed) coordinate system. $\sum\mathbf{F}$ is again the resultant of the applied forces. The velocity vector $\mathbf{V}$ and its time derivative are expressed relative to the accelerating coordinate system.

EXAMPLE 3.10

Problem Statement The Space Shuttle is powered into space with the help of the shuttle's three main engines and by two additional solid rocket boosters that are attached to the shuttle during the early stages of lift off. The following data for the space shuttle are available; the average gross weight during launch is 2,040,000 kg; each solid rocket produces 14,678,400 N of thrust; and each main engine produces 1,751,622 N. The propulsion performance of rockets and engines is traditionally quantified using the concept of the specific impulse Isp, defined as the ratio of the produced thrust T to the rate of weight of the exhaust gases $\dot{m}g$ (i.e., $Isp = T/\dot{m}g$, where g is gravity and $\dot{m}$ the mass flow rate). The shuttle's engine-specific impulse is $Isp_e = 452.1\ s$, and the solid rockets' is $Isp_r = 267.3\ s$, both measured under vacuum conditions. Note that when conditions are different, the actual Isp is less. Find the initial acceleration of the shuttle and its velocity at the early stages of the flight.

Governing Equations By selecting the shuttle as the control volume and since exhaust gases cross the boundary, we have an open system (Figure 3.12).

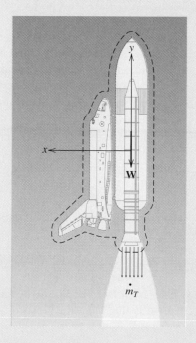

FIGURE 3.12 The Space Shuttle during liftoff.

Since the selected coordinate system is moving with the shuttle, we have an accelerating coordinate system with acceleration $\mathbf{a}_s = \dfrac{d\mathbf{V}_s}{dt}$, with $\mathbf{V}_s$ being the velocity of the shuttle relative to the ground. Therefore, for this non-inertial coordinate system the conservation of momentum is expressed as

$$\frac{d}{dt}\int_{\mathcal{V}_{os}} \mathbf{V}_g \rho\, d\mathcal{V} + \int_{A_r} \rho \mathbf{V}_g (\mathbf{V}_g \cdot \mathbf{n}) dA_r \equiv \sum \mathbf{F} - \int_{\mathcal{V}} \rho \mathbf{a}_{xyz} d\mathcal{V}, \qquad (3.14)$$

and the conservation of mass as

$$\frac{d}{dt}\int_{\mathcal{V}_{os}} \rho\, d\mathcal{V} + \int_{A_r} \rho (\mathbf{V}_g \cdot \mathbf{n}) dA_r \equiv 0,$$

where $\mathbf{V}_g$ is the velocity of the exhaust gases.

Basic Assumptions　Assume that the analysis is limited to a short time after lift off, during which the shuttle and rockets are still attached. For simplicity, we also assume that the velocity of the exhaust gases is uniform and that drag and pressure forces are negligible. Moreover, we assume that during launch the specific impulse is the same obtained under vacuum conditions (in reality, the specific impulse is less).

SOLUTION　In this problem, propulsion is produced by exhaust gases at five distinct locations: at the three engines and at the two solid rockets. The conservation of mass for the shuttle is expressed then as

$$\frac{d}{dt}\int_{\mathcal{V}} \rho_s\, d\mathcal{V} = -\sum_{k=1}^{k=3}\int_{A_{e_k}} \rho_{e_k}(\mathbf{V}_{ge_k} \cdot \mathbf{n}) dA_{e_k} - \sum_{k=1}^{k=2}\int_{A_{r_k}} \rho_{r_k}(\mathbf{V}_{gr_k} \cdot \mathbf{n}) dA_{r_k},$$

where ρ_s is the shuttle's overall density, A_{e_k} is the net engine cross-sectional area through which gases escape, and ρ_{e_k} and $\mathbf{V}_{ge_k}$ are, respectively, the gas density and the velocity there. Subscript r indicates the same quantities for the solid rockets.

Using the assumption of uniform exhaust velocity u_{e_k}, each engine mass flux term $\int_{A_{e_k}} \rho_{e_k}(\mathbf{V}_{ge_k} \cdot \mathbf{n}) dA_{e_k}$ reduces to $\rho_e u_{e_k} A_{e_k} \equiv \dot{m}_{e_k}$. The same is true for the solid rockets, for which we can write $\rho_r u_{r_k} A_{r_k} \equiv \dot{m}_{r_k}$. Now, since conditions for each engine and each rocket are identical, the conservation of mass reduces to

$$\frac{dm}{dt} = -\sum_{k=1}^{k=3} \dot{m}_{e_k} - \sum_{k=1}^{k=2} \dot{m}_{r_k} = -(3\dot{m}_e + 2\dot{m}_r) \equiv -\dot{m}_T,$$

where $\dot{m}_T$ represents the total gas mass flow rate from both the engines and the rockets. Therefore,

$$\int_{M_o}^{m} dm = -\int_0^t \dot{m}_T dt \Rightarrow m - M_o = -\dot{m}_T t,$$

where M_o is the original mass of the shuttle. The instantaneous shuttle mass as a function of time is then

$$m = M_o - \dot{m}_T t.$$

Since the coordinate system moves with the shuttle, the acceleration of the gas inside the shuttle relative to the moving coordinate system is small — that is, $\dfrac{d}{dt}\displaystyle\int_{\mathcal{V}_{os}} \rho \mathbf{V}_g$

$(\mathbf{V}_g \cdot \mathbf{n})d\mathcal{V} \approx 0$. The momentum equation then is expressed as

$$\sum_{k=1}^{k=3} \int_{A_{e_k}} \rho_{e_k} \mathbf{V}_{ge_k}(\mathbf{V}_{ge_k} \cdot \mathbf{n})dA_{e_k} + \sum_{k=1}^{k=2} \int_{A_{r_k}} \rho_{r_k} \mathbf{V}_{gr_k}(\mathbf{V}_{gr_k} \cdot \mathbf{n})dA_{r_k}$$

$$= -\int_{\mathcal{V}} \rho \frac{d\mathbf{V}_s}{dt} d\mathcal{V} + \int_{\mathcal{V}} \rho\mathbf{g}d\mathcal{V}.$$

At each engine exhaust area, we have $\mathbf{V}_{ge_k} = -u_{ge_k}\mathbf{i}$ and $\mathbf{n} = -\mathbf{i}$. Each momentum flux term there simplifies to

$$\int_{A_{e_k}} \rho_{e_k} \mathbf{V}_{ge_k}(\mathbf{V}_{ge_k} \cdot \mathbf{n})dA_{e_k} = \int_{A_{e_k}} \rho_{e_k}(-u_{ge_k}\mathbf{i})(-u_{ge_k}\mathbf{i}) \cdot (-\mathbf{i})dA_{e_k} = -\dot{m}_{ek}u_{ge_k}\mathbf{i}.$$

The corresponding term for the solid rockets reduces to $-\dot{m}_{rk}u_{gr_k}\mathbf{i}$.

Now, using the above results, the velocity of the shuttle $\mathbf{V}_s = V_s\mathbf{i}$, and the component of the gravity acting in the direction of the motion, which is $\mathbf{g} = -g\mathbf{i}$, we get

$$m\frac{dV_s}{dt} = 3\dot{m}_e u_{g_e} + 2\dot{m}_r u_{g_r} - mg.$$

From the above, it is obvious that each $\dot{m}u_g$ term represents the thrust contribution of each propulsion device: T_e for the engines and T_r for the rockets. Therefore, for identical engines and rockets,

$$3\dot{m}_e u_{g_e} + 2\dot{m}_r u_{g_r} = 3T_e + 2T_r \equiv T,$$

where T is the total thrust. The momentum equation then reduces to

$$\frac{dV_s}{dt} = \frac{T}{m} - g = \frac{T}{M_o - \dot{m}_T t} - g.$$

The initial acceleration of the shuttle then at $t = 0$ is

$$\frac{dV_s}{dt} = \frac{T}{M_o} - g. \tag{3.15}$$

The acceleration of the shuttle is proportional to the total thrust at liftoff, and it is inversely proportional to the initial total mass of the shuttle.

The velocity of the shuttle as a function of time can be obtained by integration

$$\int_0^{V_s} dV = \int_0^t \frac{T}{M_o - \dot{m}_T t} dt - \int_0^t g dt,$$

which gives

$$V_s = \frac{T}{\dot{m}_T} \ln\left(\frac{1}{1 - \frac{\dot{m}_T}{M_o}t}\right) - gt. \tag{3.16}$$

In terms of the numerical values, for the main engines

$$Isp_e = \frac{T_e}{\dot{m}_e g} \Rightarrow \dot{m}_e = \frac{T_e}{Isp_e g} = \frac{1{,}751{,}622\ N}{452.1\ s \times 9.81\ m/s^2} = 394.9\ kg/s.$$

For the solid rockets

$$Isp_r = \frac{T_r}{\dot{m}_r g} \Rightarrow \dot{m}_r = \frac{T_r}{Isp_r g} = \frac{14{,}678{,}400\ N}{267.3\ s \times 9.81\ m/s^2} = 5597.7\ kg/s.$$

The total mass flow rate $\dot{m}_T$ is

$$\dot{m}_T = 3\dot{m}_e + 2\dot{m}_r = 3 \times 394.9 \ kg/s + 2 \times 5597.7 \ kg/s = 12380.1 \ ks/s,$$

and the total thrust T is

$$T = 3T_e + 2T_r = 3 \times 1,751,622 \ N + 2 \times 14,678,400 \ N = 34,611,666 \ N.$$

For an initial gross mass of the shuttle $M_o = 2,040,000 \ kg$, according to Equation (3.15) the initial acceleration a_s is

$$a_s \equiv \frac{dV_s}{dt} = \frac{T}{M_o} - g = \frac{34,611,666 \ N}{2,040,000 \ kg} - 9.81 \ m/s^2 = 16.97 - 9.81 = 7.16 \ m/s^2,$$

which is less than the acceleration due to gravity!

By substituting the numerical values into Equation (3.16), the velocity as a function of time is

$$V_s = \frac{34,611,666 \ N}{12380.1 \ ks/s} \ln\left(\frac{1}{1 - \frac{12380.1 \ ks/s}{2,040,000 \ kg}t}\right) - 9.81 \ m/s^2 t$$

$$V_s = -2795.8 \ m/s \times \ln(1 - 6.07 \times 10^{-3}t) - 9.81 \ m/s^2 t.$$

The velocity of the shuttle as a function of time is plotted in Figure 3.3. Note that the above results are approximate since under non-vacuum conditions, Isp is less than the value obtained under vacuum conditions.

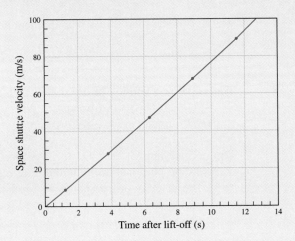

FIGURE 3.13   Velocity of the shuttle as a function of time.

3.5   Conservation of Energy

Consistent with the general Equation (3.6), the conservation of energy for an open system is expressed using $\alpha = \epsilon$ and $B = \dot{Q} - \dot{W}_s$ as

$$\frac{d}{dt}\int_{\mathcal{V}_{os}} \rho\epsilon d\mathcal{V} + \int_{A_r} \rho\epsilon(\mathbf{V} \cdot \mathbf{n})dA_r \equiv \dot{Q} - \dot{W}_s \qquad (3.17)$$

where $\epsilon = \dfrac{V^2}{2} + gz + u + \cdots + other\ forms\ of\ energy.$

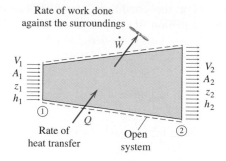

FIGURE 3.14 Open system with uniform conditions at a single inlet and a single exit.

The equation can be simplified by realizing that the mass crossing the boundary is associated with a work term defined as *flow work*,

$$\dot{W}_f = \int_{A_r} P(\mathbf{V} \cdot \mathbf{n}) dA_r = \int_{A_r} \rho P \upsilon (\mathbf{V} \cdot \mathbf{n}) dA_r,$$

where P is the pressure and υ the specific volume ($\upsilon = 1/\rho$). It is convenient, however, to separate the rate of work $\dot{W}_s$ into rate of mechanical work $\dot{W}_m$ and rate of flow work $\dot{W}_f$,

$$\dot{W}_s = \dot{W}_m + \dot{W}_f,$$

and to combine the term $P\upsilon$ with the internal energy term on the left-hand side of the equation. From this combination we get $u + P\upsilon$, which is equal to the enthalpy h. Therefore, the energy equation becomes

$$\frac{d}{dt} \int_{\mathcal{V}_{os}} \left(\frac{V^2}{2} + gz + u + \cdots \right) \rho d\mathcal{V} + \int_{A_r} \rho \left(h + \frac{V^2}{2} + gz \right) (\mathbf{V} \cdot \mathbf{n}) dA_r \equiv \dot{Q} - \dot{W}_m$$

(3.18)

Consider now a steady state problem with a single inlet and a single exit, and with uniform flow conditions across each one of them (Figure 3.14). The general form of the energy equation under these conditions reduces to

$$-(\rho \mathbf{V} \cdot \mathbf{n} A_r)_1 \left(h + \frac{V^2}{2} + gz \right)_1 + (\rho \mathbf{V} \cdot \mathbf{n} A_r)_2 \left(h + \frac{V^2}{2} + gz \right)_2 = \dot{Q} - \dot{W}_m.$$

Using mass conservation,

$$-(\rho \mathbf{V} \cdot \mathbf{n} A_r)_1 + (\rho \mathbf{V} \cdot \mathbf{n} A_r)_2 = 0 \Rightarrow (\rho \mathbf{V} \cdot \mathbf{n} A_r)_1 = (\rho \mathbf{V} \cdot \mathbf{n} A_r)_2 = \dot{m}.$$

The energy equation then simplifies to

$$\dot{m} \left[\left(h + \frac{V^2}{2} + gz \right)_2 - \left(h + \frac{V^2}{2} + gz \right)_1 \right] = \dot{Q} - \dot{W}_m,$$

which, upon collecting similar terms, simplifies further to

$$\dot{m} \left[\underbrace{(h_2 - h_1)}_{\Delta h} + \underbrace{\left(\frac{V_2^2}{2} - \frac{V_1^2}{2} \right)}_{\Delta KE} + \underbrace{g(z_2 - z_1)}_{\Delta PE} \right] = \dot{Q} - \dot{W}_m.$$

(3.19)

For the conditions stated, Equation (3.19) is a useful alternative form of the energy equation. It is often referred to as the *steady energy equation*.

Using the definition $h = u + P/\rho$, the energy equation for an incompressible fluid can be written as

$$\dot{m}\left[(u_2 - u_1) + \left(\frac{P_2}{\rho} - \frac{P_1}{\rho}\right) + \left(\frac{V_2^2}{2} - \frac{V_1^2}{2}\right) + g(z_2 - z_1)\right] = \dot{Q} - \dot{W}_m. \qquad (3.20)$$

Note that in isothermal flows (constant temperature) $u_2 - u_1 = 0$.

EXAMPLE 3.11

Problem Statement   *Flow discharged from a reservoir*. Find the rate of water being discharged from the reservoir shown in Figure 3.15. What is the velocity and pressure at point A? Consider that both the pressure at the free surface and the pressure at the delivery section is equal to the local atmospheric pressure P_a.

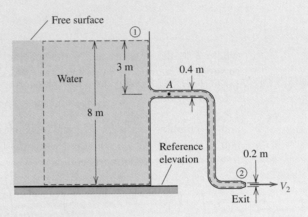

FIGURE 3.15   Schematic of Example 3.11.

Governing Equations   For the conditions of the problem, the simplified form of the energy equation (Equation (3.19)) applies here as well. Therefore, between two surfaces denoted by (1) and (2) we have

$$\left[(u_2 - u_1) + \left(\frac{P_2}{\rho} - \frac{P_1}{\rho}\right) + \left(\frac{V_2^2}{2} - \frac{V_1^2}{2}\right) + g(z_2 - z_1)\right] = \dot{Q} - \dot{W}_m.$$

Mass conservation also implies that

$$\rho V_1 A_1 = \rho V_2 A_2 = \dot{m}.$$

Basic Assumptions   The flow is assumed to be steady, isothermal, and incompressible, with negligible heat transfer to and from the surroundings.

SOLUTION   By selecting the free surface (subscript 1) and the delivery section (subscript 2), for no heat transfer ($\dot{Q} = 0$), no work ($\dot{W} = 0$), and for isothermal flow ($u_2 = u_1$), the energy equation reduces to

$$\frac{V_2^2}{2} + gz_2 + \frac{P_2}{\rho} = \frac{V_1^2}{2} + gz_1 + \frac{P_1}{\rho}.$$

At the free surface, $V_1 \approx 0$, $P_1 = P_a$, and by using the ground as a reference elevation, $z_1 = 8\ m$. At the discharge plane, the velocity is unknown, but $P_2 = P_a$ and $z_2 = 0$. Substituting the various terms,

$$\frac{V_2^2}{2} = gz_1 \Rightarrow V_2 = \sqrt{2gz_2}.$$

In terms of the numerical values

$$V_2 = \sqrt{2 \times 9.81\ m/s^2 \times 8\ m} = 12.53\ m/s.$$

The volumetric flow rate is

$$\dot{Q} = V_2 A_2 = V_2 \pi \frac{d_2^2}{4} = 12.53\ m/s \times \pi \times \frac{0.2^2}{4} m^2 = 0.394\ m^3/s,$$

and the mass flow rate is

$$\dot{m} = \rho \dot{Q} = 998\ kg/m^3 \times 0.394\ m^3/s = 393.2\ kg/s.$$

Now, if the energy equation is applied between the free surface and the plane through point A with the ground as a reference elevation, we have

$$\frac{V_A^2}{2} + gz_A + \frac{P_A}{\rho} = \frac{V_1^2}{2} + gz_1 + \frac{P_1}{\rho}.$$

The pressure at point A then is

$$P_A = P_1 + \rho \left[\frac{V_1^2}{2} - \frac{V_A^2}{2} + g(z_1 - z_A) \right].$$

Again $V_1 \approx 0$, $P_1 = P_a$, $z_1 = 8\ m$, and $z_A = 5\ m$. The velocity at A can be found by using mass conservation,

$$\dot{Q} = V_A \frac{\pi d_A^2}{4} \Rightarrow V_A = \frac{4 \times 0.394\ m^3/s}{\pi \times 0.4^2\ m^2} = 3.13\ m/s.$$

The pressure at point A then is

$$P_A = 101.3\ kPa + \frac{998\ kg/m^3}{1000\ Pa/kPa} \left[\frac{12.53^2}{2} - \frac{3.13^2}{2} + 9.81\ m/s^2 \times (8m - 5m) \right]$$

$$P_A = 204.13\ kPa.$$

EXAMPLE 3.12

Problem Statement *Pump design.* Find the pump power required to deliver water at a rate of $0.2\ m^3/s$, in a $0.8\ m$ diameter pipe at an elevation of $50\ m$ above the elevation of the suction of the pump (Figure 3.16).

Governing Equations By considering the open system as shown in the figure and for the conditions of the problem, the simplified form of the energy equation (Equation (3.19))

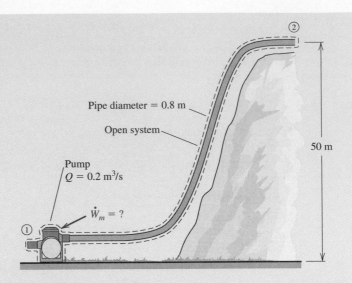

FIGURE 3.16   Pump design to meet delivery requirements.

between sections (1) and (2) applies:

$$\dot{m}\left[(u_2 - u_1) + \left(\frac{P_2}{\rho} - \frac{P_1}{\rho}\right) + \left(\frac{V_2^2}{2} - \frac{V_1^2}{2}\right) + g(z_2 - z_1)\right] = \dot{Q} - \dot{W}_m.$$

Basic Assumptions   The flow is assumed to be steady, isothermal, and incompressible, with negligible heat transfer from the pump to the surroundings.

SOLUTION   Since the cross-sectional area A_r and density ρ are constant, conservation of mass yields $V_2 = V_1$. By assuming isothermal flow $u_1 = u_2$ and with no significant heat transfer $\dot{Q} = 0$, the energy equation reduces to

$$\dot{m}\left[\left(\frac{P_2}{\rho} - \frac{P_1}{\rho}\right) + g(z_2 - z_1)\right] = -\dot{W}_m.$$

However, since the pump receives and delivers liquid at the local atmospheric pressure $P_1 = P_2 = P_a$, we finally get

$$\dot{W}_m = -\dot{m}g(z_2 - z_1).$$

The volumetric flow rate $\dot{Q}$, is defined as $\dot{Q} = VA_r = V\pi R^2$. Therefore, the fluid velocity in the pipe is

$$V = \frac{\dot{Q}}{A_r} = \frac{0.2\ m^3/s}{\pi \times 0.4^2\ m^2} = 0.4\ m/s,$$

and the mass flow rate is

$$\dot{m} = \rho\dot{Q} = 998\ kg/m^3 \times 0.2\ m^3/s = 199.6\ kg/s.$$

The power input to the pump is then

$$\dot{W}_m = -199.6\ kg/s \times 9.81\ m/s^2 \times 50\ m$$

$$\Rightarrow \dot{W}_m = -97.90\ kW.$$

The negative sign correctly indicates that in order to pump the water at the higher elevation, energy must be supplied to the system.

3.5.1 Energy Equation Along a Streamline

If a control volume is drawn around a flow line along which the velocity vector is tangent (i.e, along a streamline), in a *steady, isothermal,* and *adiabatic* flow as shown in Figure 3.17, the energy equation

$$\frac{d}{dt}\int_{\mathcal{V}_{os}}\left(\frac{V^2}{2}+gz+u+\cdots\right)\rho d\mathcal{V}+\int_{A_r}\rho\left(h+\frac{V^2}{2}+gz\right)\mathbf{V}\cdot\mathbf{n}dA_r\equiv\dot{Q}-\dot{W}_m,$$

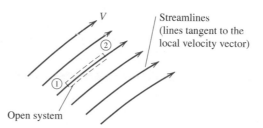

Open system

FIGURE 3.17 Open system around a section of a streamline in a steady, frictionless, and adiabatic flow.

between two arbitrary points (1) and (2) reduces to

$$\dot{m}\left[(h_2-h_1)+\left(\frac{V_2^2}{2}-\frac{V_1^2}{2}\right)+g(z_2-z_1)\right]=0.$$

By rewriting $h=u+\dfrac{P}{\rho}$, and by considering isothermal flow $u_2=u_1$, the energy equation reduces to

$$\left(\frac{P}{\rho}+gz+\frac{V^2}{2}\right)_1=\left(\frac{P}{\rho}+gz+\frac{V^2}{2}\right)_2=Constant. \qquad (3.21)$$

This equation is the well-known *steady Bernoulli equation*. This equation is re-derived in Chapter 9 using the conservation of momentum along a streamline for ideal inviscid flows that satisfy the assumptions used in deriving Equation (3.21).

The Bernoulli equation can be rearranged as

$$\underbrace{\frac{P}{\rho g}}_{Pressure\ Head}+\underbrace{\frac{V^2}{2g}}_{Velocity\ Head}+\underbrace{z}_{Elevation\ Head}=\text{Constant}\equiv\underbrace{H_T}_{Total\ Head},$$

where each term in the equation has dimensions of length. Therefore, $\dfrac{P}{\rho g}$ is traditionally defined as the *pressure head,* $\dfrac{V^2}{2g}$ as the *velocity head,* and z as the *elevation head.* The Bernoulli equation shows that along a streamline, the sum of the pressure, velocity, and elevation heads is constant, equal to the *total head* H_T (Figure 3.18).

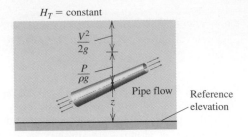

FIGURE 3.18   Schematic view of energy conservation.

Alternatively, the Bernoulli equation can be arranged as

$$\underbrace{P}_{Static\ pressure} + \underbrace{\rho\frac{V^2}{2}}_{Dynamic\ pressure} + \underbrace{\rho g z}_{Hydrostatic\ pressure} = \underbrace{P_T}_{Total\ pressure},$$

where P is the static pressure, $\rho\dfrac{V^2}{2}$ is the *dynamic pressure*, and $\rho g z$ is the *hydrostatic pressure*. In this form, the Bernoulli equation indicates that along a streamline the sum of the preceding pressure terms is constant, equal to the *total pressure* P_T.

For negligible elevation changes, the total pressure becomes

$$P + \rho\frac{V^2}{2} = P_o,$$

where P_o is designated as the *stagnation pressure*. By definition, then, the stagnation pressure is the pressure at points along a nearly horizontal streamline where the velocity is zero (i.e., at stagnation points). Both the static and stagnation pressures can be measured using specially designed manometer-type devices as shown in Figure 3.19.

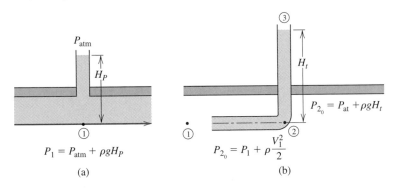

FIGURE 3.19   (a) Piezometer tube; (b) Pitot tube.

The schematic in Figure 3.19a shows a *piezometer tube*. According to the schematic, the static pressure at any point in the flow is obtained by measuring the height of the liquid z_p as

$$P_1 = \rho g z_p + P_a,$$

The schematic in Figure 3.19b is a *Pitot tube*. According to the schematic, for the streamline shown in the figure, the Bernoulli equation between points (1) and (2) gives

$$P_1 + \rho\frac{V_1^2}{2} = P_{2_o}.$$

Because of hydrostatic conditions between (2) and (3), the stagnation pressure is

$$P_{2_o} = \rho g z_t + P_a.$$

If the static pressure P_1 is measured independently by a Pitot tube

$$P_1 = \rho g z_p + P_a,$$

the velocity V_1 in the pipe can be determined as

$$V_1 = \sqrt{2\frac{P_{2_o} - P_1}{\rho}} = \sqrt{2g(z_t - z_p)}.$$

In practice, the combination of a Pitot tube and a piezometer tube is used to measure both the static pressure and the velocity in the flow.

EXAMPLE 3.13

Problem Statement   Consider a sprinter running with velocity V_o in still air, as shown in Figure 3.20a. What pressure will the sprinter experience if $V_o = 10 m/s$?

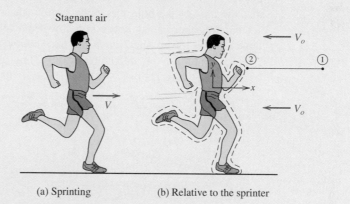

Stagnant air

(a) Sprinting   (b) Relative to the sprinter

FIGURE 3.20   Static and stagnation conditions.

Governing Equations   In the absence of heat transfer and work, the Bernoulli equation (3.20) gives the pressure distribution $P(r)$ as a function of the distance r along the streamline as

$$P(r) + \rho\frac{V(r)^2}{2} + \rho g z = \text{constant},$$

where r is the distance along the streamline $V(r)$ and z the local velocity and elevation.

Basic Assumptions   The flow is assumed to be steady, frictionless, and adiabatic.

SOLUTION   We select the sprinter as the control volume and prescribe a coordinate system attached to the sprinter (Figure 3.20b). The problem is equivalent to having a stationary sprinter and air flowing past him or her with velocity V_o at almospheric pressure P_a. The Bernoulli equation applied between points (1) and (2) along the selected streamline as shown in the figure gives

$$P_o = P_a + \rho \frac{V^2}{2},$$

where at (1) $P = P_a$ and the velocity of air is V, and at (2) the velocity of air is zero and the pressure is equal to $P = P_o$. This indicates that the sprinter experiences a pressure that increases with the square of the velocity.

For the given conditions, the stagnation pressure then is

$$P_o = 101.3 \; kPa + 1.2 \; kg/m^3 \times \frac{10^2 \; m^2/s^2}{2} \times \frac{1}{1000 \; Pa/kPa} = 101.36 \; kPa.$$

EXAMPLE 3.14

Problem Statement   Titanic *problem*. The *Titanic* sank on April 14, 1912, hitting an iceberg at 11:40 PM and sinking 160 minutes later at 2:20 AM. Recently, a sonar study of the bow of the *Titanic* on the ocean floor has revealed that the holes caused by the iceberg are much smaller than originally thought. Until this study, it was assumed that a large, 100 *m* long gash was ripped in the *Titanic*'s side, but now the sonar reveals that the area of the hole was *only* 1.4 m^2 (the size of a typical door!). The hole of the *Titanic* was approximately 6.1 *m* below sea level at the start of the sinking. Was the hole large enough to sink the *Titanic* in 160 *minutes*? The following data are available for the *Titanic*: length 269 *m*, maximum width 28 *m*, and height 30.5 *m*. [problem contributed by David J. Olinger].

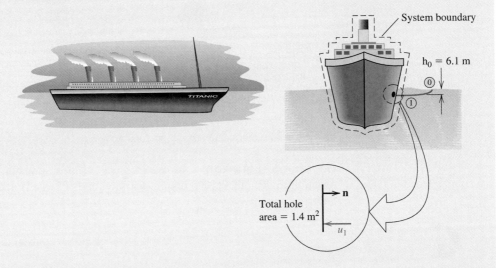

FIGURE 3.21   A simplified schematic of the *Titanic*.

Governing Equations   By selecting the entire ship as an open system, the conservation of mass is given by

$$\frac{d}{dt} \int_{\mathcal{V}} \rho d\mathcal{V} + \int_1 \rho (\mathbf{V} \cdot \mathbf{n}) dA_r = 0.$$

In this problem, however, only part of the selected volume is filled with fluid. Therefore, at any instant t, the density ρ is defined only over the volume $\mathcal{V}_f(t)$ that the fluid occupies. Therefore, the unsteady term reduces to

$$\frac{d}{dt}\int_{\mathcal{V}}\rho d\mathcal{V}=\frac{d}{dt}\int_{\mathcal{V}_f}\rho d\mathcal{V}_f.$$

For a streamline originating at the free surface (0) and terminating at the hole (1), the Bernoulli equation is expressed as

$$P_0+\rho\frac{V_0^2}{2}+\rho g z_0=P_1+\rho\frac{V_1^2}{2}+\rho g z_1,$$

where P, V, and z are respectively the pressure, velocity, and elevation from an arbitrary reference level, of the points.

Basic Assumptions   The problem can be simplified by assuming inviscid, incompressible flow, with constant uniform velocity at the hole. We also represent the *Titanic* as an empty rectangular box — a rather conservative assumption. The water inside the ship is assumed opened to the atmosphere.

SOLUTION   At the hole, we have the following conditions: $\mathbf{V}_i=-u_1\mathbf{i}$ and $\mathbf{n}_i=\mathbf{i}$. Substituting the appropriate terms in the conservation of mass, we get

$$\frac{d}{dt}\int_{\mathcal{V}_f}\rho d\mathcal{V}_f+\int_{hole}\rho\underbrace{(-u_1\mathbf{i})}_{\mathbf{V}_i}\cdot\underbrace{(\mathbf{i})}_{\mathbf{n}_i}dA_r=0,$$

$$\frac{d}{dt}\int_{\mathcal{V}_f}\rho d\mathcal{V}_f-\int_{hole}\rho u_1 dA_r=0.$$

Since sinking is defined as the point when the *Titanic* is fully filled, the upper limit of the integration is set to $\mathcal{V}_f=\mathcal{V}_t$, where $\mathcal{V}_t$ is the total volume of the *Titanic*. For uniform conditions at the hole, and constant density

$$\frac{d\mathcal{V}_f}{dt}=u_1 A_h,$$

where A_h is the area of the hole. By integration, the final result then is

$$\mathcal{V}_t=u_1 A_h t_{sink}.$$

The velocity of the water at the hole can be evaluated using the Bernoulli equation by selecting an arbitrary point at the free surface where $P_0=P_{atm}$, $V_0=0$, and $z=h_0$, and a point at the hole where $P_1=P_{atm}$, $V_1=u_1$, and $z=0$ (the reference elevation is chosen at the level of the hole). Therefore, by substitution, the water velocity at the hole is $u_1=\sqrt{2gh_0}$.

The analysis then predicts that under the assumptions and conditions stated, the *Titanic* would have sunk in

$$t_{sink}=\frac{\mathcal{V}_t}{u_1 A_h}=\frac{\mathcal{V}_t}{\sqrt{2gh_0}A_h}seconds.$$

In terms of numerical values

$$t_{sink}=\frac{269m\times 28\ m\times 30.5\ m}{\sqrt{2\times 9.81\ m/s^2\times 6.1\ m}\times 1.4\ m^2}=15000\ seconds.$$

This corresponds to 4.2 hours, which is higher than the actual time it took for the ship to sink. A more detail analysis of the same problem must consider the fact that while the *Titanic* sunk, both the depth of the hole and the level of the water inside the ship increased. The analysis must also consider that fact that the *Titanic* was not empty but partially filled with people, furniture, luggage, and cargo. Nevertheless, since the predicted 4.2 hrs from this conservative approach is not very far off from the actual time of 2.7 hrs, the analysis suggests that the *Titanic* could have sunk because of a hole that was much smaller than originally thought.

EXAMPLE 3.15

Problem Statement Consider the cylindrical reservoir with diameter D shown in Figure 3.22, where the liquid at an initial height H is draining through a small hole with diameter d at the bottom. Find the time it takes for the reservoir to empty completely.

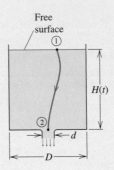

FIGURE 3.22 Draining of a reservoir.

Governing Equations For a streamline originating at the free surface (1) and terminating at the exit hole (2), the Bernoulli equation is expressed as

$$P_1 + \rho\frac{V_1^2}{2} + \rho g z_1 = P_2 + \rho\frac{V_2^2}{2} + \rho g z_2.$$

Additionally, for constant density, conservation of mass requires that

$$V_1 A_1 = V_2 A_2.$$

Basic Assumptions The flow is assumed to be frictionless and adiabatic. Moreover, we assume that since the reservoir drains slowly, the acceleration of the system is small and can be neglected.

SOLUTION Selecting one point at the free surface and another at the exit, the steady Bernoulli equation gives

$$\left(P_1 + \rho\frac{V_1^2}{2} + \rho g z_1\right)_{Free\ surface} = \left(P_2 + \rho\frac{V_2^2}{2} + \rho g z_2\right)_{Exit}.$$

Using the elevation of the draining hole as a reference elevation, $z_1 = z(t)$, $P_1 = P_a$, and $z_2 = 0$, $P_2 = P_a$. The Bernoulli equation then simplifies to

$$V_2^2 - V_1^2 = 2g z(t).$$

Mass conservation gives

$$V_2 = V_1 \frac{A_1}{A_2} = V_1 \frac{D^2}{d^2}.$$

By substitution, we get

$$V_1^2 \left(\frac{D^4}{d^4} - 1 \right) = 2gz.$$

But

$$V_1 = -\frac{dz}{dt}.$$

Therefore,

$$\sqrt{\left(\frac{D^4}{d^4} - 1 \right)} dz = \sqrt{2gz} dt,$$

or

$$-\frac{dz}{\sqrt{z}} = \sqrt{2g} \left(\frac{D^4}{d^4} - 1 \right)^{-\frac{1}{2}} dt.$$

By integration,

$$\int_H^0 -\frac{dh}{\sqrt{h}} = \int_0^{t_d} \sqrt{2g} \left(\frac{D^4}{d^4} - 1 \right)^{-\frac{1}{2}} dt,$$

and by simplification we get

$$t_d = \sqrt{2 \frac{H}{g} \left(\frac{D^4}{d^4} - 1 \right)}.$$

This is the same result provided in Chapter 1 (without proof) as the ideal analytic solution to the draining reservoir problem.

3.6 Second Law of Thermodynamics

The second law of thermodynamics for an *open isolated (or equivalently adiabatic) system* with volume $\mathcal{V}_{os}$ is expressed as

$$\frac{d}{dt} \int_{\mathcal{V}_{os}} s\rho d\mathcal{V} + \int_{A_r} \rho s (\mathbf{V} \cdot \mathbf{n}) dA_r \geq 0, \tag{3.22}$$

where s is the specific entropy, ρ is the density, $\mathbf{V}$ the velocity vector, $\mathbf{n}$ the outward pointing unit normal vector, and A_r the area where mass crosses the boundary. Again, all terms are treated exactly in the same manner as with the other laws.

For steady flow $\frac{d}{dt} \int_{\mathcal{V}_{os}} s\rho d\mathcal{V} = 0$, Equation (3.22) reduces to

$$\int_{A_r} \rho s (\mathbf{V} \cdot \mathbf{n}) dA_r \geq 0.$$

In the case of uniform flow conditions with a single inlet (i) and a single exit (e), with uniform flow conditions at both sections,

$$s_e - s_i \geq 0.$$

Finally, for isentropic flow we have $s_e = s_i$.

EXAMPLE 3.16

Problem Statement   *Heat exchange.* $2 \, kg/s$ of air at $25°\,C$ and $100 \, kPa$ enters a heat exchanger and leaves at $200°\,C$ and $400 \, kPa$, as shown in Figure 3.23. Calculate the amount of heat added and the entropy change between the inlet and the exit.

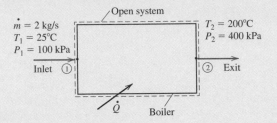

FIGURE 3.23   Energy and entropy changes in a heat exchanger.

Governing Equations   By considering the open system as shown in the figure, the conservation of energy in the form of Equation (3.18) applies

$$\dot{m}\left[(h_2 - h_1) + \left(\frac{V_2^2}{2} - \frac{V_1^2}{2}\right) + g(z_2 - z_1)\right] = \dot{Q} - \dot{W}_m;$$

the entropy change between the inlet and the exit is

$$\dot{S}_2 - \dot{S}_1 = \dot{m}\left[C_p \ln \frac{T_2}{T_1} - R \ln \frac{P_2}{P_1}\right].$$

Basic Assumptions   The flow is assumed to be steady, with no significant changes in the kinetic and potential energies between the inlet and exit sections.

SOLUTION   For steady flow, the energy equation for the control volume around the heat exchanger as shown in Figure 3.23 is given by Equation (3.18):

$$\dot{m}\left[\underbrace{(h_2 - h_1)}_{\Delta h} + \underbrace{\left(\frac{V_2^2}{2} - \frac{V_1^2}{2}\right)}_{\Delta KE} + \underbrace{g(z_2 - z_1)}_{\Delta PE}\right] = \dot{Q} - \dot{W}_m.$$

Using the assumptions of no significant changes in the kinetic and potential energies and since $\dot{W}_m = 0$, the energy equation reduces to

$$\dot{Q} = \dot{m}(h_2 - h_1) = \dot{m}C_p(T_2 - T_1).$$

By substituting the numerical values, the heat transfer is

$$\dot{Q} = 2 \, kg/s \times 1.0035 \, kJ/kg\,K \times (473.15 \, K - 298.15 \, K) = 351.23 \, kW.$$

The entropy change between the inlet and exit of the heat exchanger

$$\dot{S}_2 - \dot{S}_1 = \dot{m}\left[C_p \ln \frac{T_2}{T_1} - R \ln \frac{P_2}{P_1}\right],$$

gives

$$\dot{S}_2 - \dot{S}_1 = 2\ kg/s \times \left[1.0035 kJ/kgK \times \ln\left(\frac{473.15\ K}{298.15\ K}\right)\right.$$

$$\left. -0.287\ kJ/kgK \times \ln\left(\frac{400\ kPa}{100\ kPa}\right)\right] = 0.131\ kW/K.$$

EXAMPLE 3.17

Problem Statement   *Compressor flow.* Air enters the compressor of a gas turbine at $90\ kPa$ and $18°C$ at a rate of $40\ m^3/s$ and leaves at $600\ kPa$ (Figure 3.24). If the process is isentropic and the compressor is well insulated, find the power needed to drive the compressor.

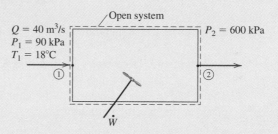

FIGURE 3.24   Isentropic and adiabatic flow in a compressor.

Governing Equations   By considering the open system as shown in the figure, the conservation of energy in the form of Equation (3.18) for the inlet (1) and exit (2) applies:

$$\dot{m}\left[(h_2 - h_1) + \left(\frac{V_2^2}{2} - \frac{V_1^2}{2}\right) + g(z_2 - z_1)\right] = \dot{Q} - \dot{W}_m.$$

For isentropic flow between (1) and (2),

$$\frac{T_2}{T_1} = \left(\frac{P_2}{P_1}\right)^{\frac{\gamma - 1}{\gamma}}.$$

Basic Assumptions   The flow is assumed to be steady and isentropic, with no significant changes in the kinetic and potential energies between the inlet and exit sections and with no heat transfer to or from the surroundings.

SOLUTION   Since the compressor is well insulated, $\dot{Q} = 0$. If we assume no significant changes in the elevation or in the velocity between the inlet and exit ports of the compressor, Equation (3.18) simplifies to

$$\dot{W}_m = -\dot{m}(h_2 - h_1) = -\dot{m}C_p(T_2 - T_1).$$

For isentropic flow,

$$\frac{T_2}{T_1} = \left(\frac{P_2}{P_1}\right)^{\frac{\gamma-1}{\gamma}} \Rightarrow T_2 = T_1 \left(\frac{P_2}{P_1}\right)^{\frac{\gamma-1}{\gamma}}.$$

By substituting the numerical values and by using $\gamma = 1.4$ for air,

$$T_2 = (273.15 \ K + 18 \ K) \times \left(\frac{600 \ kPa}{90 \ kPa}\right)^{\frac{1.4-1}{1.4}} = 500.63 \ K.$$

The mass flow rate $\dot{m}$ through the compressor is evaluated from the volumetric flow rate $\dot{Q}$ and density ρ,

$$\dot{m} = \rho \dot{Q} = \frac{P_1}{RT_1}\dot{Q} = \frac{90 \ kPa}{0.287 \ kJ/kgK \times 291.15 \ K} \times 40 \ m^3/s = 43.08 \ kg/s.$$

Returning to the energy equation, the power input to the compressor is

$$\dot{W}_m = -\dot{m}C_p(T_2 - T_1).$$

Substituting the numerical values,

$$\dot{W}_m = -43.08 \ kg/s \times 1.0035 \ kJ/kgK \times (500.63 \ K - 291.15 \ K) = -9056. \ kW.$$

The negative sign indicates that the compressor *requires* 9069.5 kW to deliver 40 m^3/s of air at the specified conditions.

EXAMPLE 3.18

Problem Statement   Consider air flow in a converging channel with inlet area of 60 cm^2 and an exit area of 40 cm^2 (Figure 3.25). At the inlet the velocity is 100 m/s, the pressure is 150 kPa, and the temperature is 300 K. The velocity at the exit is 250 m/s, and the atmospheric pressure is 101 kPA. If the flow in the channel is adiabatic and the air behaves like an ideal gas, find (a) The mass flow rate through the channel in kg/s; (b) the exit temperature; (c) the exit pressure; (d) the entropy change between the inlet and exit; (e) the net force required to keep the channel in place.

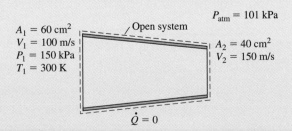

$A_1 = 60$ cm^2
$V_1 = 100$ m/s
$P_1 = 150$ kPa
$T_1 = 300$ K

$P_{atm} = 101$ kPa

Open system

$A_2 = 40$ cm^2
$V_2 = 150$ m/s

$\dot{Q} = 0$

FIGURE 3.25   Flow in a converging channel.

Governing Equations   By considering the open system as shown in the figure, the conservation of energy in the form of Equation (3.18) applied between (1) and (2) is valid

$$\dot{m}\left[(h_2 - h_1) + \left(\frac{V_2^2}{2} - \frac{V_1^2}{2}\right) + g(z_2 - z_1)\right] = \dot{Q} - \dot{W}_m.$$

The rate of entropy change with respect to time between the inlet and exit of the channel is

$$\dot{S}_2 - \dot{S}_1 = \dot{m}\left[C_p \ln \frac{T_2}{T_1} - R \ln \frac{P_2}{P_1}\right],$$

and the conservation of momentum requires that

$$\frac{d}{dt}\int_{\mathcal{V}_{os}} \mathbf{V}\rho d\mathcal{V} + \int_{A_r} \rho \mathbf{V}(\mathbf{V}\cdot\mathbf{n})dA_r = \sum \mathbf{F}.$$

The mass flow rate is calculated using the definition

$$\dot{m} = \rho V A = \frac{P}{RT}V A.$$

Basic Assumptions   It is assumed that the gas is ideal (i.e., $P = \rho R T$) and that the flow is steady, one-dimensional, and adiabatic, $\dot{Q} = 0$. It is also assumed that there are no significant changes in the kinetic and potential energies between the inlet and exit sections.

SOLUTION   The mass flow rate

$$\dot{m} = \rho V A = \frac{P}{RT}V A,$$

is

$$\dot{m} = \frac{150 \; kPa}{0.287 \; kJ/kgK \times 300 \; K} \times 100 \; m/s \times 60 \; cm^2 \times 10^{-4} \; m^2/s^2 = 1.045 \; kg/s.$$

The temperature is found by applying the energy equation

$$\dot{m} = \left[\underbrace{(h_2 - h_1)}_{\Delta h} + \underbrace{\left(\frac{V_2^2}{2} - \frac{V_1^2}{2}\right)}_{\Delta KE} + \underbrace{g(z_2 - z_1)}_{\Delta PE}\right] = \dot{Q} - \dot{W}_m.$$

For no heat transfer $\dot{Q} = 0$ and no work $\dot{W}_m = 0$ and by neglecting changes in the potential energy, it reduces to

$$(h_2 - h_1) + \left(\frac{V_2^2}{2} - \frac{V_1^2}{2}\right) = 0.$$

Substituting $(h_2 - h_1) = C_p(T_2 - T_1)$,

$$C_p(T_2 - T_1) + \left(\frac{V_2^2}{2} - \frac{V_1^2}{2}\right) = 0.$$

Therefore,

$$T_2 = T_1 - \frac{1}{2C_p}(V_2^2 - V_1^2).$$

In terms of numerical values,

$$T_2 = 300 \ K - \frac{(250^2 \ m^2/s^2 - 100^2 \ m^2/s^2)}{2 \times 1.0035 \ kJ/kgK} \times 10^{-3} \ kJ/J = 273.84 \ K.$$

The density at the exit

$$\rho = \frac{\dot{m}}{V_2 A_2} = \frac{1.045 \ kg/s}{150 \ m/s \times 40 \ cm^2 \times 10^{-4} \ cm^2/m^2} = 1.742 \ kg/m^3.$$

Using the ideal gas relation, the exit pressure is

$$P_2 = \rho R T_2 = 1.742 \ kg/m^3 \times 0.287 \ kJ/kgK \times 293.78 \ K = 146.88 \ kPa.$$

By substitution, the entropy change, between the inlet and exit of the channel

$$\dot{S}_2 - \dot{S}_1 = \dot{m} \left[C_p \ln \frac{T_2}{T_1} - R \ln \frac{P_2}{P_1} \right],$$

then gives

$$\dot{S}_2 - \dot{S}_1 = 1.045 \ kg/s \times \left[1.0035 \ kJ/kgK \times \ln \left(\frac{273.84 \ K}{300 \ K} \right) \right.$$

$$\left. - 0.287 \ kJ/kgK \times \ln \left(\frac{146.88 \ kPa}{150 \ kPa} \right) \right] = -0.0894 \ kW/K.$$

To evaluate the force in the horizontal direction we need to return to the momentum equation for the open system shown in Figure 3.6,

$$\frac{d}{dt} \int_{\mathcal{V}_{os}} \mathbf{V} \rho d\mathcal{V} + \int_{A_r} \rho \mathbf{V}(\mathbf{V} \cdot \mathbf{n}) dA_r = \sum \mathbf{F}.$$

The flow here is steady, and mass crosses the boundary of the open system at two locations. For uniform velocity across both the inlet and exit sections, $\rho V A = \dot{m} = constant$, the integral simplifies to

$$\dot{m}(V_2 - V_1) = \sum \mathbf{F}.$$

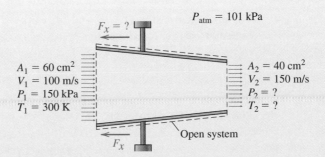

FIGURE 3.26   Conservation of momentum in a converging channel.

As shown in Figure 3.6, the appropriate forces acting along the boundary of the system are those due to the pressure at the inlet and exit sections, the atmospheric pressure, and the unknown force F_x required to hold the channel in place. Hence,

$$\dot{m}(V_2 - V_1) = P_1 A_1 - P_2 A_2 - P_{atm}(A_1 - A_2) + F_x,$$

or,

$$F_x = P_2 A_2 + P_{atm}(A_1 - A_2) - P_1 A_1 + \dot{m}(V_2 - V_1).$$

Substituting the numerical values,

$$F_x = [146.88 \; kPa \times 40 \; cm^2 + 101 \; kPa \times (60 \; cm^2 - 40 \; cm^2)$$

$$- 150 \; kPa \times 60 \; cm^2] \times 10^{-4} \; cm^2/m^2 + \frac{1.045 \; kg/s}{1000 \; N/kN} \times (250 \; m/s - 100 \; m/s)$$

$$\Rightarrow F_x = 0.0463 \; kN.$$

REFERENCES

E.A. AVALONE, and T. BAUMEISTER, *Marks' Standard Handbook for Mechanical Engineers*, 9th ed., New York, McGraw-Hill, 1987.

G.K. BATCHELOR, F.R.S., *An Introduction to Fluid Dynamics*, Cambridge, England, Cambridge University Press, 1970.

R.B. BIRD, W.E. STEWARD, and E.N. LIGHTFOOT, *Transport Phenomena*, New York, Wiley, 1960.

R.L. PANTON, *Incompressible Flow*, New York, Wiley, 1984.

R.E. SONNTAG, and G. VAN WYLEN, *Introduction to Thermodynamics, Classical and Statistical*, New York, Wiley, 1984.

PROBLEMS

1. *Entrance flow.* Consider the entrance pipe flow as shown in Figure 3.27. Due to the action of viscosity (fluid friction), after a certain distance the velocity develops into a parabolic profile. For the conditions shown in the figure, find the maximum velocity at the exit of the pipe.

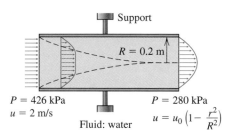

FIGURE 3.27 Schematic for Problem 1.

2. *Mixing.* Using the appropriate conservation law(s), find the maximum velocity at the exit of the pipe shown in Figure 3.28.

3. Using the appropriate conservation law(s), calculate the mass and volumetric flow rates for the flow down an inclined surface, as shown in Figure 3.29.

4. Find the average velocity through the pipe shown in Figure 3.30.

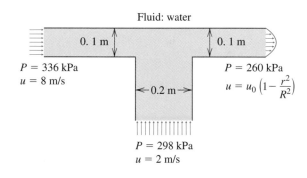

FIGURE 3.28 Schematic for Problem 2.

5. Determine the velocity as a function of x for the one-dimensional flow shown in Figure 3.31.

6. Find the velocity as a function of x for the channel flow shown in Figure 3.32, using the appropriate conservation law(s).

7. Fluid at constant volumetric flow rate Q enters a circular channel, as shown in Figure 3.33. If the thickness of the channel is H, determine the velocity as a function of distance from the origin.

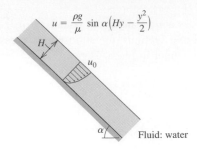

$$u = \frac{\rho g}{\mu} \sin \alpha \left(Hy - \frac{y^2}{2} \right)$$

FIGURE 3.29   Schematic for Problem 3.

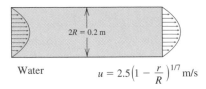

$$u = 2.5\left(1 - \frac{r}{R} \right)^{1/7} \text{ m/s}$$

FIGURE 3.30   Schematic for Problem 4.

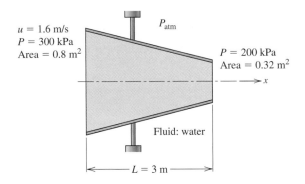

FIGURE 3.31   Schematic for Problem 5.

8. By using conservation of mass, and by assuming a spherical balloon, find an expression for the rate of inflation of the balloon as a function of time. Please state clearly your assumptions.

9. *Extrusion.* A common manufacturing method for plastic sheets is the extrusion process shown in Figure 3.34. In this process, molten material is pushed through a rectangular die and allowed to cool outside the die. Due to the dynamics of the flow, the thickness of the sheet is different from that of the die. For the conditions shown in the figure, calculate the velocity of the solidified sheet (this is of course the rate of production of solid sections).

10. Determine the rate of production of circular solid plastic rods from a melt using extrusion from a circular die as shown in Figure 3.35.

11. *Film casting.* Figure 3.36 shows a variation of the extrusion process shown previously, where the solidified material is pulled with a constant force F. If the thickness of the film

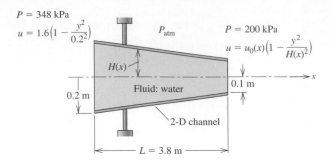

FIGURE 3.32   Schematic for Problem 6.

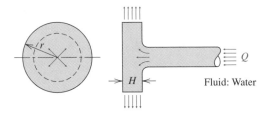

FIGURE 3.33   Schematic for Problem 7.

is $h = h_0 e^{-\frac{Fx}{4\mu \dot{Q}}}$, where x is the distance from the exit, μ is the viscosity of the melt, and $\dot{Q}$ is the volumetric rate per unit depth, find the velocity of the material outside the die as a function of x. What is the rate of production (m/s) of solidified film?

12. *Atomization by rotation.* Figure 3.37 shows a bell-shaped cone (angle θ with respect to the horizontal) rotating at high rotational speed Ω. This process is used to atomize liquids such as paint. At high rotational speeds, the velocity profile of the liquid film inside the cone (prior to atomization) is approximated by $u = \frac{\rho r \Omega \sin \theta}{\mu} \left(hy - \frac{y^2}{2} \right)$, where ρ is the density, μ the viscosity, r the local radius from the axis of rotation, and y the vertical distance from the solid surface. For a constant flow rate Q, determine the profile of the thickness of the film $h(x)$ as a function of the distance from the entrance.

13. The velocity along the height (y-direction) in a rectangular channel with $2H$ total height, when the fluid is allowed to slip along the side of the channel, is given by $u = 1/(2\mu)dP/dx(y^2 - H^2 + 2\mu H/\beta)$, where dp/dx is the pressure gradient, μ is the viscosity of the fluid, and β is known as the slip coefficient defined in terms of the shear τ_w and velocity u_w at the channel wall as $\tau_w = \beta u_w$. Find (a) the average velocity; (b) the pressure gradient needed to induce a volumetric flow rate Q.

14. *Mixing.* Consider a section of a pipe junction where three pipes merge (Figure 3.38). For the conditions shown in the figure, find the maximum velocity after the two fluid streams mix.

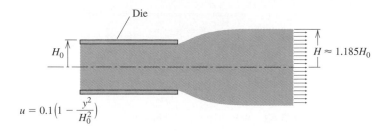

$$u = 0.1\left(1 - \frac{y^2}{H_0^2}\right)$$

FIGURE 3.34   Schematic for Problem 9.

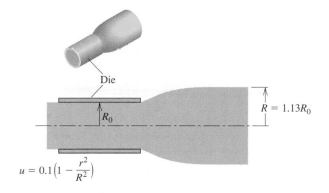

$$u = 0.1\left(1 - \frac{r^2}{R^2}\right)$$

FIGURE 3.35   Schematic for Problem 10.

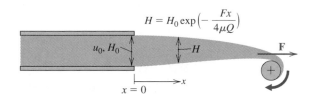

FIGURE 3.36   Schematic for Problem 11.

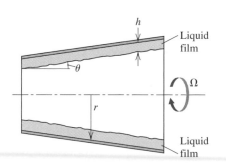

FIGURE 3.37   Schematic for Problem 12.

15. The velocity distribution in a channel where flow is induced by a combination of a pressure gradient dP/dx and motion of the upper wall with velocity V is given by $u = V/a - 1/(2\mu)dP/dx(ay - y^2)$, where a is the height of the channel. Find (a) the average velocity; (b) the pressure gradient needed to induce a volumetric flow rate Q.

16. Consider a channel at an angle θ with respect to the horizontal direction, where flow is induced by both a pressure gradient dp/dx and gravity g. The velocity distribution in the direction normal to the main direction is given by $u = 1/(2\mu)[\rho g \sin\theta - dP/dx](H^2 - y^2)$, where $2H$ is the height of the channel, and μ and ρ are, respectively, the viscosity and density of the fluid. Find (a) the average velocity; (b) the pressure gradient needed to induce a volumetric flow rate Q.

17. The velocity distribution across the thickness (y-direction) of a thin film with total thickness δ flowing down an inclined surface with and angle θ with respect to the horizontal is $u = \rho g \sin\theta/\mu(\delta y - y^2)$. Find (a) the average velocity; (b) the pressure gradient needed to induce a volumetric flow rate Q.

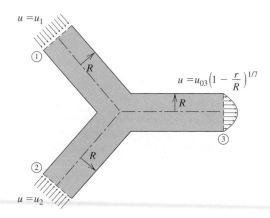

FIGURE 3.38   Schematic for Problem 14.

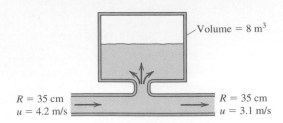

FIGURE 3.39   Schematic for Problem 19.

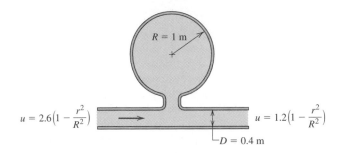

FIGURE 3.40   Schematic for Problem 20.

18. Two immiscible incompressible fluids A and B, with densities ρ_A and ρ_B ($\rho_A > \rho_B$) and viscosities μ_A and μ_B, flow between two parallel plates. Flow is induced by one plate moving with speed V while the other plate is fixed. The velocity distribution as a function of the distance in the direction normal to the main flow direction is

$$u^A = \frac{\mu_B V}{\mu_A H_B + \mu_B H_A} y \quad 0 \geq y \geq H_A,$$

$$u^B = V - \frac{\mu_A V}{\mu_A H_B + \mu_B H_A}$$

$$(H_A + H_B - y) H_A \geq y \geq H_A + H_B,$$

where H_A and H_B, are the thickness of each layer. Find (a) the average velocity; (b) the velocity V needed to induce a volumetric flow rate Q.

19. For the conditions and geometry shown in Figure 3.39, calculate the time it takes to fill the reservoir.

20. For the conditions and geometry shown in Figure 3.40, calculate the time it takes to fill the spherical reservoir.

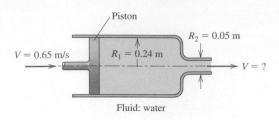

FIGURE 3.42   Schematic for Problem 24.

21. Design a supply line to fill a spherical tank with radius of 1.5 m. To deliver the water costs 80 *cents* per m^3/s. Plot the total cost as a function of the time it takes to fill the tank.

22. The velocity profile of fully developed flow of a non-Newtonian fluid (power-law fluid) in a two-dimensional channel with height H is given by

$$u = \left(\frac{1}{\mu_0} \left| \frac{\partial P}{\partial x} \right| \right)^{\frac{1}{n}} \left(\frac{n}{n+1} \right) \left[\left(\frac{H}{2} - y^{\frac{n+1}{n}} \right)^{\frac{n+1}{n}} \right].$$

Develop expressions for (a) maximum velocity; (b) average velocity; (c) the volumetric flow rate.

23. Flow enters a circular pipe as shown in Figure 3.41 and leaves through a narrow slit along the length of the pipe. If the velocity of the fluid varies linearly along the pipe, evaluate u_{r_0}.

24. A piston moving with constant velocity pushes fluid out of a circular pipe through a narrow hole. For the conditions shown in Figure 3.42, calculate the velocity of the fluid leaving the hole.

25. *Injection molding.* A plastic circular bucket is produced by the injection molding process shown in Figure 3.43. Estimate the number of buckets produced per hour. Account for the fact that each bucket takes 1.7 minutes to cool.

26. Determine whether one is better off running fast or slowly to avoid the rain. Assume no wind (i.e., the rain is falling in the vertical direction).

27. Design the manufacturing of a 2 *mm* thick aluminum film using the spray system shown in Figure 3.44. Hint: find a relation between the thickness of the film (h), the radius of the turning wheels (w), and the length of the spray section (L), and then choose reasonable processing parameters.

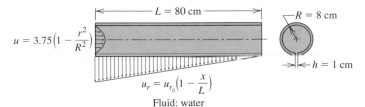

FIGURE 3.41   Schematic for Problem 23.

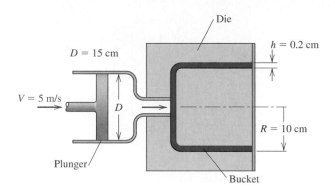

FIGURE 3.43   Schematic for Problem 25.

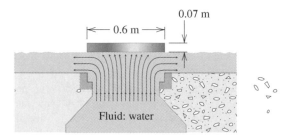

FIGURE 3.44   Schematic for Problem 27.

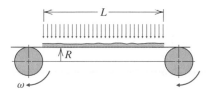

FIGURE 3.45   Schematic for Problem 30.

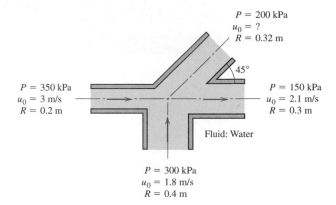

FIGURE 3.46   Schematic for Problem 32.

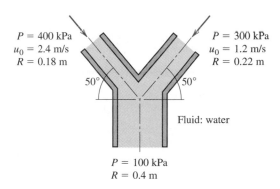

FIGURE 3.47   Schematic for Problem 33.

28. The average velocity V_a throught a pipe is defined as $V_a A_r = \int_0^R u 2\pi r \, dr$, where A_r is the total sectional area and R is the radius of the pipe. Find the average velocity through a pipe when the velocity profile is (a) $u = u_0(1 - r/R)$; (b) $u = u_0(1 - r^2/R^2)$.

29. A standing wave is set up in a pipe (diameter D) filled with gas. The density ρ of the gas as a function of time t and location x is given by $\rho = \rho_o + \delta\rho \sin(2\pi ct/\lambda) \sin(2\pi x/\lambda)$, where c is the speed of sound and λ is the wavelength. Find the net mass flow rate in a section bounded by $x = 0$ and $x = \lambda/4$. Assume that the density is uniform across the pipe.

30. During severe flooding, the manhole cover of the drainage system is lifted as shown in Figure 3.45. If the diameter of the cover is $0.6 \ m$ and the thickness is $0.07 \ m$, find the velocity of the water. Assume that the cover is made with cast iron whose density is $8000 \ kg/m^3$.

31. Consider a balloon filled with gas. If the balloon is pierced, find its initial acceleration as the gas escapes from a small hole with diameter d and velocity u_o *relative* to the balloon. Plot the velocity and acceleration of the balloon as a function of time using $\dfrac{\dot{m}}{M_o} = 0.15$, where $\dot{m}$ is the mass flow rate of air leaving the balloon and M_o is the original mass inside the balloon; also for $u_o = 100 \ m/s$.

32. Consider the pipe arrangement shown in Figure 3.46. For the conditions indicated, calculate the magnitude and direction of the force required to keep the arrangement in place by (a) assuming uniform flow conditions; (b) assuming parabolic velocity profile of the form $u = u_0\left(1 - \dfrac{r^2}{R^2}\right)$. Compare the results.

33. For the conditions shown in Figure 3.47, calculate the magnitude and direction of the force required to keep the pipe arrangement in place by (a) assuming uniform flow conditions; (b) assuming parabolic velocity profile of the form $u = u_0\left(1 - \dfrac{r^2}{R^2}\right)$. What is the required force if the flow in the $0.22 \ m$ pipe is increased by 10%?

34. Consider the pipe shown in Figure 3.48. For the conditions indicated, calculate the magnitude and direction of the force required to keep the arrangement in place by (a) assuming

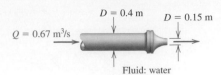

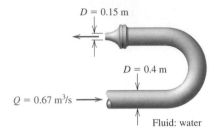

FIGURE 3.48   Schematic for Problem 34.

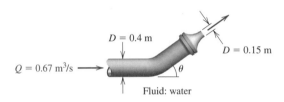

FIGURE 3.49   Schematic for Problem 35.

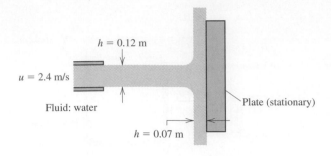

FIGURE 3.50   Schematic for Problem 36.

uniform flow conditions; (b) assuming parabolic velocity profile of the form $u = u_0 \left(1 - \dfrac{r^2}{R^2}\right)$. R is the radius of the flow sections.

35. Consider the pipe shown in Figure 3.49. For the conditions indicated, calculate the magnitude and direction of the force required to keep the arrangement in place by (a) assuming uniform flow conditions; (b) assuming parabolic velocity profile of the form $u = u_0 \left(1 - \dfrac{r^2}{R^2}\right)$. R is the radius of the flow sections.

36. Consider the pipe shown in Figure 3.50. For the conditions indicated, calculate the magnitude and direction of the force required to keep the arrangement in place as a function of angle θ by (a) assuming uniform flow conditions; (b) assuming parabolic velocity profile of the form $u = u_0 \left(1 - \dfrac{r^2}{R^2}\right)$. R is the radius of the flow sections.

37. Reconsider the arrangement shown in Problem 1. For the conditions shown in Figure 1, calculate the force required to keep the pipes in place.

38. *Force on a solid surface generated by an impinging jet of fluid.* A jet of water emerging from a pipe with cross-sectional area of $0.05\ m^2$ hits perpendicularly a stationary flat plate. If the velocity of the emerging jet is $0.1\ m/s$, calculate the force required to keep the plate in place.

39. A jet of water emerging from a pipe hits a stationary plate as shown in Figure 3.51. Calculate the force required to keep the plate in place.

40. For the schematic shown in Problem 2 and the conditions shown in Figure 3.6, calculate the force required to keep the pipes in place.

41. Consider the arrangement shown in Problem 5. For the conditions shown in Figure 3.31, calculate the force required to keep the pipes in place.

42. A propeller-driven aircraft is flying at 80 *mph*. Find the thrust produced when $220\ kg/s$ of air through the propeller are accelerated to 100 *mph*. What is the diameter of the propeller?

43. An aircraft during sea level flight at 550 *mph* produces a net thrust of $100\ kN$. What is the mass flow rate? If the velocity of the exhaust gases is constant, find the thrust produced by the engines as a function of altitude (assume that the density of the exhaust gases is equal to that of air).

44. Find the thrust produced by a rocket engine at sea level when the exhaust gas velocity is $800\ m/s$, with cross-sectional area of $0.1\ m^2$. The gas density is $1.4\ kg/m^3$.

45. Estimate the weight of a humming bird by assuming that while the bird is hovering, it induces a downward velocity of $4.5\ m/s$ through its wings, which have a span of $5\ cm$. Assume that the air accelerates to $9\ m/s$ after it goes through the wings.

46. A rocket is designed for an altitude of $10,000\ m$. At the designed conditions, the engine exit pressure is the same as the ambient. Find the thrust produced when the engine exit area is $2.8\ m^2$ and the mass flow rate is $400\ kg/s$. For the exhaust gases, use the same density as air at the same altitude.

47. In propulsion analysis, the ratio of the thrust produced T to the rate of weight of gas $\dot{w} = \dot{m}g$ (i.e., $T/\dot{m}g = I_{sp}$ is known as the specific impulse and has units of time). A rocket is designed to produce 1.6 MN of thrust and a

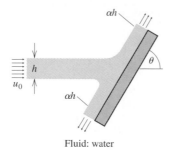

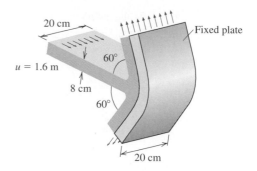

FIGURE 3.52   Schematic for Problem 50.

FIGURE 3.54   Schematic for Problem 52.

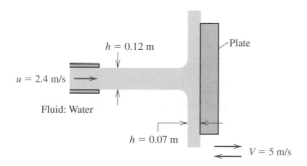

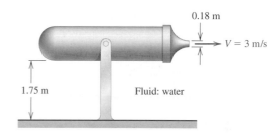

FIGURE 3.53   Schematic for Problem 51.

FIGURE 3.55   Schematic for Problem 53.

specific impulse of 400 *seconds*. Find the velocity of the exhaust gases for a 1.8 *m* diameter engine.

48. For the arrangement shown in Problem 6 and the conditions shown in the Figure 3.36, calculate the force required to keep the pipes in place.

49. Consider the arrangement shown in Problem 23. For the conditions shown in Figure 3.41, calculate the force required to keep the pipes in place.

50. Find a general expression for the magnitude and direction of the force required to keep the plate shown in Figure 3.52 in place.

51. Find the force required to keep the plate shown in Figure 3.53 in place.

52. Find the force required to keep the plate shown in Figure 3.54 in place when the plate is moving with constant velocity of (a) 5 *m/s* to the left; (b) 5 *m/s* to the right.

53. Determine the magnitude of the reaction force and moment produced at the support of the fire extinguisher shown in Figure 3.53.

54. Air flows through a converging nozzle with inlet and exits areas of 0.8 m^2 and 0.42 m^2. At the inlet, the air velocity is 80 *m/s*, pressure is 200 kPa, and absolute temperature is 400 K. If at the exit the velocity is 135 *m/s* and absolute temperature is 350 K, determine (a) the volumetric flow rate; (b) the mass flow rate; (c) the pressure and density of the air at the exit of the nozzle; (d) the net force required to keep the nozzle in place.

55. Consider air flow through a diverging nozzle with inlet and exits areas of 0.42 m^2 and 0.8 m^2. At the inlet, the air velocity is 485 *m/s*, pressure is 600 kPa, and absolute temperature is 480 K. If at the exit the absolute temperature is 430 K and the density is 40% of the inlet density determine (a) the volumetric flow rate; (b) the mass flow rate; (c) the pressure of the air at the exit of the nozzle; (d) the net force required to keep the nozzle in place.

56. Air flows through a diverging nozzle with exit to inlet area ratio of 1.5. At the inlet, the air velocity is 179 *m/s*, pressure is 800 kPa, and absolute temperature is 500 K. If at the exit the velocity is 380 *m/s* and absolute temperature is 300 K, determine (a) the pressure, density, and velocity of the air at the exit of the nozzle; and (b) the net force required to keep the nozzle in place.

57. When flow conditions in supersonic flow are not "favorable," the flow adjusts suddenly to subsonic conditions through a normal shock (Figure 3.56). Since the change takes place over a short distance, the process can be assumed to be adiabatic, but because of its abrupt nature, it is non-isentropic. Using the appropriate conservation laws, develop expressions for the velocity, temperature, and pressure at the subsonic side of the shock. Find the conditions on the subsonic side of the shock when at the supersonic side the velocity is 695 *m/s*, the pressure is 750 kPa, and the absolute temperature is 470 K.

58. If flow conditions are not favorable, supersonic flow over a very short distance changes suddenly to subsonic flow. If the velocity on the supersonic part of the shock is 695 *m/s*

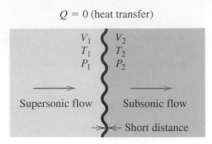

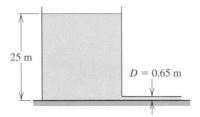

FIGURE 3.56   Schematic for Problem 57.

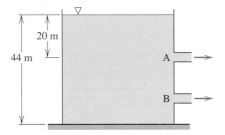

FIGURE 3.57   Schematic for Problem 59.

FIGURE 3.58   Schematic for Problem 60.

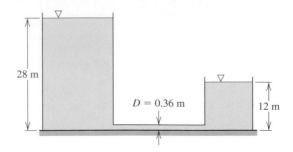

FIGURE 3.59   Schematic for Problem 62.

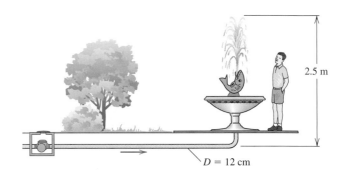

FIGURE 3.60   Schematic for Problem 63.

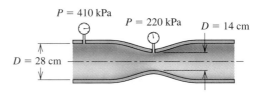

FIGURE 3.61   Schematic for Problem 65.

with pressure of 750 kPa and temperature of 470 K, find the density and pressure on the subsonic part of the shock where the velocity is 342 m/s and temperature is 692 K.

59. A reservoir is supplying water through a 0.65 m diameter pipe, as shown in Figure 3.57. If the level of the water is 25 m, evaluate the volumetric flow rate.

60. Compare the volumetric flow rates through the two identical pipes shown in Figure 3.58.

61. A 0.48 m pipe attached to a reservoir similar to that in Problem 59, delivers water at a rate of 0.24 m^3/s. Calculate the level of the water in the reservoir.

62. Calculate the volumetric flow rate through the pipe connecting the two reservoirs as shown in Figure 3.59.

63. Find the power required to run the fountain shown in Figure 3.60.

64. A pump delivers water at the rate of 0.027 m^3/s through a 0.2 m diameter pipe 20 m above the level of the suction. If the pressure right after the pump is 300 kPa and at

the delivery section is 100 kPa, find (a) the losses in the piping; (b) the power required to run the pump.

65. Calculate the volumetric flow rate of water through the pipe shown in Figure 3.61.

66. Determine the fluid velocity for the conditions shown in the arrangement in Figure 3.62.

67. An ideal two-dimensional source is represented as a mathematical point releasing fluid at a constant volumetric flow rate Q. The velocity distribution in an ideal source is given by $V = \dfrac{C}{r}$, where r is the distance from the origin. Using mass conservation, evaluate the constant C and then calculate the pressure as a function of r (note: the velocity at the center is ∞, so start from a very short distance away from the origin).

68. Find an expression for the pressure around a circular cylinder using the ideal flow velocity distribution $V = 2V_\infty \sin\theta$, where V_∞ is the velocity of the incoming flow far away from the cylinder and θ is the angle a point

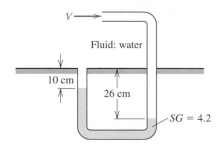

FIGURE 3.62   Schematic for Problem 66.

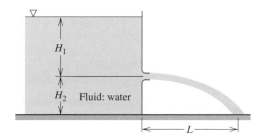

FIGURE 3.63   Schematic for Problem 70.

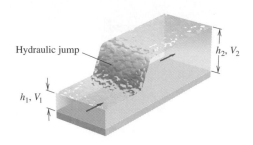

FIGURE 3.64   Schematic for Problem 74.

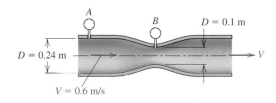

FIGURE 3.65   Schematic for Problem 76.

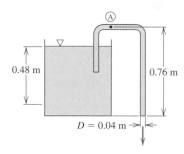

FIGURE 3.66   Schematic for Problem 78.

on the cylinder makes with respect to the horizontal. Plot the pressure distribution relative to the pressure away from the cylinder using $V_\infty = 2.4\ m/s$, $D = 0.2\ m$.

69. Using the ideal velocity distribution around a circular cylinder with $V_\infty = 2.4\ m/s$ and $D = 0.2\ m$, evaluate and plot the pressure head, velocity head, elevation head, and total head as a function of the horizontal direction. What is the stagnation pressure for this flow?

70. Develop a general expression for the range L a water jet will reach as a function of the position of the jet H_2 and the level of the water above the opening H_1 as shown in Figure 3.63.

71. Using the same reservoir arrangement as in Problem 70, find the level of water above the opening in the reservoir for $H_2 = 1.5\ m$ and $L = 3\ m$.

72. Find the power of the pump needed to deliver water at $2\ m^3/s$ in a $0.2\ m$ diameter pipe at an elevation of $50\ m$ when (a) there are no losses; (b) the losses are 18% of the kinetic energy.

73. A pump with 80% efficiency delivers water at a rate of $0.02\ m^3/s$, $8\ m$ above the suction port. If the pressure right after the pump is $320\ kPa$ and at the delivery port is $101\ kPa$, find (a) the losses in the piping system; (b) the power needed to run the pump.

74. In open channels, the flow under certain conditions undergoes a sudden jump, as shown in Figure 3.64. During this jump, there is significant energy loss. Using the appropriate conservation laws, develop expressions for the conditions after the jump (i.e., the new depth and velocity).

75. For the hydraulic jump shown in Problem 74, evaluate the energy loss by using the results of Problem 74.

76. Evaluate the pressure difference between points A and B along the pipe for the conditions shown in Figure 3.65.

77. A Pitot static tube is used to measure the speed of air flowing at standard conditions in a wind tunnel. The stagnation and static pressures in the tunnel are found to be $-1\ cm$ and $2.7\ cm$ of water (gage), respectively. Find the velocity of the flowing air.

78. Determine the volumetric flow rate through the syphon shown in Figure 3.66. What is the pressure at point A?

79. Using the control volume shown in Figure 3.67, show that the volumetric flow through a sluice gate is given by $Q = y_2 W \sqrt{\dfrac{2g(y_1 - y_2)}{1 - (y_2/y_1)^2}}$. Find the volumetric flow rate when $V_1 = 3.5\ m/s$, $y_1 = 4\ m$, and $y_2 = 0.8\ m$.

80. Develop the same expression as in Problem 79, by using the Bernoulli equation and considering (a) points 1 and 2; (b) points 3 and 4.

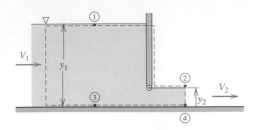

FIGURE 3.67   Schematic for Problem 79.

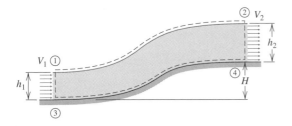

FIGURE 3.68   Schematic for Problem 81.

81. Using the control volume shown in Figure 3.68, develop a general expression for the volumetric flow rate as a function of the flow geometry. Determine the flow rate for $z_1 = 1.2\ m$, $H = 0.8\ m$, and $z_2 = 1.4\ m$.

82. Develop the same expression as in Problem 81, by using the Bernoulli equation and considering (a) points 1 and 2; (b) points 3 and 4.

83. 4 kg/s of air at $28°\ C$ and 101 kPa enter a heat exchanger and leave at $285°\ C$ and 500 kPa. How much heat was added? What is the entropy change for the process?

84. Air enters a well-insulated compressor at 100 kPa and with $20°\ C$ and with flow rate of 30 m^3/s. The air leaves at 520 kPa. If the process is isentropic, determine the power needed to drive the compressor.

85. The inlet conditions at an insulated converging nozzle are pressure 600 kPa, temperature 400 K, and velocity 50 m/s, with cross-sectional area of 0.4 m^2. The temperature at the exit of the nozzle is 300 K, and the area is 0.3 m^2. Determine (a) the pressure and velocity at the exit of the nozzle; (b) the entropy change; (c) the force required to keep the nozzle in place. Consider air that behaves as an ideal gas.

86. A can 15 cm high and 4 cm radius is filled with air at 300 kPa and a room temperature of $20°\ C$. Air is released from a 4 mm hole at a velocity equal to the sonic speed given by $V = \sqrt{\gamma RT}$, where the ratio of specific heat $\gamma = 1.4$, the gas constant for air $R = 0.287\ kJ/kgK$, and T is the absolute temperature. Calculate the rate at which the density is changing at the moment the air is first released.

87. A 2 m^3 volume tank is pressurized with air at 800 kPa and $20°\ C$. Air is released from a 6 cm hole at a speed equal to the sonic speed given by $V = \sqrt{\gamma RT}$, where the ratio of specific heat $\gamma = 1.4$, the gas constant for air $R = 0.287\ kJ/kgK$, and T is the absolute temperature. If the air inside the tank undergoes a polytropic process with $n = 1.55$, find the time it will take to empty half of the original mass of the pressurized air.

88. Fluid flows through a turbine at a rate of 3 kg/s. While the inlet velocity is assumed to be small, the exit velocity is 140 m/s. The initial enthalpy of the working fluid is 3000 kJ/kg, and the final enthalpy is 2500 kJ/kg. Find the power generated when (a) there are no thermal losses; (b) when 80 kJ/s is being lost.

89. An air compressor receives air at one atmosphere and $20°\ C$ and discharges it at 650 kPa and $200°\ C$. Heat loss from the compressor is 60 kJ/kg. Calculate the work input to the compressor. State clearly your assumptions.

90. Air at a rate of 1 kg/s is being compressed from 100 kPa, $15°\ C$, and negligible inlet velocity to 250 kPa and exit velocity of 92 m/s by a process that follows $p\mathcal{V}^{1}.35 = constant$, where P is pressure and $\mathcal{V}$ is volume. During compression, 15 kJ/kg is being lost. Find the power required for the compression. Assume air as an ideal gas.

91. Air at 600 kPa and $150°\ C$ enters an insulated nozzle at 50 m/s and leaves at 120 kPa and 550 m/s. Determine the final temperature.

92. Air flows in a well-insulated diverging nozzle. If the inlet velocity is small and the temperature drop between the inlet and exit sections is $50°\ C$, find the exit velocity.

93. A compressor receives air at ambient conditions (101 kPa and $20°\ C$) and negligible velocity. Air exits at 1.5 kPa $400°\ C$ with a velocity of 95 m/s. The power input to the compressor is 5200 kW. Find the mass flow rate.

94. A 2m^3 tank is pressurized with air at 1.8 MPa, at $20°\ C$. Air is released from a 4 cm diameter hole at a speed equal to the sonic speed V ($V = \sqrt{\gamma RT}$). If the air inside the tank undergoes changes consistent with an isentropic process, find the time it will take to empty half of the original mass of the air.

4 Differential View of Fluid Motion: Fluid Kinematics and Deformation

Finite control volume analysis (both for closed and open systems) is appropriate for the bulk (average) behavior of fluid systems. However, this kind of analysis is not always sufficient as more detailed flow information that cannot be obtained from finite volume analysis is often required. For instance, when materials are processed into complex shapes while in a molten state, the microstructure and, hence, the final quality of the product is in part, a function of *local* flow details such as pressure, velocity, and the deformation and cooling histories. Similarly, the forces on an airfoil depend on the pressure and velocity distribution all along the airfoil surface. These situations can be handled using the *differential approach*, where, with the use of differential calculus, the governing laws are derived for an infinitesimal control volume. In this context, the differential approach and differential calculus are synonymous concepts.

4.1 Differential View

Consistent with the continuum assumption, fluid "particles" are represented as points with zero dimensions but with finite density so that the integral over the entire volume accounts for the mass of the fluid. Consequently, as a volume of fluid deforms by changing location and shape, *fluid flow* is defined as the combined motion of the particles composing this volume (Figure 4.1a). This may be demonstrated by a bag of marbles, where, very much as with fluid particles, the shape of the bag changes constantly because of the mobility of the marbles, while the number of the marbles of course remains the same (Figure 4.1b).

The relationship between fluid motion (kinematics) and the accompanying forces (dynamics) is clearly articulated in the conservation of linear momentum law,

$$\underbrace{\frac{D}{Dt}\int_{\mathcal{V}}\rho\mathbf{V}\,d\mathcal{V}}_{Kinematics} \equiv \underbrace{\sum\mathbf{F}}_{Dynamics} . \qquad (4.1)$$

This chapter, using now the differential view, concentrates on the differential analog of the left-hand side of Equation (4.1). This term describes fluid kinematics or the motion of fluid particles. Subsequently, the general question of *how fluids flow* is addressed. By using purely kinematic arguments we first quantify fluid motion, and then we investigate some of its implications. Within the same kinematic framework, we also discuss the deformation of fluid particles. The differential view of fluid forces (i.e., the right-hand side of Equation (4.1)) and their relation to fluid motion are discussed in Chapter 5.

---

[1]G.A. Tokaty, *A History and Philosophy of Fluid Mechanics*.

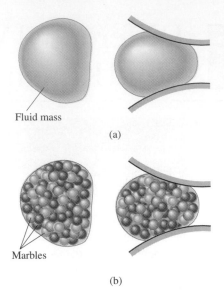

(a)

Fluid mass

Marbles

(b)

FIGURE 4.1   (a) Fluid deforming while being squeezed through a converging channel. (b) A bag of marbles squeezed through a converging channel. The two situations are similar.

4.2   Position Vector of Fluid Particles

Consider a point P representing a fluid particle. Using an appropriate coordinate system, the location of the particle at each moment in time is uniquely defined by specifying its coordinates — that is, the distance from the origin in each direction measured in *meters* (SI system of units). For example, in the three-dimensional Cartesian coordinate system shown in Figure 4.2, the coordinates $x_p(t)$, $y_p(t)$, and $z_p(t)$, at any time are sufficient to uniquely locate the particle. The coordinates can be combined to form the *position vector* $\mathbf{r}_p$

$$\mathbf{r}_p(t) = x_p(t)\,\mathbf{i} + y_p(t)\,\mathbf{j} + z_p(t)\,\mathbf{k},$$

where $\mathbf{i}$, $\mathbf{j}$, and $\mathbf{k}$ are unit vectors representing three mutually perpendicular directions. Here, the i, j, and k directions coincide with the three principal coordinate directions x, y, z. The position vector obeys all vector properties and their mathematical manipulation.

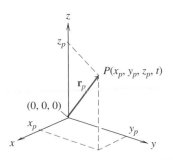

FIGURE 4.2   Representation of the position vector.

4.3   Velocity and Acceleration Fields

The velocity of a particle P, measured in *meters/second*, is defined as the rate at which it changes position with respect to time along each coordinate direction,

$$u_p = \frac{Dx_p}{Dt}, \quad v_p = \frac{Dy_p}{Dt}, \quad w_p = \frac{Dz_p}{Dt}.$$

The notation D/Dt, used here and throughout the entire textbook, signifies the rate of change with respect to time as seen by an *observer moving with the particle*. This issue is discussed in more detail in Section 4.4.

The three components of the velocity can be combined to form the velocity vector, $\mathbf{V}_p$ as

$$\mathbf{V}_p = u_p\,\mathbf{i} + v_p\,\mathbf{j} + w_p\,\mathbf{k}.$$

In more condensed form, this is expressed as

$$\mathbf{V}_p = \frac{D\mathbf{r}_p}{Dt}. \tag{4.2}$$

The magnitude of the velocity vector

$$|\mathbf{V}| = \sqrt{u_p^2 + v_p^2 + w_p^2}$$

is the *speed* of the fluid particle. This is the quantity displayed by the speedometer in a car. Consistent with the definition, the speedometer indicates a single scalar number, independent of the direction of the car's motion.

The velocity of a fluid particle can be generalized by introducing the *velocity field* $\mathbf{V} = \mathbf{V}(x, y, z, t)$. This is a smooth, differentiable function valid over the fluid domain and represents the velocity as a function of time at each *coordinate location* (x, y, z), irrespective of what particle occupies that location (Figure 4.3).

The acceleration of a fluid particle P, measured in *meters/second*, is defined as the rate of change of the particle's velocity with respect to time,

$$a_{p_x} = \frac{Du_p}{Dt} \quad a_{p_y} = \frac{Dv_p}{Dt} \quad a_{p_z} = \frac{Dw_p}{Dt}.$$

These three components represent the acceleration vector $\mathbf{a}_p$

$$\mathbf{a}_p = a_{p_x}\,\mathbf{i} + a_{p_y}\,\mathbf{j} + a_{p_z}\,\mathbf{k},$$

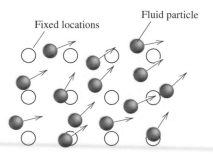

Fixed locations

Fluid particle

FIGURE 4.3   Conceptual schematic indicating the velocity field as seen from two points of view: (a) as motion of individual particles; (b) as the fluid velocity at fixed locations.

or, in a more condensed form,

$$\mathbf{a}_p = \frac{D\mathbf{V}_p}{Dt}.$$ (4.3)

Similar to the velocity field, the *acceleration field* $\mathbf{a} = \mathbf{a}(x, y, z, t)$, is a smooth, differentiable function valid over the fluid domain, representing the acceleration as a function of time at each coordinate location (x, y, z) irrespective of the identity of the particle at that location (Figure 4.3).

EXAMPLE 4.1

Problem Statement   Consider the incompressible flow through a converging nozzle shown in Figure 4.4. The cross-sectional area of the nozzle decreases linearly from an initial value of $A_1 = 0.5\ m^2$ to a final area of $A_2 = 0.1\ m^2$ over a length of 0.5 m. The volumetric flow rate is 2 m^3/s. Under these conditions

- Find the *residence time* defined as the time it takes a particle to travel the entire length of the nozzle.
- What is the location of a fluid particle originally (at $t = 0$) at the inlet, after half the residence time has elapsed?
- Find the velocity and acceleration of the particle at half the residence time.
- Find the velocity and acceleration when the particle is 0.3 m downstream from the inlet.

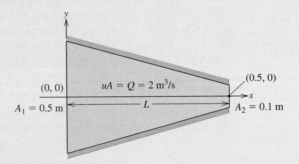

FIGURE 4.4   Schematic of Example 4.1.

Governing Equations   The distance traveled by a particle with velocity u, as a function of time, is

$$x = \int_0^t u\, dt'.$$

By definition, the acceleration a of a fluid particle is

$$a = \frac{Du}{Dt}.$$

For this one-dimensional incompressible flow, since the density is constant, mass conservation requires that $u\,A = Q$, where A is the cross-sectional area and Q the given volumetric flow rate.

Assumptions   Assume that the flow within the nozzle is one-dimensional (i.e., u is uniform at each cross section).

SOLUTION Using mass conservation, the velocity is given as

$$u = \frac{Q}{A}.$$

Since the area of the nozzle changes linearly with x, $A = c_1 - c_2 x$, where c_1 and c_2 are constants to be determined from the shape of the channel. At $x = 0$, $A = A_1$, and at $x = L$, $A = A_2$. Therefore, $c_1 = A_1$ and $c_2 = (A_1 - A_2)/L$. The velocity as a function of x is expressed as

$$u = \frac{Dx}{Dt} = \frac{Q}{A_1 - (A_1 - A_2)\frac{x}{L}}. \tag{4.4}$$

The distance x_s, traveled in time t_s, is found by integrating Equation (4.4) as

$$\int_0^{x_s} \left[\frac{A_1}{Q} - \left(\frac{A_1 - A_2}{Q} \right) \frac{x}{L} \right] Dx = \int_0^{t_s} Dt,$$

or

$$t_s = \frac{A_1}{Q} x_s - \left(\frac{A_1 - A_2}{Q} \right) \frac{x_s^2}{2L} = 0.25\, x_s - 0.2\, x_s^2. \tag{4.5}$$

A particle originally at the inlet will reach the exit in

$$t_{res} = 0.25 \times 0.5 - 0.2 \times 0.5^2 = 0.075\ s,$$

The location of a particle then as a function of time is found by solving the quadratic Equation (4.5) for $x_p = x_p(t)$,

$$x_p = L \frac{A_1}{A_1 - A_2} \left[1 - \sqrt{1 - 2\frac{A_1 - A_2}{A_1} \frac{Q}{L A_1} t} \right].$$

The $(-)$ sign in front of the square root was chosen so that at $t = 0$, $x_s = 0$. In terms of numerical values,

$$x_p = 0.625 \left[1 - \sqrt{1 - 12.8\, t} \right].$$

According to the preceding expression, at half the residence time $t = 0.0375\ s$, the particle is at

$$x_p = 0.625 \left[1 - \sqrt{1 - 12.8 \times 0.0375} \right] = 0.174\ m.$$

The velocity of the particle as a function of time $u_p = u_p(t)$ is obtained by differentiating $x_p(t)$ with respect to time

$$u_p = \frac{Dx_p}{Dt} = \frac{Q}{A_1} \left[1 - 2\left(\frac{A_1 - A_2}{A_1} \right) \frac{Q}{L A_1} t \right]^{-\frac{1}{2}},$$

or

$$u_p = \frac{4}{\sqrt{1 - 12.8\, t}}.$$

The reader can verify that by substituting t in terms of x, this expression is identical to the velocity given by Equation (4.4).

The acceleration is obtained by differentiating again with respect to time $u_p = u_p(t)$,

$$a_p = \frac{Du_p}{Dt} = \frac{Q^2}{A_1^2 L}\left(\frac{A_1 - A_2}{A_1}\right)\left[1 - 2\left(\frac{A_1 - A_2}{A_1}\right)\frac{Q}{LA_1}t\right]^{-\frac{3}{2}},$$

or

$$a_p = 25.6[1 - 12.8\, t]^{-\frac{3}{2}}.$$

Figure 4.5 shows the position x_p, velocity u_p, and acceleration a_p as a function of time of a particle entering the nozzle at $t = 0$.

At half the residence time, the velocity of the particle is

$$u_p = \frac{4}{\sqrt{1 - 12.8\, t}} = \frac{4}{\sqrt{1 - 12.8 \times 0.0375}} = 5.55\ m/s,$$

and the acceleration

$$a_p = 25.6[1 - 12.8\, t]^{-\frac{3}{2}} = 25.6[1 - 12.8 \times 0.0375]^{-\frac{3}{2}} = 68.27\ m/s^2.$$

According to Equation (4.5), a particle will arrive at $x = 0.3\ m$ in

$$t_s = 0.25\, x_s - 0.2\, x_s^2 = 0.25 \times 0.3 - 0.2 \times 0.3^2 = 0.057\ s.$$

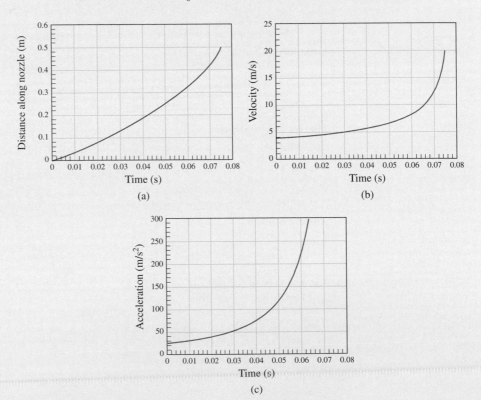

FIGURE 4.5 Results of Example 4.1 for an individual fluid particle: (a) position; (b) velocity; (c) acceleration, as a function of time.

At this time, the velocity and acceleration of the particle are respectively,

$$u_p = \frac{4}{\sqrt{1 - 12.8\ t}} = \frac{4}{\sqrt{1 - 12.8 \times 0.057}} = 7.69\ m/s,$$

and

$$a_p = 25.6[1 - 12.8\ t]^{-\frac{3}{2}} = 25.6[1 - 12.8 \times 0.057]^{-\frac{3}{2}} = 182.07\ m/s^2.$$

EXAMPLE 4.2

Problem Statement   Consider a fluid flow in which the location of individual particles is given by $x = 2e^{2t} - 2$. Find the velocity and acceleration of each particle. What are the velocity and acceleration fields for this flow?

Governing Equations   According to the definition, the velocity u and acceleration a are given as

$$u = \frac{Dx}{Dt} \quad \text{and} \quad a = \frac{Du}{Dt}.$$

SOLUTION   Using the equation of motion $x = 2e^{2t} - 2$, the velocity is readily calculated using the definition as

$$u = \frac{Dx}{Dt} = 4\ e^{2t}.$$

Similarly, the acceleration is evaluated as

$$a = \frac{Du}{Dt} = 8\ e^{2t}.$$

The expressions for the position, velocity, and acceleration of fluid particles as a function of time describe fully the kinematics of this flow (Figure 4.6).

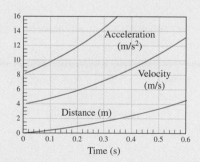

FIGURE 4.6   Results of Example 4.2 for an individual fluid particle: position, velocity, and acceleration as a function of time.

Alternatively, since $x = 2\ e^{2t} - 2$, the velocity u and acceleration a can be expressed in terms of x as

$$u = 2(x + 2), \quad a = 4(x + 2).$$

These expressions are the velocity and acceleration fields, indicating the velocity and acceleration of any particle that *happens to occupy* the position x at time t.

In summary, fluid kinematics can be studied using two equivalent points of view; one by following individual particles, and another by observing the motion at fixed locations in space. The implications of these two distinct points of view are elaborated below.

4.4   Lagrangian and Eulerian Perspectives

As stated previously, the notation D/Dt used here and throughout the entire textbook signifies the rate of change with respect to time as seen by an *observer moving with the particle* (Figure 4.7). This time derivative is known as the *material* or *substantial* derivative. This distinctive approach, in which the analysis is based on individual particles, is referred to as the *Lagrangian* point of view. Alternatively, we can use an *Eulerian* point of view when attention is instead fixed on *distinct coordinate points in space*. Obviously, at different times these fixed locations are occupied by different fluid particles. An observer then sees the effect of the particles passing through them.

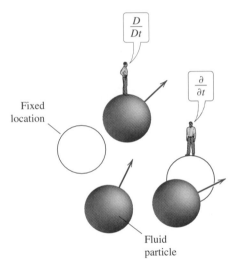

FIGURE 4.7   Schematic illustrating the difference between Eulerian and Lagrangian perspectives.

The two alternative points of view are illustrated in Figure 4.8 by the boat going down the river. In this case, the Lagrangian point of view corresponds to having the observer on the boat, while the boat is drifting with the flow. In the Eulerian approach the observer is standing by the river bank.

FIGURE 4.8   A boat drifting with the local river current. The fixed on-shore observer and the observer on the boat illustrate the difference between Eulerian and Lagrangian perspectives.

Physical phenomena though, are unaffected by the observer's point of view. For example, the acceleration of the boat shown in Figure 4.8 may *appear* different to different observers, but in reality it has a unique value. For this reason, the two points of view are not only related, but they are equivalent concepts. As established previously, in *the Lagrangian approach the velocity field is only a function of time*. However, since in the Eulerian perspective attention is focused on a particular location (x, y, z), the *velocity field is a function of x, y, z, and t*. This follows also from the fact that, at different times, different particles occupy fixed physical locations, hence, resulting in variations in both time and space.

Therefore, the acceleration of a particle which at time t occupies a fixed location can be calculated either by using the material derivative as

$$\mathbf{a} = \frac{D\mathbf{V}}{Dt},$$

or by differentiation using the chain rule as

$$\mathbf{a} = \frac{\partial \mathbf{V}}{\partial t} + \frac{\partial \mathbf{V}}{\partial x}\frac{dx}{dt} + \frac{\partial \mathbf{V}}{\partial y}\frac{dy}{dt} + \frac{\partial \mathbf{V}}{\partial z}\frac{dz}{dt}.$$

The two are equal,

$$\underbrace{\frac{D\mathbf{V}}{Dt}}_{\text{Lagrangian approach}} = \underbrace{\frac{\partial \mathbf{V}}{\partial t} + \frac{\partial \mathbf{V}}{\partial x}\frac{dx}{dt} + \frac{\partial \mathbf{V}}{\partial y}\frac{dy}{dt} + \frac{\partial \mathbf{V}}{\partial z}\frac{dz}{dt}}_{\text{Eulerian approach}},$$

when the coordinates of the particle x_p, y_p, z_p and the coordinates of the "Eulerian" point x, y, z within the flow coincide. At that instant

$$\frac{dx}{dt} = u, \quad \frac{dy}{dt} = v, \quad \frac{dz}{dt} = w,$$

where u, v, w are the components of the particle velocity. We can, therefore, write

$$\frac{D\mathbf{V}}{Dt} = \underbrace{\frac{\partial \mathbf{V}}{\partial t}}_{\text{Local change}} + \underbrace{u\frac{\partial \mathbf{V}}{\partial x} + v\frac{\partial \mathbf{V}}{\partial y} + w\frac{\partial \mathbf{V}}{\partial z}}_{\text{Fluid motion contribution}}. \tag{4.6}$$

The above discussion indicates that the acceleration of a particle is made up of two contributions: (a) the *local change of the velocity*, and (b) the acceleration due to local fluid motion known as *convective acceleration*. This is consistent with the physics of the different perspectives; when we fix attention on a particular point in space, the observed behavior is the combined effect of what is happening locally (i.e., the term $\partial/\partial t$) and the net effect due to movement or convection of particles $\left(u\frac{\partial}{\partial x} + v\frac{\partial}{\partial y} + w\frac{\partial}{\partial z}\right)$ through that point.

To generalize then, for any time-varying quantity q, the Lagrangian and Eulerian perspectives are related by

$$\frac{Dq}{Dt} = \underbrace{\frac{\partial q}{\partial t}}_{\text{Local change}} + \underbrace{u\frac{\partial q}{\partial x} + v\frac{\partial q}{\partial y} + w\frac{\partial q}{\partial z}}_{\text{Convective acceleration}}. \tag{4.7}$$

In more condensed, vector form we can write

$$\frac{Dq}{Dt} = \frac{\partial q}{\partial t} + \mathbf{V} \cdot \nabla q. \tag{4.8}$$

EXAMPLE 4.3

Problem Statement   For the Eulerian velocity field given by $u = 2(x + 2) \ m/s$, determine the acceleration field. What is the acceleration at $x = 0.2 \ m$? Verify that the same result can be obtained using a Lagrangian approach. Find the location of a particle after 1 s by considering that the particle at $t = 0$ is at the origin $x = 0$.

Governing Equations   From an Eulerian perspective, the acceleration field is given by

$$\mathbf{a} = \frac{\partial \mathbf{V}}{\partial t} + u \frac{\partial \mathbf{V}}{\partial x} + v \frac{\partial \mathbf{V}}{\partial y} + w \frac{\partial \mathbf{V}}{\partial z},$$

and from a Lagrangian point of view, it is given by

$$\mathbf{a} = \frac{D\mathbf{V}}{Dt},$$

where $\mathbf{V}$ is the velocity vector.

SOLUTION   For the given one-dimensional velocity field the acceleration field reduces to

$$a = \frac{\partial u}{\partial t} + u \frac{\partial u}{\partial x}.$$

Here, we have $u = 2(x + 2)$ and $\frac{\partial u}{\partial x} = 2$. Therefore,

$$a = 4(x + 2).$$

At $x = 0.2 \ m$, the acceleration is $a = 8.8 \ m/s^2$. Any particle passing through $x = 0.2 \ m$ then has the same acceleration.

   Using the Lagrangian viewpoint, by definition we have

$$\frac{Dx}{Dt} = u = 2(x + 2).$$

By letting $x_o = x + 2$, then

$$\frac{Dx_o}{Dt} = 2x_o,$$

which upon integration gives

$$x_o = Ce^{2t}.$$

If we assume that at $t = 0$, the particle was at the origin, then $x_o = x + 2 = 2$ and $C = 2$. The solution for x is then

$$x(t) = 2e^{2t} - 2.$$

According to this expression, a particle originally at the origin, after 1 s moves to $x = 2e - 2 = 3.44 \ m$.

   By differentiation, the acceleration of the particle is

$$a = \frac{Du}{Dt} = 8e^{2t} = 4(x + 2),$$

which is identical to the expression found previously. This illustrates clearly the equivalence of the Eulerian and Lagrangian points of view.

If fluid properties at every point in the flow are constant (i.e., they are not changing with time), the flow is known as *steady*. Consequently, in steady flows for any property q, we have

$$\frac{\partial q}{\partial t} = 0,$$

which implies that q is independent of time, but it can vary from point to point within the flow field $q = q(x, y, z)$.

In general, flows in Eulerian perspective may be steady or unsteady, but the same flows in Lagrangian perspective are *always unsteady*. This point is demonstrated by the previous example in which, while the flow in an Eulerian framework is steady $u = u(x)$, in a Lagrangian framework the same flow is *unsteady*, $u = u(t)$!

EXAMPLE 4.4

Problem Statement For the velocity field $u = 2t$, $v = z$, and $w = y$, calculate the acceleration field. Verify that the same result can be obtained by using a Lagrangian point of view.

Governing Equations Using the Eulerian perspective, the acceleration field is given by

$$\mathbf{a} = \frac{\partial \mathbf{V}}{\partial t} + u\frac{\partial \mathbf{V}}{\partial x} + v\frac{\partial \mathbf{V}}{\partial y} + w\frac{\partial \mathbf{V}}{\partial z}.$$

In the Lagrangian approach, the acceleration is given by

$$\mathbf{a} = \frac{D\mathbf{V}}{Dt},$$

where $\mathbf{V}$ is the velocity vector.

SOLUTION The acceleration in the x-direction is

$$a_x = \frac{\partial u}{\partial t} + u\frac{\partial u}{\partial x} + v\frac{\partial u}{\partial y} + w\frac{\partial u}{\partial z} = 2 + 0 + 0 + 0 = 2;$$

the acceleration in the y-direction is

$$a_y = \frac{\partial v}{\partial t} + u\frac{\partial v}{\partial x} + v\frac{\partial v}{\partial y} + w\frac{\partial v}{\partial z} = 0 + 0 + 0 + y \cdot 1 = y;$$

and the acceleration in the z-direction is

$$a_z = \frac{\partial v}{\partial t} + u\frac{\partial v}{\partial x} + v\frac{\partial v}{\partial y} + w\frac{\partial v}{\partial z} = 0 + 0 + z \cdot 1 + 0 = z.$$

Therefore, $\mathbf{a} = 2\mathbf{i} + y\mathbf{j} + z\mathbf{k}$.

Switching now to Lagrangian coordinates where all quantities are those associated with individual particles,

$$u = \frac{Dx}{Dt} = 2t \Rightarrow x = t^2 + C_0,$$

where C_0 is a constant.

Therefore, by differentiating with respect to time we get $a_x = 2$. To obtain the other acceleration components, we repeat

$$v = \frac{Dy}{Dt} = z \quad \text{and} \quad w = \frac{Dz}{Dt} = y.$$

Differentiating the first expression with respect to time,

$$\frac{D^2 y}{Dt^2} = \frac{Dz}{Dt},$$

and by substituting the expression

$$\frac{Dz}{Dt} = y,$$

we get

$$\frac{D^2 y}{Dt^2} = y.$$

The general solution to this differential equation is

$$y = C_1 e^{-t} + C_2 e^{t},$$

where C_1 and C_2 are appropriate constants to be determined by initial conditions. The velocity and acceleration are determined by differentiation as

$$v = \frac{Dy}{Dt} = -C_1 e^{-t} + C_2 e^{t} \quad \text{and} \quad a_y = \frac{D^2 y}{Dt^2} = C_1 e^{-t} + C_2 e^{t} \Rightarrow a_y = y.$$

Similarly, using

$$\frac{Dy}{Dt} = z \Rightarrow z = -C_1 e^{-t} + C_2 e^{t},$$

which, upon differentiation, yields

$$w = \frac{Dz}{Dt} = C_1 e^{-t} + C_2 e^{t} \quad \text{and} \quad a_z = \frac{D^2 z}{Dt^2} = -C_1 e^{-t} + C_2 e^{t} \Rightarrow a_z = z.$$

Obviously, all components of the acceleration vector are identical to those calculated using the Eulerian description.

4.5 Visual Kinematic Concepts

Since fluid particles can deform easily, the resulting fluid patterns can be quite complex. Therefore, the proper display of flow patterns can make fluid dynamics a highly visual science. Indeed, this possibility is often exploited in order to facilitate the study of fluid flow.

4.5.1 Streamlines, Streamfunction

The most common visual concept is the *streamline*, defined as a line tangent to the local velocity vector at all points along its length (Figure 4.9). In two dimensions, the local slope dy/dx along a streamline is given by

$$\frac{dy}{dx} = \frac{v}{u}, \tag{4.9}$$

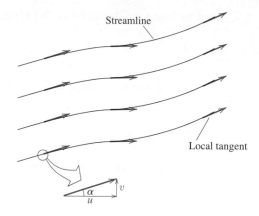

FIGURE 4.9   Definition of streamlines.

or more generally in three dimensions as

$$\frac{dx}{u} = \frac{dy}{v} = \frac{dz}{w}.$$

If u, v, w are functions of position but not of time, the resulting streamline pattern is steady. If, however, the velocities are time dependent, so are the streamlines.

In two dimensions, the local tangent vector $\mathbf{t}$ to a streamline is

$$\mathbf{t} = \frac{u\mathbf{i} + v\mathbf{j}}{\sqrt{u^2 + v^2}}.$$

Since the tangent vector is perpendicular to the normal vector $\mathbf{n}$,

$$\mathbf{n} \cdot \mathbf{t} = 0 \Rightarrow \mathbf{n} \cdot \frac{u\mathbf{i} + v\mathbf{j}}{\sqrt{u^2 + v^2}} = 0.$$

Therefore,

$$\mathbf{n} \cdot (u\mathbf{i} + v\mathbf{j}) = \mathbf{n} \cdot \mathbf{V} = 0,$$

which shows that the velocity normal to the streamline is by definition zero, or, equivalently, *no mass crosses the streamlines*.

Figure 4.9 suggests that the flow pattern can be represented well by a collection of streamlines. To make the streamline pattern more quantitative and, therefore, more useful, it is necessary to label each streamline. This is achieved by introducing the concept of the *streamfunction* ψ, defined in two dimensions as

$$u = \frac{\partial \psi}{\partial y}, \quad v = -\frac{\partial \psi}{\partial x}, \tag{4.10}$$

where u and v are the components of the velocity vector.

In polar r, θ coordinates, the definition of ψ is given as

$$v_r = \frac{1}{r}\frac{\partial \psi}{\partial \theta} \quad \text{and} \quad v_\theta = -\frac{\partial \psi}{\partial r}.$$

For two-dimensional, steady flow, $\psi = \psi(x, y)$. Then, by the use of the chain rule,

$$d\psi = \frac{\partial \psi}{\partial x}\,dx + \frac{\partial \psi}{\partial y}\,dy,$$

and by using the definition of the streamfunction, the preceding expression simplifies to

$$d\psi = -v\,dx + u\,dy.$$

Now, for a line of constant ψ, for which $d\psi = 0$, we get

$$\frac{dy}{dx} = \frac{v}{u},$$

which is identical to the definition of a streamline. Therefore, lines of constant ψ are streamlines.

As a consequence of the definition of ψ, the expression $\nabla \cdot \mathbf{V} = 0$ is satisfied automatically,

$$\nabla \cdot \mathbf{V} = \frac{\partial u}{\partial x} + \frac{\partial v}{\partial y} = \frac{\partial}{\partial x}\left(\frac{\partial \psi}{\partial y}\right) - \frac{\partial}{\partial y}\left(\frac{\partial \psi}{\partial x}\right) = \frac{\partial^2 \psi}{\partial xy} - \frac{\partial^2 \psi}{\partial yx} = 0,$$

where,

$$\frac{\partial^2 \psi}{\partial xy} = \frac{\partial^2 \psi}{\partial yx}.$$

As we will prove in later chapters, $\nabla \cdot \mathbf{V} = 0$ is the differential form of the conservation of mass for an incompressible fluid. The above result then implies that the streamfunction automatically satisfies the conservation of mass, or that any streamfunction that is a correct representation of a real incompressible flow *must* satisfy the condition $\nabla \cdot \mathbf{V} = 0$.

Consider now a general two-dimensional flow shown in Figure 4.10. The volumetric flow rate per unit depth Q crossing an arbitrary surface s, between two points 1 and 2 as shown in the figure is

$$Q = \int_1^2 \mathbf{V} \cdot \mathbf{n}\,ds,$$

where $\mathbf{V}$ is the local velocity vector, and $\mathbf{n}$ the local unit normal vector at an arbitrary point on the surface s. According to the figure, the normal vector can be constructed as $\mathbf{n} = \dfrac{dy}{ds}\mathbf{i} - \dfrac{dx}{ds}\mathbf{j}$. Therefore,

$$Q = \int_1^2 \left(u\frac{dy}{ds} - v\frac{dx}{ds}\right)ds.$$

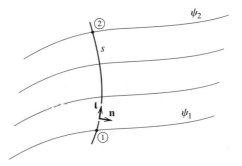

FIGURE 4.10   Volumetric flow rate between two streamlines.

By replacing the velocity components in term of the streamfunction,

$$Q = \int_1^2 \left(\frac{\partial \psi}{\partial y} \frac{dy}{ds} + \frac{\partial \psi}{\partial x} \frac{dx}{ds} \right) ds = \int_1^2 \frac{d\psi}{ds} ds = \int_1^2 d\psi = \psi_2 - \psi_1.$$

This result shows that the volumetric flow rate per unit depth Q crossing a surface is the difference in the numerical values $(\psi_2 - \psi_1)$ of the two streamlines through the endpoints of the surface.

EXAMPLE 4.5

Problem Statement Sketch the streamlines that correspond to the streamfunction $\psi = 2(x^2 - y^2)$. Determine the velocity field.

Governing Equations By definition,

$$u = \frac{\partial \psi}{\partial y} \quad \text{and} \quad v = -\frac{\partial \psi}{\partial x}.$$

SOLUTION The velocity field is determined by differentiation, as

$$u = \frac{\partial \psi}{\partial y} = -4y \quad \text{and} \quad v = -\frac{\partial v}{\partial x} = -4x.$$

The velocity field $\mathbf{V} = \mathbf{V}(x, y)$ is then

$$\mathbf{V} = -4y\,\mathbf{i} - 4x\,\mathbf{j}.$$

As shown in Figure 4.11, by sketching ψ for a number of selected streamlines, we obtain a pattern that resembles that of flow in a 90° corner oriented symmetrically around the x-axis.

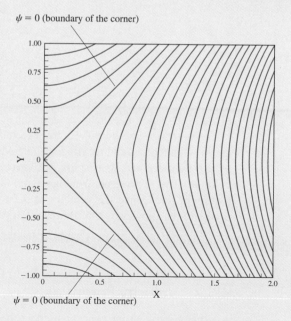

FIGURE 4.11 Streamlines for frictionless flow in a 90° corner oriented symmetrically around the x-axis.

EXAMPLE 4.6

Problem Statement   Show that the function

$$\psi = U_o y \left(1 - \frac{a^2}{x^2 + y^2} \right)$$

represents the streamline pattern of two-dimensional frictionless flow past a circular cylinder with radius a, where U_o is the approaching velocity far away from the cylinder.

Governing Equations   For the given function, to be a true streamfunction it must satisfy the condition

$$\frac{\partial u}{\partial x} + \frac{\partial v}{\partial y} = 0.$$

By definition,

$$u = \frac{\partial \psi}{\partial y} \quad \text{and} \quad v = -\frac{\partial \psi}{\partial x}.$$

SOLUTION   By differentiation, the velocity field is determined as

$$u = U_o - U_o a^2 \left[\frac{x^2 - y^2}{(x^2 + y^2)^2} \right],$$

and

$$v = -U_o a^2 \left[\frac{2xy}{(x^2 + y^2)^2} \right].$$

Therefore,

$$\frac{\partial u}{\partial x} = U_o a^2 \left[\frac{2x}{(x^2 + y^2)^2} - \frac{8xy^2}{(x^2 + y^2)^3} \right],$$

and

$$\frac{\partial v}{\partial y} = U_o a^2 \left[-\frac{2x}{(x^2 + y^2)^2} + \frac{8xy^2}{(x^2 + y^2)^3} \right],$$

the sum of which is exactly zero, hence satisfying the condition $\nabla \cdot \mathbf{V} = 0$.

As shown in Figure 4.12, by sketching a number of selected streamlines, we obtain the flow pattern around a cylinder. Observe that the streamline $\psi = 0$ arrives at the front of the cylinder and then splits into two segments, each following the contour of the cylinder. The two segments merge again at the back of the cylinder before continuing as a straight line. Since the velocity at these points is zero, they are known as *stagnation points*. Note, however, that the preceding flow field is true only under the assumption that the fluid flows freely past the cylinder without internal friction.

4.5.2 Velocity Potential

Another useful visual concept used is the *velocity potential* ϕ. In two dimensions, ϕ is defined as

$$u = \frac{\partial \phi}{\partial x}, \quad v = \frac{\partial \phi}{\partial y}. \tag{4.11}$$

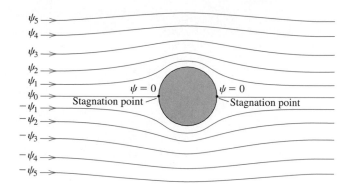

FIGURE 4.12 Streamlines for frictionless flow past a circular cylinder.

In polar coordinates, the definition of ϕ is given as

$$v_r = \frac{\partial \phi}{\partial r} \quad \text{and} \quad v_\theta = \frac{1}{r}\frac{\partial \phi}{\partial \theta}. \tag{4.12}$$

A line of constant ϕ is constructed by writing

$$d\phi = \frac{\partial \phi}{\partial x}\,dx + \frac{\partial \phi}{\partial y}\,dy = u\,dx + v\,dy.$$

The local slope along at an arbitrary point on a line of constant ϕ is

$$\frac{dy}{dx} = -\frac{u}{v}.$$

Recall that the local slope of along a streamline is $\left.\dfrac{dy}{dx}\right|_\psi = \dfrac{v}{u}$. Therefore, since $\left.\dfrac{dy}{dx}\right|_\psi$ $\left.\dfrac{dy}{dx}\right|_\phi = -1$, iso-potential lines and streamlines are orthogonal to each other.

EXAMPLE 4.7

Problem Statement Show that the velocity potential function

$$\phi = U_o x \left(1 + \frac{a^2}{x^2 + y^2}\right)$$

represents the pattern of two-dimensional frictionless flow past a circular cylinder with radius a, where U_o is the approaching velocity far away from the cylinder.

Governing Equations By definition,

$$u = \frac{\partial \phi}{\partial x} \quad \text{and} \quad v = \frac{\partial \phi}{\partial y}.$$

SOLUTION Using the definition, the velocity components are obtained by differentiation as

$$u = U_o - U_o a^2 \left[\frac{x^2 - y^2}{(x^2 + y^2)^2}\right],$$

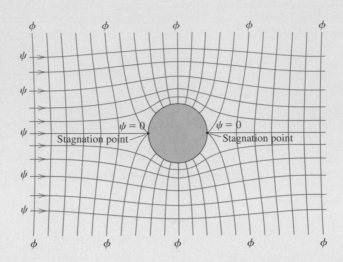

FIGURE 4.13 Superimposed iso-velocity potential lines and streamlines for frictionless flow past a circular cylinder.

and

$$v = -U_o a^2 \left[\frac{2xy}{(x^2 + y^2)^2} \right],$$

which are identical to those obtained in Example 4.5 using the concept of the stream-function.

Figure 4.13 shows the sketch of a number of selected lines of constant velocity potential. By superimposing the streamline pattern, we observe the orthogonality of the two patterns.

EXAMPLE 4.8

Problem Statement   Using the streamfunction for a two-dimensional incompressible flow $\psi = 2(x^2 - y^2)$, construct the velocity potential function.

Governing Equations   By definition,

$$u = \frac{\partial \psi}{\partial y} = \frac{\partial \phi}{\partial x} \quad \text{and} \quad v = -\frac{\partial \psi}{\partial x} = \frac{\partial \phi}{\partial y}.$$

SOLUTION   Using the definition, the velocity components are obtained by differentiating ψ as

$$u = \frac{\partial \psi}{\partial y} = -4y,$$

and

$$v = -\frac{\partial \psi}{\partial x} = -4x.$$

However, by definition

$$u = \frac{\partial \phi}{\partial x} = -4y,$$

which by integration yields,

$$\phi = -4xy + f(y).$$

Similarly,

$$v = \frac{\partial \phi}{\partial y} = -4x,$$

which, upon integration gives

$$\phi = -4xy + f(x).$$

Therefore, the only way for the two functions to be identical is when $f(x) = f(y) = constant = C$. By arbitrarily selecting the constant $C = 0$, the velocity potential function ϕ is then,

$$\phi = -4xy.$$

Figure 4.14, shows the sketch of a number of selected lines of constant velocity potential. By superimposing the streamline pattern, we verify the orthogonality of the two patterns (Figure 4.15).

4.5.3   Pathlines, Streaklines

Other useful visual concepts are the *pathlines* and *streaklines*. By definition, a pathline is the actual path followed by a fluid particle. This is analogous to the lines in a long-exposure photograph of a busy road taken at night. Very much like fluid particles, these lines represent the trajectories of individual cars.

A *streakline* is the locus of all particles that at an earlier instant of time, passed through a prescribed point in space. This is similar to smoke released from a chimney

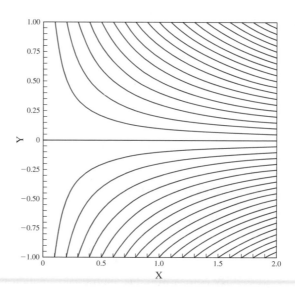

FIGURE 4.14   Velocity potential lines for frictionless flow in a 90° corner oriented symmetrically around the x-axis.

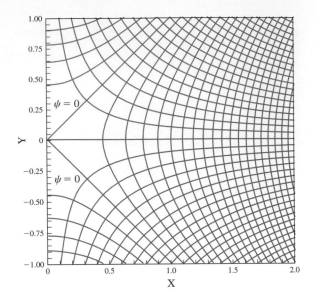

FIGURE 4.15 Superimposed iso-velocity potential lines and streamlines for frictionless flow in a 90° corner oriented symmetrically around the *x*-axis.

(or a fire): at any instant of time, the smoke particles form a streakline because they are released (at different times, of course) from the same point (i.e., the chimney or fire). In steady flows, streamlines, pathlines, and streaklines are identical.

EXAMPLE 4.9

Problem Statement Consider the two-dimensional flow field given by $\mathbf{V} = 2t\,\mathbf{i} + \mathbf{j}$, where t is time. For this flow, determine the pattern of *streamlines*, *pathlines*, and *streaklines*.

Governing Equations By definition,

$$u = \frac{\partial \psi}{\partial y} \quad \text{and} \quad v = -\frac{\partial \psi}{\partial x}.$$

SOLUTION For this flow

$$u = 2t \quad \text{and} \quad v = 1.$$

According to the definition

$$u = \frac{\partial \psi}{\partial y} = 2t \Rightarrow \psi = 2ty + f(x),$$

and

$$v = -\frac{\partial \psi}{\partial x} = 1 \Rightarrow \psi = -x + f(y).$$

By comparison, the appropriate solution is

$$\psi = 2ty - x.$$

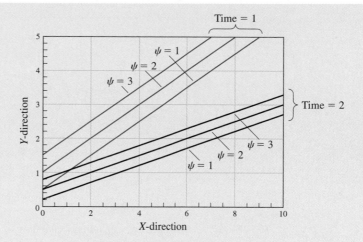

FIGURE 4.16   Instantaneous streamlines corresponding to the flow in Example 4.9.

Selected values of ψ at different times t are sketched in Figure 4.16. The figure shows that the streamlines are straight lines whose slope decreases with time.

Alternatively, the streamlines could be constructed using the slope of the local streamlines

$$\frac{dy}{dx}\bigg|_{\psi} = \frac{v}{u} = \frac{1}{2t},$$

which, upon integration,

$$\int_{y_0}^{y} 2t\, dy = \int_{x_0}^{x} dx,$$

where y_0 and x_0 are the coordinates of a known point along the streamline, gives

$$2ty - x = 2ty_0 - x_0 = \psi.$$

The trajectory of a single particle which, at time $t = t_0$, is at X_o, Y_o, can be evaluated using the definition of the velocity,

$$u = \frac{Dx}{Dt} = 2t \Rightarrow \int_{X_o}^{X} Dx = \int_{t_o}^{t} 2t\, Dt,$$

giving

$$X - X_o = t^2 - t_0^2. \tag{4.13}$$

Similarly, the velocity in the y-direction gives,

$$v = \frac{Dy}{Dt} = 1 \Rightarrow \int_{Y_o}^{Y} Dy = \int_{t_o}^{t} Dt \Rightarrow Y - Y_o = t - t_0. \tag{4.14}$$

By eliminating time from the foregoing equations, the trajectory of the particle (i.e., the pathline) is obtained as

$$(X - X_o) = (Y - Y_o)^2 + 2t_o(Y - Y_o). \tag{4.15}$$

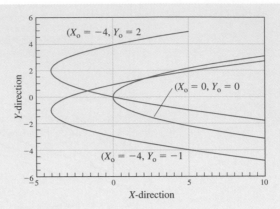

FIGURE 4.17   Pathlines corresponding to the flow in Example 4.9.

Recall that implicit in the above equation is the fact that the particle at $t = t_o$ was originally at X_o, Y_o. Figure 4.17 shows the pathlines for a number of such particles. Consistent with the above equation, the pathlines are a set of parabolas.

A streakline is the locus of all particles that at some earlier time, passed through a particular reference point. For illustration, we will select the origin as the reference point ($X_o = Y_o = 0$). Arbitrarily, we will select three earlier times when particles passed through the reference point. These times are $t = 0, 1, 2 \ s$. The locus of these particles, at a subsequent time $t > 2$ then, represent streaklines.

First, consider particle A, which at $t = t_0 = 0$ was at the origin ($X_0 = 0, Y_0 = 0$). According to Equations (4.13) and (4.14), then, the particle location as a function of time is

$$X = t^2 \quad \text{and} \quad Y = t.$$

Particle B was at the origin ($X_0 = 0, Y_0 = 0$) at $t = t_0 = 1$. Again, according to Equations (4.13) and (4.14), the particle location as a function of time is

$$X = t^2 - 1 \quad \text{and} \quad Y = t - 1.$$

Finally, because particle C was at the origin ($X_0 = 0, Y_0 = 0$) at $t = t_0 = 2$, according to Equations (4.13) and (4.14), the location of this particle as a function of time is

$$X = t^2 - 4 \quad \text{and} \quad Y = t - 2.$$

The location of the particles at the different times is summarized in the following table.

Figure 4.18 shows the streakline at time $t = 2$. From the above example, it is clear that streamlines, pathlines, and streaklines are not necessarily the same lines but can represent three distinctly different visualization perspectives. In steady flows, though (flows independent of time), the three lines are identical.

TABLE 4.1   Summary of the Location of Particles A, B, C at Different Times.

Particle	At time = 0	At time = 1	At time = 2
A	$X = 0, Y = 0$	$X = 1, Y = 1$	$X = 4, Y = 2$
B	$X = -1, Y = -1$	$X = 0, Y = 0$	$X = 3, Y = 1$
C	$X = -4, Y = -2$	$X = -3, Y = -1$	$X = 0, Y = 0$

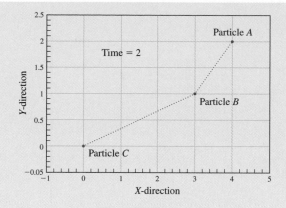

FIGURE 4.18 Streaklines corresponding to the flow in Example 4.9.

4.6 Deformation of Fluid Elements: Translation, Rotation, Strain

Fluid particles in motion deform, and, therefore, can conform to complex geometries and shapes. In practical terms, deformation represents the different ways in which particles can change shape and/or position under the influence of external forces. This situation is similar to the different ways a dancer in a densely packed dance floor is affected by other dancers. For instance, a dancer can be pushed, forced to spin around, or squeezed. Similarly, football players along the line of scrimmage experience similar types of deformation(!).

Fluid particles undergo similar deformations. Consider an infinitesimal, two-dimensional fluid element in a general flow field, as shown in Figure 4.19a. At time $t = t_o$, the element is assumed to be a perfect square. For reference, the corner A at the origin of the coordinate system is assumed to move with velocity u, v. With respect to point A, point B moves with velocity $u_B = u + \dfrac{\partial u}{\partial x} dx$ and $v_B = v + \dfrac{\partial v}{\partial x} dx$. Similarly, point D moves with velocity $u_D = u + \dfrac{\partial u}{\partial y} dy$ and $v_D = v + \dfrac{\partial v}{\partial y} dy$. In the definitions that follow, point C does not participate.

At an instant of time later, $t = t_o + \Delta t$; because of the ability of fluid particles to deform, this fluid element deforms as shown in Figure 4.19b. The particle has simultaneously moved, rotated, and changed its shape.

Geometrically, the final deformation is the result of having the four corners move with different velocities along each coordinate direction. For example, the distance ea is given as

$$ea = dx + (u_B - u)\Delta t = dx + \frac{\partial u}{\partial x} dx \Delta t.$$

Similarly, we can write

$$be = (v_B - v)\Delta t = \frac{\partial v}{\partial x} dx \Delta t,$$

$$fa = dy + (v_D - v)\Delta t = dy + \frac{\partial v}{\partial y} dy \Delta t,$$

$$fd = (u_D - u)\Delta t = \frac{\partial u}{\partial y} dy \Delta t.$$

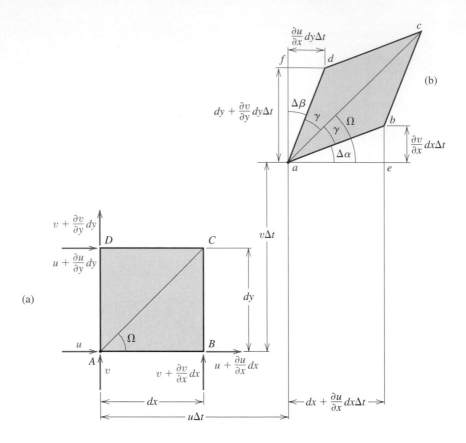

FIGURE 4.19 Schematic of an originally square-shaped fluid particle (a); its shape after deformation (b).

Formally, the modes of deformation shown above are defined either as (a) *simple translation*, (b) *simple rotation*, or (c) *strain*. Strain is further identified as *extensional strain* or *shear strain*, depending on the mechanism responsible for the deformation. In fluid dynamics, it is customary to define deformation in relative terms by expressing the time rate of change with respect to an earlier configuration. Therefore, the expressions developed subsequently are all in rate form.

Rate of Translation

Fluid particles change position according to their local fluid velocity. Therefore, in the time interval Δt the reference corner A moves in each coordinate direction a distance of $\Delta x = u\Delta t$, $\Delta y = v\Delta t$ and, if we consider the third direction, $\Delta z = w\Delta t$. The rate of translation, then, along each coordinate direction is defined as,

$$lim_{\Delta t \to 0} \frac{\Delta x}{\Delta t}, \frac{\Delta y}{\Delta t}, \frac{\Delta z}{\Delta t},$$

which is simply the velocity u, v, and w of the particle in each direction.

Rate of Rotation

If we define Ω_z as the angle that the diagonal, AC, makes with respect to the horizontal axis, the rate of rotation around the z-axis is defined as the rate at which Ω_z changes with time,

$$Rate\ of\ Rotation = lim_{\Delta t \to 0} \frac{\Delta \Omega_z}{\Delta t} = \frac{d\Omega_z}{dt}.$$

The rate of rotation is considered positive when the diagonal angle increases in the counterclockwise direction.

Using the geometry of Figure 4.19, $\Delta \Omega_z = \Omega_{zt+\Delta t} - \Omega_{zt}$. But $\Omega_{zt+\Delta t} = \Delta \alpha + \gamma$. However, $2\gamma + \Delta \alpha + \Delta \beta = 90$ and $\Omega_{zt} = 45$. After simple manipulation, we can write $\Delta \Omega_z = \dfrac{\Delta \alpha - \Delta \beta}{2}$. Using again the geometry of the new configuration, we get

$$\Delta \alpha = \tan^{-1} \frac{(be)}{(ea)} = \tan^{-1} \left[\frac{\dfrac{\partial v}{\partial x} dx \Delta t}{dx + \dfrac{\partial u}{\partial x} dx \Delta t} \right],$$

in the limit of $\Delta t \to 0$

$$lim_{\Delta t \to 0} \frac{\Delta \alpha}{\Delta t} = \frac{d\alpha}{dt} = \frac{\partial v}{\partial x}.$$

Similarly,

$$\Delta \beta = \tan^{-1} \frac{(fd)}{(fa)} = \tan^{-1} \left[\frac{\dfrac{\partial u}{\partial y} dy \Delta t}{dy + \dfrac{\partial v}{\partial y} dy \Delta t} \right],$$

and in the limit of $\Delta t \to 0$,

$$lim_{\Delta t \to 0} \frac{\Delta \beta}{\Delta t} = \frac{d\beta}{dt} = \frac{\partial u}{\partial y}.$$

By substituting the expressions for $\Delta \alpha$ and $\Delta \beta$ into the definition, the rate of rotation around the z-axis is obtained as

$$\frac{d\Omega_z}{dt} = \frac{1}{2} \left(\frac{\partial v}{\partial x} - \frac{\partial u}{\partial y} \right) \equiv \frac{1}{2} \omega_z. \tag{4.16}$$

The same procedure can be repeated for the rate of rotation of fluid elements around the other two coordinate axes

$$\frac{d\Omega_x}{dt} = \frac{1}{2} \left(\frac{\partial w}{\partial y} - \frac{\partial v}{\partial z} \right) \equiv \frac{1}{2} \omega_x. \tag{4.17}$$

$$\frac{d\Omega_y}{dt} = \frac{1}{2} \left(\frac{\partial u}{\partial z} - \frac{\partial w}{\partial x} \right) \equiv \frac{1}{2} \omega_y. \tag{4.18}$$

The above quantities are the components of a vector that, for obvious reasons, is called the *angular velocity vector*. A new quantity $\boldsymbol{\omega}$, defined for convenience as twice the angular velocity vector (to avoid the 1/2 in the angular velocity)

$$\boldsymbol{\omega} = \omega_x\, \mathbf{i} + \omega_y\, \mathbf{j} + \omega_z\, \mathbf{k} = 2\frac{d\boldsymbol{\Omega}}{dt} \tag{4.19}$$

is known as the *vorticity vector*. The vorticity indicates the tendency of the fluid to rotate around each coordinate axis.

Observe that the vorticity vector can be represented in a mathematically compact form by writing

$$\boldsymbol{\omega} = \nabla \times \mathbf{V}, \tag{4.20}$$

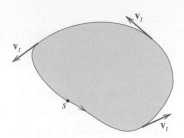

FIGURE 4.20   Definition of circulation.

where, $\mathbf{V}$ is the velocity vector. The operator ∇ stands for

$$\nabla = \frac{\partial}{\partial x}\mathbf{i} + \frac{\partial}{\partial y}\mathbf{j} + \frac{\partial}{\partial z}\mathbf{k},$$

while the operator $\times$ denotes the cross product.

The vorticity is a very important concept in fluid mechanics. Of particular importance is the special flow where the fluid elements do not rotate (i.e., $\boldsymbol{\omega} = 0$). In such case, the flow is called *irrotational*.

If we substitute the definition of ϕ, Equation (4.11), in the expression for vorticity, we get

$$\omega_z = \frac{\partial v}{\partial x} - \frac{\partial u}{\partial y} = \frac{\partial}{\partial x}\left(\frac{\partial \phi}{\partial y}\right) - \frac{\partial}{\partial y}\left(\frac{\partial \phi}{\partial x}\right) = 0,$$

which indicates that ϕ is defined *only* for irrotational flows. By definition, then, the velocity potential *must* satisfy $\boldsymbol{\omega} = 0$, if it is truly a velocity potential representing an actual flow.

The rotation of the particles is also expressed in terms of a new quantity known as the *circulation* Γ. The circulation is defined as the integral of the tangent velocity along a closed boundary s, as shown in Figure 4.20.

$$\Gamma = \oint_s \mathbf{V} \cdot \mathbf{t}\, ds. \tag{4.21}$$

In the SI system, Γ, is expressed in units of m^2/s. According to the Stokes theorem, the line integral can be converted into a surface integral over the area A enclosed by s as

$$\oint_s \mathbf{V} \cdot \mathbf{t}\, ds = \int_A (\nabla \times \mathbf{V}) \cdot \mathbf{n}\, dA,$$

where $\mathbf{n}$ is the outward pointing unit normal vector to the surface. But since $\boldsymbol{\omega} = \nabla \times \mathbf{V}$, the circulation can be alternatively expressed in terms of the vorticity vector as

$$\Gamma = \int_A \boldsymbol{\omega} \cdot \mathbf{n}\, dA. \tag{4.22}$$

If the flow is irrotational ($\boldsymbol{\omega} = 0$), the circulation, then, is zero. Chapter 9 explores further the implications of both the rotation and the irrotationality of fluid flow.

Strain, Rate of Strain

Strain describes the deformation of fluid particles and *rate of strain* the rate of deformation of the particles. Strain is further defined in terms of the mechanisms responsible for the deformation: (a) a particle extended in one or more directions deforms in what is known as *extensional strain*, and (b) a particle under the influence of a lateral or *shear* force deforms in what is known as *shear strain*.

Extensional Strain: Extensional strain is defined as the relative change of a linear dimension of a fluid element in the direction of the element. Positive strain is associated with a fluid line increasing in length. By returning to the schematic of Figure 4.19, the *extension* or *extensional strain* in the x-direction, $\Delta\epsilon_{xx}$ is defined as the relative change of ab with respect to the original length, dx. Despite the fact that extensional strain is associated with a single direction, it is usually identified by a double subscript—for instance, xx. As shown subsequently, this is done for notational convenience, as all forms of strain are mathematically combined into a tensor quantity, which of course requires double subscripts.

The increase in length is due to the difference of the velocity of points A and B,

$$\Delta\epsilon_{xx} = \frac{dx + \frac{\partial u}{\partial x}dx\,\Delta t - dx}{dx} = \frac{\partial u}{\partial x}\Delta t.$$

The rate of extensional strain in the limit of $\Delta t \to 0$ is given as

$$lim_{\Delta t \to 0}\frac{\Delta\epsilon_{xx}}{\Delta t} = \frac{d\epsilon_{xx}}{dt} = \dot{\epsilon}_{xx} = \frac{\partial u}{\partial x}.$$

Similarly, we can define the rate of extensional strain in the other two principal directions as

$$\frac{d\epsilon_{yy}}{dt} = \dot{\epsilon}_{yy} = \frac{\partial v}{\partial y},$$

$$\frac{d\epsilon_{zz}}{dt} = \dot{\epsilon}_{zz} = \frac{\partial w}{\partial z}.$$

Shear Strain: Shear strain, $\Delta\epsilon_{xy}$, is defined as the average decrease in the angle between two originally perpendicular lines. Positive shear strain is associated with *decreasing angle*. The double subscript xy indicates the directions associated with each line, and, consequently, the plane formed by them. For instance, the two perpendicular lines AB and AD shown in Figure 4.19a form the xy plane, which includes also the angle between them. Because of the action of shear in the deformed configuration, the angle between AB and AD is no longer 90° but less.

The average decrease of the enclosed angle is $\Delta\epsilon_{xy} = (\Delta\alpha + \Delta\beta)/2$. Using earlier results, we can write

$$\Delta\epsilon_{xy} = \frac{1}{2}\left(\frac{\partial v}{\partial x} + \frac{\partial u}{\partial y}\right)\Delta t.$$

The rate of change of ϵ_{xy} with respect to time, in the limit of $\Delta t \to 0$, is defined as the *rate of shear strain*. Therefore,

$$lim_{\Delta t \to 0}\frac{\Delta\epsilon_{xy}}{\Delta t} = \frac{d\epsilon_{xy}}{dt} = \dot{\epsilon}_{xy} = \frac{1}{2}\left(\frac{\partial v}{\partial x} + \frac{\partial u}{\partial y}\right).$$

Similarly, for a general three-dimensional flow, we can write the rate of shear strain for the other planes as

$$\frac{d\epsilon_{yz}}{dt} = \dot{\epsilon}_{yz} = \frac{1}{2}\left(\frac{\partial w}{\partial y} + \frac{\partial v}{\partial z}\right),$$

$$\frac{d\epsilon_{zx}}{dt} = \dot{\epsilon}_{zx} = \frac{1}{2}\left(\frac{\partial u}{\partial z} + \frac{\partial w}{\partial x}\right).$$

Since each component of the shear strain is associated with a particular plane, the indices associated with ϵ are interchangeable — that is,

$$\dot{\epsilon}_{xy} = \dot{\epsilon}_{yx},$$

$$\dot{\epsilon}_{yz} = \dot{\epsilon}_{zy},$$

$$\dot{\epsilon}_{zx} = \dot{\epsilon}_{xz}.$$

Often, the various measures of rate of strain are mathematically combined, and they are simply referred to as *rate of strain*. For convenience, the components of strain are usually depicted in tensor form as

$$\dot{\epsilon} = \begin{bmatrix} \dot{\epsilon}_{xx} & \dot{\epsilon}_{xy} & \dot{\epsilon}_{xz} \\ \dot{\epsilon}_{xy} & \dot{\epsilon}_{yy} & \dot{\epsilon}_{yz} \\ \dot{\epsilon}_{xz} & \dot{\epsilon}_{yz} & \dot{\epsilon}_{zz} \end{bmatrix} = \frac{1}{2}\begin{bmatrix} \left(2\frac{\partial u}{\partial x}\right) & \left(\frac{\partial u}{\partial y} + \frac{\partial v}{\partial x}\right) & \left(\frac{\partial u}{\partial z} + \frac{\partial w}{\partial x}\right) \\ \left(\frac{\partial u}{\partial y} + \frac{\partial v}{\partial x}\right) & \left(2\frac{\partial v}{\partial y}\right) & \left(\frac{\partial w}{\partial y} + \frac{\partial v}{\partial z}\right) \\ \left(\frac{\partial u}{\partial z} + \frac{\partial w}{\partial x}\right) & \left(\frac{\partial w}{\partial y} + \frac{\partial v}{\partial z}\right) & \left(2\frac{\partial w}{\partial z}\right) \end{bmatrix} \qquad (4.23)$$

Implicit in the above tensor is the distinction between the two kinds of strain: diagonal terms represent the rate of extentional strain, and off-diagonal terms the rate of shear strain.

In practice, we define a new tensor Δ, known as the *rate of deformation* tensor, as $\Delta = 2\dot{\epsilon}$. Conveniently, this definition removes the factor $1/2$ from $\dot{\epsilon}$. Therefore,

$$\Delta = \begin{bmatrix} \Delta_{xx} & \Delta_{xy} & \Delta_{xz} \\ \Delta_{xy} & \Delta_{yy} & \Delta_{yz} \\ \Delta_{xz} & \Delta_{yz} & \Delta_{zz} \end{bmatrix} = \begin{bmatrix} \left(2\frac{\partial u}{\partial x}\right) & \left(\frac{\partial u}{\partial y} + \frac{\partial v}{\partial x}\right) & \left(\frac{\partial u}{\partial z} + \frac{\partial w}{\partial x}\right) \\ \left(\frac{\partial u}{\partial y} + \frac{\partial v}{\partial x}\right) & \left(2\frac{\partial v}{\partial y}\right) & \left(\frac{\partial w}{\partial y} + \frac{\partial v}{\partial z}\right) \\ \left(\frac{\partial u}{\partial z} + \frac{\partial w}{\partial x}\right) & \left(\frac{\partial w}{\partial y} + \frac{\partial v}{\partial z}\right) & \left(2\frac{\partial w}{\partial z}\right) \end{bmatrix}. \qquad (4.24)$$

The rate of deformation can be expressed in a more condensed form as

$$\Delta = (\nabla \mathbf{V} + \nabla \mathbf{V}^T),$$

where $\nabla \mathbf{V}$ is the velocity gradient tensor and $\nabla \mathbf{V}^T$ the transpose of $\nabla \mathbf{V}$.

EXAMPLE 4.10

Problem Statement   *Fluid deformation in Couette flow*. The velocity distribution $u = u(y)$, between two plates at distance H apart, when one plate is moving with velocity V relative to the other plate, is given by

$$u = V\frac{y}{H}.$$

Describe how fluid particles will deform in this flow.

Governing Equations   For this two-dimensional flow, the possible types of deformation are linear motion, given by the velocity $u = u(y)$, and the rate of rotation around the z-axis defined as,

$$\omega_z = \frac{\partial v}{\partial x} - \frac{\partial u}{\partial y}.$$

The particles may also deform because of extension or shear, depending on the rate of deformation tensor Δ

$$\Delta = \begin{bmatrix} \left(2\dfrac{\partial u}{\partial x}\right) & \left(\dfrac{\partial u}{\partial y} + \dfrac{\partial v}{\partial x}\right) \\ \left(\dfrac{\partial u}{\partial y} + \dfrac{\partial v}{\partial x}\right) & \left(2\dfrac{\partial v}{\partial y}\right) \end{bmatrix}.$$

SOLUTION   Using the given flow kinematics with the proper definition of the various deformation modes, we have:

Linear Motion   For this flow, the velocity distribution $v = 0$, which implies that fluid particles move only in the x-direction, with the given u velocity.

Rate of Rotation   For the rate of rotation we have

$$\frac{\partial u}{\partial x} = \frac{\partial v}{\partial x} = \frac{\partial v}{\partial y} = 0,$$

and

$$\frac{\partial u}{\partial y} = \frac{V}{H}.$$

Particles, then, rotate around the z-axis at a constant rate of

$$\omega_z = \frac{\partial v}{\partial x} - \frac{\partial u}{\partial y} = -\frac{V}{H},$$

Since this is a negative quantity, by definition, the particles rotate in the clockwise direction as shown in Figure 4.21.

Rate of Deformation   The rate of deformation tensor for this flow simplifies to

$$\Delta = \begin{bmatrix} 0 & \dfrac{V}{H} \\ \dfrac{V}{H} & 0 \end{bmatrix}.$$

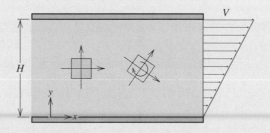

FIGURE 4.21   Rate of rotation of fluid particles in Couette flow.

Since $\Delta_{xx} = \Delta_{yy} = 0$, there is no extensional deformation. However, fluid particles deform under shear according to

$$\Delta_{xy} = \frac{V}{H}.$$

Again, since because is a positive quantity, according to the definition, fluid particles deform as shown in Figure 4.22.

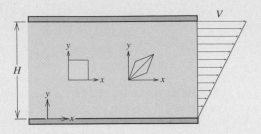

FIGURE 4.22   Rate of shear strain of fluid particles in Couette flow.

EXAMPLE 4.11

Problem Statement   For the same Couette flow described in Example 4.10, calculate the circulation around an arbitrary closed loop, *ABCD*, as shown in Figure 4.23.

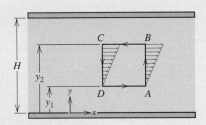

FIGURE 4.23   Circulation around an arbitrary closed loop in Couette flow.

Governing Equations   According to the definition, the circulation is calculated using either

$$\Gamma = \oint \mathbf{V} \cdot \mathbf{t}\, ds,$$

where the integral is over a closed loop in the counterclockwise direction, or

$$\Gamma = \int_s \omega_z\, ds,$$

with *S* being the area enclosed by the loop.

SOLUTION   According to the definition,

$$\Gamma = \oint \mathbf{V} \cdot \mathbf{t}\, ds = \int_A^B \mathbf{V} \cdot \mathbf{t}\, ds + \int_B^C \mathbf{V} \cdot \mathbf{t}\, ds + \int_C^D \mathbf{V} \cdot \mathbf{t}\, ds + \int_D^A \mathbf{V} \cdot \mathbf{t}\, ds.$$

Using the figure:

$$\text{Along } AB: \quad u = y\frac{V}{H}\mathbf{i}, \quad \mathbf{t} = \mathbf{j}, \quad \text{and} \quad ds = dy.$$

$$\text{Along } BC: \quad u = y_2\frac{V}{H}\mathbf{i}, \quad \mathbf{t} = -\mathbf{i}, \quad \text{and} \quad ds = -dx.$$

$$\text{Along } CD: \quad u = y\frac{V}{H}\mathbf{i}, \quad \mathbf{t} = -\mathbf{j}, \quad \text{and} \quad ds = -dy.$$

$$\text{Along } DA: \quad u = y_1\frac{V}{H}\mathbf{i}, \quad \mathbf{t} = \mathbf{i}, \quad \text{and} \quad ds = dx.$$

By substitution,

$$\Gamma = y\frac{V}{H}\int_{x_A}^{x_B} \underbrace{(\mathbf{i}\cdot\mathbf{j})}_{=0}\,dy - y_2\frac{V}{H}\int_{x_B}^{x_C} \mathbf{i}\cdot(-\mathbf{i})\,dx -$$

$$y\frac{V}{H}\int_{x_C}^{x_D} \underbrace{\mathbf{i}\cdot(-\mathbf{j})}_{=0}\,dy + y_1\frac{V}{H}\int_{x_D}^{x_A} (\mathbf{i}\cdot\mathbf{i})\,dx.$$

Simplifying the above,

$$\Gamma = y_2\frac{V}{H}(x_C - x_B) + y_1\frac{V}{H}(x_A - x_D).$$

But, since $x_A = x_B$ and $x_C = x_D$,

$$\Gamma = -\frac{V}{H}(y_2 - y_1)(x_A - x_D).$$

Alternatively, the circulation can be calculated using

$$\Gamma = \int_S \omega_z\,dS.$$

From the previous example, $\omega_z = -\dfrac{V}{H}$,

$$\Gamma = -\frac{V}{H}\int_{y_1}^{y_2}\int_{x_D}^{x_A} dx\,dy = -\frac{V}{H}(y_2 - y_1)(x_A - x_D),$$

which is identical to the result found previously.

REFERENCES

G.A. TOKATY, *A History and Philosophy of Fluid Mechanics*, New York, Dover, 1994.

M. VAN DYKE, *An Album of Fluid Motion*, Stanford, California, The Parabolic Press, 1982.

G.K. BATCHELOR, F.R.S., *An Introduction to Fluid Dynamics*, Cambridge, England, Cambridge University Press, 1970.

R.L. PANTON, *Incompressible Flow*, New York, Wiley, 1984.

PROBLEMS

1. The velocity field for fully developed flow in a two dimensional channel can be expressed as $u = u_o \left[1 - \left(\frac{y}{h}\right)^2\right]$ where u_o is the maximum velocity at the center of the channel. The width of the channel is $2h$. For this flow let $u_0 = 1.5 \ m/s$ and $h = 0.2 \ m$.

   • Find the location of a particle originally with position vector $0 \ m\mathbf{i} + 0.1 \ m\mathbf{j}$, after $t = 0.3 \ s$.
   • Find the time needed for a particle at $x = 0.0$ and $y = 0.12 \ m$ to move to $x = 2.3 \ m$.
   • Find the distance between two particles A, B originally at $x_A = 0.0$, $y_A = 0.0$, $x_B = 0.0$, $y_B = 0.15$, after $0.4 \ s$.

2. The velocity profile for fully developed two dimensional pipe flow can be expressed as $u = u_o \left[1 - \left(\frac{r}{R}\right)^2\right]$ where u_o is the maximum velocity at the center of the pipe and R the radius of the pipe. For $u_0 = 6.4 \ m/s$ and $R = 0.25 \ m$.

   • Find the location of a particle after $t = 0.3 \ s$. The particle's original position vector was $x = 0 \ m$ and $r = 0.14 \ m$.
   • Find the time needed for a particle at $x = 0$ and $r = 0.12 \ m$ to move to $x = 2.3 \ m$.

3. Find the time it takes a particle to travel the length of the channel for the one-dimensional flow shown in Figure 4.24. What is the acceleration of a particle at the exit of the channel? Where is the particle after $1/3$ of the resident time? Find the velocity and acceleration of the particle at that location.

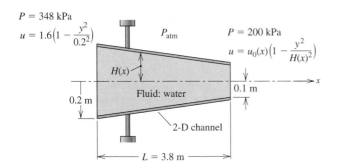

$P = 348 \ kPa$
$u = 1.6\left(1 - \dfrac{y^2}{0.2^2}\right)$

P_{atm}

$P = 200 \ kPa$
$u = u_0(x)\left(1 - \dfrac{y^2}{H(x)^2}\right)$

$H(x)$

$0.2 \ m$

Fluid: water

$0.1 \ m$

2-D channel

$L = 3.8 \ m$

FIGURE 4.24   Schematic for Problem 3.

4. Find the time it takes a particle along the centerline to travel the length of the channel as shown in Figure 4.25. What is the acceleration of a particle at the exit of the channel?

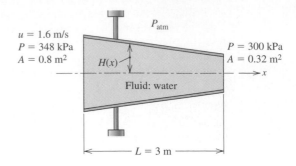

$u = 1.6 \ m/s$
$P = 348 \ kPa$
$A = 0.8 \ m^2$

$H(x)$

P_{atm}

$P = 300 \ kPa$
$A = 0.32 \ m^2$

x

Fluid: water

$L = 3 \ m$

FIGURE 4.25   Schematic for Problem 4.

5. Reconsider the above problem and find the distance between two particles A, B after $0.4 \ s$. The particles at $t = 0$ were at $x_A = 0.0$, $y_A = 0.0$, $x_B = 0.0$, $y_B = 0.15$.

6. The flow between two infinite plates where the lower plate is stationary and the upper plate is moving with velocity u_o in the direction of increasing x-axis is expressed as $u = u_o \dfrac{y}{h}$ where h is the distance between the plates. If $u_o = 1.5 \ m/s$ and $h = 0.2 \ m$, repeat the same calculations as in Problem 1.

7. The velocity profile of fully developed flow of a Non-Newtonian fluid (power-law fluid) in a two-dimensional channel with height H is given by

$$u = u_0 \left[1 - \left(\frac{2y}{H}\right)^{\frac{n+1}{n}}\right].$$

For a volumetric flow rate $Q = 2.2 \ m^3/s$ and $H = 0.6 \ m$, find the effect of the parameter n on the time it takes a particle along the centerline to travel $1.2 \ meters$.

8. The velocity distribution across the thickness H, of a layer of fluid flowing under the influence of gravity g, down an inclined surface with angle θ with respect to the horizontal is given by $u = \dfrac{1}{\mu} \rho g \sin\theta \left(Hy - \dfrac{y^2}{2}\right)$. For this flow consider two particles, one at $x_1 = 0.0$ and $y_A = H/2$ and another at $x_1 = 0.0$ and $y_A = H$. If $H = 8 \ cm$ and $\theta = 24°$, find: (a) the position vector of each particle after $0.2 \ seconds$, (b) the distance between them, (c) the acceleration of each particle. How much time it takes each particle to travel $40 \ cm$?

9. The ideal velocity around a circular cylinder with radius R is given by $u = 2U \sin\theta$ where U is the velocity away from the cylinder. For $R = 0.6 \ m$ and $U = 3.6 \ m/s$ find the time it takes for a particle to travel around the upper side of the cylinder.

10. Determine the velocity and acceleration fields for an one dimensional flow where the particles' position vector is

defined by $r(t) = \mathbf{i}x(t) = \mathbf{i}x_0 e^{at}$, where t is time, x_0 is the position at time $t = 0$ and a is a constant.

11. Find the acceleration vector corresponding to the velocity $u = U_o e^{-at} \cos(x - \omega t)\mathbf{i}$, where t is time and x is the distance along the horizontal direction from the origin of Cartesian coordinate system.

12. Sketch the following flow fields and derive general expressions for the corresponding acceleration fields: (a) $u = 6$, $v = 2$ (b) $u = 6$, $v = 2x$ (c) $u = 2x$, $v = -2y$, (d) $u = xy$, $v = 0$.

13. For the general flow given by $\mathbf{V} = 3.5xe^{-t}\mathbf{i} - 3.5ye^{-t}\mathbf{j}$, determine the acceleration fields using: (a) a Lagrangian description and (b) an Eulerian description.

14. Consider the flow field $u = Ux$ and $v = -Uy$. Find the acceleration fields using: (a) a Lagrangian description and (b) an Eulerian description.

15. Compute the acceleration field for the following velocity fields: (a) $u = U\cos(x - 2t)$, (b) $u = 5.1(3x^2 - 2y)$, (c) $u = 3.4(1 - e^{-3t})$.

16. Consider the flow field $u = xU\cos\alpha t$ and $v = yU\sin\alpha t$, where α is a constant and t is time. Find the acceleration field using: (a) a Lagrangian description and (b) an Eulerian description.

17. When fluid friction is neglected, the ideal velocity distribution around a two-dimensional cylinder with radius a, due to a uniform flow with velocity U_0 is given by

$$u = U_0 - U_0 a^2 \left[\frac{x^2 - y^2}{(x^2 + y^2)^2} \right]$$

and

$$v = -U_0 a^2 \frac{2xy}{(x^2 + y^2)^2}.$$

For this velocity distribution calculate and plot the acceleration of fluid particles along the surface of the cylinder.

18. For the two dimensional particle flow given by the particle velocities $u = 2xt^2$, $v = 5yt$, determine the velocity and acceleration fields using (a) a Lagrangian description and (b) an Eulerian description.

19. Consider a general flow where, at time $t = 0$ a particle A crosses the origin with velocity 1 m/s. One second later the particle is 1.5 meters away with velocity of 5 m/s. At that moment another particle, B crosses the origin with velocity 3 m/s. What is the average velocity of particle A? If the average velocity of particle A is 2 m/s, what then must be the velocity of particle B the moment it crosses the origin?

20. Due to the action of wind-induced shear at a free surface of a lake, the water is suddenly set to motion with velocity u_o. By reaction, due to the action of viscosity the shear is transmitted to the lower fluid layers. The velocity as a function of time is given by

$$u = u_0(1 - erf\,\eta) = u_0\left(1 - erf\frac{y}{2\sqrt{vt}}\right).$$

The integral $\int_0^\eta e^{-\eta^2} d\eta$, which appears in many engineering applications, forms the basis for the well-known *error function* defined as

$$erf\,\eta = \frac{2}{\sqrt{\pi}} \int_0^\eta e^{-\eta^2} d\eta.$$

For this flow find and plot the acceleration as a function of location and time.

21. Sketch the streamlines that correspond to the velocity field $u = x + 3y$, $v = 3x - y$, $w = 0$.

22. The streamfunction

$$\psi = V_0 \left[y + \frac{(\delta(x) - y)^3}{3\delta^2(x)} \right] \quad 0 \le y \le \delta(x)$$

approximates the streamfunction in a laminar boundary layer for a flat plate, where V_0 is the velocity away from the plate, y is the coordinate direction normal to the main flow and $\delta(x)$ is the thickness boundary layer. (a) Sketch selected streamlines, (b) determine the velocity field, (c) determine the rate of deformation tensor and discuss the way fluid particles will deform.

23. For the streamfunction $\psi = 2/3r^{3/2}\sin(3/2\theta)$, find the corresponding velocity potential function.

24. For the velocity field given by $u = xe^{-\alpha y}$ and $v = \frac{1}{\alpha}e^{-\alpha y}$, where α is a constant, find the acceleration of a particle at $x = 2$ and $y = 3$. What is the pathline of a particle originally at the origin. Determine the streakline for selected particles that at an earlier instance passed through the origin.

25. For the velocity field $u = x/(1 + 3t)$, $v = y/(1 + 2t)$, and $w =$, (a) find the streamlines, ((b) pathlines, and (c) the streaklines at $t = 3$ $seconds$ for all particles that at $t = 0$ where at $x = y = 1$.

26. Consider the two dimensional flow field $u = 3.2t$, $v = 4.8t$, where t is time find: (a) the velocity and acceleration fields (b) determine the streamline, pathline and streakline patterns.

27. Find the streamlines, pathlines and streaklines for the flow described by $u = 2x(1 + t)$ and $v = 2y$, where t is time.

28. For the velocity field given by $u = xe^{-\alpha y}$ and $v = \frac{1}{\alpha}e^{-\alpha y}$, determine the proper expression for the streamfunction.

29. For steady flow find the equation of the streamline passing through the point $(1, 2, 3)$ if $u = 3.5x$, $v = 4y$, and $w = -7.5z$.

30. Is it possible to construct the streamfunction and the velocity potential for Problem 1? If yes, sketch selected streamlines and iso-potential lines. What is the total flow through a section normal to the plane of the flow defined by $y_1 = -0.15$ m and $y_2 = 0.15$ m?

31. For the flow field $\mathbf{V} = \mathbf{i} + 3t\mathbf{j}$, where t is time determine the streamline, pathline and streakline patterns.

32. Given $\psi = 3x^2 y$. Is this a valid streamfunction expression? If yes, sketch selected streamlines. What is the corresponding velocity potential function?

33. Given $\psi = 5xy$. Is this a valid streamfunction expression? If yes, sketch selected streamlines. What is the corresponding velocity potential function?

34. Given $\psi = -u(x, y)y + v(x, y)x$ where u, v are the components of the two dimensional velocity vector. Is this a valid streamfunction expression?

35. Sketch the streamfunction $\psi = 1.5(x^2 - y^2)$ and analyze the associated velocity field.

36. Given $\psi = Q \ln(x^2 + y^2)^{0.5}$ Is this a valid streamfunction expression? If yes, sketch selected streamlines. What is the corresponding velocity potential function?

37. Given $\phi = 4 \ln(x^2 + y^2)^{0.5}$ Is this a valid velocity potential expression? If yes, sketch selected iso-velocity potential lines. What is the corresponding streamfunction?

38. Construct the velocity potential function in polar coordinates given the streamfunction $\psi = Ar^n \sin(n\theta)$.

39. Consider the streamfunction $\psi = U\left(r - \dfrac{R^2}{r}\right) \sin\theta - \dfrac{\Gamma}{2\pi} \ln \dfrac{r}{R}$, where R, U and Γ are constants and r the coordinate in cylindrical coordinates. Plot selected streamlines for $\Gamma = 3\pi RU$ $\Gamma = 4\pi RU$ and $\Gamma = 5\pi RU$. Consider $R = 1\ m$ and $U = 2\ m/s$. Using the definition, calculate the velocity components. Show by integration that Γ is the circulation around the cylinder.

40. For steady compressible flow the continuity equation is expressed as
$$\frac{\rho u}{x} + \frac{\rho v}{y} = 0.$$
Consider a new streamfunction defined as $\rho u = \dfrac{\chi}{y}$ and $\rho v = -\dfrac{\chi}{x}$. Show that χ is constant along streamlines, and that the numerical difference $\Delta \chi$ between two streamlines is the mass flow rate per unit depth.

41. *Stokes' streamfunction.* The continuity equation for steady axisymmetric problems is expressed as
$$\frac{1}{r}\frac{ur}{r} + \frac{\rho w}{z} = 0.$$
Show that an appropriate streamfunction for such flows can be defined as $ur = -\dfrac{\psi}{z}$ and $wr = \dfrac{\psi}{r}$.

42. Find and plot the Stokes' streamfunction defined as shown in the above problem that represents uniform flow in the z-direction. Are the streamlines equally spaced? Why?

43. Show that the velocity potential
$$\phi = Uz + \frac{R^3 Uz}{2(r^2 + z^2)^{1.5}},$$
represents flow past a sphere of radius R. U is the approaching velocity away from the sphere.

44. Which of the following functions represent a true velocity potential function; (a) $f = Ax$, (b) $f = Ay$, (c) $f = Ax^2$, (d) $f = A \cos(x + y)$; where A is a constant.

45. Consider two infinite plates at a distance $2H$ apart; the top plate moving to the right with velocity U_1 and the bottom plate moving to the left with velocity U_2. The velocity distribution for this flow is $u = \left(\dfrac{U_1 + U_2}{2}\right)\dfrac{y}{H} - \left(\dfrac{U_2 - U_1}{2}\right)$. For this flow: (a) sketch the streamlines, (b) find the rate of deformation for the flow and (c) describe the deformation of particles at $y = 0$, $y = H$ and $y = -H$.

46. Reconsider the two dimensional, fully developed channel flow with $u = u_o\left[1 - \left(\dfrac{y}{h}\right)^2\right]$ where u_o is the maximum velocity at the center of the channel. The width of the channel is $2h$. For this flow let $u_0 = 1.5\ m/s$ and $h = 0.2\ m$. Describe how a fluid particle at $x = 1.2\ m$ and $y = 0.08\ m$ will deform.

47. For the flow described in Problem 2, develop a general expression for the rate of deformation tensor. Described the deformation of particles at $x_A = x_B = 0$, $y_A = R/2$ $y_B = R$ (use again, $u_0 = 6.4\ m/s$ and $R = 0.25\ m$).

48. For the flow described in Problem 8, develop a general expression for the rate of deformation tensor. Described the deformation of the particles at $x_A = x_B = 0$, $y_A = H/2$ $y_B = H$ (use again, $H = 8\ cm$ and $\theta = 24°$).

49. Extensional strain $\dot{\epsilon}$ is defined as the rate of change with time of the relative change of the length of a fluid element. For a constant $\dot{\epsilon}$, find the final length of a a fluid element (originally with length l_0) after a time of Δt.

50. Show that the general flow $u = -\dfrac{1}{2}\dot{\epsilon}(1 + b)x$, $u = -\dfrac{1}{2}\dot{\epsilon}(1 - b)y$, $w = +\dot{\epsilon}z$ is a shear-free flow (i.e., the fluid particles in this flow do not deform under shear). $\dot{\epsilon}$ is a constant. For this flow, describe the deformation of an initially cubic fluid element (side= l) for: (a) $b = 0$ and $\dot{\epsilon} > 0$, (b) $b = 0$ and $\dot{\epsilon} < 0$, and $b = 1$.

51. The Maxwell orthogonal rheometer consists of two parallel rotating eccentric disks (different axis of rotation) with a gap b between them. Both disks are rotating with the same angular velocity ω. For this rheometer if the distance between the two centers of rotation is a the velocity distribution is $u = -\omega\left(y - \dfrac{a}{b}z\right)$, $v = \omega x$ and $W = 0$. Is this a shearfree flow? Describe the deformation of fluid particles across the gap as function of x, and y.

52. For the flow described in Problem 2, find the circulation for the closed loops shown in Figure 4.26 by using: (a) the velocity around the boundary and (b) the vorticity enclosed within the prescribed area.

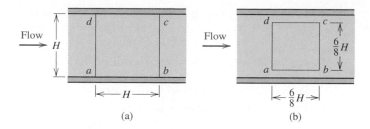

FIGURE 4.26 Schematic for Problem 52.

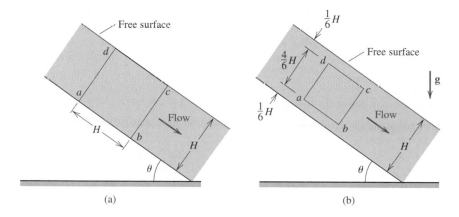

FIGURE 4.27 Schematic for Problem 53.

53. For the flow described in Problem 8, find the circulation for the closed loops shown in Figure 4.27 by using: (a) the velocity around the boundary and (b) the vorticity enclosed within the prescribed area.

54. The radial velocity distribution of an ideal viscous vortex is given by $v_r = \dfrac{C}{r}\left(1 - e^{-\frac{r^2}{4vt}}\right)$, where r is the radial distance, v is the kinematic viscosity, t is time and C is a constant. Using the definition, evaluate the circulation around a circle with radius R. From the result, evaluate the vorticity enclosed within the same circle.

55. Derive the rate of deformation tensor and the vorticity vector for the flow field described by $u = \epsilon x$, $v = -\epsilon y$ and $w = 0$, where ϵ is a constant. How do you interpret this flow?

56. Derive the rate of deformation tensor and the vorticity vector for the flow field described by $u = 1 - y$, $v = w = 0$. How do you interpret this flow?

57. For the flow described in Problem 8, find and plot (a) the vorticity as a function of time and location, (b) the relevant components of the rate of deformation. Discuss the way fluid particles deform (select 2 locations for the plots).

58. For the Non-Newtonian channel flow and conditions discussed in Problem 7, develop a general expression for the

rate of deformation tensor. Discuss the effect of the shear thinning (thickening) parameter n on the deformation of particles at the centerline $y = 0$, at $y = H/2$ and at $y = H$.

59. A velocity field is given as $u = x^2 + y$, $v = -2xy$. Determine the circulation around a square contour defined by the segments: (a) $0 \le x \le 4$, $y = 0$; (b) $0 \le y \le 4$, $x = 4$, (c) $0 \le x \le 4$, $y = 4$; $0 \le y \le 4$, $x = 0$. Verify the results by integrating the vorticity over the area of square.

60. The velocity along the height (y-direction) in a rectangular channel with $2H$ total height when the fluid is allowed to slip along the side of the channel is given by $u = 1/(2\mu)\,dP/dx(y^2 - H^2 + 2\mu H/\beta)$, where dp/dx is the pressure gradient, μ is the viscosity of the fluid, and β is known as the slip coefficient defined in terms of the shear τ_w and velocity u_w at the channel wall as $\tau_w = \beta u_w$. Describe the deformation tensor for this flow. Also describe the way a particle (a) along the centerline, (b) along the channel wall and (c) at a distance $H/2$ from the centerline, will deform under shear and extension.

61. For the velocity distribution given above find the circulation around a hypothetical square loop in the flow with its center along the centerline and side of $H/2$.

62. The velocity distribution in a channel where flow is induced by a combination of a pressure gradient dP/dx and motion

of the upper wall with velocity V is given by $u = V/a - 1/(2\mu)\,dP/dx(ay - y^2)$, where a is the height of the channel. Describe the deformation tensor for this flow. Also describe the way a particle (a) along the lower wall (b) along the upper channel wall and (c) at a distance $a/2$ from the lower wall, will deform under shear and extension.

63. For the velocity distribution given above find the circulation around a hypothetical square loop in the flow with one side along the lower wall and total side length of $a/2$.

64. Consider a channel at an angle θ with respect to the horizontal direction, where flow is induced by both a pressure gradient dp/dx and gravity g. The velocity distribution in the direction normal to the main direction is given by $u = 1/(2\mu)[\rho g \sin\theta - dP/dx](H^2 - y^2)$, where $2H$ is the height of the channel μ and ρ are respectively the viscosity and density of the fluid. Describe the deformation tensor for this flow. Also describe the way a particle (a) along the centerline, (b) along the channel wall and (c) at a distance $H/2$ from the centerline, will deform under shear and extension.

65. For the velocity distribution given above find the circulation around a hypothetical square loop in the flow with its center along the centerline and side of $H/2$.

66. The velocity distribution across the thickness (y-direction) of a thin film with total thickness δ flowing down an inclined surface with and angle θ with respect to the horizontal is $u = \rho g \sin\theta/\mu(\delta y - y^2)$. Describe the deformation tensor for this flow. Also describe the way a particle (a) along the wall, (b) along the free surface and (c) at a distance $\delta/2$ from the lower surface.

67. The velocity field near a stagnation point can be approximated by $u = Ax\mathbf{i}$ and $v = -Ay\mathbf{j}$. (a) Sketch selected streamlines, (b) developed appropriate expressions for the velocity potential function, (c) sketch selected pathlines.

68. For the velocity field $\mathbf{u} = C(x^2\mathbf{i} + z^2\mathbf{j} - 2xz\mathbf{k})$, find the rate of deformation tensor; what are the proper units of C. Discuss the way fluid particles will deform.

69. For the velocity field $\mathbf{u} = C(z^2\mathbf{i} + y^2\mathbf{j} - 2yz\mathbf{k})$, find the rate of deformation tensor; what are the proper units of C. Discuss the way fluid particles will deform.

70. For the velocity distribution given above find the circulation around a hypothetical square loop in the flow with its center along the solid surface and side of $\delta/2$.

71. Two immiscible incompressible fluids A and B with densities ρ_A and ρ_B ($\rho_A > \rho_B$) and viscosities μ_A and μ_B flow between two parallel plates. Flow is induced by one plate moving with speed V while the other plate is fixed. The velocity distribution as a function of the distance in the direction normal to the main flow direction is

$$u^A = \frac{\mu_B V}{\mu_A H_B + \mu_B H_A} y \quad 0 \geq y \geq H_A,$$

$$u^B = V - \frac{\mu_A V}{\mu_A H_B + \mu_B H_A}(H_A + H_B - y)$$

$$H_A \geq y \geq H_A + H_B,$$

where H_A and H_B the thickness of each layer. Describe the deformation tensor for this flow. Also describe the way a particle (a) along each wall, (b) along the interface free surface.

72. For the velocity distribution given above find the circulation around a hypothetical square loop in the flow with its center along the common interface surface and side of δ ($\delta < H_A, H_B$).

73. The velocity field near a stagnation point can be approximated by $u = Ax\mathbf{i}$ and $v = -Ay\mathbf{j}$. (a) evaluate the vorticity vector, (b) determine the rate of deformation tensor and discuss the way fluid particles will deform near the stagnation point.

5 Differential Form of the Conservation Laws

> *If fluid parts are continuous and uniformly distributed, then that of them which is the least compressed is driven along by that which is more compressed.*[1]
>
> —Archimedes

5.1 Differential View of the Dynamics of Flow: Internal State of Stress

The main objective in this chapter is to develop the differential form of the governing laws. Very much as in Chapter 4, the equations in this form can be used to analyze detailed fluid motions not possible with a finite control volume approach. Below we reconsider the general expression for the conservation of linear momentum for a finite control volume (Equation (5.1))

$$\underbrace{\frac{D}{Dt} \int_{\mathcal{V}} \rho \mathbf{V} d\mathcal{V}}_{Kinematics} \equiv \underbrace{\sum \mathbf{F}}_{Dynamics} \tag{5.1}$$

and focus attention on the right-hand side term of the equation. This term describes the dynamics of fluid particles, or equivalently, the forces associated with the fluid motion (kinematics) discussed in Chapter 4. To a large extent, these forces are similar to those introduced in earlier chapters such as the surface $\mathbf{f}_s$ and body $\mathbf{f}_b$ forces,

$$\sum \mathbf{F} = \underbrace{\int_A \mathbf{f}_s \, dA}_{Surface\ Forces} + \underbrace{\int_{\mathcal{V}} \mathbf{f}_b \, d\mathcal{V}}_{Body\ Forces} . \tag{5.2}$$

However, their definition must be adjusted to reflect the fact that they are associated with fluid particles and, therefore, are internal to the fluid. A new concept here then, is the *state of stress* at a point in the fluid.

Stress State at a Fluid Point: Stress Tensor

During flow, fluid particles are constantly deforming and moving under the influence of externally applied forces. For instance, recall that according to the conservation of linear momentum, motion (acceleration) can be induced from rest only by a net resultant force. By reaction, as applied forces are transmitted throughout the material, internal forces develop between fluid particles. In an exact analogy, when we squeeze a bag of marbles, individual marbles are set in motion and internal forces develop as the marbles press against each other (Figure 5.1). For convenience, internal forces are expressed in terms of stresses denoted by σ. By definition, stress is defined as the force per unit area along which the force is acting on. Consequently, stresses and forces are equivalent concepts.

---

[1] G.A. Tokaty, *A History and Philosophy of Fluid Mechanics.*

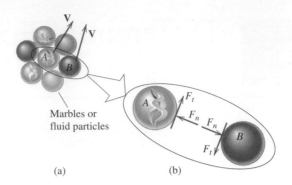

(a)   (b)

FIGURE 5.1 Interparticle force interactions between two marbles or fluid particles.

Consistent with the above physical picture, the free body diagram for an infinites-imal two dimensional fluid element is represented as shown in Figure 5.2. Notice that stresses are distinguished as *normal* and *tangential*. The subscripts xy or yx on the tangential stresses indicate respectively the face and the direction the stresses are applied to. For example, the first subscript x in σ_{xy} shows the coordinate direction (here x direction) which is normal to the face under consideration, and the second subscript y is the actual direction of the stress (here y direction).

By definition, *tensile* normal stresses are considered positive, and *compressive* normal stresses are considered negative. The *tangential stresses* are considered positive when the unit normal vector to a face, and the stresses acting on that face are either *both* pointing toward positive directions, or *both* are pointing toward negative directions. Following the definitions, all terms depicted in Figure 5.2 are positive quantities.

To guarantee static equilibrium for the free body diagram in Figure 5.2 and, therefore, to ensure that $\sum \mathbf{F} = 0$ and $\sum \mathbf{M} = 0$, where $\mathbf{F}$ and $\mathbf{M}$ are respectively the force and moment vectors, we must have

$$\sigma_{xy} = \sigma_{yx}.$$

By including the third direction, the stress state at a particular point in a three-dimensional flow is given by the tensor

$$\sigma = \begin{bmatrix} \sigma_{xx} & \sigma_{xy} & \sigma_{xz} \\ \sigma_{yx} & \sigma_{yy} & \sigma_{yz} \\ \sigma_{zx} & \sigma_{zy} & \sigma_{zz} \end{bmatrix}.$$

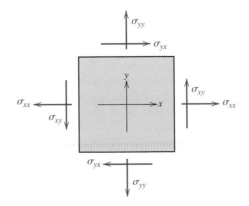

FIGURE 5.2 Stress state at a point in the fluid.

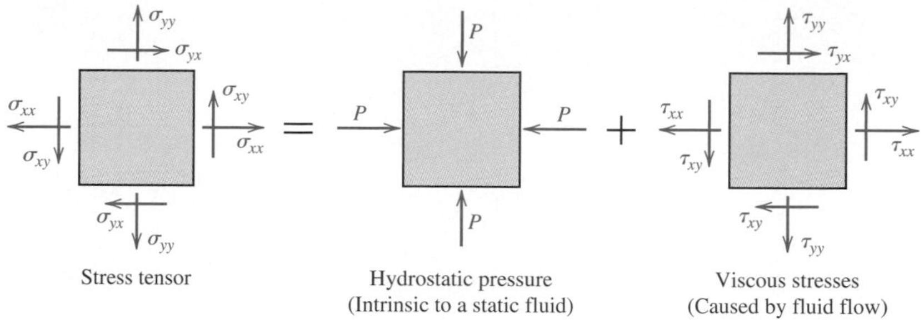

FIGURE 5.3 Stress state σ as a combination of the hydrostatic pressure P and viscous (flow) stresses τ.

Again, for static equilibrium $\sigma_{xy} = \sigma_{yx}$, $\sigma_{xz} = \sigma_{zx}$, $\sigma_{yz} = \sigma_{zy}$, which makes the stress tensor *symmetric* (i.e., off-diagonal terms are equal).

When the fluid is static, in the absence of fluid motion the fluid's internal stress state is identical to the local hydrostatic pressure which, by definition, acts in a compressive direction as shown in Figure 5.3. During flow, additional internal forces develop as a result of particle/particle and solid surface/particle friction. These flow-related stresses are known as *viscous stresses*. Therefore, the total internal *state of stress* at a point is typically represented as

$$\underbrace{\sigma}_{\textit{Total Stress}} = \underbrace{-P}_{\textit{Hydrostatic pressure}} + \underbrace{\tau}_{\textit{Stresses due to motion}} . \tag{5.3}$$

The discussion can be generalized to three dimensions using *tensor notation* as

$$\sigma = \begin{bmatrix} \sigma_{xx} & \sigma_{xy} & \sigma_{xz} \\ \sigma_{xy} & \sigma_{yy} & \sigma_{yz} \\ \sigma_{xz} & \sigma_{yz} & \sigma_{zz} \end{bmatrix} = \begin{bmatrix} -P + \tau_{xx} & \tau_{xy} & \tau_{xz} \\ \tau_{xy} & -P + \tau_{yy} & \tau_{yz} \\ \tau_{xz} & \tau_{yz} & -P + \tau_{zz} \end{bmatrix} .$$

Since pressure is a compressive force acting normal to the face of a particle, it is represented as a negative quantity along the diagonal of the stress tensor.

Stress State for a Differential Control Volume

The concept of the stress at a point is generalized to represent the stress state in the entire flow by introducing the *stress field* $\sigma = \sigma(x, y, z, t)$. This is a smooth differentiable function valid over the entire fluid domain. Since $\sigma = \sigma(x, y, z, t)$ is differentiable, the stress state in the neighborhood of a point M can be constructed using Taylor's expansion as a function of the stress at M $\sigma(M)$ as shown in Figure 5.4.

5.1.1   Body Forces on a Differential Control Volume

The best known body force is the force due to gravity. For the differential control volume shown in Figure 5.5, the infinitesimal force due to gravity $d\mathbf{F}_b$ is expressed as

$$d\mathbf{F}_b = \mathbf{g}dm = \rho\mathbf{g}dxdydz = \rho d\mathcal{V},$$

where $\mathbf{g}$ is the local acceleration vector due to gravity, $d\mathcal{V} = dxdydz$, and ρ is the local density. The body force per unit volume is

$$\frac{d\mathbf{F}_b}{d\mathcal{V}} = \mathbf{f}_b = \rho\mathbf{g}.$$

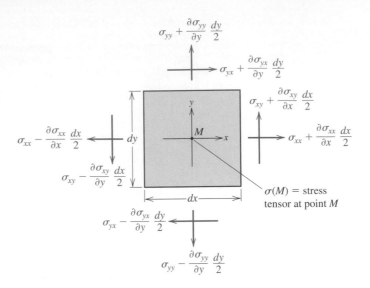

FIGURE 5.4 Stress state in the neighborhood of a fluid point M.

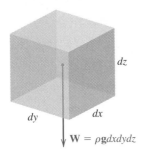

FIGURE 5.5 Body force (weight) for a differential fluid element.

Since $\mathbf{f}_b$ is a vector, it must be defined using an appropriate coordinate system. Other body forces that may be included are *electro-magnetic forces*. Because of the introductory nature of the material, however, only the gravitational forces are considered here.

5.1.2   Force on an Arbitrary Fluid Surface

The stress tensor represents the internal state of stress at a point. However, a shown in Equation (5.2) to develop the differential form of the momentum equation we need the force per unit surface area acting on an arbitrary fluid surface. Hence, for an arbitrary plane passing through a fluid point M, we need to evaluate $\mathbf{f}_s$ (Figure 5.6). This force is known as the *traction force vector*.

For static equilibrium we must have

$$\sum \mathbf{F} \equiv \mathbf{0},$$

where $\mathbf{F}$ represents all forces acting on the fluid element. As seen in the schematic, the outward unit normal vector $\mathbf{n}$ on the surface at point M is $\mathbf{n} = \cos\alpha\,\mathbf{i} + \sin\alpha\,\mathbf{j}$, and the projections of the fluid area A, along each coordinate direction are $A_x = A\sin\alpha$, and

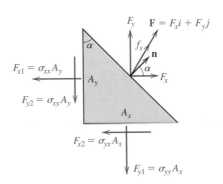

FIGURE 5.6 Traction force vector: the force per unit fluid surface area.

$A_y = A \cos \alpha$. Following the definition of the local stress tensor we have

$$F_x - F_{x_1} - F_{x_2} = 0 \Rightarrow$$

$$F_x = \underbrace{\underbrace{\sigma_{xx}}_{Stress} \underbrace{A \cos \alpha}_{Projected\ area}}_{F_{x_1}} + \underbrace{\underbrace{\sigma_{yx}}_{Stress} \underbrace{A \sin \alpha}_{Projected\ area}}_{F_{x_2}} \ ;$$

similarly,

$$F_y - F_{y_1} - F_{y_2} = 0 \Rightarrow$$

$$F_y = \underbrace{\underbrace{\sigma_{xy}}_{Stress} \underbrace{A \cos \alpha}_{Projected\ area}}_{F_{y_1}} + \underbrace{\underbrace{\sigma_{yy}}_{Stress} \underbrace{A \sin \alpha}_{Projected\ area}}_{F_{y_2}} \ .$$

The surface force per unit surface area $\mathbf{f}_s$, in condensed form is

$$\mathbf{f}_s = \frac{F_x}{A}\mathbf{i} + \frac{F_y}{A}\mathbf{j} = (\cos \alpha \mathbf{i} + \sin \alpha \mathbf{j}) \cdot \begin{bmatrix} \sigma_{xx} & \sigma_{xy} \\ \sigma_{yx} & \sigma_{yy} \end{bmatrix},$$

and in three-dimensional *vector notation*

$$\mathbf{f}_s = \mathbf{n} \cdot \sigma. \tag{5.4}$$

5.2 Conservation Laws

In this section we combine the differential view of the kinematics (motion), and dynamics (forces) of fluids to develop the differential form of the conservation of mass and momentum laws. Conservation of mass is also known as the *continuity equation*.

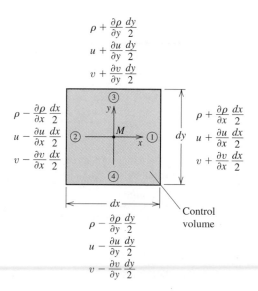

FIGURE 5.7 Flow conditions around a differential fluid element.

5.2.1   Conservation of Mass

Consider a fixed infinitesimal control volume ($\mathcal{V} \to 0$) around a point M in the flow whose velocity and density are respectively u, v, and ρ (Figure 5.7). Since flow crosses the boundary of the control volume, this choice corresponds to an open system for which the conservation of mass is written as

$$\underbrace{\frac{\partial}{\partial t} \int_{\mathcal{V}} \rho \, d\mathcal{V}}_{open\ system} + \underbrace{\int_{A} \rho \mathbf{V} \cdot \mathbf{n} \, dA}_{surface\ integral} \equiv 0. \tag{5.5}$$

The mass within the control volume is $dm = \rho d\mathcal{V} = \rho \, dx dy W$, where W is the width of the control volume. The integrals in Equation 5.5 are evaluated by assuming constant conditions along each face of the control volume as depicted in Figure 5.7 and are summarized as follows.

Face 1

Area $= dy\,W$

$\rho_1 = \rho + \dfrac{\partial \rho}{\partial x}\dfrac{dx}{2}$

$u_1 = u + \dfrac{\partial u}{\partial x}\dfrac{dx}{2}$

Outward unit normal vector $\mathbf{n} = \mathbf{i}$

Face 2

Area $= dy\,W$

$\rho_2 = \rho - \dfrac{\partial \rho}{\partial x}\dfrac{dx}{2}$

$u_2 = u - \dfrac{\partial u}{\partial x}\dfrac{dx}{2}$

Outward unit normal vector $\mathbf{n} = -\mathbf{i}$

Face 3

Area $= dx\,W$

$\rho_3 = \rho + \dfrac{\partial \rho}{\partial y}\dfrac{dy}{2}$

$v_3 = v + \dfrac{\partial v}{\partial y}\dfrac{dy}{2}$

Outward unit normal vector $\mathbf{n} = \mathbf{j}$

Face 4

Area $= dx\,W$

$\rho_4 = \rho - \dfrac{\partial \rho}{\partial y}\dfrac{dy}{2}$

$v_4 = v - \dfrac{\partial v}{\partial y}\dfrac{dy}{2}$

Outward unit normal vector $\mathbf{n} = -\mathbf{j}$

Substituting the above into Equation (5.5) we get

$$\frac{\partial \rho}{\partial t}dx\,dy\,W + \left(\rho + \frac{\partial \rho}{\partial x}\frac{dx}{2}\right)\left(u + \frac{\partial u}{\partial x}\frac{dx}{2}\right)dy\,W - \left(\rho - \frac{\partial \rho}{\partial x}\frac{dx}{2}\right)\left(u - \frac{\partial u}{\partial x}\frac{dx}{2}\right)dy\,W$$

$$+ \left(\rho + \frac{\partial \rho}{\partial y}\frac{dy}{2}\right)\left(v + \frac{\partial v}{\partial x}\frac{dy}{2}\right)dx\,W - \left(\rho - \frac{\partial \rho}{\partial y}\frac{dy}{2}\right)\left(v - \frac{\partial v}{\partial x}\frac{dy}{2}\right)dx\,W \equiv 0$$

By multiplying through, and neglecting small terms such as $\dfrac{(dx)^2}{2}$ etc., the conservation of mass becomes

$$\frac{\partial \rho}{\partial t} + \rho\frac{\partial u}{\partial x} + \rho\frac{\partial v}{\partial y} + u\frac{\partial \rho}{\partial x} + v\frac{\partial \rho}{\partial y} = 0.$$

By combining the various terms and by including the third dimension, the final expression for the conservation of mass in differential form is expressed as

$$\frac{\partial \rho}{\partial t} + \frac{\partial \rho u}{\partial x} + \frac{\partial \rho v}{\partial y} + \frac{\partial \rho w}{\partial z} \equiv 0. \tag{5.6}$$

In vector notation, Equation (5.6) is further condensed into

$$\frac{\partial \rho}{\partial t} + \nabla \cdot (\rho \mathbf{V}) = 0. \tag{5.7}$$

For steady flow, the conservation of mass reduces to

$$\nabla \cdot (\rho \mathbf{V}) = 0. \tag{5.8}$$

When the density is constant, as in *incompressible fluids*, Equation (5.7) simplifies to

$$\nabla \cdot \mathbf{V} = 0, \tag{5.9}$$

which is valid for both steady and unsteady flows. In Cartesian coordinates, the three-dimensional form of the conservation of mass for incompressible fluids is expressed as

$$\frac{\partial u}{\partial x} + \frac{\partial v}{\partial y} + \frac{\partial w}{\partial z} = 0. \tag{5.10}$$

Equation (5.6) in cylindrical coordinates with velocity components $u_r, u_\theta,$ and u_z is given as

$$\frac{\partial \rho}{\partial t} + \frac{1}{r}\frac{\partial \rho r u_r}{\partial r} + \frac{1}{r}\frac{\partial \rho u_\theta}{\partial \theta} + \frac{\partial \rho u_z}{\partial z} = 0. \tag{5.11}$$

Alternative Derivation of Mass Conservation

The conservation of mass can be alternatively derived using the approach outlined in the following text. Consider again a fixed control volume with mass crossing the boundaries as shown in Figure 5.8. For this open system, the conservation of mass is

$$\frac{\partial}{\partial t} \int_{\mathcal{V}} \rho \, d\mathcal{V} + \int_{A} \rho (\mathbf{V} \cdot \mathbf{n}) \, dA \equiv 0.$$

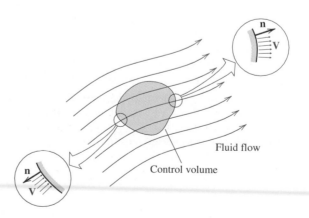

FIGURE 5.8 Fixed differential control volume in the flow.

For constant limits of integration, the time derivative can be brought inside the integral giving

$$\int_{\mathcal{V}} \frac{\partial \rho}{\partial t}\, d\mathcal{V} + \int_A \rho(\mathbf{V} \cdot \mathbf{n})\, dA = 0.$$

Recall from calculus (see Appendix E) that the Gauss divergence theorem transforms a surface integral into a volume integral according to

$$\int_A \mathbf{G} \cdot \mathbf{n}\, dA = \int_{\mathcal{V}} \nabla \cdot \mathbf{G}\, d\mathcal{V},$$

where $\mathbf{G}$ is a vector function. Therefore, by defining $\rho\mathbf{V} = \mathbf{G}$, the surface integral becomes

$$\int_A \rho(\mathbf{V} \cdot \mathbf{n})\, dA = \int_{\mathcal{V}} (\nabla \cdot \rho\mathbf{V})\, d\mathcal{V},$$

which, upon substitution into the conservation of mass, yields

$$\int_{\mathcal{V}} \left[\frac{\partial \rho}{\partial t} + \nabla \cdot (\rho\mathbf{V}) \right] d\mathcal{V} = 0. \tag{5.12}$$

Since the choice of the volume is arbitrary, the term inside the integral must be zero everywhere in the flow — that is,

$$\frac{\partial \rho}{\partial t} + \nabla \cdot (\rho\mathbf{V}) = 0.$$

This is also equivalent to reducing the size of the control volume to a mathematical point as defined by continuum theory. Note, however, that the final result is identical to the expression derived using the infinitesimal element approach, Equation (5.7).

EXAMPLE 5.1

Problem Statement   Show that the velocity field $u = 4\cos x - 3\sin y$, $v = 2y\sin x - 2xzt$, $w = 3tx + 2z\sin x$, satisfies mass conservation for a three-dimensional incompressible flow.

Governing Equations   The general form of mass conservation is given by Equation (5.6)

$$\frac{\partial \rho}{\partial t} + \frac{\partial \rho u}{\partial x} + \frac{\partial \rho v}{\partial y} + \frac{\partial \rho w}{\partial z} \equiv 0.$$

Basic Assumptions   The flow is assumed to be incompressible (i.e., $\rho = $ constant).

SOLUTION   Using the incompressibility assumption, the conservation of mass simplifies to

$$\frac{\partial u}{\partial x} + \frac{\partial v}{\partial y} + \frac{\partial w}{\partial z} \equiv 0.$$

For the given velocity field

$$\frac{\partial u}{\partial x} = -4\sin x,$$

$$\frac{\partial v}{\partial y} = 2\sin x,$$

and

$$\frac{\partial w}{\partial z} = 2 \sin x.$$

By substitution,

$$\frac{\partial u}{\partial x} + \frac{\partial v}{\partial y} + \frac{\partial w}{\partial z} = 4 \sin x - 4 \sin x = 0,$$

which shows that mass is indeed conserved.

EXAMPLE 5.2

Problem Statement   Consider an one-dimensional unsteady flow where $u = u(x, t)$. If the density ρ, changes according to $\rho = \rho_o e^{-at}$, where ρ_o is the density at some original time $t = 0$, find the velocity field.

Governing Equations   The general form of mass conservation is given by

$$\frac{\partial \rho}{\partial t} + \frac{\partial \rho u}{\partial x} + \frac{\partial \rho v}{\partial y} + \frac{\partial \rho w}{\partial z} \equiv 0.$$

Basic Assumptions   The flow is one-dimensional, $v = w = 0$.

Boundary Conditions   The velocity is assumed to be constant at all times at $x = 0$ (i.e., $u(x = 0, t) = u_0$).

SOLUTION   Using the basic assumption, mass conservation reduces to

$$\frac{\partial \rho}{\partial t} + \frac{\partial \rho u}{\partial x} = 0.$$

Therefore,

$$\frac{\partial \rho u}{\partial x} = -\frac{\partial \rho}{\partial t}.$$

But the rate of change of the density is

$$\frac{\partial \rho}{\partial t} = -a\rho_o e^{-at}.$$

Hence,

$$\frac{\partial \rho u}{\partial x} = a\rho_o e^{-at}.$$

By integration,

$$\rho u = ax\rho_o e^{-at} + C(t),$$

where $C(t)$ is a constant to be determined from the boundary condition. After rearranging, the velocity field is given by

$$u = ax + \frac{C(t)}{\rho_o} e^{at}.$$

Using the condition at $x = 0$, $u = u_0$,

$$u_0 = \frac{C(t)}{\rho_o} e^{at},$$

from which we get $C(t) = \rho_0 u_0 e^{-at}$. By substitution, the velocity is finally given by

$$u = u_0 + ax.$$

5.2.2   Conservation of Linear Momentum

The conservation of momentum for the same open system as in the schematic of Figure 5.9 can be expressed as:

$$\underbrace{\int_{\mathcal{V}} \frac{\partial \rho v_i}{\partial t}\, d\mathcal{V} + \int_A \rho v_i\, (\mathbf{V} \cdot \mathbf{n})\, dA}_{Rate\ of\ Change\ of\ Momentum} \equiv \underbrace{\sum \mathbf{F}_i}_{Resultant\ Force} = \underbrace{\int \mathbf{f}_{s_i}\, dA}_{Surface\ Forces} + \underbrace{\int \mathbf{f}_{b_i}\, d\mathcal{V}}_{Body\ Forces}, \quad (5.13)$$

where v_i, $\mathbf{f}_{s_i}$, and $\mathbf{f}_{b_i}$ are respectively the components of the velocity vector, the surface force per unit surface area, and the body force per unit volume.

The conservation of momentum can be readily derived by using either a differential control volume or the alternative approach demonstrated in Section 5.2.1. Below we use the latter approach.

The left-hand side of Equation (5.13) can be transformed into a volume integral using the Gauss divergence theorem to yield

$$\int_{\mathcal{V}} \frac{\partial \rho v_i}{\partial t}\, d\mathcal{V} + \int_A \rho v_i\, (\mathbf{V} \cdot \mathbf{n})\, dA = \int_{\mathcal{V}} \left[\frac{\partial \rho v_i}{\partial t} + \nabla \cdot (\rho v_i \mathbf{V}) \right] d\mathcal{V},$$

which, by differentiation and by using conservation of mass (Equation (5.7)), can be simplified to

$$\int_{\mathcal{V}} \rho \left(\frac{\partial v_i}{\partial t} + \mathbf{V} \cdot \nabla v_i \right) d\mathcal{V}.$$

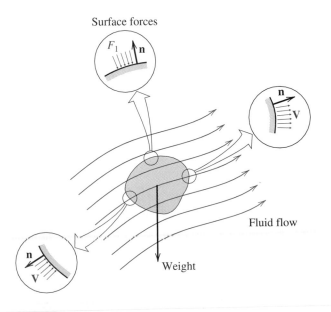

FIGURE 5.9   Fixed differential control volume in the flow.

Subsequently, by substituting $\mathbf{f}_b = \rho\mathbf{g}$ and $\mathbf{f}_s = \mathbf{n} \cdot \sigma$, as defined in Section 5.1, into Equation (5.13), we get

$$\int_{\mathcal{V}} \rho \left(\underbrace{\frac{\partial v_i}{\partial t}}_{Local\ Accel.} + \underbrace{\mathbf{V} \cdot \nabla v_i}_{Convective\ Accel.} \right) d\mathcal{V} = \int_{\mathcal{V}} \underbrace{\rho\mathbf{g}_i}_{Body\ force}\ d\mathcal{V} + \int_A \underbrace{(\mathbf{n} \cdot \sigma)_i}_{Internal\ force}\ d\mathcal{A}.$$

Applying the divergence theorem to the surface integral, we get

$$\int_{\mathcal{V}} \rho \left(\frac{\partial v_i}{\partial t} + \mathbf{V} \cdot \nabla v_i \right) d\mathcal{V} = \int_{\mathcal{V}} \rho\mathbf{g}_i\ d\mathcal{V} + \int_{\mathcal{V}} (\nabla \cdot \sigma)_i\ d\mathcal{V}.$$

Again, since the choice of volume is arbitrary, the final form of the momentum equation can be expressed as:

$$\rho\frac{Dv_i}{Dt} = \rho\left(\frac{\partial v_i}{\partial t} + \mathbf{V} \cdot \nabla v_i \right) = \rho\mathbf{g}_i + (\nabla \cdot \sigma)_i. \tag{5.14}$$

In Cartesian coordinates and by using $\sigma = -P + \iota$, the three components of the momentum equation are

$$\rho\left(\frac{\partial u}{\partial t} + u\frac{\partial u}{\partial x} + v\frac{\partial u}{\partial y} + w\frac{\partial u}{\partial z}\right) = \rho g_x - \frac{\partial P}{\partial x} + \frac{\partial \tau_{xx}}{\partial x} + \frac{\partial \tau_{xy}}{\partial y} + \frac{\partial \tau_{xz}}{\partial z}$$

$$\rho\left(\frac{\partial v}{\partial t} + u\frac{\partial v}{\partial x} + v\frac{\partial v}{\partial y} + w\frac{\partial v}{\partial z}\right) = \rho g_y - \frac{\partial P}{\partial y} + \frac{\partial \tau_{yx}}{\partial x} + \frac{\partial \tau_{yy}}{\partial y} + \frac{\partial \tau_{yz}}{\partial z} \tag{5.15}$$

$$\underbrace{\rho\left(\frac{\partial w}{\partial t} + u\frac{\partial w}{\partial x} + v\frac{\partial w}{\partial y} + w\frac{\partial w}{\partial z}\right)}_{Kinematics} = \underbrace{\rho g_z - \frac{\partial P}{\partial z} + \frac{\partial \tau_{zx}}{\partial x} + \frac{\partial \tau_{zy}}{\partial y} + \frac{\partial \tau_{zz}}{\partial z}}_{Dynamics}$$

where g_x, g_y, g_z are the components of the gravity vector with respect to the coordinate axes. The conservation of linear momentum in cylindrical, and spherical coordinates is given in Appendix C.

EXAMPLE 5.3

Problem Statement Show that the momentum equation in differential form predicts the familiar hydrostatic pressure distribution for liquid at rest. Consider that gravity is acting downwards along the negative $\mathbf{j}$ direction of a three-dimensional Cartesian coordinate system and that the depth of the liquid is H.

Governing Equations The momentum equation in Cartesian coordinates is as shown in Equation (5.15).

Basic Assumptions The fluid is assumed to be at rest, $u = v = w = 0$, with the gravity vector given by $\mathbf{g} = -g\mathbf{j}$.

Boundary Conditions The pressure at the free surface is equal to the atmospheric pressure. Therefore, $P(y = H) = P_{atm}$, where H is the height of the liquid with respect to the origin placed at the bottom of the liquid.

SOLUTION For static fluid the velocity is zero, $u = v = w = 0$. Therefore, the entire left-hand side and the viscous stresses τ on the right-hand side of Equation (5.15) are zero. Hence,

$$0 = \rho g_x - \frac{\partial P}{\partial x},$$

$$0 = \rho g_y - \frac{\partial P}{\partial y},$$

$$0 = \rho g_z - \frac{\partial P}{\partial z}.$$

For this problem, $g_x = g_z = 0$, and $g_y = -g$. Therefore,

$$\frac{\partial P}{\partial x} = 0 \Rightarrow P = P(y, z),$$

$$\frac{\partial P}{\partial z} = 0 \Rightarrow P = P(y, x).$$

The combination of the above implies that the pressure is only a function of y

$$P = P(y).$$

The component of the momentum equation in the y direction is

$$\frac{\partial P}{\partial y} = -\rho g.$$

Upon integration it gives

$$P = -\rho g y + C.$$

The constant C is found from the boundary condition $P(y = H) = P_{atm}$ as

$$P_{atm} = -\rho g H + C \Rightarrow C = P_{atm} + \rho g H.$$

Finally, the pressure distribution is obtained as

$$P = P_{atm} + \rho g (H - y).$$

But since $(H - y)$ is the depth from the free surface h, the differential form of the momentum equation predicts the same hydrostatic pressure variation as in Chapter 2

$$P = P_{atm} + \rho g h.$$

5.2.3   Differential View of Fluid Statics

According to Example 5.3, the momentum equation for a liquid at rest in two dimensions reduces to

$$\rho g_x - \frac{\partial P}{\partial x} = 0,$$

$$\rho g_y - \frac{\partial P}{\partial y} = 0, \tag{5.16}$$

where g_x, and g_y are respectively the components of the gravity vector in the x, and y directions. In vector form the foregoing equations are equivalent to

$$\frac{\partial P}{\partial x}\mathbf{i} + \frac{\partial P}{\partial y}\mathbf{j} = \rho g_x\,\mathbf{i} + \rho g_y\,\mathbf{j}.$$

In a more condensed form, the above expression is written as

$$\nabla P = \rho \mathbf{g}.$$

Consider now a two-dimensional free surface exposed to a constant pressure P_0. Here, the overall pressure distribution is a function of x and y, $P = P(x, y)$. Using the identity along the free surface

$$dP = \frac{\partial P}{\partial x}\,dx + \frac{\partial P}{\partial y}\,dy,$$

where $dP = 0$, we have,

$$\frac{\partial P}{\partial x}\,dx + \frac{\partial P}{\partial y}\,dy = 0,$$

which simplifies to

$$\frac{dy}{dx} = -\left(\frac{dP/dx}{dP/dy}\right).$$

Substituting now the result obtained in Equation (5.16), the slope of the free surface for a stationary fluid is given by

$$\frac{dy}{dx} = -\frac{g_x}{g_y}. \tag{5.17}$$

As expected, this result shows that, when gravity is acting in the vertical direction, with $g_x = 0$, and $g_y = -g$, the surface is flat ($dy/dx = 0$).

Accelerating Fluid Masses

Fluids subjected to the same acceleration throughout their mass cannot develop relative motion, and, hence, they do not deform. In such cases the entire fluid mass moves as a solid body, or is in a *rigid-body rotation*. Provided then that the acceleration of the fluid mass is taken into consideration, the equations and concepts developed for stationary fluids are applicable here as well.

By selecting the moving mass of fluid as a control volume and by prescribing a coordinate system that it is attached to and moving with it, we have a non-inertial coordinate system. The gravity term, then, must be modified to include the acceleration of the coordinate system as discussed in Chapter 2.

Therefore, if $\mathbf{a}$ is the acceleration of the mass, and $\mathbf{g}$ the local gravity, the effective gravitational acceleration is defined as

$$\mathbf{g}_{eff} = \mathbf{g} - \mathbf{a} = (g_x - a_x)\mathbf{i} + (g_y - a_y)\mathbf{j}.$$

Consequently, when the mass of fluid accelerates, the slope of the free surface is

$$\frac{dy}{dx} = -\left(\frac{g_x - a_x}{g_y - a_y}\right). \tag{5.18}$$

Consider now a reservoir filled with liquid as shown in Figure 5.10. When the reservoir is stationary, the slope of the free surface defined by

$$\frac{dy}{dx} = -\frac{g_x}{g_y} = -\left(\frac{-g\sin\alpha}{-g\cos\alpha}\right) = -\tan\alpha,$$

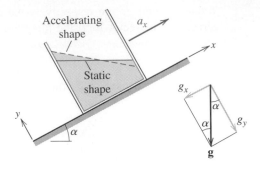

FIGURE 5.10   Mass of fluid with constant acceleration.

is parallel to the ground (i.e., it is horizontal). However, when the reservoir is accelerating in the x direction with acceleration a_x, the slope of the free surface relative to the horizontal is now given by

$$\frac{dy}{dx} = -\left(\frac{g_x - a_x}{g_y}\right) = -\left(\frac{-g\sin\alpha - a_x}{-g\cos\alpha}\right) = -\left(\frac{g\sin\alpha + a_x}{g\cos\alpha}\right),$$

which shows that it increases and, as shown in the figure, the free surface dips down at its forward edge.

EXAMPLE 5.4

Problem Statement   Consider a container filled with liquid at rest in an elevator (Figure 5.11). Find the pressure at the bottom of the container when the elevator is accelerating (a) upward, against the local gravity, with acceleration $a = 5\ m/s^2$; (b) downward, in the direction of the local gravity with $a = 5\ m/s^2$; and (c) downward, again in the direction of the local gravity with $a = g$, where g is the local gravity. Assume that the height of the water in the container is $H = 0.2\ m$.

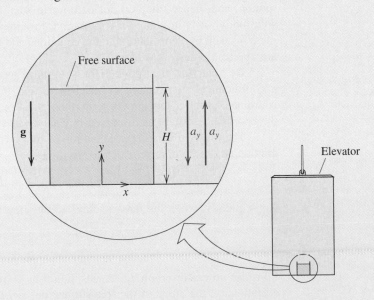

FIGURE 5.11   Container moving with constant vertical acceleration.

Governing Equations   Since the liquid moves as a solid body, the pressure distribution is given by the same equations obtained for stationary fluid conditions,

$$\frac{\partial P}{\partial x} = \rho g_{eff_x},$$

$$\frac{\partial P}{\partial y} = \rho g_{eff_y}, \qquad (5.19)$$

where $\mathbf{g}_{eff}$ is the effective gravity, corrected for the fact that the mass of fluid accelerates and moves as a rigid body with constant acceleration a_x and a_y.

$$\mathbf{g}_{eff} = (g_x - a_x)\mathbf{i} + (g_y - a_y)\mathbf{j}.$$

g_x, and g_y are the components of the local gravitational acceleration vector.

Basic Assumptions   The acceleration is acting in the general direction as does the local gravity, chosen here to coincide with the y-axis. Therefore, we have $g_x = a_x = 0$. We are assuming also that the free surface ($y = H$) is open to the atmosphere and $P(y = H) = P_{atm} = 101.3 \; kPa$.

SOLUTION   Using the assumptions, in all three cases

$$\frac{dP}{dx} = 0.$$

This implies that the pressure varies only along the y direction according to

$$\frac{dP}{dy} = \rho g_{eff_y} = \rho(g_y - a_y),$$

or,

$$P = \rho(g_y - a_y)y + C,$$

where $g_y = -g = -9.81 \; m/s^2$ and C is a constant to be determined from the boundary condition $P(y = H) = P_{atm}$. Therefore,

$$P = P_{atm} + \rho(g_y - a_y)(y - H).$$

When the fluid is at rest $a_y = 0$, the pressure at the bottom of the container $y = 0$ is

$$P = 101.3 \; kPa - \frac{998 \; kg/m^3}{1000 \; Pa/kPa} \times (-9.81 \; m/s^2) \times 0.2 \; m = 103.26 \; kPa.$$

(a) When $a_y = 5 \; m/s^2$, the pressure at the bottom of the container at $y = 0$ is

$$P = 101.3 \; kPa - \frac{998 \; kg/m^3}{1000 \; Pa/kPa} \times (-9.81 \; m/s^2 - 5 \; m/s^2) \times 0.2 \; m = 104.26 \; kPa.$$

We observe that because of the acceleration, the pressure at the bottom of the container is larger than the pressure obtained under static conditions.

(b) When $a_y = -5 \; m/s^2$, the pressure at the bottom of the container is

$$P = 101.3 \; kPa - \frac{998 \; kg/m^3}{1000 \; Pa/kPa} \times (-9.81 \; m/s^2 + 5 \; m/s^2) \times 0.2 \; m = 102.26 \; kPa.$$

In this case the resulting pressure is lower than the static pressure.

(c) When $a_y = -g \ m/s^2$, $dP/dy = 0$ and, hence, the pressure is constant, equal to atmospheric $P = P_{atm} = 101.3 \ kPa$. This shows that, when the fluid is accelerating with the same acceleration as the local gravity, the effects of gravity disappear! Obviously, this result has many practical implications, such as in creating artificial weightless environments for space applications.

EXAMPLE 5.5

Problem Statement   Consider a cylindrical container as shown in Figure 5.12, filled with liquid and rotating around the z-axis with constant angular velocity ω. Determine the shape of the free surface, and the pressure distribution inside the container.

Basic Assumptions   Again, the fluid is assumed to rotate without deforming.

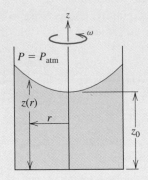

FIGURE 5.12   Mass of fluid under constant rotation.

Governing Equations   For the cylindrical geometry shown in Figure 5.12, it is more convenient to use the momentum equation expressed in cylindrical coordinates given in Appendix C. It can be easily shown that for stationary fluid in two dimensions, the momemntum equation reduces to

$$\frac{\partial P}{\partial r} = \rho g_r,$$

$$\frac{\partial P}{\partial z} = \rho g_z,$$

where g_r, and g_z are the components of the acceleration along the r and z directions respectively.

SOLUTION   In this problem, we have $g_z = -g$, and because of the rotation, $g_r = \omega^2 r$. Since the pressure along the free surface is uniform, $dP = 0$. Therefore,

$$dP = \frac{\partial P}{\partial r} \, dr + \frac{\partial P}{\partial z} \, dz = 0.$$

The slope of the free surface then is

$$\frac{dz}{dr} = -\frac{dP/dr}{dP/dz} = \frac{\omega^2 r}{g}.$$

By integrating, the height of the free surface z_s as a function of r is

$$z_s = \frac{\omega^2 r^2}{2g} + C.$$

Using the boundary condition that the minimum height of the free surface at $r = 0$, is $z_s = z_0$, we get $C = z_0$. The actual shape, then, of the free surface is

$$z_s = z_0 + \frac{\omega^2 r^2}{2g}.$$

According to

$$\frac{\partial P}{\partial z} = -\rho g,$$

the pressure in the vertical direction at any radius r is

$$P = -\rho g z + C(r),$$

where the function $C(r)$ is evaluated using the fact that, at the given radius r, the pressure at the free surface is equal to the external pressure P_0. Therefore,

$$P_0 = -\rho g\, z_s(r) + C(r),$$

or

$$C(r) = P_0 + \rho g \left(z_0 + \frac{\omega^2 r^2}{2g} \right).$$

The pressure distribution then is obtained as

$$P = P_0 + \rho g(z_0 - z) + \rho \frac{\omega^2 r^2}{2}.$$

At this point, the constant z_0 is still unknown. In practical problems, z_0 can be evaluated from the total volume of fluid V_c in the container by using

$$V_c = \int_0^R 2\pi r z\, dr \Rightarrow z_0 = \frac{V_c}{\pi R^2} - \frac{\omega^2 R^2}{4g},$$

where R is the radius of the container. An interesting question here is to find the maximum rotational speed ω_{max} to make $z_0 = 0$. From the above relation we get

$$\omega_{max} = \sqrt{4 \frac{V_c}{\pi} \frac{g}{R}}.$$

5.2.4 Conservation of Mechanical Energy

Consistent with the definition of mechanical work, the conservation of mechanical energy is obtained from the scalar product of the momentum equation Equation (5.14) with the velocity vector

$$\mathbf{V} \cdot \left(\rho \frac{D\mathbf{V}}{Dt} \right) \equiv \mathbf{V} \cdot (\rho \mathbf{g} \mid \nabla \cdot \sigma),$$

which, after simplification, it reduces to

$$\underbrace{\rho \frac{D}{Dt}\left(\frac{V^2}{2}\right)}_{\text{Change of Kinetic energy}} \equiv \underbrace{\rho \mathbf{V} \cdot \mathbf{g}}_{\text{Change of potential energy}} + \underbrace{\mathbf{V} \cdot (\nabla \cdot \sigma)}_{\text{Change of mechanical work}}. \tag{5.20}$$

The conservation statement relates the rate of change of mechanical work to the rate of change of the kinetic and potential energies. The equation, however, *does not* include the thermal energy. As discussed in Chapter 2, conservation of thermal energy is included in the first law of thermodynamics.

Since the conservation of mechanical energy is derived from the momentum equation, the two equations are obviously not independent. For analysis, then, we can use either the conservation of momentum or the conservation of mechanical energy, but not both.

5.3   Constitutive Relations

The natural laws govern the material's internal response to external effects such as forces and heat transfer. As expected, different materials respond differently to these effects: water flow is quantitatively different from air flow, and heat transfer in wood is quite different from heat flow in metal. This type of material behavior is introduced into the theory using *constitutive relations*. These are experimentally based relations which, by means of material constants, introduce the material behavior into the conservation laws. The need for substance-specific constitutive relations is by no means a violation of the universality of the conservation laws; the laws of nature are universal statements about how physical phenomena, and not material constants, are related to each other.

The need for constitutive relations and its implications can be shown mathematically as well by considering the conservation of mass and momentum for a general isothermal flow

$$\frac{\partial \rho}{\partial t} + \nabla \cdot (\rho \mathbf{V}) \equiv 0,$$

$$\rho\left(\frac{\partial v_i}{\partial t} + \mathbf{V} \cdot \nabla v_i\right) \equiv \rho \mathbf{g}_i + (\nabla \cdot \sigma)_i.$$

The proceding four equations — three components of the momentum equation and the continuity equation — are in terms of thirteen unknowns (ρ, u, v, w, σ_{xx}, σ_{xy}, σ_{xz}, σ_{yy}, σ_{yz}, σ_{zz}, σ_{yx}, σ_{zx}, σ_{zy}). Since there are more unknowns than equations, the problem as formulated has no solution. The introduction of constitutive relations achieves the necessary *closure*, resulting then in the same number of equations as unknowns.

The development of the mathematical form of the conservation principles, as demonstrated in Section 5.2, is rather straightforward. The quantitative generalization of the material behavior, however, is not as easy, and at present remains one of the most critical (if not the most critical) issue in modeling fluid flow. On one hand, this difficulty is due to the complexity of the behavior of real fluids and the uncertainty in describing them. On the other hand, however, some physical phenomena are not as well understood and, therefore, cannot be adequately described. This is the case of turbulent flow, which at present may be the single most difficult issue facing fluid dynamicists.

Laminar vs. Turbulent Flow

Many fluid systems under certain critical flow conditions become unstable. Beyond these critical conditions, the flow develops unsteady random fluctuations, resulting in a highly

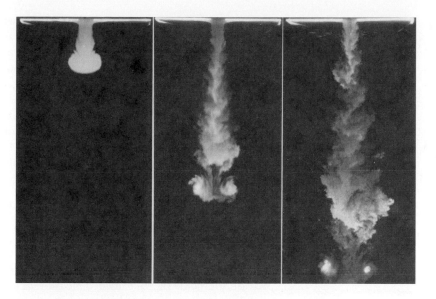

FIGURE 5.13 Developement of a starting turbulent jet at 0.27, 0.6, and 1.0 seconds after initiation of the flow. The flow was visualized by the laser-induced fluorescence technique where the jet fluid contained a fluorescence dye and a laser sheet was used to illuminate the jet cross section. Note the initially smooth interface in the first image, and the convulated turbulent interfaces in the next two time instants. The starting vortex ring of the jet is clearly visible in these images. [Courtesy of H. Johari].

recirculating flow characterized by eddies that have a wide range of sizes. Under these conditions, the flow is no longer *laminar* and, after a transition, it becomes *turbulent*. Figure 5.13 shows a typical flow structure of a turbulent jet. Contrary to turbulent flow, under laminar conditions small perturbations of the flow field decay, thus resulting in a stable flow field. Because of random fluctuations and other complicated turbulent phenomena, it is impossible to constitutively describe turbulent flow behavior with absolute certainty.

The understanding of turbulence and its theoretical description remains the most important challenge for fluid dynamicists. Given its importance, turbulent phenomena have been studied extensively for a number of problems, producing a substantial body of information. However, despite the volume of information produced and the number of available theories, there is no single theoretical framework that can explain all of the phenomena observed in experiments.

Laminar Flow: Newtonian Fluids

Flow is characterized by the continuous deformation of fluid particles and a corresponding internal state of stress. As argued previously, the precise relationship between the two depends on the material properties. For instance, honey and water are different fluids; under the influence of the same external force, they deform and, hence, flow differently. Obviously then, the corresponding stress distribution within each fluid is different.

However, any relationship between deformation and stresses must be general and applicable to all possible flow situations. In reality, the definition of the various forms of deformation and stresses may allow for cases, where either deformation, or stresses can exist without the other. For instance, while the stress distribution in liquid at rest is finite and equal to the hydrostatic pressure, the fluid particles are not moving, and, hence, they are not deforming. Therefore, the pressure and fluid deformation cannot be related to each other through a general material property. Similarly, when fluids are subjected to the same acceleration throughout their mass, fluid particles cannot develop relative motion, and, hence, they do not deform. The entire fluid mass then moves as a rigid body. This is the case, for instance, of a bucket filled with liquid on a rotating platform (e.g., Example 5.2.3). Therefore, since fluids can rotate without deforming, these two also cannot be related to each other by a general material property.

By excluding the pressure and the rate of rotation, the above observations indicate that only the remaining terms, such as the rate of deformation Δ and the viscous stress tensor τ, can be "constitutively" related to each other. Indeed, consistent with the preceding physical arguments, it is well *established experimentally* that under laminar conditions, the rate of deformation for many fluids is proportional to the local viscous stress state

$$Viscous\ Stresses = \mu(Rate\ of\ Deformation). \tag{5.21}$$

The proportionality constant μ, which has units of $kg/m\ sec$, is the *v*iscosity, a characteristic property of the fluid. A more general definition for μ is that

Viscosity is a measure of how easily the fluid deforms or flows under an applied stress. Alternatively, viscosity is a measure of the fluid's internal resistance to deformation or the internal friction between fluid layers.

For most fluids the viscosity is a function of temperature, and the viscosity typically decreases with increasing temperature. Fluids for which the viscosity at constant temperature remains constant, independently of τ and Δ, are known as *Newtonian*. Typical Newtonian fluids are water and air. Equation (5.21) is known as "Newton's law of viscosity." Again, it must be emphasized that this is a constitutive relation and not a law. The characterization as a law is due to the wide acceptance and applicability of the relation.

Viscosity is the single most important fluid property, and it is responsible for a number of important fluid phenomena. The ratio $\nu = \dfrac{\mu}{\rho}$ in units of m^2/s is known as the *kinematic viscosity*. Experimentally, μ is measured using specially designed *viscometers*, whose operation is based on simple flow between one plate fixed and another moving with constant velocity V. If the gap H between the plates is small compared to the length of the plates, the flow can be assumed with reasonable accuracy to be one-dimensional. For this type of flow, the velocity is constant in the direction of flow but varies in the normal direction. For instance, if the flow is in the x direction, the velocity is a function of y, $u = u(y)$. The reader can easily show that under these conditions the constitutive relation reduces to the expression used, without derivation, in Chapter 1

$$\tau_{xy} = \mu\frac{\partial u}{\partial y} = \mu\frac{V}{H}.$$

According to the above relation, the viscosity can be determined by measuring the necessary force per unit area required to maintain the plate moving with constant velocity V. As discussed in Chapter 12, most other viscometers operate on similar principles.

In an arbitrary, three-dimensional laminar flow, the relation is generalized as

$$\begin{bmatrix} \tau_{xx} & \tau_{xy} & \tau_{xz} \\ \tau_{xy} & \tau_{yy} & \tau_{yz} \\ \tau_{xz} & \tau_{yz} & \tau_{zz} \end{bmatrix} = \mu \begin{bmatrix} \left(2\dfrac{\partial u}{\partial x}\right) & \left(\dfrac{\partial u}{\partial y}+\dfrac{\partial v}{\partial x}\right) & \left(\dfrac{\partial u}{\partial z}+\dfrac{\partial w}{\partial x}\right) \\ \left(\dfrac{\partial u}{\partial y}+\dfrac{\partial v}{\partial x}\right) & \left(2\dfrac{\partial v}{\partial y}\right) & \left(\dfrac{\partial w}{\partial y}+\dfrac{\partial v}{\partial z}\right) \\ \left(\dfrac{\partial u}{\partial z}+\dfrac{\partial w}{\partial x}\right) & \left(\dfrac{\partial w}{\partial y}+\dfrac{\partial v}{\partial z}\right) & \left(2\dfrac{\partial w}{\partial z}\right) \end{bmatrix},$$

or, in more condensed form, as

$$\tau = \mu\Delta. \tag{5.22}$$

Laminar Flow Generalized Newtonian Fluids

The viscosity of a large number of fluids at constant temperature is not constant. In the simplest case, the viscosity is a function of the rate of deformation that is, $\mu = \mu(\Delta)$. Fluids exhibiting such behavior are known as *non-Newtonian*. Figure 1.14 shows typical results from a simple viscometric τ vs. Δ experiment for non-Newtonian fluids. As shown in the figure, for some fluids the viscosity is decreasing with increasing rate of deformation. In this case, the fluids are known as *shear thinning*. Paint, for instance, is a typical shear thinning fluid; the faster one moves the brush, the easier paint flows. When the viscosity is increasing with rate of deformation, the fluid is *shear thickening*. Typical shear thickening fluids are two-phase mixtures, such as semisolid materials, drilling fluids, mud, etc.

The behavior of these fluids, known as *Generalized Newtonian Fluids* or simply as *non-Newtonian* fluids, is expressed as

$$\tau = \mu(\Delta)\Delta. \tag{5.23}$$

Many non-Newtonian fluids can be described using a "power law" constitutive relation. For a one-dimensional problem, this relation reduces to

$$\tau_{xy} = \underbrace{\mu\left(K\left|\frac{\partial u}{\partial y}\right|\right)^{n-1}}_{\text{Effective Viscosity}}\frac{\partial u}{\partial y}. \tag{5.24}$$

K is a constant appended for consistency in the units. Comparing this relation to that of a Newtonian fluid, the term that multiplies the velocity gradient is essentially the effective viscosity of the fluid. Unlike Newtonian fluids, the effective viscosity changes according to local flow conditions. The power-law designation relates to the exponent $(n-1)$ in the effective viscosity. Depending on the magnitude of n, a number of behaviors such as shear thinning ($n < 1$) or shear thickening ($n > 1$) can be described. The Newtonian behavior is recovered when $n = 1$.

For multi-dimensional problems, the power law is expressed in terms of the second invariant of Δ, II, which is defined as

$$II = \Delta_{xx}^2 + \Delta_{yy}^2 + \Delta_{zz}^2 + 2\Delta_{xy}^2 + 2\Delta_{xz}^2 + 2\Delta_{yz}^2.$$

In such a case,

$$\tau = \mu(K|\Delta|)^{n-1}\Delta,$$

where $|\Delta| = \sqrt{0.5\,II}$. The justification for using this particular expression may be beyond the level of this introductory material. However, the discussion here and the solution to a typical non-Newtonian flow problem in Chapter 7 are included for completeness.

Turbulent Flow: Reynolds Stresses

In fluid flow, there are two competing physical phenomena: (1) the convection of fluid, which is directly related to the velocity, and (2) the resistance to flow as expressed by the viscous stresses and the viscosity of the fluid. For each effect, we can identify characteristic time scales that indicate the time associated with physical events related to each effect. For example, events associated with the convective velocity U and a characteristic length of the flow geometry L, on average, take place in $t_\alpha = L/U$ seconds. In other words, it takes a fluid particle t_α seconds to travel a distance L with velocity U. Similarly, events associated with viscosity μ are characterized by the time $t_\upsilon = L^2/\nu$. This follows from the definition $\tau_{xy} = \mu\partial u/\partial y$, by substituting the estimates $\tau_{xy} \approx \rho L^2/t_\upsilon^2$ and $\partial u/\partial y \approx 1/t_\upsilon$.

A nondimensional quantity known as the Reynolds number is defined as the ratio of these two length scales $Re = t_\upsilon/t_\alpha = UL/\nu$. At low Re, where t_υ is small, the effects of viscosity dominate the flow; any disturbances in the flow are quickly dissipated by fluid friction (i.e., viscosity). The viscosity, therefore, acts to stabilize the flow. At high *Re*, the opposite is true, and the flow is dominated by the effects of convection. In such case, as the flow becomes unstable, disturbances develop and grow before viscosity can dissipate them. This point can be easily demonstrated by considering water flow from a faucet; when the faucet is wide open, the velocity of the jet is high, resulting in a chaotic and, hence, turbulent flow. Evidence of turbulence in the flow is the relatively high noise of the jet. When the faucet is closed slowly, the velocity of the water decreases and the flow becomes progressively more ordered and, hence, quieter.

Using the faucet analogy again, when the faucet is partially closed, the reverse sequence of events take place; the flow starts as laminar, it progressively undergoes a series of changes, and it eventually becomes fully turbulent. For a general flow then, as *Re* increases, flow changes from laminar, undergoes a transition, and finally becomes fully turbulent.

The eddies in fully turbulent flow have a wide range of sizes. The large eddies depend on the geometry of the flow for, instance, the diameter of the faucet in the above example. The small-scale eddies, however, unlike the larger ones, appear to have a universal structure, independent of the geometric configuration of the problem. The "universality" ideas were exploited by Kolmogorov with the concept of an energy cascade; the energy is stored in eddies with characteristic dimensions that are related to each other by a constant factor. The energy stored in such eddies is dissipated there by viscosity. Using Kolmokorov's postulate, the smallest length scale η can be related to the rate of dissipation per unit mass ϵ and the kinematic viscosity through

$$\eta \approx \left(\frac{\nu^3}{\epsilon}\right)^{\frac{1}{4}}.$$

Furthermore, if *isotropy* and *universality* are assumed, this information can be eventually related to the velocity field. However, the universal picture is not quite complete since turbulent phenomena can behave in unusual ways not predicted by *universal* arguments. For example, near solid surfaces turbulence appears to be intermittent as *turbulent bursts* appear occasionally.

The classical approach to theoretically describe turbulent flow is to decompose the flow properties into locally time-averaged (e.g., $\bar{u}$, $\bar{P}$, $\bar{\rho}$) and timewise fluctuating components, u', P', ρ',

$$u = \bar{u} + u_i' \quad v = \bar{v} + v_i' \quad w = \bar{w} + w_i' \quad P = \bar{P} + p' \quad \rho = \bar{\rho} + \rho'.$$

The average quantities are obtained over a sufficiently long time that they are independent of time. According to this averaging, the average of the fluctuating components is equal to zero. However, this is not true for the product of the fluctuations. Using standard averaging concepts, the averaged governing equations are expressed as

$$\nabla \cdot \bar{\mathbf{V}} = 0. \tag{5.25}$$

$$\rho\left(\frac{D\bar{V}}{Dt}\right) = -\nabla\bar{P} + \mu\nabla^2\bar{\mathbf{V}} \underbrace{-\rho\left[\nabla \cdot \overline{v_i' v_j'}\right]}_{Reynolds\ Stresses}. \tag{5.26}$$

The additional stresses, known as Reynolds stresses, that result from the turbulent fluctuations encompass all phenomena associated with turbulence. Moreover, these stresses represent the essence of the difficulties associated with the modeling of turbulence because their behavior is not completely understood. The literature on this issue is rich, and numerous books and scientific studies are devoted to this topic. From a practical standpoint, turbulence can be both advantageous and harmful. Turbulence can promote mixing, enhance heat transfer, and delay the onset of separation. It can, however, also lead to undesired effects such as energy loss due to excessive dissipation.

Within the context of this book, it is impossible to satisfactorily review all relevant issues related to turbulence; therefore, only basic themes are presented.

5.4 Flow Boundary Conditions

The governing equations, along with the appropriate constitutive relations (whenever available), describe fully the fluid flow within a given geometry. However, the mathematical model cannot be solved unless we specify the boundary conditions for the problem. Physically, the need for boundary conditions is not surprising, because the boundary conditions provide information on the way the fluid interacts with its surroundings. Mathematically, this is expected, because upon integration, the differential equations yield constants that are fixed using boundary conditions.

5.4.1 No-Slip Boundary Condition

In the vast majority of problems, we have fluids flowing along solid boundaries. Consistent with continuum theory, the *no-slip* boundary condition states that there is no relative motion between fluid particles and the solid boundaries with which they are in contact — that is,

$$\mathbf{V}_{fluid} = \mathbf{V}_{wall}. \tag{5.27}$$

If the wall is not moving, then the fluid velocity at the wall is

$$\mathbf{V}_{fluid} = 0. \tag{5.28}$$

Despite its apparent simplicity, the no-slip boundary condition leads to some physical inconsistencies that are not yet resolved completely. For instance, the no-slip condition cannot explain the motion of a liquid interface in contact with a solid boundary; according to this condition, the liquid interface in a partially filled glass must remain stationary with respect to the glass when the glass is moved! At the macroscopic level at least, we know that this is not the case. Consider also the flow near an exit, as shown in Figure 5.14. According to the no-slip condition, the velocity of fluid particles near the exit is mathematically singular as the velocity changes instantaneously from zero to a finite value.

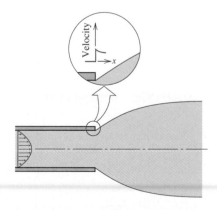

FIGURE 5.14 Velocity singularity at the exit of a die.

The physical inconsistencies can be removed mathematically by assuming that fluid particles in contact with solid surfaces *slip*. This means that fluid particles in contact with the wall can slide, resulting in an apparent fluid velocity. As a matter of fact, there is strong evidence that this occurs in many non-Newtonian fluids. In such cases, the *slip velocity* $\mathbf{V}_s$ is assumed to be proportional to the shear along the wall.

$$\mathbf{V}_s \propto \tau_w \Rightarrow \mathbf{V}_s = \lambda \tau_w, \tag{5.29}$$

where λ is a proportionality constant to be determined experimentally. However, for most Newtonian fluids there is no real experimental evidence to support the presence of slip.

EXAMPLE 5.6

Problem Statement   Fully developed, laminar flow of a Newtonian, incompressible fluid in a two-dimensional rectangular channel is given by the general parabolic expression

$$u = a\,y^2 + b\,y + c,$$

where $a, b,$ and c are constants to be determined by the boundary conditions (see Chapter 7 for the derivation). Determine the actual velocity distribution by considering (a) no-slip and (b) slip boundary conditions.

Governing Equations   For the case of slip boundary condition

$$u_s = \lambda\,\tau_w,$$

where u_s is the velocity of the fluid along the contacting surface, τ_w is the local wall shear stress, and λ an appropriate experimental constant.

Basic Assumptions   The flow is assumed to be symmetric around the centerline, with the maximum velocity u_0 occurring along that line, as shown in Figure 5.15.

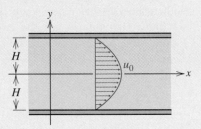

FIGURE 5.15   Fully developed flow in a two-dimensional channel.

SOLUTION   Because of the symmetry in the flow,

$$\left.\frac{\partial u}{\partial y}\right|_{y=0} = 0,$$

which implies that $b = 0$. Moreover, since the maximum velocity occurs at $y = 0$, $c = u_0$, the velocity then takes the form

$$u = u_0 + a\,y^2. \tag{5.30}$$

When the no-slip condition $u = 0$ at $y = \pm H$ is applied, we get $a = -u_0/H^2$, which gives

$$u = u_0 \left(1 - \frac{y^2}{H^2} \right).$$

u_0 can be related either to the total volumetric flow rate or to the pressure gradient responsible for inducing this flow. These issues are explored fully in Chapter 7.

In the case of finite slip, the wall shear stress τ_w is given by

$$\tau_w = -\tau_{xy} = -\mu \frac{\partial u}{\partial y},$$

where τ_{xy} is the shear stress on the fluid along the wall. Therefore,

$$u_s = -\lambda \tau_{xy} = -\lambda \mu \left. \frac{\partial u}{\partial y} \right|_{y=H}.$$

According to Equation (5.30)

$$u_s = u_0 + aH^2, \quad \text{and} \quad \left. \frac{\partial u}{\partial y} \right|_{y=H} = 2ay = 2aH.$$

Therefore,

$$u_0 + aH^2 = -\lambda \mu 2 a H,$$

or

$$a = - \left(\frac{u_0}{H^2 + 2\lambda \mu H} \right).$$

The velocity profile is given by

$$u = u_0 \left(1 - \frac{y^2}{H^2 + 2\lambda \mu H} \right).$$

When $\lambda = 0$, we recover the profile obtained using the no-slip velocity boundary condition.

5.4.2 Applied Surface Forces

In many applications, an amount of fluid is squeezed by an external force (Figure 5.16). In injection molding, for instance, plastic material in molten state is compressed by a press into a die to form the final part. For such problems, the boundary condition at the interface is obtained by considering the balance of forces there

$$\mathbf{f}_a = \frac{d\mathbf{F}}{dA} = \mathbf{n} \cdot \sigma, \tag{5.31}$$

where $\mathbf{f}_a$ is the externally applied force $\mathbf{F}$ per unit surface area A. $\mathbf{n} = n_x \mathbf{i} + n_y \mathbf{j} + n_z \mathbf{k}$ is the outward pointing unit normal vector, and σ is the local stress tensor defined by $\sigma = -P + \tau$, with τ being the viscous stress tensor. The total applied force $\mathbf{F}$ is recovered

FIGURE 5.16   External force applied to a fluid domain.

by integration over A as

$$\mathbf{F} = \int_A \mathbf{f}_a dA = \int_A \mathbf{n} \cdot \sigma dA.$$

In Cartesian coordinates, the boundary condition in component form is expressed as

$$f_{a_x} = n_x(-P + \tau_{xx}) + n_y \tau_{yx} + n_z \tau_{zx},$$
$$f_{a_y} = n_x \tau_{xy} + n_y(-P + \tau_{yy}) + n_z \tau_{zy},$$
$$f_{a_z} = n_x \tau_{xz} + n_y \tau_{yz} + n_z(-P + \tau_{zz}).$$

FIGURE 5.17   Special cases of externally applied forces: (a) normal, compressive force, (b) tangential, shear force.

Special Cases

When a force is applied in the normal direction $\mathbf{f}_{a_y} = -f_{a_y}\mathbf{j}$, as shown in Figure 5.17a and $\mathbf{n} = \mathbf{j}$, the boundary condition reduces to

$$-f_{a_y} = -P + \tau_{yy} \quad \text{or,} \quad f_{a_y} = P - 2\mu\frac{\partial v}{\partial y}.$$

When a shear force is applied in the horizontal direction $\mathbf{f}_{a_x} = f_{a_x}\mathbf{i}$, as shown in Figure 5.17b and $\mathbf{n} = \mathbf{j}$, the boundary condition reduces to

$$f_{a_x} = \tau_{yx} = \mu\left(\frac{\partial u}{\partial y} + \frac{\partial v}{\partial x}\right).$$

EXAMPLE 5.7

Problem Statement   Consider a balloon with an initial diameter of 0.2 *m* subjected to a squeezing force $\mathbf{F}_s$ applied by a press as shown in Figure 5.18. If $\mathbf{F}_s = 40$ *N* and the contact area between the press and the balloon at the deformed state is a circle with 0.08 *m* diameter, find the pressure inside the balloon.

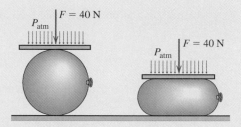

FIGURE 5.18 Balloon deformed by an applied compressive force.

Governing Equations The conservation of mass Equation (5.6) and momentum Equation (5.14) in differential form describe fully this problem.

Boundary Conditions At the press, the boundary condition is given as

$$f_{a_x} = n_x(-P + \tau_{xx}) + n_y\tau_{yx} + n_z\tau_{zx},$$

$$f_{a_y} = n_x\tau_{xy} + n_y(-P + \tau_{yy}) + n_z\tau_{zy},$$

$$f_{a_z} = n_x\tau_{xz} + n_y\tau_{yz} + n_z(-P + \tau_{zz}),$$

where P is the pressure, τ the viscous stress tensor, and $\mathbf{n}$ the outward pointing unit normal vector.

Basic Assumptions Assume that at the final deformed state, the gas inside the balloon is stationary, $\mathbf{V} = \mathbf{0}$, and the gravitational effects are negligible.

SOLUTION Under static conditions, the conservation of mass $\nabla \cdot \mathbf{V} = 0$ is satisfied automatically. Also, since $\mathbf{V} = \mathbf{0}$, the momentum equation in each coordinate direction reduces to the simple expressions,

$$\frac{\partial P_g}{\partial x} = 0, \quad \frac{\partial P_g}{\partial y} = 0, \quad \frac{\partial P_g}{\partial x} = 0,$$

which, upon integration, show that the pressure within the balloon is constant

$$P_g = P_g(x, y, z) = \text{constant}.$$

The externally applied force per unit surface area is due to the force on the press and the ambient atmospheric pressure,

$$f_{a_y} = \frac{\mathbf{F}_a}{A_{press}} = -\left(\frac{F_{press}}{A_{press}} + P_{atm}\right)\mathbf{j}.$$

At the press surface, the outward unit normal vector is $\mathbf{n} = \mathbf{j}$. Additionally, since the gas inside the balloon is static, $\tau = 0$. The boundary condition then

$$f_{a_y} = n_x\tau_{xy} + n_y(-P + \tau_{yy}) + n_z\tau_{zy},$$

at the press simplifies to

$$f_{a_y} = -P_g,$$

where P_g is the pressure inside the balloon. By substitution,

$$-\left(\frac{F_{press}}{A_{press}} + P_{atm}\right) = -P_g,$$

or, by using a standard value for the atmospheric pressure,

$$P_g = 101.3\ kPa + \frac{4 \times 0.04\ kN}{\pi \times 0.08^2\ m^2} = 109.26\ kPa.$$

5.4.3 Surface Tension Forces

Surface tension is a surface force that develops at the interface between two immiscible fluids and lies in the plane of the surface. Surface tension forces are expressed as

$$d\mathbf{F} = -\kappa \beta \mathbf{n}\, dA, \tag{5.32}$$

where β is the surface energy per unit area, κ is the local curvature, $\mathbf{n}$ is the outward pointing unit normal vector, and A the area. The curvature is considered positive when the surface is concave. If the interface is part of a circular arc with radius R,

$$d\mathbf{F} = -\frac{\beta}{R}\mathbf{n}\, dA.$$

This boundary condition is treated in the same manner as previously with the applied surface forces Equation (5.31).

5.4.4 Free Surface Boundary Conditions

The dynamics of free surfaces are very important in fluid mechanics. The evolution of these boundaries depends on the internal material dynamics, and the external influence from the surrounding material. A free surface is described by two boundary conditions a kinematic condition that signifies that the particles at the free surface move with the local fluid velocity, and a dynamic condition that expresses force balance at the surface.

The schematic in Figure 5.19 shows a typical free surface. During deformation, the height of the free surface h moves with the vertical fluid velocity v as

$$\frac{Dh}{Dt} = v.$$

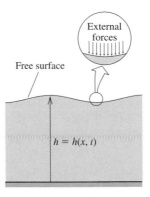

FIGURE 5.19   Free surface boundary.

By considering a stationary observer the *kinematic condition* is expressed as

$$\frac{\partial h}{\partial t} + u\frac{\partial h}{\partial x} + w\frac{\partial h}{\partial z} = v. \tag{5.33}$$

The *dynamic condition* expresses force balance at the free surface

$$\frac{d\mathbf{F}_a}{dA} = f_t\mathbf{t} + f_n\,\mathbf{n} = \mathbf{n}\cdot\sigma, \tag{5.34}$$

where $\mathbf{t}$ and $\mathbf{n}$ represent unit vectors in the tangential and normal directions respectively, and f_t and f_n the components of $\mathbf{f}$ along these directions. This expression is treated in exactly the same way as elaborated in Section 5.4.2. For instance, if the only external force per unit area is the atmospheric pressure P_{atm}, we have

$$\mathbf{n}\cdot\sigma = -P_{atm}\,\mathbf{n}, \tag{5.35}$$

or in case of a tangential force f_t, we write

$$f_t = (\mathbf{n}\cdot\sigma)\cdot\mathbf{t}. \tag{5.36}$$

When no external forces are applied, we have the "traction free" boundary condition

$$\mathbf{n}\cdot\sigma = \mathbf{0}. \tag{5.37}$$

5.5 Navier-Stokes Equations

The equations of motion for general isothermal laminar flow of a Newtonian incompressible fluid were originally derived by Navier (1785–1836) and then by Stokes (1819–1903). The complete derivation of these equations was the culmination of centuries of philosophical, empirical, and scientific observation and discussion of fluid phenomena. A partial list of famous contributors to the theoretical foundations of fluid mechanics include Aristotle, Archimedes, Hero of Alexandria, Leonardo da Vinci, Evangelista Torricelli, Blaise Pascal, Isaac Newton, Daniel Bernoulli, Leonhart Euler, Louis Lagrange, Jean le Rond d' Alembert, and many others. These equations form the basis for studying most fundamental phenomena in fluid flow.

In Cartesian coordinates $\mathbf{V} = \mathbf{V}(x, y, z, t)$, the equations are given as

Continuity

$$\frac{\partial u}{\partial x} + \frac{\partial v}{\partial y} + \frac{\partial w}{\partial z} = 0. \tag{5.38}$$

x-momentum

$$\rho\left(\frac{\partial u}{\partial t} + u\frac{\partial u}{\partial x} + v\frac{\partial u}{\partial y} + w\frac{\partial u}{\partial z}\right) = \rho g_x - \frac{\partial P}{\partial x} + \mu\left[\frac{\partial^2 u}{\partial x^2} + \frac{\partial^2 u}{\partial y^2} + \frac{\partial^2 u}{\partial z^2}\right].$$

y-momentum

$$\rho\left(\frac{\partial v}{\partial t} + u\frac{\partial v}{\partial x} + v\frac{\partial v}{\partial y} + w\frac{\partial v}{\partial z}\right) = \rho g_y - \frac{\partial P}{\partial y} + \mu\left[\frac{\partial^2 v}{\partial x^2} + \frac{\partial^2 v}{\partial y^2} + \frac{\partial^2 v}{\partial z^2}\right] \tag{5.39}$$

z-momentum

$$\rho\left(\frac{\partial w}{\partial t} + u\frac{\partial w}{\partial x} + v\frac{\partial w}{\partial y} + w\frac{\partial w}{\partial z}\right) - \rho g_z - \frac{\partial P}{\partial z} + \mu\left[\frac{\partial^2 w}{\partial x^2} + \frac{\partial^2 w}{\partial y^2} + \frac{\partial^2 w}{\partial z^2}\right].$$

In vector notation, the equations are expressed as

$$\nabla \cdot \mathbf{u} = 0, \tag{5.40}$$

$$\rho \left(\frac{\partial u_i}{\partial t} + \mathbf{u} \cdot \nabla u_i \right) = \rho g_i - \nabla P + \mu \nabla^2 u_i. \tag{5.41}$$

The preceding equations are obtained by assuming constant ρ and by substituting the constitutive relations $\tau = \mu \Delta$ into the continuity and momentum equations. The Navier-Stokes equations in cylindrical and spherical coordinates are provided in Appendix C.

EXAMPLE 5.8

Problem Statement   *Couette flow: Flow between two parallel plates.* Consider fluid between two long parallel plates of width W. A tangential force per unit area $\tau_{xy} = f_t$ causes the upper plate to move to the right while the lower plate is fixed. This flow is known as *Couette flow.* Determine the induced velocity as shown in the Figure 5.20.

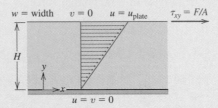

FIGURE 5.20   *Couette flow.* Flow induced by a shear force.

Governing Equations   For this two-dimensional problem, the flow is governed by the conservation of mass,

$$\frac{\partial u}{\partial x} + \frac{\partial v}{\partial y} = 0,$$

and momentum

$$\rho \left(\frac{\partial u}{\partial t} + u \frac{\partial u}{\partial x} + v \frac{\partial u}{\partial y} \right) = \rho g_x - \frac{\partial P}{\partial x} + \frac{\partial \tau_{xx}}{\partial x} + \frac{\partial \tau_{xy}}{\partial y},$$

$$\rho \left(\frac{\partial v}{\partial t} + u \frac{\partial v}{\partial x} + v \frac{\partial v}{\partial y} \right) = \rho g_y - \frac{\partial P}{\partial y} + \frac{\partial \tau_{xy}}{\partial x} + \frac{\partial \tau_{yy}}{\partial y}.$$

Assumptions   We assume steady state, laminar flow of a Newtonian fluid driven by shear, and negligible gravity effects.

Constitutive Relations   For a Newtonian fluid under laminar conditions,

$$\tau = \mu \Delta.$$

Boundary Conditions   Because of the no-slip boundary condition, at $y = 0$ we have $u = v = 0$. At $y = H$, $v = 0$, and $\tau_{xy} = f_t$.

SOLUTION   Since no flow crosses the plate, the vertical velocity is zero ($v = 0$) everywhere. The velocity vector then is $\mathbf{V} = u\,\mathbf{i}$. For this flow, the continuity equation predicts

$$\frac{\partial u}{\partial x} + \frac{\partial v}{\partial y} = 0 \Rightarrow \frac{\partial u}{\partial x} = 0 \Rightarrow u = u(y).$$

As the flow is induced by shear and not by pressure, $\dfrac{dP}{dx} = \dfrac{dP}{dy} = 0$. Additionally, $\tau_{xx} = \tau_{yy} = 0$ and $v = w = 0$. Finally, for steady flow and negligible gravity effect, the two-dimensional momentum equations reduce to

$$\frac{\partial \tau_{xy}}{\partial y} = 0 \Rightarrow \tau_{xy} = \text{Constant}.$$

Since the shear stress at the top plate is equal to the applied stress f_t,

$$\tau_{xy} = f_t.$$

The above verifies that the externally applied stress is transmitted throughout the extent of the fluid. As a result of this stress, the fluid deforms continuously in a manner consistent with shear deformation,

$$\Delta_{xy} = \frac{\partial u}{\partial y}.$$

For a Newtonian fluid,

$$\tau_{xy} = \mu \Delta_{xy} = \mu \frac{\partial u}{\partial y}.$$

By applying the boundary condition,

$$f_t = \mu \frac{\partial u}{\partial y},$$

after integration it gives

$$u = \frac{f_t}{\mu} y + C_1.$$

The constant $C_1 = 0$ is obtained from the boundary condition $u = 0$ at $y = 0$. The velocity distribution, then, within the gap is

$$u = \frac{f_t}{\mu} y.$$

The upper plate (at $y = H$) moves with velocity u_{plate}

$$u_{plate} = \frac{f_t}{\mu} H.$$

Using $f_t/\mu = u_{plate}/H$, the velocity u can be expressed also as

$$u = u_{plate} \frac{y}{H}.$$

This demonstrates the fact that the applied shear stress f_t induces internal stresses (τ_{xy}), which continuously deform the fluid (Δ). This continuous deformation is the resulting fluid flow (u).

The same problem could be posed in the reverse order by considering the plate moving to the right with velocity u_{plate}. Using the geometry and physics of the problem, the momentum equation then yields the same result (left to the reader to verify)

$$u = u_{plate} \frac{y}{H}.$$

Because of the motion of the plate and the induced deformation, the shear within the flow is a constant given by

$$\tau_{xy} = \mu \frac{\partial u}{\partial y} = \mu \frac{u_{plate}}{H} = f_t.$$

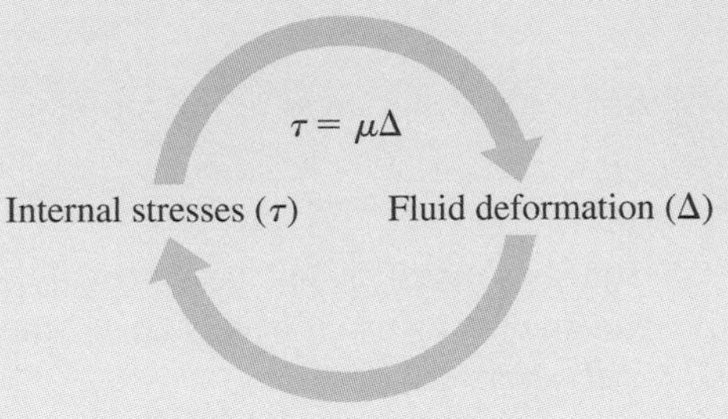

$$\tau = \mu \Delta$$

Internal stresses (τ)   Fluid deformation (Δ)

FIGURE 5.21 Cyclic relationship between fluid deformation and internal fluid stresses.

The above arguments clearly indicate the cyclic relationship between the shear stresses and the rate of deformation (Figure 5.21).

5.6 Non-Isothermal Flows[2]

In non-isothermal problems, the conservation of mass and momentum laws must be augmented by including the conservation of energy law.

5.6.1 Conservation of Total Energy

The conservation of total energy (thermal and mechanical energy) or the first law of thermodynamics is expressed as

$$\underbrace{\int_{\mathcal{V}} \frac{\partial \rho e}{\partial t} d\mathcal{V} + \int_{A} \rho e \mathbf{V} \cdot \mathbf{n} dA}_{\textit{Rate of Change of Energy}} \equiv \underbrace{\int \dot{Q} \, dA}_{\text{Rate of Heat Transfer}} + \underbrace{\int \dot{W}_s dA + \int \dot{W}_b d\mathcal{V}}_{\textit{Rate of Work due to all applied Forces}},$$

(5.42)

where $e = \varepsilon + \dfrac{V^2}{2}$, with ε being the internal energy per unit mass and V the magnitude of the velocity vector. Heat transfer $\dot{Q}$ is expressed in terms of the heat flux vector per unit area $\mathbf{q}$ as $\dot{Q} = -\mathbf{q} \cdot \mathbf{n}$. The negative sign is added for physical consistency, because $\mathbf{q} \cdot \mathbf{n}$ is a positive quantity when the flux is leaving the volume. The rate of work due to surface forces $\dot{W}_s$ is given by $\dot{W}_s = \mathbf{n} \cdot (\sigma \cdot \mathbf{V})$. The rate of work due to body forces $\dot{W}_b$ is expressed as $\rho \mathbf{g} \cdot \mathbf{V}$. Using the same procedure as in Sections 5.2.1 and 5.2.2, the energy equation reduces to

$$\rho \frac{D}{Dt} \left(\varepsilon + \frac{V^2}{2} \right) \equiv -\nabla \cdot \mathbf{q} + \nabla \cdot (\sigma \cdot \mathbf{V}) + \rho \mathbf{g} \cdot \mathbf{V}.$$

(5.43)

---

[2]This section can be skipped without loss of continuity.

5.6.2 Conservation of Thermal Energy

The conservation of thermal energy is easily derived by subtracting the mechanical energy (Equation (5.20)) from the total energy (Equation (5.43)).

$$\underbrace{\rho \frac{D\varepsilon}{Dt}}_{\text{Rate of Internal Energy Change}} = \underbrace{-\nabla \cdot \mathbf{q}}_{\text{Diffusion of Heat}} \underbrace{- P\nabla \cdot V}_{\text{Flow Work}} + \underbrace{\tau : \nabla\mathbf{V}}_{\text{Viscous Dissipation}} \,.$$

However, $d\varepsilon = c\,dT$, where c_v, is the specific heat coefficient at constant volume and T the temperature. For a liquid $c_v \approx c_p \approx c$. Then, by assuming constant c

$$\rho c \frac{DT}{Dt} = -\nabla \cdot \mathbf{q} - P\nabla \cdot V + \tau : \nabla\mathbf{V}.$$

For an incompressible liquid $\nabla \cdot V = 0$, the conservation of thermal energy reduces to

$$\rho c \frac{DT}{Dt} = -\nabla \cdot \mathbf{q} + \tau : \nabla\mathbf{V}. \tag{5.44}$$

In Cartesian coordinates, the energy equation for an incompressible liquid is given as

$$\rho c \left(\frac{\partial T}{\partial t} + u\frac{\partial T}{\partial x} + v\frac{\partial T}{\partial y} + w\frac{\partial T}{\partial z} \right) = -\left(\frac{\partial q_x}{\partial x} + \frac{\partial q_y}{\partial y} + \frac{\partial q_z}{\partial z} \right) + \Phi_v, \tag{5.45}$$

where Φ_v is the heat generated due to the action of the viscous forces.

$$\Phi_v = \tau_{xx}\frac{\partial u}{\partial x} + \tau_{yy}\frac{\partial v}{\partial y} + \tau_{zz}\frac{\partial w}{\partial z} + \tau_{xy}\left(\frac{\partial u}{\partial y} + \frac{\partial v}{\partial x} \right) + \tau_{yz}\left(\frac{\partial v}{\partial z} + \frac{\partial w}{\partial y} \right) + \tau_{zx}\left(\frac{\partial w}{\partial x} + \frac{\partial u}{\partial z} \right). \tag{5.46}$$

5.6.3 Constitutive Description of Heat Transfer

The conservation of thermal energy, Equation (5.44), addresses directly the phenomenon of heat transfer in liquids: the local energy storage ($\rho c \frac{\partial T}{\partial t}$) is determined by local transfer of heat, by means of *convection* ($\mathbf{V} \cdot \nabla T$), *conduction* ($-\nabla \cdot \mathbf{q}$), and *heat generation*, which may be due to viscous dissipation ($\tau : \nabla\mathbf{V}$), electromagnetic heating, etc.

However, Equation (5.44) remains incomplete since the intrinsic mechanism by which heat conducts has not been defined. This can be achieved using "Fourier's Law" of conduction

$$\mathbf{q} = -\kappa\nabla T, \tag{5.47}$$

or, in vector form,

$$\mathbf{q} = -\kappa \left(\frac{\partial T}{\partial x}\mathbf{i} + \frac{\partial T}{\partial y}\mathbf{j} + \frac{\partial T}{\partial z}\mathbf{k} \right), \tag{5.48}$$

where κ is the thermal conductivity, and ∇T is the temperature gradient. The conductivity is an experimentally determined property and shows the relative ease by which heat conducts in a particular material. The negative sign is inserted to denote the fact that heat conducts in the direction of decreasing temperature. Again, "Fourier's Law" is not a law but a constitutive relation. Similar to Newton's law of viscosity, the characterization of this relation as as law is due to its wide acceptance and applicability.

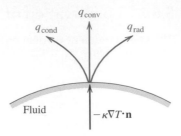

FIGURE 5.22 Thermal boundary conditions.

The conservation of thermal energy equation in Cartesian coordinates is

$$\rho c \left(\frac{\partial T}{\partial t} + u \frac{\partial T}{\partial x} + v \frac{\partial T}{\partial y} + w \frac{\partial T}{\partial z} \right) = \kappa \left(\frac{\partial^2 T}{\partial x^2} + \frac{\partial^2 T}{\partial y^2} + \frac{\partial^2 T}{\partial z^2} \right)$$

$$+ \tau_{xx} \frac{\partial u}{\partial x} + \tau_{yy} \frac{\partial v}{\partial y} + \tau_{zz} \frac{\partial w}{\partial z} + \tau_{xy} \left(\frac{\partial u}{\partial y} + \frac{\partial v}{\partial x} \right)$$

$$+ \tau_{yz} \left(\frac{\partial v}{\partial z} + \frac{\partial w}{\partial y} \right) + \tau_{zx} \left(\frac{\partial w}{\partial x} + \frac{\partial u}{\partial z} \right).$$

5.6.4 Thermal Boundary Conditions

A physical system in contact with fluids can exchange heat by conduction, convection, or by radiation as shown in Figure 5.22. By defining everything external to the flow as the "surroundings," we are able to describe the appropriate boundary conditions at the interface, where heat arrives (or leaves) at a rate of $\mathbf{q}$ per unit area.

When heat arrives at or leaves the fluid by conduction, we have

$$\mathbf{q} = -\kappa_f \nabla T \cdot \mathbf{n}, \tag{5.49}$$

and by convection,

$$\mathbf{q} = h_c (T - T_\infty), \tag{5.50}$$

where κ_f is the conductivity, T is the temperature of the fluid at the boundary, and T_∞ the ambient temperature. h_c is the heat transfer coefficient.

When fluid exchanges heat by radiation with a body at temperature T_∞, we have

$$\mathbf{q} = \sigma \epsilon F_{12} (T^4 - T_\infty^4), \tag{5.51}$$

where σ is the Stefan-Boltzmann constant, T_1 is again the temperature of the fluid, ϵ is the emissivity, and F_{12} is a geometric factor known as the *view factor*.

REFERENCES

G.A. TOKATY, *A History and Philosophy of Fluid Mechanics*, New York, Dover, 1994.

R.B. BIRD, W.E. STEWART, and E.N. LIGHTFOOT, *Transport Phenomena*, New York, Wiley, 1960.

R.B. BIRD, R.C. ARMSTRONG, and O. HASSAGER, *Dynamics of Polymeric Liquids*, 2nd Ed., New York, Wiley-Interscience, 1987.

R.L. PANTON, *Incompressible Flow*, New York, Wiley, 1984.

Y.S. TOULOUKIAN, S.C. SAXENA, and P. HESTERMANS, "Thermophysical Properties of Matter," The TPRC Data Series, Vol. **11** — Viscosity, New York, Plenum Press, 1975.

PROBLEMS

1. A surface force per unit area of $1.35\ kN/m^2$ is acting the fluid surface as shown in Figure 5.23. Find the stress state of a particle at the surface.

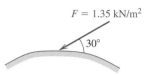

$F = 1.35\ \text{kN/m}^2$

$30°$

FIGURE 5.23 Schematic for Problem 1.

2. A surface force per unit area of $760\ N/m^2$ is acting on a fluid surface as shown in Figure 5.24. Find the stress state of a particle at the surface.

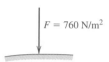

$F = 760\ \text{N/m}^2$

FIGURE 5.24 Schematic for Problem 2.

3. A surface force per unit area of $2.2\ kN/m^2$ is acting on a fluid surface as shown in Figure 5.25. For each case, find the stress state of a particle at the surface.

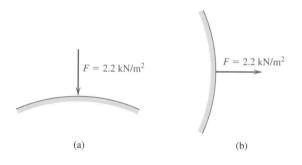

$F = 2.2\ \text{kN/m}^2$ $F = 2.2\ \text{kN/m}^2$

(a)   (b)

FIGURE 5.25 Schematic for Problem 3.

4. Describe the local stress state 12 m below the surface of a fresh water lake. What is the rate of deformation tensor? Are fluid particles deforming?

5. What are the stresses that a submarine 100 m below the free surface will experience? What is the rate of deformation tensor? Are fluid particles deforming?

6. The stress state at a point in the flow in Cartesian coordinates is given by $\sigma_{xx} = 80\ Pa$, $\sigma_{xy} = \sigma_{yx} = 160\ Pa$, $\sigma_{yy} = 0$. Express the state in tensor form and then calculate the force per unit area acting on a plane: (a) aligned with the x-axis (b) aligned with the y-axis and (c) at 45° with respect to the horizontal direction.

7. The stress state at a point in the flow in Cartesian coordinates is given by $\sigma_{xx} = 60\ Pa$, $\sigma_{xy} = \sigma_{yx} = 0$, $\sigma_{yy} = -50\ Pa$. Express the state in tensor form and then calculate the force per unit area acting on a plane: (a) aligned with the x-axis (b) aligned with the y-axis and (c) at 45° with respect to the horizontal direction.

8. The stress tensor at a point $\mathbf{r} = \sqrt{3}\mathbf{j}$ in the flow is given by

$$\mathbf{T}\begin{bmatrix} 2 & 1 & 0 \\ 1 & 4 & 1 \\ 0 & 1 & 0 \end{bmatrix}$$

(a) Find the traction force per unit area $\mathbf{f} = \mathbf{n} \cdot \mathbf{T}$ on a surface whose outward unit normal vector is $1/\sqrt{3}(\mathbf{i}+\mathbf{j}+\mathbf{k})$.
(b) Find the moment $\mathbf{M} = \mathbf{r} \times \mathbf{f}$ produce by $\mathbf{f}$.

9. Given the stress state at a point in the flow is Cartesian coordinates given by $\sigma_{xx} = 1000\ Pa$, $\sigma_{xy} = \sigma_{yx} = 0$, $\sigma_{yy} = 1000\ Pa$: (a) find the force per unit area acting on a plane passing through this point as a function of the angle the plane makes with respect to the horizontal, (b) find the plane direction where the shear stress is maximum.

10. Develop an expression for the local state of stress when the fluid in a shear free flow develops a uniform force given by $1500\ N\mathbf{i} - 800\ N\mathbf{j}$ on an inclined surface 120° degrees with respect to the horizontal and total area of $0.1\ m^2$.

11. The stress state at a point in a general flow in Cartesian coordinates is given by $\sigma_{xx} = 750\ kPa$, $\sigma_{xy} = \sigma_{yx} = 1500\ kPa$, $\sigma_{yy} = -2750\ kPa$, $\sigma_{xz} = \sigma_{zx} = 2200\ kPa$, $\sigma_{zz} = 1250\ kPa$. If the local hydrostatic pressure is $250\ kPa$, determine the local viscous stress tensor.

12. Which expressions may represent steady fluid flow with constant density: (a) $u = x$, $v = 3y$, $z = -4z$, (b) $u = \cos \alpha x$, $v = \sin \alpha y$, $w = Az$, (c) $u = axy$, $v = ayz$, $w = azx$, (d) $u = -e^x$, $v = yz$, $w = ze^x$.

13. Find the simplest expression for the density variation in the one dimensional, steady state flow with velocity $u = \dfrac{20}{5 - 4x^{3.15}}\ m/s$.

14. Find the simplest expression for the velocity for an one-dimensional problem with density variation given by $\rho = Axe^{-\alpha t}$ where A and α are constants and t is time.

15. Consider the functions $u = 3x^2 + 5$ and $v = Ay^2 + 3$. Find the constant A for which u, v could be functions representing 2D incompressible flow.

16. Which expressions represent 2D incompressible flow? (a) $u = u_o \cos(\omega x)$, $v = u_o \sin(\omega x)$, (b) $u = u_o \cos(\omega x)$, $v = u_o \cos(\omega x)$, (c) $u = u_o \cos(\omega x)$, $v = u_o y \sin(\omega x)$, (d) (a) $u = u_o \sin(\omega y)$, $v = u_o \sin(\omega y)$.

17. Find the simplest expression for the density variation in the one dimensional, steady state flow with velocity $u = 3.5e^{1.2x}\ m/s$.

18. For the steady flow of incompressible fluid where $u = 2x$, $v = 3y$ and $w = az$, find the constant a.

19. In steady, two-dimensional flow the component of the velocity in the x-direction is given by $u = 2(x^2 - y^2)$. Find an appropriate expression for the other component of the velocity.

20. For the steady, two-dimensional incompressible flow with axial velocity field $u = 2(x + 2)$, find the simplest expression for the velocity in the other direction.

21. Show that the sum of the diagonal terms of the viscous stress tensor for an incompressible fluid is zero.

22. Using the one dimensional differential element shown in Figure 5.26 develop the proper expression for the conservation of linear momentum in one dimension.

23. Show that the solution, $\dfrac{dP}{dx} = -10\ Pa/m$, $v = 0$, and $u = \dfrac{1}{2\mu}\dfrac{dP}{dx}[y^2 - (\dfrac{H}{2})^2]$ m/s satisfies the conservation laws (i.e conservation of mass and linear momentum) for laminar flow through a two dimensional channel with height $H = 0.2\ m$. P is pressure, and y is the distance measured from the centerline.

24. If the maximum velocity in laminar flow of a Newtonian fluid through a two dimensional channel is $1.5\ m/s$, using the conservation of linear momentum find the pressure gradient to sustain this flow. The velocity distribution for this flow is $v = 0$, $u = 3.2[1 - (\dfrac{2y}{H})^2]$ m/s, with $H = 0.2\ m$, and y is the distance measured from the centerline.

25. Show that the solution, $\dfrac{dP}{dx} = 0$, $v = 0$, and $u = \dfrac{1}{\mu}\rho g \sin\alpha$ $(Hy - \dfrac{y^2}{2})$ satisfies the conservation laws (i.e., conservation of mass and linear momentum) for laminar flow down an inclined surface due to gravity g with angle $\alpha = 30°$ degrees with respect to the horizontal direction, depth of $0.4\ m$, and ρ is the density, and μ the viscosity of the fluid.

26. Show that the solution, $\dfrac{dP}{dz} = -8\ Pa/m$, $v = 0$, and $u = \dfrac{1}{4\mu}\dfrac{\partial P}{\partial z}(r^2 - R^2)$ m/s satisfies the conservation laws in cylindrical form (given in Appendix C) for laminar flow through a pipe with radius $R = 0.1\ m$. P is pressure, μ is the viscosity, and r is the radial distance.

27. Show that the solution, $\dfrac{dP}{dx} = -12$ Pa/m, $v = 0$, and $u = u_{max}[1 - (\dfrac{2y}{H})^{\frac{n+1}{n}}]$ m/s satisfies the conservation laws (i.e., conservation of mass and linear momentum) for laminar flow of a Power-Law Non-Newtonian fluid through a two dimensional channel with height $H = 0.2\ m$, and $n = 1.2$.

28. If the maximum velocity in laminar flow of a Power-Law Non-Newtonian fluid through a two dimensional channel is $2.2\ m/s$, using the conservation of linear momentum find the pressure gradient to sustain this flow. The velocity distribution for this flow is $v = 0$, and $u = u_{max}[1 - (\dfrac{2y}{H})^{\frac{n+1}{n}}]$ m/s, y is the distance measured from the centerline. Use $H = 0.18\ m$ and $n = 0.8$. Repeat the calculation for $n = 1.2$. What is the conclusion of your calculations?

29. Show that the solution, $\dfrac{dP}{dx} = \dfrac{dP}{dy} = 0$, $v = 0$, and $u = u_0(1 - erf\dfrac{y}{2\sqrt{vt}})$ satisfies the conservation laws (i.e conservation of mass and linear momentum) for laminar flow of liquid set in motion impulsively by a velocity u_0, applied to the free surface. t is time, v is the kinematic viscosity, y is the coordinate normal to the free surface and erf the error function defined as $erf\eta = \dfrac{2}{\sqrt{\pi}}\int_0^\eta e^{-\eta^2}d\eta$.

30. The streamfunction

$$\psi = V_0\left[y + \frac{(\delta(x) - y)^3}{3\delta^2(x)}\right] \quad 0 \le y \le \delta(x)$$

approximates the streamfunction in a laminar boundary layer for a flat plate, where V_0 is the velocity away from the plate, y is the coordinate direction normal to the main flow and $\delta(x)$ is the thickness boundary layer. Find the stress as a function of the distance along the plate.

31. Find the maximum acceleration the container shown in Figure 5.27, can be moved without spilling its contents.

32. Determine the acceleration with which the container shown in Figure 5.28 is accelerated.

33. Find the fastest time the container filled with liquid ($SG = 0.88$) as shown in Figure 5.29 can be accelerated from rest

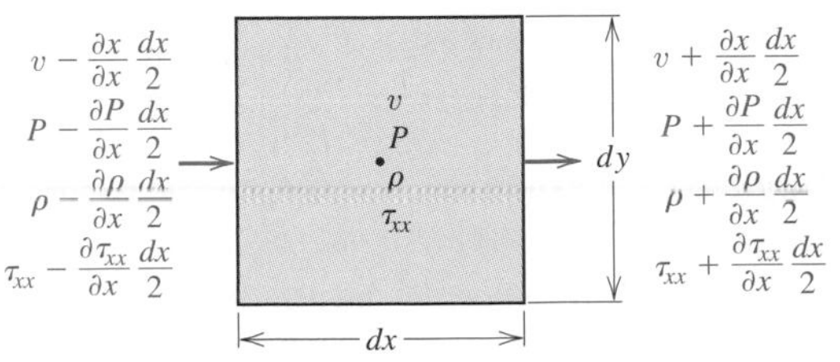

$v - \dfrac{\partial x}{\partial x}\dfrac{dx}{2}$

$P - \dfrac{\partial P}{\partial x}\dfrac{dx}{2}$

$\rho - \dfrac{\partial \rho}{\partial x}\dfrac{dx}{2}$

$\tau_{xx} - \dfrac{\partial \tau_{xx}}{\partial x}\dfrac{dx}{2}$

v
P
ρ
τ_{xx}

$v + \dfrac{\partial x}{\partial x}\dfrac{dx}{2}$

$P + \dfrac{\partial P}{\partial x}\dfrac{dx}{2}$

$\rho + \dfrac{\partial \rho}{\partial x}\dfrac{dx}{2}$

$\tau_{xx} + \dfrac{\partial \tau_{xx}}{\partial x}\dfrac{dx}{2}$

dy

dx

FIGURE 5.26 Schematic for Problem 22.

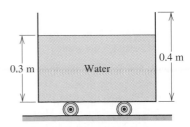

0.3 m   Water   0.4 m

FIGURE 5.27 Schematic for Problem 31.

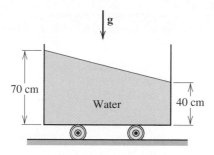

FIGURE 5.28   Schematic for Problem 32.

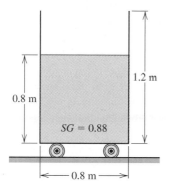

FIGURE 5.29   Schmeatic for Problem 33.

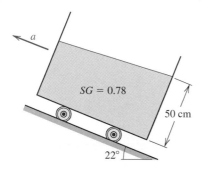

FIGURE 5.30   Schematic for Problem 34.

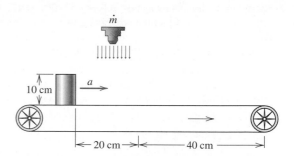

FIGURE 5.31   Schematic for Problem 36.

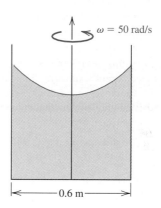

FIGURE 5.32   Schematic for Problem 37.

36. In food processing a cubic can (side $= 0.1\ m$) starting from rest is filled with soup (density $= 1200\ kg/m^3$) as shown in Figure 5.31. Calculate the maximum flow rate with which the can is filled without spilling the soup. If the process is complete when the can is at the end of the conveyor belt, find the time of the process.

37. The water in the 0.6 m-diameter cylindrical tank shown in Figure 5.32 is rotating as a rigid body at a rate of 20 rad/s. If the volume of the water is 0.38 m^3, find: (a) the shape of the free surface and (b) the pressure distribution along the bottom of the tank.

38. The free surface of water in a rotating 0.6 m-diameter cylindrical tank is displaced by 15 cm from the lowest point of 0.1 m measured from the bottom of the tank. Find: (a) the rotating speed and (b) the volume of the water in the cylinder.

39. A 0.6 m-diameter cylindrical tank with height of 0.4 m is 60% filled. Find the maximum rotating speed with which the tank can be rotated without spilling its contents.

40. The bottom of a 1 m-diameter cylindrical container is held together by 8, 1 cm-diameter bolts. If the container is filled with 1 m^3 of water, and rotating at a constant speed of 1.5 revolutions per second, find the stress on each bolt.

41. A 50 kg rigid container is dropped from a certain height. During the fall the container experiences a drag force F_d,

at a constant acceleration a, and moved 100 m without spilling the contents.

34. Find the maximum acceleration the cubic container shown in Figure 5.30, can be moved in the direction shown without spilling its contents. Assume that 65% of the container volume is filled and that the specific gravity of the liquid is 0.78.

35. Find the fastest time t, a cubic container (side $= H$) partially filled with liquid (density ρ) can be accelerated from rest at a constant acceleration a, and moved a distance L without spilling the contents. Plot the results for $H = 0.4\ m$, $L = 10\ m$ and $\rho = 850\ kg/m^3$ as a function of the original height of the liquid in the container.

proportional to its instantaneous velocity V, $(F_d = CV)$. For $C = 1.2$, find and plot the effective gravity of contents inside the container as a function of time. Find the hydrostatic pressure at the bottom of a cubic box (side $= 0.4$ m) filled with water.

42. The axial velocity distribution across the height (y-direction) in steady, laminar flow of air through a two dimensional channel is $u = 3.2[1 - (2\frac{y}{H})^2]$ m/s. By assuming a channel width $w = 1$ m, height $H = 1.$ m, and length $L = 4.$ m, find: (a) the stress state at $y = 0$, $y = H/2$ and $y = H$, (b) the total force on the walls of the channel and (c) the pressure difference across a section of the channel 0.5 m long (Hint: Select a finite control volume 0.5 m long and apply conservation of linear momentum.)

43. The axial velocity as a function of the distance across the thickness H of a film of liquid flowing down an inclined surface is given by the general expression $u = a+by+cy^2$, and $v = 0$. For a constant volumetric flow rate Q, by applying appropriate boundary conditions find: (a) the actual velocity distribution and (b) the total force induced by the fluid on the solid surface. State clearly your assumptions.

44. The velocity distribution across the thickness (y-direction) of a fluid film with density ρ flowing down an inclined surface due to gravity g is $u = \frac{1}{\mu}\rho g \sin\alpha(Hy - \frac{y^2}{2})$. If the fluid is water, angle $\alpha = 30°$ degrees with respect to the horizontal direction and depth of 0.4 m for laminar flow find: (a) the stress state at $y = 0$, $y = H/2$ and $y = H$, (b) the total force on the walls of the surface.

45. For the Couette flow described in Example 5.5, and shown in Figure 5.5 the velocity profile between the two plates is given by the general expression $u = ay + b$. Find the actual velocity distribution for a modified problem where: (a) both plates are moving to the right with the same velocity U, (b) both plates are moving to the right with the top plate moving with half the velocity of the bottom plate and (c) the top plate is moving to the right with velocity U while the bottom plate is moving to the left with the same velocity. For the above cases, find the stress state of fluid particles along the surface when $U = 5$ m/s, the distance between the plates is 0.28 m and the fluid is water.

46. Reconsider the Couette flow described in Example 5.5 by replacing the no-slip boundary condition with a slip condition as explained in Section 5.4.1. Is this possible? Explain your answer.

47. A Newtonian liquid of density ρ and viscosity μ, is bounded by two long plates separated by a distance H. The liquid is initially at rest, and motion is induced by moving the lower plate with constant velocity V, while the upper plate remains stationary. (a) Show that the Navier-Stokes equations for this problem reduces to

$$\frac{\partial u}{\partial t} = \nu \frac{\partial^2 u}{\partial y^2},$$

where u is the velocity in the direction of flow and y the direction normal to the direction of u. (b) Show that the velocity distribution

$$u = V\left(1 - \frac{y}{H}\right) - 2\frac{V}{\pi}\sum_{k=1}^{\infty}\frac{1}{k}\sin\left(\frac{k\pi y}{H}\right)e^{-\frac{k^2\pi^2\nu}{H^2}t}$$

is a solution to the problem, i.e., it satisfies both the governing equations and boundary conditions.

48. A semi-infinite $(0 \leq x < \infty)$ Newtonian fluid is set in motion by a plate at $y = 0$ oscillating with velocity $V = V_0\cos\omega t$. Show that the Navier-Stokes equations reduce to

$$\frac{\partial u}{\partial t} = \nu\frac{\partial^2 u}{\partial y^2},$$

where u is the velocity in the direction of flow and y the direction normal to the direction of u, and show that the velocity distribution

$$u(y,t) = V_0 exp\left(-\sqrt{\frac{\omega}{2\nu}}y\right)\cos\left(\omega t - \sqrt{\frac{\omega}{2\nu}}y\right),$$

satisfies both the governing equations and boundary conditions.

49. For the problem described above, find the total force on the plate per unit width, as a function of time.

50. Show that the velocity profile

$$u_x = -\frac{1}{36\mu}\frac{dP}{dx}\frac{1}{a}(\sqrt{3}z + 3y + a)$$

$$(\sqrt{3}z + 3y - a)(\sqrt{3}z - 3y - a),$$

represents steady flow of a Newtonian fluid through an infinite channel with a cross section that it is an equilateral triangle with side a. Consider the origin at the centroid of the cross-section and that the flow is in the x-direction. Also show that the volumetric flow rate Q is

$$Q = -\frac{\sqrt{3}}{320}\mu\frac{dP}{dx}a^4.$$

51. Creeping flow is defined a very slow flow such that the convective terms compared to the viscous and pressure forces of the equations of motion become small and can be neglected. Under steady state conditions demonstrate that creeping flow past a sphere of radius R is given by the velocity field $u_\phi = 0$, $u_r = \frac{U}{2}[2 - 3\frac{R}{r} + \frac{R^3}{r^3}]\cos\theta$, and

$u_\theta = \frac{U}{4}[4 - 3\frac{R}{r} - \frac{R^3}{r^3}]\sin\theta$, where U is the velocity of the flow far away from the cylinder.

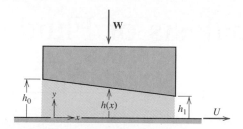

FIGURE 5.33   Schematic for Problem 54.

52. Show that the general expression for the temperature T, $T = ay + b$ satisfies the conservation of the thermal energy for a two dimensional problem with stagnant fluid between two plates held at two different temperatures T_1 and T_2. Assume that the side normal to the x-direction is insulated.

53. Using the expression $T = ay + b$, for the temperature T distribution between two infinite plates. Determine the actual temperature distribution using: (a) constant temperature T_1 and T_2 respectively at the bottom and top plates, (b) constant temperature at the bottom and heat loss/gain by convection at the top plate, (c) constant temperature at the bottom and heat loss/gain by radiation.

54. *Lubrication Theory.* In lubrication flows, fluid flows through a narrow gap of almost constant height as shown in Figure 5.33, while one surface is moving (here, the lower surface is moving with velocity U). When $(h_0 - h_1)/L \ll 1$ and $h \ll L$, (a) Show that the flow is described by

$$\mu \frac{\partial^2 u}{\partial y^2} = \frac{dP}{dx},$$

(b) using the appropriate boundary conditions find the velocity distribution, and determine the volumetric flow rate.

55. For the lubrication problem described above, use the fact the the volumetric flow rate Q is constant and, hence, $dQ/dx = 0$, to show that (a) the pressure equation satisfies

$$\frac{d}{dx}\left[\frac{h^3}{\mu}\frac{dP}{dx}\right] = -6U\frac{dh}{dx}.$$

(b) Using a linear gap profile $h(x) = h_0 - \dfrac{h_0 - h_1}{L}x$, show that the pressure distribution is

$$P(x) = P_0 + \frac{6\mu U L (h - h_0)(h_1 - h)}{(h_1^2 - h_0^2)h^2}.$$

56. Using the results from above find the load W that the lubrication fluid will support by integrating $W = \int_0^L (P - P_0)dx$.

57. Using the basic flow results for the lubrication problems described above find the total viscous force the bearing will experience by integrating $F_v = -\int_0^L \tau_w dx$, where τ_w is the shear at the bearing.

58. The Navier-Stokes equation for steady Couette flow between two plates reduces to $\partial \tau_{xy}/\partial y = 0$. Consider a narrow gap of constant height $H = 4\ cm$ filled with two immiscible fluids; water and glycerin each with equal height. If the lower plate is fixed, and the upper plate is moving to the right with velocity of $20\ m/s$, find the velocity distribution across the gap.

59. Reconsider the geometry and conditions of the previous problem. Find the velocity distribution across the gap thickness when the upper plate is moving to the right with velocity of $10\ m/s$ and the lower plate is moving to the left with velocity of $10\ m/s$.

6 Dimensional Analysis of Fluid Systems

Unlike earlier chapters, attention in this chapter is focused on a less rigorous mathematical representation of fluid flow phenomena. In this chapter, fluid systems are examined from a purely dimensional point of view. We will show that in a physical problem, dimensional arguments can help predict the relative importance of individual effects. This knowledge can be used to develop relevant assumptions with which to simplify problems. In addition, scaling and similarity arguments, which follow naturally from dimensional analysis, are important in the design of experiments and the interpretation of the results.

At the onset, mathematical models of fluid flow are valid only when the parameters are expressed in consistent dimensions and use the same system of units. The model then must be *dimensionally homogeneous* or *dimensionally consistent*. By convention, we use $[p]$ to represent the dimensions of a variable p. For example, if D is a length term such as the diameter of a pipe or cylinder, $[D]$ indicates the dimensions of D. If we use the symbols L, t, and M to denote the three primary dimensions of length, time, and mass, then $[D] = L$, which reads: "D has dimensions of length." Of course in the SI system, L is expressed in units of meters. Velocity V, is expressed as $[V] = L/T$, which again reads as "V has dimensions of length divided by time," or in units of meters per second. Note that from a dimensional point of view all parameters can be expressed in terms of the fundamental dimensions of time T, space L, mass M, and temperature Θ.

6.1 Functional Relations Established by Dimensional Analysis

Dimensional analysis is based on two simple observations:

- Since natural phenomena obey a set of universal conservation principles, the relevant variables in a problem $q_1, q_2 \ldots q_n$ must be related to each other — that is,

$$F(q_1, q_2, \ldots q_n) = 0. \tag{6.1}$$

Clearly, the exact relationship between these variables is the solution to the problem, which may be obtained by one or a combination of the methods discussed in Chapter 1.

- A number of physical variables with appropriate dimensions can be combined to form *groups of variables* that have no dimensions (i.e., they are dimensionless).

---

[1] Sir Thomas Heath, *A History of Greek Mathematics*.

Using these observations then, the solution, as sought by the functional relationship given by Equation (6.1), can now be written in the general form

$$F(\Pi_1, \Pi_2, \ldots \Pi_k), \tag{6.2}$$

where $\Pi_1, \Pi_2, \ldots \Pi_k$ are nondimensional groups formed using the variables $q_1, q_2, \ldots q_n$.

The expression given by Equation (6.2) is a significant result and it points to the important conclusion that all physical problems can be formulated and studied as a function of a group of nondimensional variables irrespective of the magnitude of individual variables. An obvious application of this observation is in experiments where, instead of obtaining results for individual variables, they can be obtained for nondimensional groups of variables. Since the groups of variables are less than the number of variables, the number of experiments, the time required to run them, and the overall experimentation cost are reduced. Additionally, dimensional analysis provides a convenient method to organize important variables providing thus a better insight to the physics of the problem.

In the following sections, we illustrate the use of dimensional analysis for a two-dimensional flow past a circular cylinder (Figure 6.1). This is a classical fluid dynamics problem, and since it represents most of the salient features of flow past bluff bodies, it usually serves as a prototype flow for such problems.

In this case, because of viscous effects, the flow separates by moving away from the cylinder, thus forming a highly unsteady vortical region behind the cylinder. This unsteady flow induces a net unsteady force $\mathbf{F}(t)$ on the cylinder, the magnitude and direction of which depend on the cylinder diameter D, the magnitude of the flow away from the cylinder V, and the viscosity μ and density ρ of the fluid. In practice, the object is to determine the relationship between the force $\mathbf{F}(t)$ and the flow parameters

$$\mathbf{F} = \mathbf{F}(V, D, \mu, \rho, t).$$

Note that dimensional analysis cannot (and does not) provide quantitative answers to a given problem and, therefore, it does not eliminate the need to actually solve the problem. Dimensional analysis, however, provides a qualitative understanding of the functional relationship between variables. In other words, the analysis makes possible to *a priori* represent the solution in terms of nondimensional parameters, in a way that both helps to obtain the solution and provides insight into the physics of the problem.

Before proceeding further, it is essential to formalize the procedure of obtaining the dimensionless groups. This formal procedure is provided by the *Buckingham-Π theorem* described in the following subsection.

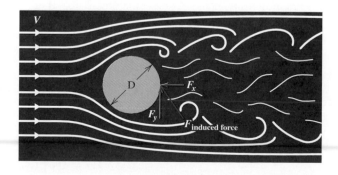

FIGURE 6.1   Schematic of a two-dimensional flow past a circular cylinder.

6.1.1 Buckingham-Π Theorem

According to the theorem, m independent physical parameters, which collectively contain n primary dimensions (such as dimensions of length L, time T, mass M, temperature Θ, etc.), can form $m - n$ nondimensional groups. To obtain these groups, we select r parameters that, taken together, contain all n primary dimensions but do not themselves form a dimensionless group. These parameters are called "repeating" parameters. The nondimensional groups are formed by combining each of the remaining $m - r$ parameters with the repeating parameters. The exact arrangement of the parameters within a group is obtained either by inspection or by using simple algebraic relations by requiring that the arrangement be nondimensional.

For most problems, the number of repeating parameters r is the same as the number of primary dimensions n. However, in some cases this may not be the case, and $r < n$. This problem arises when the repeating parameters are *dependent* (i.e., a combination of the dimensions of some of the repeating parameters yields the dimensions of another repeating parameter). For instance, in capillary rise h due to surface tension, one cannot select a length term such as the diameter of a tube $D[L]$, the specific weight of a fluid $\gamma = \rho g [M/(L^2 T^2)]$, and surface tension $\sigma [M/T^2]$ as repeating parameters because by combining D and γ $(D^2 \gamma)$, we can get the dimensions of σ. In such a case we must select as repeating parameters the minimum number of *independent* parameters (e.g., D and γ in the cited example). Additionally, since the repeating parameters appear in all nondimensional groups the dependent variables whose functional relation we wish to establish must not be chosen as repeating parameters.

EXAMPLE 6.1

The force F on a circular cylinder depends on the freestream velocity V; the diameter of the cylinder D; the density and viscosity of the fluid ρ and μ, respectively; and time t — that is, $(m = 6)$. These six parameters collectively represent dimensions of length L, time T, and mass M $(n = 3)$. In this example, we select $V[L/t]$, $D[L]$ and $\rho[M/L^3]$ as repeating parameters $(r = 3)$. This is an appropriate set for repeating parameters because together they represent all primary dimensions for the flow problem (e.g., L, T, M), and they are independent (i.e, the dimensions of one parameter cannot be obtained as a combination of the other two).

According to the Buckingham-Π theorem, we need to determine only three nondimensional groups Π_1, Π_2, and Π_3 $(m - r = 3)$ as

$$\Pi_1 = \Pi_1 \left(F, \underbrace{V, D, \rho}_{repeating\ parameters} \right),$$

$$\Pi_2 = \Pi_2 \left(\mu, \underbrace{V, D, \rho}_{repeating\ parameters} \right),$$

and

$$\Pi_3 = \Pi_3 \left(t, \underbrace{V, D, \rho}_{repeating\ parameters} \right).$$

F was not selected as a repeating parameter because it is the main variable whose functional dependence we wish to establish.

For Π_1, we write then

$$[F][V]^\alpha[D]^\beta[\rho]^\gamma = L^0 t^0 M^0,$$

where $[p]$ denotes the dimensions of variable p. Here, $[F] = ML/T^2$, $[V] = L/T$, $[D] = L$, and $[\rho] = M/L^3$. Therefore,

$$\left[\frac{ML}{T^2}\right]\left[\frac{L}{T}\right]^\alpha [L]^\beta \left[\frac{M}{L^3}\right]^\gamma = L^0 t^0 M^0.$$

Upon collecting terms pertaining to each fundamental dimension, we get

$$\begin{aligned} \text{For } M: \quad & 1 + \gamma && = 0 \\ \text{For } T: \quad & -2 - \alpha && = 0 \\ \text{For } L: \, & 1 + \alpha + \beta - 3\gamma && = 0. \end{aligned}$$

Solving the preceding equations, we get $\gamma = -1$, $\alpha = -2$, and $\beta = -1 - \alpha + 3\gamma = -1 + 2 - 3 = -2$. The nondimensional group then is

$$\Pi_1 = \frac{F}{\rho V^2 D^2}.$$

For the second nondimensional group Π_2, we write

$$[\mu][V]^\alpha[D]^\beta[\rho]^\gamma = L^0 t^0 M^0.$$

But $[\mu] = M/LT$; therefore

$$\left[\frac{M}{LT}\right]\left[\frac{L}{T}\right]^\alpha [L]^\beta \left[\frac{M}{L^3}\right]^\gamma = L^0 t^0 M^0,$$

which, upon collecting terms, gives

$$\begin{aligned} \text{For } M: \quad & 1 + \gamma && = 0 \\ \text{For } T: \quad & -1 - \alpha && = 0 \\ \text{For } L: \, & -1 + \alpha + \beta - 3\gamma && = 0. \end{aligned}$$

Solving the preceding equations we get $\gamma = -1$, $\alpha = -1$, and $\beta = 1 - \alpha + 3\gamma = 1 + 1 - 3 = -1$. The nondimensional group then is

$$\Pi_2 = \frac{\mu}{\rho V D}.$$

Since Π_2 is nondimensional, its inverse $\dfrac{\rho V D}{\mu}$ is also a nondimensional group, which is the well-known Reynolds number (Re). As shown in Chapter 8, Re is a very important parameter in fluid mechanics.

By inspection, we easily determine that for Π_3

$$\Pi_3 = \frac{Vt}{D}.$$

Therefore, from simple dimensional arguments we conclude (without solving the problem) that the nondimensional force is a function of the Reynolds number and time.

$$\frac{F}{\rho V^2 D^2} = f\left(Re, \frac{Vt}{D}\right).$$

Of course, the actual functional relationship is still unknown and can be established only by solving the problem.

This result is, however, significant, because the simple rearrangement of the variables *a priori* identifies the *Re* as the most significant parameter. The *Re* number then helps guide the solution procedure; below a critical value, the flow is laminar—hence, offering the possibility for computational and/or experimental solutions. Above this critical value, the flow becomes turbulent, and most likely, the solution will have to be obtained experimentally.

In practice, the steady drag and lift forces for bluff bodies of various shapes are indeed expressed as

$$\frac{F_D}{\frac{1}{2}\rho V^2 D^2} = C_D,$$

and

$$\frac{F_L}{\frac{1}{2}\rho V^2 D^2} = C_L,$$

where C_D and C_L are experimentally determined coefficients which depend on Re.

EXAMPLE 6.2

An open-topped tank with diameter D is filled with a liquid with density ρ and viscosity μ, to a height h. Under the influence of gravity g, the tank is drained through a hole with diameter d (Figure 6.2). Using dimensional analysis, find the time t it takes to empty the tank.

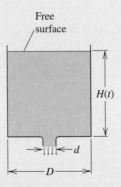

FIGURE 6.2   Schematic of a draining tank.

In this problem, we have seven parameters t, D, d, h, ρ, μ, g ($m = 7$) that collectively represent three primary dimensions, length L, time T, and mass M ($n = 3$). By selecting the independent parameters ρ, D, and g as repeating parameters ($r = 3$), according to the Buckingham-Π theorem, we can establish four independent nondimensional groups $(m - r = 4)$ as

$$\Pi_1 = \Pi_1 \left(h, \underbrace{\rho, D, g}_{repeating\ parameters} \right), \quad \Pi_2 = \Pi_2 \left(d, \underbrace{\rho, D, g}_{repeating\ parameters} \right),$$

$$\Pi_3 = \Pi_3 \left(t, \underbrace{\rho, D, g}_{repeating\ parameters} \right), \quad \Pi_4 = \Pi_4 \left(\mu, \underbrace{\rho, D, g}_{repeating\ parameters} \right).$$

By observation, the first two parameters are simply $\Pi_1 = \dfrac{h}{D}$ and $\Pi_2 = \dfrac{d}{D}$; in some cases, a simple arrangement of the variables is sufficient to form the groups, and the repeating parameters method is not necessary. For instance, in forming Π_1 and Π_2, there is no way to combine either ρ or g with any other variable to yield a nondimensional group (mass M and time T do not appear in any other variable). Therefore, ρ and g do not appear in Π_1 and Π_2.

For Π_3 we have $[t] = T$, $[\rho] = M/L^3$, $[D] = L$, and $[g] = L/T^2$. Therefore,

$$[T] \left[\frac{M}{L^3} \right]^\alpha [L]^\beta \left[\frac{L}{T^2} \right]^\gamma = L^0 T^0 M^0,$$

which, upon collecting terms

$$\begin{aligned}
\text{For } M: & \quad \alpha & = 0 \\
\text{For } T: & \quad 1 - 2\gamma & = 0 \\
\text{For } L: & \ -3\alpha + \beta + \gamma & = 0
\end{aligned}$$

By solving the above equations, we get $\alpha = 0$, $\gamma = 1/2$, and $\beta = -\gamma = -1/2$. The nondimensional group then is

$$\Pi_3 = t \sqrt{\frac{g}{D}}.$$

Since $[\mu] = M/LT$, for the nondimensional group Π_4 we write

$$\left[\frac{M}{LT} \right] \left[\frac{M}{L^3} \right]^\alpha [L]^\beta \left[\frac{L}{T^2} \right]^\gamma = L^0 T^0 M^0,$$

which, upon collecting terms

$$\begin{aligned}
\text{For } M: & \quad 1 + \alpha & = 0 \\
\text{For } T: & \quad -1 - 2\gamma & = 0 \\
\text{For } L: & \ -1 - 3\alpha + \beta + \gamma & = 0.
\end{aligned}$$

Solving the above equations, we get $\alpha = -1$, $\gamma = -1/2$, and $\beta = 1 + 3\alpha - \gamma = 1 - 3 + 1/2 = -3/2$. The nondimensional group then is

$$\Pi_4 = \frac{\mu}{\rho D^{\frac{3}{2}} g^{\frac{1}{2}}}.$$

Therefore, the time it takes for the tank to empty is functionally related to

$$t \sqrt{\frac{g}{D}} = f \left(\frac{h}{D}, \frac{d}{D}, \frac{\mu}{\rho D^{\frac{3}{2}} g^{\frac{1}{2}}} \right).$$

This problem is identical to the one discussed in Chapter 1, where, by neglecting μ, the solution was obtained as

$$t\sqrt{\frac{g}{D}} = \sqrt{2\frac{h}{D}\left(\frac{D^4}{d^4} - 1\right)}.$$

Therefore, the simple dimensional analysis predicts correctly the important nondimensional groups that appear in the exact solution to the ideal problem (i.e., when viscosity (and, hence, viscous losses) are neglected).

EXAMPLE 6.3

Steady flow in a pipe is sustained by applying a pressure difference ΔP between the inlet and exit of a horizontal pipe of length L. In addition to the pressure difference, the induced velocity V in the pipe depends on the diameter D, the roughness e of the pipe, and the viscosity μ and density ρ of the fluid. Using dimensional analysis, find the proper form of the functional relationship between ΔP and the other flow parameters.

For this problem, we have seven parameters ($m = 7$), which collectively represent dimensions of length L, time T, and mass M ($n = 3$). By selecting V, D, and ρ as repeating parameters ($r = 3$), according to the Buckingham-Π theorem we need to determine only four nondimensional groups Π_i and $i = 1\ldots4$ ($m - r = 4$) as

$$\Pi_1 = \Pi_1\left(e, \underbrace{V, D, \rho}_{repeating\ parameters}\right),$$

$$\Pi_2 = \Pi_2\left(L, \underbrace{V, D, \rho}_{repeating\ parameters}\right),$$

$$\Pi_3 = \Pi_3\left(\mu, \underbrace{V, D, \rho}_{repeating\ parameters}\right),$$

and

$$\Pi_4 = \Pi_4\left(\Delta P, \underbrace{V, D, \rho}_{repeating\ parameters}\right).$$

ΔP was not selected as a repeating parameter because it is the main variable whose functional dependence we wish to establish.

By inspection, we get

$$\Pi_1 = \frac{e}{D},$$

and

$$\Pi_2 = \frac{L}{D}.$$

Moreover, from Example 6.1, Π_3 yields the familiar Re as

$$\Pi_3 = \frac{\rho V D}{\mu} = Re.$$

For Π_4 we write

$$[\Delta P][V]^{\alpha}[D]^{\beta}[\rho]^{\gamma} = L^0 t^0 M^0.$$

Again, $[p]$ denotes the dimensions of variable p. Here, $[\Delta P] = M/LT^2$, $[V] = L/T$, $[D] = L$, and $[\rho] = M/L^3$. Note that the three repeating parameters are independent (i.e, the dimensions of one parameter cannot be obtained as a combination of the other two). Therefore,

$$\left[\frac{M}{LT^2}\right]\left[\frac{L}{T}\right]^{\alpha} [L]^{\beta} \left[\frac{M}{L^3}\right]^{\gamma} = L^0 T^0 M^0,$$

which, upon collecting terms pertaining to each fundamental dimension, is equivalent to

$$
\begin{aligned}
\text{For } M: && 1 + \gamma &= 0 \\
\text{For } T: && -2 - \alpha &= 0 \\
\text{For } L: -1 + \alpha + \beta - 3\gamma &= 0.
\end{aligned}
$$

Solving the above equations, we get $\gamma = -1$, $\alpha = -2$, and $\beta = 1 - \alpha + 3\gamma = 1 + 2 - 3 = 0$. The nondimensional group then is,

$$\Pi_4 = \frac{\Delta P}{\rho V^2}.$$

The pressure difference is expressed in the general functional relationship

$$\frac{\Delta P}{\rho V^2} = \frac{\Delta P}{\rho V^2}\left(Re, \frac{e}{D}, \frac{L}{D}\right).$$

At this point, the actual functional relationship can be established only by solving the problem using any appropriate solution method.

If surface roughness is neglected, in Chapter 7 the pressure difference for laminar, fully developed flow in a pipe is found to be

$$\frac{dP}{dz} = -128\mu \frac{Q}{\pi D^4},$$

which, after rearrangement is expressed as

$$\frac{\Delta P}{\rho V^2} = -\frac{L}{D}\frac{32}{Re}.$$

Dimensional analysis then predicts correctly the pressure's functional dependence on L, D, and Re.

6.2 Scaling and Order-of-Magnitude Analysis

The discussion here revolves around the use of *order-of-magnitude analysis* as a way to help analyze fluid systems. As the name suggests, the analysis establishes the relative importance of individual flow parameters by comparing them with each other. This procedure lacks the mathematical rigor of earlier chapters and, instead, relies heavily on experience and on the ability of the analyst to identify and compare critical flow parameters.

When comparing flow parameters one must decide whether they are "large" or "small," and whether to include them in the analysis. Obviously, this decision is rather subjective. Nevertheless, the ability to accurately invoke assumptions based on sound physical arguments may be the single most important attribute of a successful engineer.

Regardless of the apparent subjectivity of the procedure, there are some distinct steps one can follow, such as (a) to *scale* the variables by nondimensionalization, using the magnitude of parameters characteristic of the problem; and (b) to complete the scaling by *nondimensionalizing* the governing equations, constitutive relations, and boundary conditions.

For illustration, consider again the two-dimensional flow past a circular cylinder. If we assume that the flow is isothermal and laminar, the two-dimensional Navier-Stokes equations describe the problem as

Continuity

$$\frac{\partial u}{\partial x} + \frac{\partial v}{\partial y} = 0.$$

x-momentum

$$\rho \left(\underbrace{\frac{\partial u}{\partial t}}_{\text{unsteady effects}} + \underbrace{u\frac{\partial u}{\partial x} + v\frac{\partial u}{\partial y}}_{\text{convection}} \right) = \underbrace{\rho g_x}_{\text{gravity forces}} - \underbrace{\frac{\partial P}{\partial x}}_{\text{pressure forces}} + \underbrace{\mu \left[\frac{\partial^2 u}{\partial x^2} + \frac{\partial^2 u}{\partial y^2} \right]}_{\text{viscous forces}}.$$

y-momentum

$$\rho \left(\underbrace{\frac{\partial v}{\partial t}}_{\text{unsteady effects}} + \underbrace{u\frac{\partial v}{\partial x} + v\frac{\partial v}{\partial y}}_{\text{convection}} \right) = \underbrace{\rho g_y}_{\text{gravity forces}} - \underbrace{\frac{\partial P}{\partial y}}_{\text{pressure forces}} + \underbrace{\mu \left[\frac{\partial^2 v}{\partial x^2} + \frac{\partial^2 v}{\partial y^2} \right]}_{\text{viscous forces}}.$$

We note that the conservation of linear momentum in each direction relates the unsteady effects and the effects due to convection, gravity, pressure, and viscosity. Scaling and order-of-magnitude analysis help establish the relative importance of each one of these effects. Often our ability to simplify the problem may determine whether a solution can be obtained or not.

Selection of Scaling Quantities

The first step in the process is to scale the flow variables using quantities characteristic of the flow. These are parameters which, when varied independently (for instance in an experiment), have the greatest effect on the flow. In the example problem, the obvious choice for characteristic length and velocity scales are respectively the diameter D of the cylinder and the freestream velocity V. Velocity components such as u and v are considered large or small, depending on the magnitude of the ratios $\frac{u}{V}$ and $\frac{v}{V}$, respectively. Similarly, the magnitude-of-length parameters, such as x and y, are compared to D. Since $|V| = L/T$, once we introduce the velocity V and the diameter D as characteristic flow parameters, we cannot independently introduce a characteristic time t_{ref}. Instead, we derive one using the definition of velocity as $t_{ref} = D/V$.

The second step is to extend the scaling procedure to all the terms comprising the governing equations, a procedure that leads naturally into the *nondimensionalization* of these equations.

Nondimensionalization

By definition the ratios $\frac{u}{V}$, $\frac{v}{V}$, $\frac{x}{D}$, and $\frac{y}{D}$ have no dimensions and, therefore, are *dimensionless* or *nondimensional* quantities. Using V, D, and $t_{ref} = D/V$ to define new variables as $u^* = \frac{u}{V}$, $v^* = \frac{v}{V}$, $x^* = \frac{x}{D}$, $y^* = \frac{y}{D}$, $t^* = \frac{tV}{D}$, the entire set of governing equations can be expressed in nondimensional form. This is achieved by replacing each dimensional term with its nondimensional counterpart. For instance,

$$\frac{\partial u}{\partial x} = \frac{\partial (u^* V)}{\partial (x^* D)},$$

and since V and D are constant numbers,

$$\frac{\partial u}{\partial x} = \frac{V}{D} \frac{\partial u^*}{\partial x^*}.$$

Similarly, it can be shown that

$$\frac{\partial u}{\partial t} = \frac{V^2}{D} \frac{\partial u^*}{\partial t^*} \quad \text{and} \quad \frac{\partial^2 u}{\partial x^2} = \frac{V}{D^2} \frac{\partial^2 u^*}{\partial x^{*2}}.$$

Using the same procedure, the governing equations can be expressed in dimensionless form as

$$\frac{V}{D} \frac{\partial u^*}{\partial x^*} + \frac{V}{D} \frac{\partial v^*}{\partial y^*} = 0 \Rightarrow \frac{\partial u^*}{\partial x^*} + \frac{\partial v^*}{\partial y^*} = 0,$$

and

$$\rho \left(\frac{V^2}{D} \frac{\partial u^*}{\partial t^*} + \frac{V^2}{D} u^* \frac{\partial u^*}{\partial x^*} + \frac{V^2}{D} v^* \frac{\partial u^*}{\partial y^*} \right) = \rho g_x - \frac{1}{D} \frac{\partial P}{\partial x^*} + \mu \frac{V}{D^2} \left[\frac{\partial^2 u^*}{\partial x^{*2}} + \frac{\partial^2 u^*}{\partial y^{*2}} \right],$$

$$\rho \left(\frac{V^2}{D} \frac{\partial v^*}{\partial t^*} + \frac{V^2}{D} u^* \frac{\partial v^*}{\partial x^*} + \frac{V^2}{D} v^* \frac{\partial v^*}{\partial y^*} \right) = \rho g_y - \frac{1}{D} \frac{\partial P}{\partial y^*} + \mu \frac{V}{D^2} \left[\frac{\partial^2 v^*}{\partial x^*} + \frac{\partial^2 v^*}{\partial y^{*2}} \right].$$

Since the term $\frac{\mu V}{D}$ has dimensions of pressure, it can be used to scale the pressure term as $P^* = \frac{PD}{\mu V}$. Simplifying, then,

$$\frac{\rho D V}{\mu} \left(\frac{\partial u^*}{\partial t^*} + u^* \frac{\partial u^*}{\partial x^*} + v^* \frac{\partial u^*}{\partial y^*} \right) = \frac{\rho D^2 g_x}{\mu V} - \frac{\partial P^*}{\partial x^*} + \left[\frac{\partial^2 u^*}{\partial x^{*2}} + \frac{\partial^2 u^*}{\partial y^{*2}} \right],$$

$$\frac{\rho D V}{\mu} \left(\frac{\partial v^*}{\partial t^*} + u^* \frac{\partial v^*}{\partial x^*} + v^* \frac{\partial v^*}{\partial y^*} \right) = \frac{\rho D^2 g_y}{\mu V} - \frac{\partial P^*}{\partial y^*} + \left[\frac{\partial^2 v^*}{\partial x^{*2}} + \frac{\partial^2 v^*}{\partial y^{*2}} \right].$$

Note that the nondimensional parameter $\frac{\rho D V}{\mu}$ is the Reynolds number Re. The nondimensional term $\frac{\rho D^2 g}{\mu V}$ is known as the Stokes number St. The fact that these nondimensional parameters carry the names of famous fluid dynamicists is not coincidental but is because these persons have made special contributions to the fluid flow phenomena governed by these dimensionless groups.

Simplification of Governing Equations

At this point, through purely dimensional arguments the problem is expressed solely in terms of nondimensional variables. Moreover, each term in the conservation of momentum (e.g., the viscous, the inertia, and the gravity forces) is normalized with variables whose magnitude characterizes the problem.

The governing equations in their nondimensional forms are

$$\nabla \cdot \mathbf{u}^* = 0,$$

$$Re \left(\frac{\partial u_i^*}{\partial t^*} + \mathbf{u}^* \cdot \nabla u_i \right) = -\nabla P^* \cdot \mathbf{I} + \nabla^2 \mathbf{u}^* + St = 0,$$

where $\mathbf{I}$ is the unit tensor. These equations may be simplified by comparing the relative magnitude of each nondimensional term:

- When the Re is small (i.e., $Re \approx 0$), the momentum equations simplify by neglecting the inertia terms as

$$0 = -\nabla P^* \cdot \mathbf{I} + \nabla^2 \mathbf{u}^* + St.$$

  For small Re, the flow is called *creeping flow*. St can be also neglected when gravity effects are not important.

- When the Re is large, inertia effects in the flow are important. In the case of small (negligible) viscous and gravity forces, the momentum equations can be simplified to

$$Re \left(\frac{\partial u_i^*}{\partial t^*} + \mathbf{u}^* \cdot \nabla u_i \right) = -\nabla \cdot P^* \mathbf{I}.$$

  Together with the continuity equation, they constitute the *Euler equations*. Since the viscosity is neglected, these equation are also known as the *inviscid equations*.

However, simplifications must be used with caution, since by neglecting terms from the mathematical problem, we completely eliminate their influence on the solution. For instance, the viscosity represents the internal resistance to flow, and it is responsible for the development of large velocity gradients near solid walls. These large velocity gradients can lead to flow separation, development of recirculating zones, and energy dissipation. Therefore, by neglecting the viscosity, the solution to the Euler equations will not predict these phenomena.

6.2.1 Significance of Dimensionless Groups

In a typical flow problems, we are dealing with forces that include: inertia F_I, pressure F_P, viscous F_V, surface tension F_T, and gravity F_G. However, these forces are not arbitrary but must satisfy the conservation of momentum—that is,

$$F_I = F_P + F_V + F_T + F_G. \tag{6.3}$$

Using their definitions, the magnitude of these forces may be *estimated* from variables characteristic of various scales in the flow such as velocity V, and length L and from fluid properties such as the density ρ, the viscosity μ, and the surface tension σ of the fluid as shown below

$$F_I \equiv M\frac{dV}{dt} \approx \rho L^3 \frac{V^2}{L} = \rho V^2 L^2 \quad (\text{mass } M \text{ is expressed as } \rho L^3)$$

$$F_P \equiv \Delta P A \approx \Delta P L^2 \quad (A \text{ is the surface area})$$

$$F_G \equiv Mg \approx \rho L^3 g$$

$$F_V \equiv \mu A\frac{\partial u}{\partial y} \approx \mu L^2 \frac{V}{L} = \mu L V$$

$$F_T \equiv \sigma L.$$

Now, in a general problem the relative importance of these forces can be established by comparison. For instance, the ratio of the inertia to viscous forces

$$\frac{F_{inertia}}{F_{viscous}} = \frac{\rho L V}{\mu} \equiv Re,$$

indicates the relative importance of convection or inertia forces ($\rho V^2 L^2$) to the viscous forces ($\mu V L$). Therefore, Re is not just a simple consequence of the nondimensionalization, but it has physical significance as well; at small Re, the viscous effects dominate the flow, while at large Re, the inertia effects are more important. It is not surprising, then, that the Re indicates also whether the flow is laminar or turbulent. The definition of Re is also consistent with the argument in Chapter 5 where turbulent flow was introduced.

Similarly, the ratio of the gravity forces to viscous forces

$$\frac{F_{gravity}}{F_{viscous}} = \frac{\rho L^2 g}{\mu V} \equiv St,$$

indicates the relative importance of gravity forces to viscous forces. This nondimensional group is known as the Stokes number St, named after Stokes, a famous fluid dynamicist.

Dimensionless groups can also be obtained as a combination of other groups such as

$$St = \frac{\rho L^2 g}{\mu V} = \frac{\rho V L}{\mu}\frac{Lg}{V^2} = \frac{Re}{Fr^2},$$

where Fr is the Froude number defined as $\frac{V}{\sqrt{gL}}$. The Froude number (actually Fr^2) indicates the relative importance of inertia ($\rho V^2 L^2$) to gravity forces ($\rho g L^3$). All well founded nondimensional groups have similar physical significance. Table 6.1 lists a number

TABLE 6.1   List of Nondimensional Groups and their Significance.

Name	Formula	Quantities Represented
Reynolds, Re	$\dfrac{LV\rho}{\mu}$	inertia force/viscous force
Froude, Fr	$\dfrac{V}{\sqrt{gL}}$	inertia force/gravity force
Euler, Eu	$\dfrac{\Delta P}{0.5\rho V^2}$	pressure force/inertia force
Weber, We	$\dfrac{\rho L V^2}{\sigma}$	inertia force/surface tension force

of such nondimensional groups, which are obtained by taking the ratios indicated in the last column of the table.

EXAMPLE 6.4

Problem Statement   Consider air flow past a 0.1 m-diameter circular cylinder at a velocity 100 m/s. For this flow, establish the importance of the viscous and gravity forces relative to the inertia forces.

SOLUTION   The forces in this problem are those due to inertia F_I, gravity F_G, and viscosity F_V. Following the procedure shown in Section 6.2.1, their magnitude can be *estimated* by using the magnitude of variables characteristic of the problem. Therefore, by using $M \approx \rho D^3$ we have

$$F_I \equiv M \frac{dV}{dt} \approx \rho V^2 D^2, \quad F_G \equiv Mg \approx \rho D^3 g, \quad F_V \equiv \mu A \frac{\partial u}{\partial y} \approx \mu D V,$$

where the characteristic length was taken to be the diameter of the cylinder D and velocity V the free stream velocity. A is the surface area.

By comparison, then, the relative importance of gravity to inertia forces is

$$\frac{F_G}{F_I} = \frac{\rho D^3 g}{\rho V^2 D^2} = \frac{gD}{V^2} = \frac{9.81 \ m/s^2 \times 0.1 \ m}{100^2 \ m^2/s^2} = 9.81 \times 10^{-5}.$$

Similarly, by using the viscosity of air $\mu = 1.82 \times 10^{-5} \ N \cdot s/m^2$, and density of $\rho = 1.204 \ kg/m^3$, the relative importance of viscous to inertia forces is given by

$$\frac{F_V}{F_I} = \frac{\mu D V}{\rho V^2 D^2} = \frac{\mu}{\rho D V} = \frac{1.82 \times 10^{-5} \ N \cdot s/m^2}{1.204 \ kg/m^3 \times 0.1 \ m \times 100 \ m/s} = 1.51 \times 10^{-6}.$$

According to this analysis, both viscous and gravity forces appear to be small compared with inertia forces. Therefore, one is tempted to remove them from consideration, which for many problems is the correct approach to take. However, for some problems this may be misleading and may lead to wrong conclusions; in flows past solid surfaces, for instance, the viscous effects are always important irrespective of the Reynolds number (see Chapter 8 for a more detailed discussion of the topic). This is because the relevant length scale for such flows is the thickness of the boundary layer δ, which turns out to be very small ($\delta \ll D$), making $F_V/F_I = \mu/\rho\delta V$ larger than the ratio obtained above. Note also that by removing the effects of viscosity in such flows, we are also eliminating the possibility of predicting viscous-related phenomena, (e.g., flow separation and vortex shedding).

6.3   Similitude, Design of Experiments

Since turbulent flow, material constants, and constitutive parameters are primarily determined by experiments, experimentation is central to the study of fluid mechanics. However, as fluid problems are typically characterized by a large number of variables that must be varied and studied independently, experiments can be both expensive and time consuming. Additionally, the physical size of problems often introduces serious difficulties in the design of experiments. Fortunately, scaling and dimensional analysis, as introduced previously, become pivotal in the proper design and organization of experiments.

In this section, we summarize and expand on earlier observations in order to show what is probably the most important aspect of dimensional analysis — the design of experiments and the proper organization of the results. Key to this discussion is the fact that as a consequence of the nondimensionalization, the solution to the dimensionless equations represents an infinite family of *similar* solutions.

For illustration, consider the flow between two long plates with the upper plate moving to the right with velocity u_{plate}. This is the same geometry, boundary conditions, and physics as Example 5.8, discussed in Chapter 5. According to the solution, the momentum equation in the x direction simplifies to

$$\mu \frac{\partial^2 u}{\partial y^2} = 0.$$

Introducing the nondimensional variables

$$u^* = \frac{u}{u_{plate}}, \quad y^* = \frac{y}{H},$$

where the characteristic velocity and length scales are chosen to be the velocity of the plate u_{plate} and the distance H, the momentum equation reduces to

$$\frac{\partial^2 u^*}{\partial y^{*2}} = 0.$$

The nondimensional boundary conditions now are at $y^* = 0$, $u^* = 0$ and at $y^* = 1$, $u^* = 1$.

By integrating the momentum equation, we get

$$u^* = C_1 y^* + C_2,$$

where the constants C_1 and C_2 are obtained from the boundary conditions. Finally, then, the velocity profile is expressed as

$$u^* = y^*.$$

Since we can have an infinite number of combinations of u_{plate} and H, the preceding single nondimensional solution represents the solution to an infinite number of problems of laminar flow between two plates.

The above implications are useful in experiments where, instead of experimenting on real prototype problems that might be expensive to build or too large to install in laboratories, it is possible to perform the experiments on *models* (i.e., on physically scaled setups). For instance, instead of studying flow past an airplane wing (or the entire plane) as shown in Figure 6.3a, the same results can be obtained from a scaled-down model (Figure 6.3b). The two are equivalent only when the model and the prototype problems are *geometrically*, *kinematically*, and *dynamically* similar.

The model is said to be *geometrically* similar to the prototype when ratios between corresponding geometric parameters are the same. For instance, if t is the thickness and c the chord length of the airfoil, the two airfoils are geometrically similar when

$$\frac{t_p}{t_m} = \frac{c_p}{c_m} \equiv L_R,$$

where subscripts p and m correspond to quantities related to the prototype and the model respectively, and L_R is the ratio of the linear dimensions. If linear dimensions are similar,

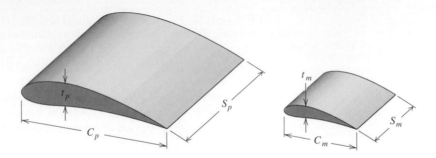

FIGURE 6.3 Flow past an airfoil.

corresponding surface areas (A_p, A_m) are then proportional to the square of L_R.

$$\frac{A_p}{A_m} = \frac{c_p^2}{c_m^2} = L_R^2.$$

Corresponding volumes then are proportional to the cube of the ratio of L_R.

$$\frac{\mathcal{V}_p}{\mathcal{V}_m} = \frac{c_p^3}{c_m^3} = L_R^3.$$

The resulting flow field in the model problem is said to be *kinematically* similar when the prototype and model consist of geometrically similar streamlines. Therefore, as some streamlines coincide with geometric surfaces, kinematic similarity implies also geometric similarity (but not the reverse). Kinematic similarity also implies that the acceleration vectors at corresponding points are parallel and have the same magnitude ratio, $a_p/a_m = L_a$.

For dynamic similarity, forces at corresponding points must have the same ratios. However, kinematic similarity and dynamic similarity are related. For instance, the acceleration a multiplied by the fluid mass m yields the inertia force. Also, since the acceleration vectors are parallel, according to conservation of momentum, the resultant force that balances this acceleration must point in the same direction and must have the same magnitude ratio as inertia forces. Therefore, not only inertia force similarity, but also the resultant force similarity is implied by kinematic similarity.

Now, given kinematic similarity $(a_p/a_m = L_a)$, and dynamic similarity for the inertia force $(F_{I_p}/F_{I_m} = L_{F_I})$, we have

$$\frac{F_{I_p}}{F_{I_m}} = L_{F_I} = \frac{m_p a_p}{m_m a_m} = L_a \frac{m_p}{m_m} \Rightarrow \frac{m_p}{m_m} = \frac{L_{F_I}}{L_a} \equiv L_m,$$

we also get similar mass distribution. The above discussion shows that for dynamic similarity, we must have both kinematic (and, hence, geometric) similarity and similar mass distribution (L_m).

Typical forces in a general flow may be the inertia forces F_I, the pressure forces F_P, the viscous forces F_V, the surface tension forces F_T, and the force due to gravity F_G. As shown in Section 6.2.1, the magnitude of these forces may be estimated using quantities representative of the fluid (ρ, μ, σ), the geometry of the problem (L), and the operating conditions (V) as

$$F_I \approx \rho V^2 L^2$$

$$F_P \approx \Delta P L^2$$

$$F_G \approx \rho L^3 g \tag{6.4}$$

$$F_V \approx \mu L^2 \frac{V}{L} = \mu L V$$

$$F_T \approx \sigma L.$$

These forces, however, in the prototype and the model must satisfy conservation of linear momentum. Since these forces as shown in Equation (6.4) are not exact but are estimates of their magnitude, for both the model and prototype we can write only

$$\mathcal{O}\left(\frac{F_G}{F_I}\right) + \mathcal{O}\left(\frac{F_P}{F_I}\right) + \mathcal{O}\left(\frac{F_V}{F_I}\right) + \mathcal{O}\left(\frac{F_T}{F_I}\right) = \mathcal{O}(1), \tag{6.5}$$

where $\mathcal{O}()$ indicates the order of the term inside the parenthesis. For instance, $\mathcal{O}(1)$ implies a quantity of order 1.

Note that Equation (6.5) is in terms of four ratios. Therefore, once three of them are specified and fixed by imposing similarity, the fourth is also fixed through conservation of linear momentum (in other words, it is specified automatically by nature). Similarity then is required for three of the four ratios shown above. In a general problem then, dynamic similarity between the prototype and the model needs to be enforced for one less force than the total number of forces relevant to the problem.

By definition, for dynamic similarity the model and the conditions of the model experiment must satisfy relations of the form

$$\frac{F_{G_p}}{F_{G_m}} = \frac{F_{P_p}}{F_{P_m}} = \frac{F_{V_p}}{F_{V_m}} = \frac{F_{T_p}}{F_{T_m}} = \frac{F_{I_p}}{F_{I_m}},$$

which can be rearranged as a set of simultaneous relations given by

$$\frac{F_{G_p}}{F_{I_p}} = \frac{F_{G_m}}{F_{I_m}}, \quad \frac{F_{V_p}}{F_{I_p}} = \frac{F_{V_m}}{F_{I_m}},$$

and

$$\frac{F_{P_p}}{F_{I_p}} = \frac{F_{P_m}}{F_{I_m}}, \quad \frac{F_{T_p}}{F_{I_p}} = \frac{F_{T_m}}{F_{I_m}}.$$

Now, in terms of the characteristic flow parameters these ratios are expressed in nondimensional form as

$$\frac{F_{G_p}}{F_{I_p}} = \left(\frac{\rho L^3 g}{\rho V^2 L^2}\right)_p = \left(\frac{Lg}{V^2}\right)_p = \frac{1}{Fr_p^2} = \frac{St_p}{Re_p},$$

$$\frac{F_{V_p}}{F_{I_p}} = \left(\frac{L\mu V}{\rho V^2 L^2}\right)_p = \left(\frac{\mu}{\rho L V}\right)_p = \frac{1}{Re_p},$$

$$\frac{F_{P_p}}{F_{I_p}} = \left(\frac{\Delta P L^2}{\rho V^2 L^2}\right)_p = \left(\frac{\Delta P}{\rho V^2}\right)_p = Eu_p,$$

and

$$\frac{F_{T_p}}{F_{I_p}} = \left(\frac{\sigma L}{\rho V^2 L^2}\right)_p = \left(\frac{\sigma}{\rho V^2 L}\right)_p = \frac{1}{We_p},$$

where St is the Stokes number, Re the Reynolds number, Fr the Froude number, Eu the Euler number, and We the Weber number. Therefore, dynamic similarity is achieved when the significant nondimensional force ratios between the prototype and the model are the same

$$Fr_p = Fr_m, \quad Re_p = Re_m,$$

and

$$Eu_p = Eu_m, \quad We_p = We_m.$$

For instance, in flow past an airfoil the relevant forces are those due to inertia F_I, the viscosity F_V, and the drag force F_D. In principle, F_D is a function of the pressure and viscous forces along the body. Here we are assuming that the drag force is primarily due to the pressure forces. In addition to geometric similarity, if we select the flow parameters such that two of the forces, say F_I and F_V are similar — that is,

$$\frac{F_{I_m}}{F_{I_p}} = \frac{F_{V_m}}{F_{V_p}} \Rightarrow Re_p = Re_m,$$

we automatically get

$$C_{D_p} = \left(\frac{F_D}{\rho V^2 L^2}\right)_p = \left(\frac{F_D}{\rho V^2 L^2}\right)_m = C_{D_m}.$$

This indicates that the measured drag coefficient C_D is identical to the one expected in the prototype problem.

Similarly, in pipe flow the relevant forces are the pressure F_P, inertia F_I, and viscous F_V forces. Therefore, if through proper choice of the parameters, we ensure geometric similarity and that $Re_p = Re_m$, we will automatically get $Eu_p = Eu_m$, or

$$\left(\frac{\Delta P}{\rho V^2}\right)_p = \left(\frac{\Delta P}{\rho V^2}\right)_m.$$

6.3.1   Incomplete Similarity

Complete similarity requires that similarity be enforced for one less force than the total number of forces in the problem. Often, however, this may not be possible since the model parameters may lead to unreasonable requirements for the equipment and/or working fluid. For instance, if the Re is matched between the model and the prototype

$$\left(\rho \frac{VL}{\mu}\right)_p = \left(\rho \frac{VL}{\mu}\right)_m,$$

for the same working fluid we get

$$\frac{V_m}{V_p} = \frac{L_p}{L_m} > 1.$$

Hence, $V_m > V_p$, making the velocities in the model much higher than those in the prototype. It is possible, then, that existing equipment may not be possible to handle the conditions of the model. Additionally, if the primary fluid is a gas, at high enough V_m the compressibility effects may become significant (whereas the prototype might be incompressible), in which case the results from the model will not be equivalent to the prototype problem.

These difficulties arise often in the design of experiments. In such cases, the general practice is to seek to achieve similarity for the most critical forces in the problem and obtain measurements for the primary nondimensional groups. The remaining forces for which similarity was not possible are also measured, and their effect on the prototype problem is estimated by extrapolation, analytic methods, and the like.

However, nondimensionalization allows various degrees of flexibility; the prototype and model problems can be made equivalent by using fluids other than the fluid in the prototype. It is possible then to substitute the working fluid from gas to liquid and vice versa. Additionally, this flexibility might allow the use of already existing experimental facilities (facilities that use the equivalent working fluid), hence, reducing the cost of the experiment (or even making possible the experiment). Additionally, different fluids may allow the use of experimental methods not possible for other fluids.

EXAMPLE 6.5

Problem Statement   A $0.8\ m$ airfoil is designed to be used in flight at operating conditions such that $Re = 10^7$. The performance of the airfoil is evaluated experimentally using a 10:1 scaled-down model. Find the air velocity required in the experiment in order to reproduce the conditions of the prototype. What is the velocity if the experiment is performed in water? Find the ratio between the drag force obtained in the experiment and the force expected in the prototype. Condider air conditions at $20°C$.

SOLUTION   For the given $Re = 10^7$, air viscosity $\mu = 1.82 \times 10^{-5}\ N \cdot s/m^2$, and density $\rho = 1.204\ kg/m^3$, the velocity in the prototype problem is

$$Re = \frac{\rho L V}{\mu} \Rightarrow V = Re\frac{\mu}{\rho L} = 10^7 \times \frac{1.82 \times 10^{-5}\ N \cdot s/m^2}{1.204\ kg/m^3 \times 0.8\ m} = 188.95\ m/s.$$

Forces measured experimentally using the airfoil model will be *similar* to those in the actual airfoil when the prototype and model are geometrically, kinematically, and dynamically similar. In this problem, geometric similarity is achieved by constructing the model with linear reduction of 10:1.

The relevant forces for this problem are the drag F_D, inertia F_I, and viscous forces F_V. In principle, F_D is a function of the pressure and viscous forces along the body. Here, we are assuming that the drag force is primarily due to the pressure forces. As discussed previously, by matching $Re_p = Re_m$ we are ensuring that

$$\left.\frac{F_D}{F_I}\right|_p = \left.\frac{F_D}{F_I}\right|_m,$$

where subscripts m and p refer to model and prototype quantities. Therefore,

$$Re_p = Re_m \Rightarrow \left.\frac{\rho V L}{\mu}\right|_m = \left.\frac{\rho V L}{\mu}\right|_p.$$

When the working fluid is air for both cases, we have

$$\left.\rho\frac{V L}{\mu}\right|_m = \frac{1.204\ kg/m^3 \times V_m \times L_m}{1.82 \times 10^{-5} N \cdot s/m^2} = \left.\rho\frac{V L}{\mu}\right|_p = \frac{1.204\ kg/m^3 \times V_p \times L_p}{1.82 \times 10^{-5} N \cdot s/m^2} \Rightarrow$$

$$\frac{V_m}{V_p} = \frac{L_p}{L_m} = 10.$$

The experiment in air then must be performed at velocity $V_m = 10 \times 188.95\ m/s = 1889.5\ m/s$, which is obviously very large (speed of sound at this temperature is only about $350\ m/s$)! As discussed in Chapter 11, at this velocity the compressibility effects will both dominate and alter the flow so that the prototype and model flows will no longer be comparable.

If the working fluid is water, the viscosity is $\mu = 1.002 \times 10^{-3} \ N \cdot s/m^2$, and the density is $\rho = 998.2 \ kg/m^3$, we have

$$\rho \frac{VL}{\mu}\bigg|_m = \frac{998.2 \ kg/m^3 \times V_m \times L_m}{1.002 \times 10^{-3} \ N \cdot s/m^2} = \rho \frac{VL}{\mu}\bigg|_p = \frac{1.204 \ kg/m^3 \times V_p \times L_p}{1.82 \times 10^{-5} \ N \cdot s/m^2} \Rightarrow$$

$$\frac{V_m}{V_p} = 6.64 \times 10^{-2} \times \frac{L_p}{L_m} = 0.664.$$

Under these conditions, the velocity of the model is $V_m = 0.664 \times 188.95 \ m/s = 125.46 \ m/s$, which is still very large for a water tunnel! Therefore, the only way to reduce this velocity further is to use a larger model airfoil, which of course may not be possible because of the size of the water channel.

The drag force obtained from the model airfoil is related to the actual forces on the prototype through

$$\frac{F_D}{F_I}\bigg|_p = \frac{F_D}{F_I}\bigg|_m \Rightarrow \frac{F_{D_p}}{F_{D_m}} = \frac{F_{I_p}}{F_{I_m}} = \frac{\rho_p}{\rho_m} \frac{V_p^2}{V_m^2} \frac{L_p^2}{L_m^2} \Rightarrow$$

$$\frac{F_{D_p}}{F_{D_m}} = \frac{1.204}{998.2} \times \left(\frac{188.95}{125.46}\right)^2 \times \left(\frac{10}{1}\right)^2 = 0.274.$$

Under these conditions, the force on the model is larger than the actual force on the prototype. This problem exemplifies some of the practical difficulties encountered in the design of experiments and the proper matching of the parameters.

REFERENCES

T. HEATH, *A History of Greek Mathematics*, Vol. I, p. 67, New York, Dover, 1981.

R.L. PANTON, *Incompressible Flow*, New York, Wiley, 1984.

PROBLEMS

1. Show that the velocity distribution $u = \frac{1}{\mu}\rho g \sin \alpha \left(Hy - \frac{y^2}{2}\right)$, which represents laminar flow down an inclined surface due to gravity g, with angle α with respect to the horizontal direction and depth of H is dimensionally consistent. ρ and μ are the density and viscosity of the fluid.

2. *Film casting.* Show that the velocity $u = u_0 e^{\frac{F}{4\mu \dot{Q}}x}$ of a thin film as a function of x, drawn from a die by pulling it with a constant force per unit width F is dimensionally consistent. u_o is the velocity at the exit of the die and $\dot{Q}$ the volumetric flow per unit width.

3. A solid body quenched in liquid cools according to

$$mC\frac{dT}{dt} = h(T - T_L),$$

where T is the temperature, m is the mass of the body, C the solid's specific heat, h is a heat transfer coefficient, T_L is the temperature of the liquid and t is time. For aluminum $\rho = 2707 \ kg/m^3$ and $C = 0.896 \ kJ/kg°C$. What are the units of h?

4. Capillary waves develop at the interface of two immiscible fluids due to the dominant effect of surface tension σ. The propagation speed c depends on the density ρ, surface tension σ, the capillary wavelength λ. Using dimensional analysis, express c as a function of the relevant variables.

5. The torque T that must be applied to a shaft to overcome friction in a journal bearing is a function of the diameter D, the bearing length L, the clearance between the bearing and the shaft d, the angular velocity ω, the oil viscosity μ, the load of the shaft W and the volumetric flow rate Q. Using the Buckingham-Π Theorem express T in terms of non-dimensional terms.

6. The period T of a pendulum depends on the mass m, the length l, gravity g and the swing angle θ; using dimensional analysis organize the variables in terms of non-dimensional groups.

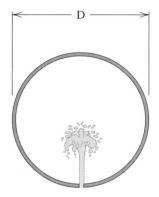

FIGURE 6.4 Schematic for Problem 13.

7. Show that the velocity $u = u_0 \left(1 - erf\, \dfrac{y}{2\sqrt{vt}}\right)$ for laminar flow of liquid set impulsively in motion by a velocity u_0, applied to the free surface is dimensionally consistent. t is time, v is the kinematic viscosity, y is the coordinate normal to the free surface, and erf the error function defined as $erf(\eta) = \dfrac{2}{\sqrt{\pi}} \displaystyle\int_0^{\eta} e^{-\eta^2}\,d\eta$.

8. A liquid droplet suspended in space once deformed slightly it will oscillate with frequency ω, which appears to be a function of the density ρ, the viscosity μ, and radius R. Using dimensional analysis, find the proper non-dimensional parameter for the frequency of oscillation, and the time τ it takes for the oscillations to die-down as a function of the flow properties.

9. The general form of the fluid velocity in a two-dimensional channel is given by $u = a + by^2$. Using dimensional analysis find the proper dimensions of the constants a and b.

10. A solid particle with diameter d and density ρ_p in a liquid with viscosity μ and density ρ settles with due to gravity. Using dimensional analysis find the settling velocity as a function of parameters of the flow.

11. Find the approximate size of a liquid droplet that can remain attached to a vertical wall due to surface tension.

12. The pressure P, within a container filled with liquid rotating at a constant angular velocity ω, is a function of the distance from the axis of rotation r, the local depth z, gravity g, and the density of the fluid ρ. Using dimensional analysis find the proper functional relationship for the pressure in terms of the appropriate non-dimensional parameters.

13. *Mixing in a confined domain.* Using dimensional analysis find a characteristic time to represent turbulent mixing of a mass of fluid with density ρ, injected with an initial impulse I (expressed as force × time), into a circular container with diameter D as shown in Figure 6.4.

14. The viscosity of a fluid is often determined in a parallel disk viscometer where liquid is placed in the gap h, between two circular disks with diameter D, with the upper plate rotating at a constant angular speed ω. The viscosity is evaluated my measuring the torque $\mathcal{T}$, required to sustained the rotation of the plate. Using dimensional analysis, find the functional relationship of the viscosity μ, to the parameters of the instrument.

15. The viscosity of a fluid can be measured using a falling-ball viscometer where a cylindrical slug with radius R_1, and density ρ_1, is allowed to drop at a constant velocity V, due to the action of gravity g, in a cylinder with radius R_2, filled with the liquid with density ρ, whose viscosity μ, is being measured. Find the functional relationship between the relevant parameters of the experiment using dimensional analysis.

16. Using dimensional analysis derive an expression for the terminal velocity of a smooth sphere falling under gravity in an incompressible fluid. Select the appropriate parameters that affect the terminal velocity.

17. In flow through a constriction both the velocity and pressure change with the reduction of area. This configuration is often used to calculate the velocity V_1, of the fluid with density ρ, in a constant diameter pipe D_1, by inserting in the pipe a short section with a constriction with diameter D_2, and by measuring the resulting pressure drop $P_1 - P_2$. Using dimensional analysis find the proper functional relationship for the velocity.

18. As discussed in earlier chapters surface tension σ, can support the weight a column of liquid with density ρ, and height h, in a small-diameter tube. Using dimensional analysis, find the proper non-dimensional groups in this problem. Repeat the same non-dimensionalization by using the specific weight $\gamma = \rho g$ as a variable instead of ρ and g separately. Hint: select repeating parameters that are independent.

19. A Bingham fluid is a fluid which will not flow (i.e deform) unless the local stress exceeds a finite stress. These fluids are found in many industrial applications, especially in two-phase (fluid and particle) flows. Arrange the variables in non-dimensional groups for an experiment where the objective is to establish the relationship between the density ρ, viscosity μ, and the finite yield stress τ_o, on the pressure drop in flow of a Bingham fluid through a sudden expansion where the fluid flows from a section with diameter d, to a larger diameter D.

20. A hydraulic jump is a phenomenon in free surface (gravity g is acting normal to the surface) flows where liquid with depth h_1, and velocity V_1, slows down with a sudden increase in the depth to h_2, and a decrease in the velocity. Using dimensional analysis find the functional dependence of the new depth on the flow parameters.

21. The force on the mast of a Flettner-rotor ship depends on the rotational speed ω of the cylindrical masts with diameter D, height H, the velocity V and density ρ of the wind, the speed of the ship. Using dimensional analysis find the

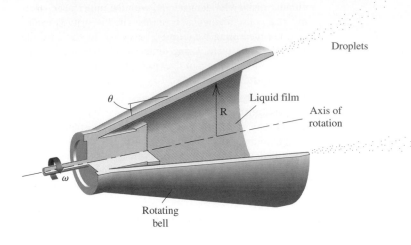

FIGURE 6.5   Schematic for Problem 24.

functional dependence of the force F to the problem parameters.

22. In a viscous coupling device where parts with different rotational speeds ω_1 and ω_2, are coupled together, the torque T developed is a function of the density ρ, diameter D, the viscosity μ, the volume of fluid $\mathcal{V}$. Using dimensional analysis, find the proper functional dependence of the torque to the problem variables.

23. In gravity assisted flow of a thin film, in addition to gravity g, the important parameters are the thickness of the film h, surface tension σ, and density of the fluid ρ. Using dimensional analysis express the velocity of the film as a function of the problem variables.

24. One method of atomizing liquids is the rotating bell shown in Figure 6.5. The velocity distribution across the thickness is a function of the angular velocity ω, the volumetric flow rate $\dot{Q}$, the angle of the bell θ, the local radius from the center of rotation R, the density ρ, and viscosity μ, of the liquid. Find the proper functional dependence of the velocity and thickness on flow parameters.

25. The important operational parameters in pumps are the head rise H, and the rate of shaft work $\dot{W}$. These parameters depend on the effective diameter of the impeller D, the roughness e, the rotational speed ω, the viscosity μ, and density ρ of the fluid and other geometric parameters denoted by L_i. Using dimensional analysis, find the functional dependence of H and $\dot{W}$ on the flow parameters.

26. The lift force F_L of a supersonic jet is a function of the length L, velocity V, angle of attack β, speed of sound α, the density ρ and viscosity μ f the surrounding air. Using non-dimensional analysis, find the functional dependence of F_L on the parameters of the problem.

27. The angular velocity Ω of a windmill depends on the wind speed V, the diameter D, the number of blades N and the density ρ of the surrounding air. Using dimensional analysis, find the functional dependence of Ω on the flow parameters.

28. The well known Blasius' similarity transformation for laminar boundary layer flow, is based on a non-dimensional streamfunction f constructed using the velocity U, away from the edge of the boundary layer, the distance x from the point of the initial growth of the layer, and the density ρ and viscosity μ of the fluid. Using dimensional analysis, construct the proper definition of f.

29. The total viscous force F_v on a flat rectangular plate with length L and width W due to a uniform velocity U is, in addition to L, W and U, a function of the density ρ and viscosity μ of the fluid. Using dimensional analysis find F_v as a function of the appropriate non-dimensional groups.

30. In the flow past a circular cylinder the frequency f with which vortices are shed is a function of the bulk velocity U, the diameter D, and roughness e of the cylinder and the density ρ and viscosity μ of the fluid. Using dimensional analysis find the frequency as a function of the appropriate non-dimensional variables.

31. A small-amplitude surface wave travels with speed c which is a function of gravity g, the depth h, and the wavelength λ. Using dimensional analysis, find the non-dimensional speed as a function of the non-dimensional wavelength.

32. A small-amplitude surface wave in the case when surface tension is important travels with a speed c which is a function of gravity g, the depth h, and the wavelength λ and the density of the liquid. Using dimensional analysis, find the non-dimensional speed as a function of the appropriate non-dimensional groups.

33. The interface between two immiscible fluids with densities ρ_1 and ρ_2 and moving with two different velocities U_1 and U_2 with gravity g acting normal to the interface is unstable. If c is the speed of the instability and λ the wavelength, find

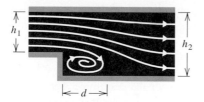

FIGURE 6.6 Schematic for Problem 35.

the functional dependence of the non-dimensional speed on the appropriate non-dimensional groups.

34. In open channel flows the height of the liquid above a weir is a function of the gravity g, the volumetric flow rate per unit length $\dot{Q}$, density ρ and viscosity μ. Using dimensional analysis develop the appropriate non-dimensional groups.

35. Flow over a backward-facing step as shown in Figure 6.6 is encountered in many applications. Due to the geometry, the flow develops a recirculation zone near the step. The length of this zone d is a function of the velocity in the section prior to the step, and the heights h_1 and h_2 before and after the step as well as the viscosity and density of the fluid and gravity. Find the proper functional form of the length of the recirculation zone in term of the parameters of the problem.

36. Due to friction (viscosity), flow past a flat plate develops a thin layer of slower fluid near the surface of the plate. The thickness of this layer is a function of the viscosity of the fluid μ, the density ρ, the distance x from the edge of the flat plate and the approaching velocity U. Using dimensional analysis, find the dependence of the thickness of the layer as a function of the flow parameters.

37. The momentum equation for fully developed flow in a channel due to an applied pressure gradient across a segment of the channel is given by

$$\frac{\partial^2 u}{\partial y^2} = \frac{1}{\mu}\frac{dP}{dx},$$

where μ is the viscosity and P the pressure. Using P_0, u_0 and the height of the channel H as characteristic quantities, non-dimensionalize the equation and interpret the physical significance of the non-dimensional parameter that results from the non-dimensionalization.

38. The free surface of an initially stagnant fluid with infinite depth is suddenly set in motion with velocity V_s. The velocity distribution across the thickness of the fluid as a function of time is given by the solution of the conservation of linear momentum

$$\rho\frac{\partial u}{\partial t} = \mu\frac{\partial^2 u}{\partial y^2},$$

where ρ is the density, and μ the viscosity of the fluid. Using appropriate characteristic quantities (note

that in this problem there is no characteristic length !) non-dimensionalize the equation and interpret physically any non-dimensional groups that emerge.

39. Disturbances in compressible fluids travel as pressure waves within the fluid. In the case of a weak (small disturbance) one dimensional wave the conservation of mass and momentum laws can be written as

$$\frac{\partial \rho}{\partial t} + \rho\frac{\partial u}{\partial x} + u\frac{\partial \rho}{\partial x} = 0$$

$$\rho\left(\frac{\partial u}{\partial t} + u\frac{\partial u}{\partial x}\right) = -c^2\frac{\partial \rho}{\partial x},$$

where c is the speed of the disturbance. Using as characteristic quantities the conditions of the undisturbed fluid ρ_0, u_0, and a length L, express the equations in non-dimensional form.

40. Find the functional dependence of the power in a hydraulic machine if it depends on the angular velocity of the rotating shaft, the size of the machine, and the density and viscosity of the working fluid.

41. The gap between two cylinders of radii R_1 and R_2 ($R_2 > R_1$) is filled with an incompressible fluid whose viscosity and density are respectively μ and ρ. Using dimensional analysis find the functional dependence of the torque required to maintain a constant rotational speed.

42. A thin water film with thickness $H = 0.01\ m$ flows under the influence of gravity g with velocity $V = 0.05\ m/s$. For this problem find the importance of the viscous, gravity and surface tension forces relative to inertia forces. What if the fluid is glycerin.

43. For the flow past a backward-facing step shown in Figure 6.6 use an average velocity in the section prior to the step of $10\ m/s$, and $h_1 = 0.5\ m$ and $h_2 = 1\ m$ to establish the importance of the viscous and gravity forces relative to inertia forces. Consider water as the fluid.

44. In the film casting problem described in Problem 2, the inlet velocity and die height are respectively $u_o = 0.1\ m/s$ and $h_0 = 0.04\ m$. The length of the film is $L_x = 0.18\ m$ and the width $W = 0.3\ m$, with a final velocity of $u_f = 0.2\ m/s$. The fluid is a melt of Low-density polyethylene with properties $\mu = 10^3\ Pa \cdot s$, $\rho = 782\ kg/m^3$ and surface tension $\sigma = 0.025\ N/m$. Establish the importance of the various forces relative to the pulling force F.

45. Consider an extrusion process where round rods are extruded from a circular die with diameter $D = 0.04\ m/s$ at a rate such that the Re based on the diameter and the bulk velocity is 100. By assuming the fluid to be a melt of Low-density polyethylene with properties $\mu = 10^3\ Pa \cdot s$, $\rho = 782\ kg/m^3$ and surface tension $\sigma = 0.025\ N/m$. Establish the importance of the various forces in the process.

46. The pressure variation in a partially filled container with liquid rotating with angular velocity ω is a function of the distance r from the center of rotation, the local depth z

from a free surface exposed to a pressure P_0, the density of the liquid ρ, and the gravitational acceleration g. Using dimensional analysis, express this functional relationship in terms of non-dimensional groups.

47. A 1/8 scaled-down model of an aircraft is tested both in air and in water. Find the required speed in both experiments in order to reproduce the results of 60 m/s in the actual aircraft.

48. The test section of a wind tunnel is a 0.45 m square. The tunnel can be operated at a maximum speed of 50 m/s. Find, and plot the range of actual airfoil sizes and operating velocities that can be tested in the tunnel.

49. A wind tunnel can deliver air at a maximum speed of 80 m/s. What is the span of a model plane when the prototype plane has a span of 8 m and it is designed to fly at 90 mph?

50. A 1/6 scale model of an aircraft is tested in a water and a wind tunnel. What is the ratio of the speeds in the tunnels?

51. A ship 100 m long moves in fresh water at 30 mph. Find the viscosity of a fluid to use in testing of a 3.5 m long model if dynamic similarity is to be enforced.

52. A ship 80 m long moves in fresh water at 25 mph. Find the viscosity of a fluid to use in testing of a 2.5 m long model if dynamic similarity is to be enforced.

53. The test section of a wind tunnel is a 0.56 m square. The tunnel can be operated at a maximum speed of 1 m/s. Find, and plot the range of airfoil sections of an actual aircraft, and the corresponding operating velocities that can be tested in the tunnel.

54. The drag on a surface ship is due to two effects: (a) the viscous forces (skin friction) and (b) the gravity forces (wave resistance). Develop the appropriate dimensionless groups for the drag of a boat with characteristic length L traveling at velocity V.

55. Find the ratio of the drag forces between the model and prototype for a 25 m long boat designed for a maximum speed of 3.5 m/s, by considering: (a) only the wave resistance and (b) only the skin friction. Show whether exact similarity where both components of the drag force are considered is possible.

56. A plane is designed to take off at 95 mph. However, the maximum available speed for testing is 5 m/s. What is the largest model scale that can be used?

57. A 1.2 long model of a 45 m long ship is tested in a towing tank at a speed of 0.7 m/s. Find the velocity of the ship that corresponds to the experimental conditions.

58. A spillway of a dam is designed by testing a one-tenth scaled model. The flow was found experimentally to be 0.66 m/s and the volumetric flow rate 0.08 m^3/s. What is the expected velocity of the prototype?

59. In flow through porous media the velocity of the fluid, known as the *superficial velocity* v_0 is defined as the volumetric flow rate per unit normal cross-sectional area that includes both the solid and the liquid. Using Darcy's law v_0 is found as: $v_0 = -\kappa/\mu(\nabla p - \rho g)$, where μ and ρ are respectively the viscosity and density of the fluid, p is the density and g gravity. The parameter κ is known as the permeability of the porous material. Using dimensional analysis find the dimension of κ.

Exact Analytic Solutions

Science may set limits to knowledge, but should not set limits to imagination.[1]
— Bertrand Russell

7.1 Mathematical Modeling

The first part of this book dealt with the mathematical framework for describing fluid flow. As established, all natural phenomena, and by extension, all fluid flows are governed by a set of universal laws. Consequently, the conservation of mass and momentum form the theoretical basis for *modeling* isothermal flow problems. In non-isothermal problems these equations are augmented by the energy equation. A mathematical model is considered complete when the number of equations, such as the *governing laws* and *constitutive relations*, is the same as the number of unknown parameters.

Because of limitations in the available constitutive models to describe general turbulent flows, mathematically "complete" models are possible only for laminar flows and for a limited number of turbulent flows. Moreover, even under laminar conditions, the models may still be incomplete as a result of unknown fluid behavior. This is the case of complex fluids such as polymers, two-phase mixtures, and the like, whose behavior is still not very well understood.

Finally, since the governing equations in their original form are nonlinear, exact analytic solutions are possible only for a limited number of simple problems. Nevertheless, despite the simplicity of these problems, they are still worth studying, as they illuminate the basic physics of more complicated flows. In the following text, we present exact analytic solutions for a selected number of classical laminar flows. Implicit, then, is the fact that the Reynolds number (Re) is low enough to ensure that the flow is laminar. For a 2-D channel flow with total height H, experiments show that the flow remains laminar provided that $Re_H <\sim 1000$, whereas in pipe flow with diameter D, the limit is $Re_D < 2100 \sim 2500$. The subscripts on Re denote the length based on which the number is evaluated.

7.2 Poiseuille Flow: Fully Developed Channel Flow

Consider isothermal, laminar flow ($Re_H <\sim 1000$) of an incompressible Newtonian fluid in a long two-dimensional channel with height H as shown in Figure 7.1. The flow is established and maintained by an applied pressure gradient between the inlet and the exit of the channel. This flow is known as the *plane Hagen-Poiseuille flow* or simply the *plane Poiseuille flow*. The objective here is to determine the flow field and other flow parameters.

Governing Equations

The flow is governed by the continuity and momentum equations. For incompressible flow, using the Cartesian coordinate system shown in Figure 7.1, these equations are

---

[1]Bertrand Russell, *A History of Western Philosophy.*

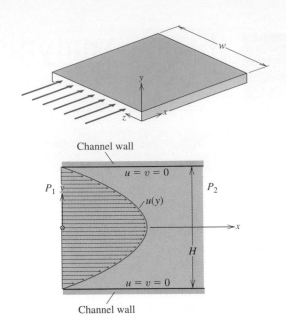

FIGURE 7.1   Fully developed flow in a two-dimensional channel.

Continuity

$$\frac{\partial u}{\partial x} + \frac{\partial v}{\partial y} = 0.$$

x-momentum

$$\rho \left(\frac{\partial u}{\partial t} + u \frac{\partial u}{\partial x} + v \frac{\partial u}{\partial y} \right) = \rho g_x - \frac{\partial P}{\partial x} + \frac{\partial \tau_{xx}}{\partial x} + \frac{\partial \tau_{xy}}{\partial y}.$$

y-momentum

$$\rho \left(\frac{\partial v}{\partial t} + u \frac{\partial v}{\partial x} + v \frac{\partial v}{\partial y} \right) = \rho g_y - \frac{\partial P}{\partial y} + \frac{\partial \tau_{xy}}{\partial x} + \frac{\partial \tau_{yy}}{\partial y}.$$

Constitutive Relations

The constitutive behavior of a Newtonian fluid for two-dimensional laminar flow is described using Newton's law of viscosity as

$$\underbrace{\begin{bmatrix} \tau_{xx} & \tau_{xy} \\ \tau_{xy} & \tau_{yy} \end{bmatrix}}_{\text{Viscous stress } (\tau)} = \mu \underbrace{\begin{bmatrix} \left(2\frac{\partial u}{\partial x} \right) & \left(\frac{\partial u}{\partial y} + \frac{\partial v}{\partial x} \right) \\ \left(\frac{\partial u}{\partial y} + \frac{\partial v}{\partial x} \right) & \left(2\frac{\partial v}{\partial y} \right) \end{bmatrix}}_{\text{Rate of deformation } (\Delta)}$$

Boundary Conditions

The no-slip boundary condition applied at the channel wall requires that the velocities $u(x, y = \pm H/2) = 0$, and $v(x, y = \pm H/2) = 0$.

Assumptions

In this problem, we assume steady conditions (i.e., $\dfrac{\partial u}{\partial t} = \dfrac{\partial v}{\partial t} = 0$), and that the flow is *fully developed*. The fully developed assumption implies that the velocity profile does not change along the direction of the flow. For this reason, there is no need to specify inlet (at $x = 0$) velocity boundary conditions. Additionally, body forces due to gravity are neglected (i.e., $g_x = g_y = 0$).

Solution

The fully developed assumption implies that the velocity components u and v are only functions of y and, therefore, they can be only of the form $u = u(y)$, $v = v(y)$. The assumption also implies that $\dfrac{\partial u}{\partial x} = \dfrac{\partial v}{\partial x} = 0$. The continuity equation

$$\frac{\partial u}{\partial x} + \frac{\partial v}{\partial y} = 0,$$

then reduces to

$$\frac{\partial v}{\partial y} = 0.$$

As v cannot be a function of x, upon integration, it must be a constant number. Also, since at the channel wall we have $v = 0$, the continuity equation is satisfied when v is zero everywhere in the flow. This fact, along with the assumption of steady, fully developed flow and in the absence of body forces, reduce the momentum equations to

$$0 = -\frac{\partial P}{\partial x} + \frac{\partial \tau_{xy}}{\partial y},$$

$$0 = -\frac{\partial P}{\partial y} + \frac{\partial \tau_{yy}}{\partial y},$$

and the constitutive relation to

$$\begin{bmatrix} \tau_{xx} & \tau_{xy} \\ \tau_{xy} & \tau_{yy} \end{bmatrix} = \mu \begin{bmatrix} 0 & \dfrac{\partial u}{\partial y} \\ \dfrac{\partial u}{\partial y} & 0 \end{bmatrix}.$$

By combining the constitutive relations with the momentum equations we get

$$0 = -\frac{\partial P}{\partial x} + \mu \frac{\partial^2 u}{\partial y^2},$$

$$0 = -\frac{\partial P}{\partial y} + 0.$$

Upon integration, the momentum equation in the y-direction gives

$$\frac{\partial P}{\partial y} = 0 \Rightarrow P = P(x) \Rightarrow \frac{\partial P}{\partial x} = \frac{dP}{dx},$$

indicating that the pressure is only a function of x. This means that at a fixed x the pressure is uniform.

Returning to the momentum equation in the x-direction

$$\underbrace{\mu \frac{\partial^2 u}{\partial y^2}}_{function\ of\ y} = \underbrace{\frac{dP}{dx}}_{function\ of\ x} \ .$$

Since the left-hand side is only a function of y and the right-hand side is only a function of x, the two are equal when they are both constants. Therefore, in channel flow dP/dx is constant. Note also, that the pressure gradient is always a negative quantity. This is because a net pressure force in the direction of flow in required to maintain the flow (i.e., to overcome the resistance and induce flow). Therefore, the upstream pressure must be larger than that downstream.

By integration, we get

$$\frac{\partial u}{\partial y} = \frac{1}{\mu}\frac{dP}{dx}y + C_1.$$

By integrating once more,

$$u = \frac{1}{2\mu}\frac{dP}{dx}y^2 + C_1 y + C_2.$$

The constants of integration C_1, C_2 are evaluated by satisfying the boundary conditions that the velocity at the channel walls is zero ($u = 0$ at $y = -H/2$, and $y = H/2$),

$$0 = \frac{1}{2\mu}\frac{dP}{dx}\left(\frac{-H}{2}\right)^2 + C_1\left(\frac{-H}{2}\right) + C_2,$$

$$0 = \frac{1}{2\mu}\frac{dP}{dx}\left(\frac{H}{2}\right)^2 + C_1\left(\frac{H}{2}\right) + C_2.$$

By adding the two equations, we get

$$C_2 = -\frac{1}{2\mu}\frac{dP}{dx}\left(\frac{H}{2}\right)^2,$$

and by subtracting them, we get

$$C_1 = 0.$$

The velocity profile then is

$$u = \frac{1}{2\mu}\frac{dP}{dx}y^2 - \frac{1}{2\mu}\frac{dP}{dx}\left(\frac{H}{2}\right)^2,$$

$$u = \frac{1}{2\mu}\frac{dP}{dx}\left[y^2 - \left(\frac{H}{2}\right)^2\right]. \tag{7.1}$$

As shown in Figure 7.2, this is a parabolic profile with the maximum velocity u_{max} occurring at the centerline $y = 0$ — that is,

$$u_{max} = -\frac{1}{2\mu}\frac{dP}{dx}\left(\frac{H}{2}\right)^2.$$

Since $\dfrac{dP}{dx}$ is negative, the $(-)$ sign in the preceding expression correctly indicates motion along the positive x direction.

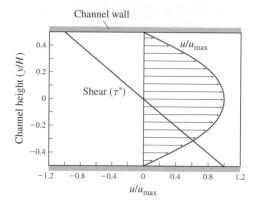

FIGURE 7.2   Velocity and shear distribution for fully developed flow in a two-dimensional channel.

The viscous shear stress distribution is

$$\tau_{xy} = \mu \frac{\partial u}{\partial y} = \frac{dP}{dx}\, y,$$

The absolute value of the shear then is zero at the centerline and maximum at the walls as shown in Figure 7.2. This is expected, since the flow is symmetric along the centerline and the velocity gradient is the largest at the walls.

The velocity in the channel is proportional to the pressure gradient. This implies that once the pressure gradient is fixed, the velocity cannot be varied independently. In other words, the pressure gradient determines the exact amount of flow through the channel. This is not surprising, because the pressure gradient is responsible for the induced flow. The flow is also inversely proportional to the viscosity: the higher is the viscosity, the less is the induced flow. This is again consistent with the fact the viscosity is a measure of the fluid's resistance to flow.

The velocity profile (Equation (7.1)) can be manipulated further by determining the necessary pressure gradient required to establish a given volumetric flow rate $\dot{Q}$ per unit depth. By definition

$$\dot{Q} = \int_{-\frac{H}{2}}^{\frac{H}{2}} u\, dy = \int_{-\frac{H}{2}}^{\frac{H}{2}} \frac{1}{2\mu} \frac{dP}{dx} \left[y^2 - \left(\frac{H}{2} \right)^2 \right] dy,$$

then

$$\dot{Q} = \frac{1}{2\mu} \frac{dP}{dx} \left[\frac{y^3}{3} - \left(\frac{H}{2} \right)^2 y \right]_{-\frac{H}{2}}^{\frac{H}{2}} = -\frac{1}{\mu} \frac{dP}{dx} \frac{H^3}{12}.$$

Therefore, the required pressure gradient is

$$\frac{dP}{dx} = -12 \frac{\mu \dot{Q}}{H^3}.$$

Note the importance of the height H of the channel. Since the pressure gradient is inversely proportional to H^3, as H decreases, the required pressure gradient to sustain a given $\dot{Q}$ increases rapidly.

The velocity profile can be written in terms of $\dot{Q}$ as

$$u = \frac{3}{2}\frac{\dot{Q}}{H}\left[1 - \left(\frac{2y}{H}\right)^2\right]. \tag{7.2}$$

The maximum velocity which occurs at $y = 0$ is

$$u_{max} = \frac{3}{2}\frac{\dot{Q}}{H}.$$

In terms of the average velocity defined as $u_{ave} = \dfrac{\dot{Q}}{H}$, the maximum velocity is expressed as

$$u_{max} = \frac{3}{2}u_{ave}.$$

7.3   Poiseuille Flow: Fully Developed Pipe Flow

The velocity distribution for steady, laminar ($Re_D <\sim 2100$), fully developed flow through a circular pipe with diameter D as shown in Figure 7.3 can be determined by using the conservation of mass and linear momentum in cylindrical coordinates $\mathbf{V} = \mathbf{V}(r, \theta, z)$ (given in Appendix C), and the same boundary conditions and assumptions used in the plane Poiseuille flow. Additionally, we assume that the flow is axisymmetric (i.e., the flow is independent of the azimuthal direction θ). Therefore, all quantities that depend on θ are set to zero. This flow is known as the *Hagen-Poiseuille flow* or simply the *Poiseuille flow*.

The equations of continuity and conservation of linear momentum for the two-dimensional laminar flow problem for a Newtonian fluid in the cylindrical coordinates shown in Figure 7.3 are

Governing Equations

Continuity

$$\frac{1}{r}\frac{\partial}{\partial r}(rv_r) + \frac{\partial v_z}{\partial z} = 0.$$

r-momentum

$$\rho\left(v_r\frac{\partial v_r}{\partial r} + v_z\frac{\partial v_r}{\partial z}\right) = \rho g_r - \frac{\partial P}{\partial r} + \mu\left[\frac{\partial}{\partial r}\left(\frac{1}{r}\frac{\partial}{\partial r}(rv_r)\right) + \frac{\partial^2 v_r}{\partial z^2}\right].$$

z-momentum

$$\rho\left(v_r\frac{\partial v_z}{\partial r} + v_z\frac{\partial v_z}{\partial z}\right) = \rho g_r - \frac{\partial P}{\partial z} + \mu\left[\frac{1}{r}\frac{\partial}{\partial r}\left(r\frac{\partial}{\partial r}(v_z)\right) + \frac{\partial^2 v_z}{\partial z^2}\right].$$

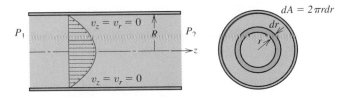

FIGURE 7.3   Fully developed flow in a pipe.

Solution

The fully developed assumption implies that $v_r = v_r(r)$, $v_z = v_z(r)$, and that $\dfrac{\partial v_r}{\partial z} = \dfrac{\partial v_z}{\partial z} = 0$. The continuity equation

$$\frac{1}{r}\frac{\partial}{\partial r}(r v_r) + \frac{\partial v_z}{\partial z} = 0,$$

then reduces to

$$\frac{1}{r}\frac{\partial}{\partial r}(r v_r) = 0 \Rightarrow v_r = \text{constant},$$

where, since v_r cannot be a function of z, it must be a constant number. Moreover, since $v_r = 0$ at the pipe wall, the continuity equation is satisfied only when v_r is zero everywhere in the flow.

Again, using the assumption of steady, fully developed flow with negligible body forces, the momentum equations reduce to

$$0 = -\frac{\partial P}{\partial r},$$

$$0 = -\frac{\partial P}{\partial z} + \mu \frac{1}{r}\frac{\partial}{\partial r}\left(r\frac{\partial}{\partial r}(v_z)\right).$$

Upon integration, the momentum equation in the r-direction gives

$$\frac{\partial P}{\partial r} = 0 \Rightarrow P = P(z),$$

indicating that the pressure is only a function of z. Consequently,

$$\frac{\partial P}{\partial z} = \frac{dP}{dz}.$$

As discussed in Section 7.1, the pressure gradient is always a negative number.

Returning to the momentum in the z-direction

$$\underbrace{\mu\frac{1}{r}\frac{\partial}{\partial r}\left(r\frac{\partial}{\partial r}(v_z)\right)}_{\textit{function of } r} = \underbrace{\frac{dP}{dz}}_{\textit{function of } z}.$$

Since the left-hand side is only a function of r and the right-hand side is only a function of z, the two are equal when they are both constants.

By rearranging,

$$\frac{\partial}{\partial r}\left(r\frac{\partial}{\partial r}(v_z)\right) = r\frac{1}{\mu}\frac{dP}{dz},$$

and by integrating

$$\frac{\partial u}{\partial r} = \frac{1}{\mu}\frac{dP}{dz}\frac{r}{2} + \frac{C_1}{r}.$$

By integrating once more

$$u = \frac{1}{\mu}\frac{r^2}{4}\frac{dP}{dz} + C_1 \ln r + C_2.$$

Since the velocity at the center of the cross section $r = 0$ is finite, C_1 must be zero. The other constant of integration C_2 is evaluated by satisfying the boundary condition $v_z = 0$

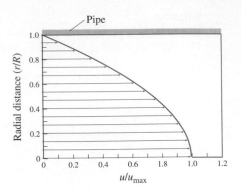

FIGURE 7.4   Velocity distribution in fully developed flow in a pipe.

at $r = R$ as

$$C_2 = -\frac{1}{\mu}\frac{R^2}{4}\frac{dP}{dz}.$$

The velocity profile then is

$$v_z = \frac{1}{4\mu}\frac{dP}{dz}(r^2 - R^2). \tag{7.3}$$

As shown in Figure 7.4, this is a parabolic profile with the maximum velocity $v_{z_{max}}$ occurring at the centerline $r = 0$—that is,

$$v_{z_{max}} = -\frac{1}{4\mu}\frac{dP}{dz}R^2.$$

The velocity profile can also be expressed in terms of the volumetric flow rate

$$\dot{Q} = \int_0^R u\,2\pi r\,dr = \int_0^R \frac{1}{4\mu}\frac{dP}{dz}(r^2 - R^2)2\pi r\,dr,$$

$$\dot{Q} = -\frac{\pi}{8\mu}\frac{dP}{dz}R^4.$$

Therefore, the required pressure gradient to induce $\dot{Q}$ is

$$\frac{dP}{dz} = -\frac{8\mu\dot{Q}}{\pi R^4}.$$

Since the pressure gradient is inversely proportional to R^4, in order to sustain a given $\dot{Q}$, the required dP/dx increases rapidly with decreasing R.

The velocity profile in terms of $\dot{Q}$ is given as

$$v_z = \frac{2\dot{Q}}{\pi R^2}\left[1 - \left(\frac{r}{R}\right)^2\right]. \tag{7.4}$$

The maximum velocity occurs at the centerline $r = 0$. Hence,

$$v_{z_{max}} = \frac{2\dot{Q}}{\pi R^2}.$$

In terms of the average velocity defined as $v_{z_{ave}} = \dfrac{\dot{Q}}{\pi R^2}$, the maximum velocity is

$$v_{z_{max}} = 2v_{z_{ave}}.$$

7.4 Gravity Flow: Flow Down an Inclined Plane

Consider laminar liquid flow flowing steadily down an inclined plane under the influence of gravity, as shown in Figure 7.5. The objective here is to determine the flow parameters as a function of the inclination angle α.

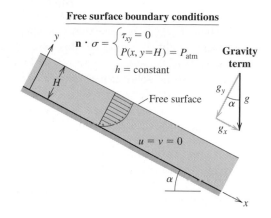

FIGURE 7.5 Fully developed flow down an inclined surface.

Governing Equations

For this two-dimensional isothermal flow, the continuity and momentum equations in the Cartesian coordinate system shown in Figure 7.5 are

Continuity

$$\frac{\partial u}{\partial x} + \frac{\partial v}{\partial y} = 0.$$

x-momentum

$$\rho\left(u\frac{\partial u}{\partial x} + v\frac{\partial u}{\partial y}\right) = \rho g_x - \frac{\partial P}{\partial x} + \frac{\partial \tau_{xx}}{\partial x} + \frac{\partial \tau_{xy}}{\partial y}.$$

y-momentum

$$\rho\left(u\frac{\partial v}{\partial x} + v\frac{\partial v}{\partial y}\right) = \rho g_y - \frac{\partial P}{\partial y} + \frac{\partial \tau_{xy}}{\partial x} + \frac{\partial \tau_{yy}}{\partial y}.$$

where g_x and g_y are the components of the gravity vector, **g**. With respect to the chosen coordinate system $g_x = g\,\sin\alpha$ and $g_y = -g\,\cos\alpha$.

Constitutive Relations

The constitutive relation for laminar flow of a Newtonian fluid is

$$\begin{bmatrix} \tau_{xx} & \tau_{xy} \\ \tau_{xy} & \tau_{yy} \end{bmatrix} = \mu \begin{bmatrix} \left(2\dfrac{\partial u}{\partial x}\right) & \left(\dfrac{\partial u}{\partial y} + \dfrac{\partial v}{\partial x}\right) \\ \left(\dfrac{\partial u}{\partial y} + \dfrac{\partial v}{\partial x}\right) & \left(2\dfrac{\partial v}{\partial y}\right) \end{bmatrix}.$$

Boundary Conditions

As discussed in Chapter 5, the free surface is described by two boundary conditions: a kinematic condition that indicates that fluid particles at the free surface move with the local velocity — that is,

$$\frac{\partial h}{\partial t} + u\frac{\partial h}{\partial x} = v,$$

and a dynamic condition that expresses a force balance at the free surface

$$\textit{Traction Vector} = \mathbf{n} \cdot \sigma = \mathbf{F}_{applied},$$

where $\mathbf{F}_{applied}$ is the externally applied force per unit area along the free surface with local height $h = h(x)$. In this case, the applied force is the atmospheric pressure acting normal to the surface as shown in Figure 7.5.

At the solid surface, we are using the classical no-slip boundary condition $u(x, y = 0) = 0$ and $v(x, y = 0) = 0$.

Assumptions

Again, we assume that the flow is *fully developed* (i.e., derivatives of the velocity with respect to the x-direction are zero).

Solution

The fully developed assumption implies that the velocity components u and v can only be of the form $u = u(y)$, $v = v(y)$ and that $\dfrac{\partial u}{\partial x} = \dfrac{\partial v}{\partial x} = 0$.

The continuity equation

$$\frac{\partial u}{\partial x} + \frac{\partial v}{\partial y} = 0,$$

then reduces to

$$\frac{\partial v}{\partial y} = 0 \Rightarrow v = \text{ constant},$$

where, very much as in the earlier examples, since v cannot be a function of x and that at the channel wall $v = 0$, the continuity equation is satisfied when v is zero everywhere in the flow.

Under steady conditions, the kinematic free surface boundary condition

$$\frac{\partial h}{\partial t} + u\frac{\partial h}{\partial x} = v,$$

reduces to

$$u\frac{\partial h}{\partial x} = v,$$

and since $v = 0$,

$$\frac{\partial h}{\partial x} = 0 \Rightarrow h = \text{ constant}.$$

As a consequence of the above assumptions, the depth of the liquid is constant (i.e., the free surface is parallel to the inclined plane).

For constant free surface elevation, the outward unit normal vector is $\mathbf{n} = \mathbf{j}$. The dynamic boundary condition along the free surface then is given as

$$\underbrace{(0\mathbf{i} + \mathbf{j})}_{\text{normal vector}} \cdot \underbrace{\begin{bmatrix} -P + \tau_{xx} & \tau_{xy} \\ \tau_{xy} & -P + \tau_{yy} \end{bmatrix}}_{\text{stress tensor}} = -P_a\mathbf{j}.$$

By performing the vector-tensor product, we obtain the following relations:

x-direction

$$\tau_{xy} = 0.$$

y-direction

$$(-P + \tau_{yy}) = -P_a.$$

However, since $v = 0$ we have $\tau_{yy} = 2\mu \dfrac{\partial v}{\partial y} = 0$. Therefore, the dynamic boundary conditions reduce to

$$\tau_{xy} = 0 \text{ and } P(x, y = H) = P_a.$$

Using the fully developed assumption and $v = 0$, the momentum equations reduce to

$$0 = -\frac{\partial P}{\partial x} + \rho g \, \sin \alpha + \frac{\partial \tau_{xy}}{\partial y},$$

$$0 = -\frac{\partial P}{\partial y} - \rho g \, \cos \alpha + \frac{\partial \tau_{yy}}{\partial y}.$$

and the constitutive relations to

$$\begin{bmatrix} \tau_{xx} & \tau_{xy} \\ \tau_{xy} & \tau_{yy} \end{bmatrix} = \mu \begin{bmatrix} 0 & \dfrac{\partial u}{\partial y} \\ \dfrac{\partial u}{\partial y} & 0 \end{bmatrix}.$$

By combining the constitutive relations with the momentum equations

$$0 = -\frac{\partial P}{\partial x} + \rho g \, \sin \alpha + \mu \frac{\partial^2 u}{\partial y^2},$$

$$0 = -\frac{\partial P}{\partial y} - \rho g \, \cos \alpha.$$

Upon integration, then, the momentum equation in the y-direction gives

$$\frac{\partial P}{\partial y} = -\rho g \, \cos \alpha \Rightarrow P = -\rho g \, \cos \alpha y + C(x),$$

which indicates that the pressure varies linearly with the depth. This is not surprising, because as there is no flow in this direction, the pressure variation must be the same as the hydrostatic pressure distribution.

The constant $C(x)$ is determined by satisfying the boundary condition at the free surface $y = H$, $P = P_a$,

$$C(x) = P_a + \rho g \, \cos \alpha H.$$

However, since P_a is independent of x, $C(x)$ cannot be a function of x and, hence, it is a constant number. Therefore, the pressure varies only along the y-direction as

$$P = P_a + \rho g \, \cos \alpha (H - y).$$

Returning to the x-momentum equation,

$$\mu \frac{\partial^2 u}{\partial y^2} = \frac{\partial P}{\partial x} \quad \rho g \, \sin \alpha.$$

But, as the pressure varies only in the y-direction, $\dfrac{\partial P}{\partial x} = 0$ and then

$$\mu \frac{\partial u}{\partial y} = -\rho g \, \sin \alpha \, y + C_1.$$

Using the boundary condition that the shear stress is zero at the free surface $y = H$,

$$\tau_{xy} = \mu \frac{\partial u}{\partial y} = -\rho g \, \sin \alpha \, H + C_1 = 0.$$

Therefore, $C_1 = \rho g \, \sin \alpha \, H$, and

$$\mu \frac{\partial u}{\partial y} = \rho g \, \sin \alpha (H - y),$$

which shows that the shear stress assumes its maximum value at the solid surface, and then decreases linearly to zero at the free surface.

Integrating, the foregoing relation once more,

$$u = \frac{1}{\mu} \rho g \, \sin \alpha \left(Hy - \frac{y^2}{2} \right) + C_2,$$

and by satisfying the no-slip boundary condition $u = 0$ at $y = 0$, $C_2 = 0$. The velocity profile is obtained as

$$u = \frac{1}{\mu} \rho g \, \sin \alpha \left(Hy - \frac{y^2}{2} \right). \tag{7.5}$$

This is a parabolic velocity profile, and, as shown in Figure 7.6, the maximum velocity u_{max} occurs at the free surface

$$u_{max} = \frac{H^2}{2\mu} \rho g \, \sin \alpha.$$

The total flow rate per unit width $\dot{Q}$ then is

$$\dot{Q} = \int_0^H u \, dy = \int_0^H \frac{1}{\mu} \rho g \, \sin \alpha \left(Hy - \frac{y^2}{2} \right) dy = \frac{1}{\mu} \rho g \, \sin \alpha \left(H \frac{H^2}{2} - \frac{H^3}{6} \right)$$

$$\dot{Q} = \frac{H^3}{3\mu} \rho g \, \sin \alpha. \tag{7.6}$$

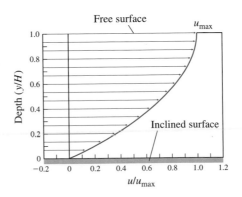

FIGURE 7.6   Velocity distribution in fully developed flow down an inclined surface.

7.5 Film Drawing

Thin films are usually produced using the method shown in Figure 7.7 where molten material is drawn out of a die and, upon solidification, is rolled into large rolls. Because of the action of the roll, the material is effectively pulled out of the die with force $\mathbf{F}$. The analysis seeks to determine the relation between the force $\mathbf{F}$ per unit film width, the volumetric flow rate per unit width $\dot{Q}$, and the thickness of the film (i.e., $h = h(x)$).

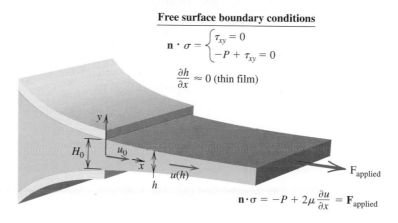

Free surface boundary conditions

$$\mathbf{n} \cdot \sigma = \begin{cases} \tau_{xy} = 0 \\ -P + \tau_{xy} = 0 \end{cases}$$

$$\frac{\partial h}{\partial x} \approx 0 \text{ (thin film)}$$

$$\mathbf{n} \cdot \sigma = -P + 2\mu \frac{\partial u}{\partial x} = \mathbf{F}_{\text{applied}}$$

FIGURE 7.7 Film drawing from a rectangular die.

Governing Equations

Again, the governing laws are

Continuity

$$\frac{\partial u}{\partial x} + \frac{\partial v}{\partial y} = 0.$$

x-momentum

$$\rho\left(\frac{\partial u}{\partial t} + u\frac{\partial u}{\partial x} + v\frac{\partial u}{\partial y}\right) = \rho g_x - \frac{\partial P}{\partial x} + \frac{\partial \tau_{xx}}{\partial x} + \frac{\partial \tau_{xy}}{\partial y}.$$

y-momentum

$$\rho\left(\frac{\partial v}{\partial t} + u\frac{\partial v}{\partial x} + v\frac{\partial v}{\partial y}\right) = \rho g_y - \frac{\partial P}{\partial y} + \frac{\partial \tau_{yx}}{\partial x} + \frac{\partial \tau_{yy}}{\partial y}.$$

where g_x and g_y are the components of the gravity vector $\mathbf{g}$.

Constitutive Relations

For laminar flow of a Newtonian fluid, the constitutive relation is

$$\begin{bmatrix} \tau_{xx} & \tau_{xy} \\ \tau_{xy} & \tau_{yy} \end{bmatrix} = \mu \begin{bmatrix} \left(2\dfrac{\partial u}{\partial x}\right) & \left(\dfrac{\partial u}{\partial y} + \dfrac{\partial v}{\partial x}\right) \\ \left(\dfrac{\partial u}{\partial y} + \dfrac{\partial v}{\partial x}\right) & \left(2\dfrac{\partial v}{\partial y}\right) \end{bmatrix}.$$

Boundary Conditions

The free surface is described by two boundary conditions: a kinematic condition and a dynamic condition. The kinematic condition is

$$\frac{\partial h}{\partial t} + u\frac{\partial h}{\partial x} = v,$$

and the dynamic condition is

$$Traction\ Vector = \mathbf{n} \cdot \sigma = F_{applied},$$

where $F_{applied}$ is the external force per unit area along the free surface.

Since the film is pulled with a force $\mathbf{F} = F\mathbf{i}$ in the x-direction, at any plane normal to the horizontal direction

$$\mathbf{i} \cdot \underbrace{(\mathbf{i} + 0\mathbf{j})}_{\mathbf{n}} \cdot \sigma = \frac{F}{h}\mathbf{i}.$$

Assumptions

We assume that the flow is steady (i.e., $\frac{\partial u}{\partial t} = \frac{\partial v}{\partial t} = 0$). Furthermore, we assume that the film is very thin compared to the length L (i.e., $h \ll L$) and that the vertical velocity v is small compared to the horizontal velocity u. Moreover, we assume that the overall process is slow and, hence, the convective effects are negligible. The free surface is treated as a traction free boundary (i.e., $\mathbf{n} \cdot \sigma = \mathbf{0}$).

Solution

The analysis that follows is valid for $h \ll L$ by assuming $\frac{dh}{dx} \approx 0$. Therefore, along with the assumptions of steady flow and $v \approx 0$, the kinematic equation is satisfied automatically.

From the description of the problem, the significant parameters are the horizontal velocity u and thickness h of the film. The horizontal velocity changes along the length of the film x and, hence, $\frac{\partial u}{\partial x}$ cannot be neglected. According to the continuity equation, though, $\frac{\partial u}{\partial x} = -\frac{\partial v}{\partial y}$. Therefore, despite the fact that v is assumed to be small, the derivative $\frac{\partial v}{\partial y}$ is not small and must be retained in the equation.

The dynamic boundary condition is obtained by assuming that the normal vector is essentially in the y-direction, i.e., $\mathbf{n} = \mathbf{j}$,

$$\underbrace{(0\mathbf{i} + \mathbf{j})}_{normal\ vector} \cdot \underbrace{\begin{bmatrix} -P + \tau_{xx} & \tau_{xy} \\ \tau_{xy} & -P + \tau_{yy} \end{bmatrix}}_{stress\ tensor} = \mathbf{0}.$$

By multiplying the normal vector and the stress tensor the boundary condition in the x-direction is

$$\tau_{xy} = 0 \Rightarrow \left.\frac{\partial u}{\partial y}\right|_{y=\pm h/2} = 0. \tag{7.7}$$

and in the y-direction

$$(-P + \tau_{yy})|_{y=\pm h/2} = \left.\left(-P + 2\mu\frac{\partial v}{\partial y}\right)\right|_{y=\pm h/2} = 0. \tag{7.8}$$

Using the assumptions, the momentum equations reduce to

$$0 = -\frac{\partial P}{\partial x} + \frac{\partial \tau_{xx}}{\partial x} = \frac{\partial}{\partial x}(-P + \tau_{xx}),$$

$$0 = -\frac{\partial P}{\partial y} + \frac{\partial \tau_{yy}}{\partial y} = \frac{\partial}{\partial y}(-P + \tau_{yy}),$$

and the constitutive relations to

$$\begin{bmatrix} \tau_{xx} & \tau_{xy} \\ \tau_{xy} & \tau_{yy} \end{bmatrix} = \mu \begin{bmatrix} 2\dfrac{\partial u}{\partial x} & 0 \\ 0 & 2\dfrac{\partial v}{\partial y} \end{bmatrix}.$$

Upon integration, the momentum equation in the y-direction gives

$$\frac{\partial}{\partial y}\left(-P + 2\mu\frac{\partial v}{\partial y}\right) = 0 \Rightarrow -P + 2\mu\frac{\partial v}{\partial y} = C_1(x).$$

According to the boundary condition (Equation (7.8)), along the free surface independently of x, the constant is zero, $C_1(x) = 0$. Therefore, across the entire thickness

$$-P + 2\mu\frac{\partial v}{\partial y} = 0.$$

Using the continuity equation $\dfrac{\partial v}{\partial y} = -\dfrac{\partial u}{\partial x}$, we have

$$-P + 2\mu\frac{\partial v}{\partial y} = -P - 2\mu\frac{\partial u}{\partial x} = 0 \Rightarrow P = -2\mu\frac{\partial u}{\partial x}.$$

Returning to the momentum in the x-direction,

$$\frac{\partial}{\partial x}\left(-P + 2\mu\frac{\partial u}{\partial x}\right) = 0,$$

upon integration we get

$$-P + 2\mu\frac{\partial u}{\partial x} = C_2(x).$$

But according to the boundary condition (Equation (7.7)),

$$\mathbf{i} \cdot (\mathbf{i} \cdot \sigma) = -P + 2\mu\frac{\partial u}{\partial x} = \frac{F}{h}.$$

Therefore,

$$C_2(x) = \frac{F}{h}.$$

However, since $P = -2\mu\dfrac{\partial u}{\partial x}$,

$$4\mu\frac{\partial u}{\partial x} = \frac{F}{h} \Rightarrow \frac{\partial u}{\partial x} = \frac{F}{4\mu h}. \tag{7.9}$$

Since v is small, mass conservation is satisfied globally using

$$\dot{Q} = u\,h,$$

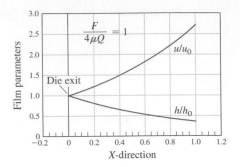

FIGURE 7.8   Velocity distribution and film thickness profile as a function of the distance from the die.

where $\dot{Q}$ is the volumetric flow rate per unit width. At the exit of the die ($x = 0$), we have $\dot{Q} = u_0 H_o$. Differentiating the preceding equation,

$$u\frac{dh}{dx} + h\frac{du}{dx} = 0,$$

and by substituting for $\dfrac{\partial u}{\partial x}$ the result in Equation (7.9),

$$u\frac{dh}{dx} + \frac{F}{4\mu} = 0.$$

Since $u = \dot{Q}/h$, we get

$$\frac{dh}{dx} + \frac{hF}{4\mu\dot{Q}} = 0.$$

Therefore, the film thickness decreases exponentially as

$$h = h_0 e^{-\frac{F}{4\mu\dot{Q}}x}, \tag{7.10}$$

where h_o is the thickness of the film at $x = 0$. The velocity increases according to

$$u = u_0 e^{\frac{F}{4\mu\dot{Q}}x}, \tag{7.11}$$

where u_o is the velocity at the exit of the die. The x-direction velocity distribution and film profile are shown in Figure 7.8.

7.6   Fully Developed Non-Newtonian Channel Flow

Consider isothermal, steady laminar flow ($Re_H <\sim 1000$) of an incompressible non-Newtonian fluid in a long two-dimensional channel with height H, as shown in Figure 7.1. The flow is established and maintained by an applied pressure gradient between the ends of the channel. The objective here is to calculate the flow field in the channel.

Governing Equations

The flow is governed by the incompressible continuity and momentum equations. In the Cartesian coordinate system shown in Figure 7.1, these equations are

Continuity

$$\frac{\partial u}{\partial x} + \frac{\partial v}{\partial y} = 0.$$

x-momentum

$$\rho \left(u \frac{\partial u}{\partial x} + v \frac{\partial u}{\partial y} \right) = \rho g_x - \frac{\partial P}{\partial x} + \frac{\partial \tau_{xx}}{\partial x} + \frac{\partial \tau_{xy}}{\partial y}.$$

y-momentum

$$\rho \left(u \frac{\partial v}{\partial x} + v \frac{\partial v}{\partial y} \right) = \rho g_y - \frac{\partial P}{\partial y} + \frac{\partial \tau_{xy}}{\partial x} + \frac{\partial \tau_{yy}}{\partial y}.$$

Constitutive Relations

The behavior of the non-Newtonian fluid in laminar flow is assumed to be described by a "power-law" constitutive law of the form

$$\underbrace{\begin{bmatrix} \tau_{xx} & \tau_{xy} \\ \tau_{xy} & \tau_{yy} \end{bmatrix}}_{\text{Viscous stress } (\tau)} - \mu \begin{bmatrix} \underbrace{\left(2\frac{\partial u}{\partial x} \right) \quad \left(\frac{\partial u}{\partial y} + \frac{\partial v}{\partial x} \right)}_{} \\ \underbrace{\left(\frac{\partial u}{\partial y} + \frac{\partial v}{\partial x} \right) \quad \left(2\frac{\partial v}{\partial y} \right)}_{\text{Rate of deformation } (\Delta)} \end{bmatrix},$$

where the viscosity μ is assumed to be a function of the rate of deformation $\wedge$, given by the power-law relation

$$\mu = \mu_o (0.5\Delta_{II})^{\left(\frac{n-1}{2} \right)}.$$

Δ_{II} is the second invariant of the rate of deformation tensor defined as

$$\Delta_{II} = \Delta_{11}\Delta_{11} + \Delta_{12}\Delta_{21} + \Delta_{21}\Delta_{12} + \Delta_{22}\Delta_{22}.$$

Because of the symmetry of Δ, we have

$$\Delta_{II} = \Delta_{11}^2 + 2\Delta_{12}^2 + \Delta_{22}^2 \Rightarrow \Delta_{II} = 4\left(\frac{\partial u}{\partial x} \right)^2 + 2\left(\frac{\partial u}{\partial y} + \frac{\partial v}{\partial x} \right)^2 + 4\left(\frac{\partial v}{\partial y} \right)^2.$$

Note that the term μ_o combines with $(\Delta_{II})^{(n-1)/2}$ to yield the effective viscosity μ at a given rate of deformation. Therefore, μ_o by itself is not a viscosity term and, hence, does not have units of viscosity.

Boundary Conditions

The no-slip boundary condition applied at the channel wall requires that $u(x, y = \pm H/2) = 0$ and $v(x, y = \pm H/2) = 0$.

Assumptions

The flow is assumed to be fully developed, which implies that the velocity profile does not change along the direction of the flow (i.e., derivatives of the velocity with respect to the x-direction are zero). Additionally, the body forces due to gravity are neglected (i.e., $g_x = g_y = 0$).

Solution

The fully developed assumption implies that the velocity components u and v are of the form $u = u(y)$, $v = v(y)$ and that $\dfrac{\partial u}{\partial x} = \dfrac{\partial v}{\partial x} = 0$. The continuity equation

$$\frac{\partial u}{\partial x} + \frac{\partial v}{\partial y} = 0,$$

then reduces to

$$\frac{\partial v}{\partial y} = 0 \Rightarrow v = \text{ constant.}$$

Since at the channel wall $v = 0$, the continuity equation is satisfied when v is zero everywhere in the flow. This fact, along with the assumption of steady, fully developed flow and in the absence of body forces, reduces the momentum equations to

$$0 = -\frac{\partial P}{\partial x} + \frac{\partial \tau_{xy}}{\partial y},$$

$$0 = -\frac{\partial P}{\partial y} + \frac{\partial \tau_{yy}}{\partial y},$$

and the constitutive relations to

$$\begin{bmatrix} \tau_{xx} & \tau_{xy} \\ \tau_{xy} & \tau_{yy} \end{bmatrix} = \mu \begin{bmatrix} 0 & \dfrac{\partial u}{\partial y} \\ \dfrac{\partial u}{\partial y} & 0 \end{bmatrix}.$$

Since Δ_{II} simplifies to

$$\Delta_{II} = 2\left(\frac{\partial u}{\partial y}\right)^2,$$

the viscosity μ reduces to

$$\mu = \mu_o \left(\frac{\partial u}{\partial y}\right)^{(n-1)}.$$

By combining the constitutive relations with the momentum equations

$$0 = -\frac{\partial P}{\partial x} + \frac{\partial}{\partial y}\underbrace{\left[\mu_0 \left(\frac{\partial u}{\partial y}\right)^n\right]}_{\tau_{xy}},$$

$$0 = -\frac{\partial P}{\partial y} + 0.$$

The simplified form of the conservation of linear momentum is similar to the form obtained for the plane Poiseuille flow. The only difference is the constitutive expression used to describe τ_{xy}.

Upon integration, the momentum in the y-direction gives

$$\frac{\partial P}{\partial y} = 0 \Rightarrow P = P(x),$$

indicating that the pressure is only a function of x. Therefore,

$$\frac{\partial P}{\partial x} = \frac{dP}{dx}.$$

Returning to the momentum in the x-direction

$$\underbrace{\frac{\partial}{\partial y}\left(\mu_0\left(\frac{\partial u}{\partial y}\right)^n\right)}_{function\ of\ y} = \underbrace{\frac{dP}{dx}}_{function\ of\ x}.$$

Since the left-hand side of the equation is a function of y and the right-hand side is a function of x, they must be both constants. By integration,

$$\underbrace{\mu_0\left(\frac{\partial u}{\partial y}\right)^n}_{\tau_{xy}} = \frac{dP}{dx}y + C_1.$$

Because of the symmetry of the fully developed flow, along the centerline $\tau_{xy} = 0$ there. Therefore, $C_1 = 0$ and

$$\frac{\partial u}{\partial y} = \left(\frac{1}{\mu_0}\frac{dP}{dx}y\right)^{\frac{1}{n}} = \left(\frac{1}{\mu_0}\frac{dP}{dx}\right)^{\frac{1}{n}}y^{\frac{1}{n}}.$$

In this expression, the sign of $\dfrac{\partial u}{\partial y}$ is always opposite to that of y; for positive y, the velocity *decreases* with increasing y, and $\dfrac{\partial u}{\partial y} < 0$, but with the negative y, we have $\dfrac{\partial u}{\partial y} > 0$. Typically, the proper sign is taken care of automatically by the sign of the pressure gradient, which, as discussed in Section 7.2, is always negative. However, for even-power exponent $1/n$, the correct sign is lost. Therefore, to guard against this, the pressure term is more accurately expressed as

$$-\left(\frac{1}{\mu_0}\left|\frac{dP}{dx}\right|\right)^{\frac{1}{n}}.$$

Integrating once more

$$u = -\left(\frac{1}{\mu_0}\left|\frac{dP}{dx}\right|\right)^{\frac{1}{n}}\left(\frac{n}{n+1}\right)y^{\frac{n+1}{n}} + C_2.$$

The constant C_2 is calculated using the no-slip boundary condition $u = 0$ at $y = h/2$

$$C_2 = \left(\frac{1}{\mu_0}\left|\frac{dP}{dx}\right|\right)^{\frac{1}{n}}\left(\frac{n}{n+1}\right)\left(\frac{H}{2}\right)^{\frac{n+1}{n}}.$$

The velocity profile then is

$$u = -\left(\frac{1}{\mu_0}\left|\frac{dP}{dx}\right|\right)^{\frac{1}{n}}\left(\frac{n}{n+1}\right)\left[y^{\frac{n+1}{n}} - \left(\frac{H}{2}\right)^{\frac{n+1}{n}}\right]. \tag{7.12}$$

The volumetric flow rate per unit width is

$$\dot{Q} = 2\int_0^{\frac{H}{2}} u\,dy = 2\int_0^{\frac{H}{2}} -\left(\frac{1}{\mu_0}\left|\frac{dP}{dx}\right|\right)^{\frac{1}{n}}\left(\frac{n}{n+1}\right)\left[y^{\frac{n+1}{n}} - \left(\frac{H}{2}\right)^{\frac{n+1}{n}}\right]dy$$

$$\dot{Q} = -2\left(\frac{1}{\mu_0}\left|\frac{dP}{dx}\right|\right)^{\frac{1}{n}}\left(\frac{n}{n+1}\right)\int_0^{\frac{H}{2}}\left[y^{\frac{n+1}{n}} - \left(\frac{H}{2}\right)^{\frac{n+1}{n}}\right]dy$$

$$\dot{Q} = -2\left(\frac{1}{\mu_0}\left|\frac{dP}{dx}\right|\right)^{\frac{1}{n}}\left(\frac{n}{n+1}\right)\left[\frac{n}{2n+1}\left(\frac{H}{2}\right)^{\frac{2n+1}{n}} - \left(\frac{H}{2}\right)^{\frac{n+1}{n}}\frac{H}{2}\right]$$

$$\dot{Q} = -2\left(\frac{1}{\mu_0}\left|\frac{dP}{dx}\right|\right)^{\frac{1}{n}}\left(\frac{n}{n+1}\right)\left(\frac{H}{2}\right)^{\frac{2n+1}{n}}\left[\frac{n}{2n+1} - 1\right]$$

$$\Rightarrow \dot{Q} = \left(\frac{2n}{2n+1}\right)\left(\frac{1}{\mu_0}\left|\frac{dP}{dx}\right|\right)^{\frac{1}{n}}\left(\frac{H}{2}\right)^{\frac{2n+1}{n}}.$$

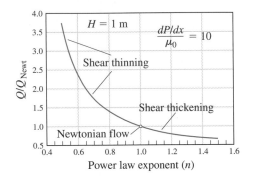

FIGURE 7.9 Volumetric flow rate as a function of the power-law exponent n for a non-Newtonian fluid through a two-dimensional channel (constant pressure gradient).

As shown in Figure 7.9, for the same applied pressure gradient and μ_o, the volumetric flow rate decreases with increasing n due to the increased viscosity (i.e., due to the increased resistance to flow).

The velocity profile can be expressed in terms of $\dot{Q}$

$$u = \left(\frac{2n+1}{n+1}\right)\frac{\dot{Q}}{H}\left(1 - \left(\frac{2y}{H}\right)^{\frac{n+1}{n}}\right), \tag{7.13}$$

the maximum velocity occurs along the centerline $y = 0$,

$$u_{max} = \left(\frac{2n+1}{n+1}\right)\frac{\dot{Q}}{H}.$$

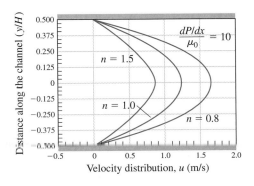

FIGURE 7.10 Velocity profile as a function of the power-law exponent n for fully developed flow of a non-Newtonian fluid through a two-dimensional die (constant pressure gradient).

The velocity profile in terms of u_{max} is

$$u = u_{max} \left(1 - \left(\frac{2y}{H} \right)^{\frac{n+1}{n}} \right). \tag{7.14}$$

Equation (7.14) is plotted in Figure 7.10. As shown in the figure, because of the non-Newtonian ($n \neq 1$) behavior, the velocity profile deviates from the purely parabolic profile ($n = 1$). For the same pressure gradient and μ_o, the induced velocity decreases with increasing n, because of the increased resistance to flow.

7.7   Transient Flow: Impulsively Started Flow

Consider a large body of water as shown in Figure 7.11. Because of the action of wind-induced shear, the water at the surface is suddenly set to motion with velocity u_o. By reaction, because of the action of viscosity, the shear is transmitted to the lower fluid layers. The objective, then, is to find the velocity of the fluid as a function of time and depth.

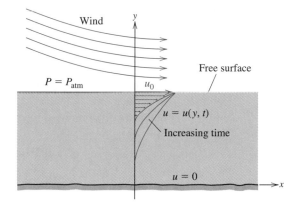

FIGURE 7.11   Impulsively, wind-induced flow of an initially stagnant free surface.

Governing Equations

The flow is governed by the continuity and momentum equations. For incompressible flow, in the Cartesian coordinate system shown in Figure 7.11, these equations are

Continuity

$$\frac{\partial u}{\partial x} + \frac{\partial v}{\partial y} = 0.$$

x-momentum

$$\rho \left(\frac{\partial u}{\partial t} + u \frac{\partial u}{\partial x} + v \frac{\partial u}{\partial y} \right) = \rho g_x - \frac{\partial P}{\partial x} + \frac{\partial \tau_{xx}}{\partial x} + \frac{\partial \tau_{xy}}{\partial y}.$$

y-momentum

$$\rho \left(\frac{\partial v}{\partial t} + u \frac{\partial v}{\partial x} + v \frac{\partial v}{\partial y} \right) = \rho g_y - \frac{\partial P}{\partial y} + \frac{\partial \tau_{xy}}{\partial x} + \frac{\partial \tau_{yy}}{\partial y}.$$

Constitutive Relations

The constitutive behavior of the Newtonian fluid in laminar flow is given in terms Newton's law of viscosity.

$$
\underbrace{\begin{bmatrix} \tau_{xx} & \tau_{xy} \\ \tau_{xy} & \tau_{yy} \end{bmatrix}}_{Viscous\ stress\ (\tau)} = \mu \underbrace{\begin{bmatrix} \left(2\dfrac{\partial u}{\partial x}\right) & \left(\dfrac{\partial u}{\partial y} + \dfrac{\partial v}{\partial x}\right) \\ \left(\dfrac{\partial u}{\partial y} + \dfrac{\partial v}{\partial x}\right) & \left(2\dfrac{\partial v}{\partial y}\right) \end{bmatrix}}_{Rate\ of\ deformation\ (\Delta)}.
$$

Initial and Boundary Conditions

Initially the flow is assumed to be static $\mathbf{u}\,(x, y, t = 0) = 0$. At the free surface $y = 0$, the horizontal velocity $u\,(x, y = 0, t) = u_0$. Far away from the free surface, the velocity is undisturbed (i.e., $u = 0$).

Assumptions

The major assumption is that the free surface is considered flat (i.e., $\dfrac{\partial h}{\partial x} = 0$).

Solution

In this problem, the u velocity is specified at the free surface. Therefore, there is no need to satisfy the dynamic boundary condition. This condition can be used after the final solution is obtained to determine the force exterted by the constant-velocity wind on the free surface.

Using the assumption of a flat free surface (i.e., $\dfrac{\partial h}{\partial t} = \dfrac{\partial h}{\partial x} = 0$), the free-surface kinematic condition

$$
\frac{\partial h}{\partial t} + u\frac{\partial h}{\partial x} = v,
$$

implies that $v = 0$. Therefore, the continuity equation

$$
\frac{\partial u}{\partial x} + \frac{\partial v}{\partial y} = 0,
$$

simplifies to

$$
\frac{\partial u}{\partial x} = 0 \Rightarrow u = u(y).
$$

The constitutive relation reduces to

$$
\begin{bmatrix} \tau_{xx} & \tau_{xy} \\ \tau_{xy} & \tau_{yy} \end{bmatrix} = \mu \begin{bmatrix} 0 & \dfrac{\partial u}{\partial y} \\ \dfrac{\partial u}{\partial y} & 0 \end{bmatrix}.
$$

The momentum equation in the y-direction simplifies to

$$
0 = \rho g - \frac{\partial P}{\partial y},
$$

which, upon integration, reduces to the well-known hydrostatic pressure variation

$$
P = P_a + \rho g y.
$$

Since this is a shear-driven flow, the pressure gradient in the x-direction is zero. The momentum equation in the x-direction then simplifies to

$$\rho \frac{\partial u}{\partial t} = \frac{\partial \tau_{xy}}{\partial y} \Rightarrow \frac{\partial u}{\partial t} = \nu \frac{\partial^2 u}{\partial y^2}.$$

From a physical point of view, the dynamic response of the fluid at a certain location y is governed by the viscosity ν and time t. Using dimensional analysis, we can define a new nondimensional independent variable as

$$\eta = \frac{y}{2\sqrt{\nu t}},$$

where the factor 2 was included for convenience in the solution procedure. With respect to this new variable, the boundary conditions remain the same: at the free surface $\eta = 0$, $u(x, y = 0, t) = u_0$, and at the bottom of the water, $\eta = \infty$, $u = 0$.

The momentum equation then can be expressed in terms of η

$$\frac{\partial u}{\partial \eta} \frac{\partial \eta}{\partial t} = \nu \frac{\partial}{\partial \eta}\left(\frac{\partial u}{\partial y}\right) \frac{\partial \eta}{\partial y},$$

but

$$\frac{\partial \eta}{\partial t} = -\frac{\eta}{2t} \quad \text{and} \quad \frac{\partial \eta}{\partial y} = \frac{1}{2\sqrt{\nu t}}.$$

Using the chain rule $\dfrac{\partial u}{\partial y} = \dfrac{\partial u}{\partial \eta} \dfrac{\partial \eta}{\partial y}$,

$$\frac{\partial u}{\partial \eta} \frac{\partial \eta}{\partial t} = \nu \frac{\partial^2 u}{\partial \eta^2}\left(\frac{\partial \eta}{\partial y}\right)^2 \Rightarrow -\frac{\eta}{2t} \frac{\partial u}{\partial \eta} = \nu \frac{\partial^2 u}{\partial \eta^2}\left(\frac{1}{2\sqrt{\nu t}}\right)^2,$$

which, since $u = u(\eta)$, simplifies to

$$\frac{d^2 u}{d\eta^2} + 2\eta \frac{du}{d\eta} = 0.$$

Therefore, using this convenient coordinate transformation, the final equation is expressed in terms of a single independent variable that combines the effects of time and space. Such variables are called *similarity variables*, and the solution of the final system is called a *similarity solution*.

By defining a new variable as $F = \dfrac{du}{d\eta}$, the preceding equation simplifies to a first-order ordinary differential equation in F

$$\frac{dF}{d\eta} + 2\eta F = 0,$$

which by separating the terms as

$$\frac{1}{F} \frac{dF}{d\eta} = -2\eta,$$

can be integrated to yield

$$\ln F = \ln \frac{du}{d\eta} = -\eta^2 + C,$$

or

$$\frac{du}{d\eta} = C_1 e^{-\eta^2}.$$

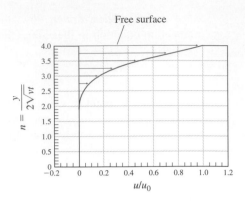

FIGURE 7.12   Nondimensional velocity distribution in an impulsively started free surface flow.

Integrating once more

$$u = C_1 \int_0^\eta e^{-\eta^2} \, d\eta + C_2,$$

the constants of integration can be evaluated using the boundary conditions: at $\eta = 0$, $u = u_0$ (i.e., $C_2 = u_0$), and at at $\eta = \infty$ $u = 0$—that is,

$$C_1 = -\frac{u_0}{\int_0^\infty e^{-\eta^2} \, d\eta}.$$

The velocity profile then is

$$u = u_0 \left(1 - \frac{\int_0^\eta e^{-\eta^2} \, d\eta}{\int_0^\infty e^{-\eta^2} \, d\eta} \right).$$

The integral $\int_0^\eta e^{-\eta^2} \, d\eta$, which appears in many engineering applications, forms the basis of the well-known *error function* defined as

$$erf \ \eta = \frac{2}{\sqrt{\pi}} \int_0^\eta e^{-\eta^2} \, d\eta.$$

Mathematically, $erf \ (\infty) = 1$; therefore,

$$\int_0^\infty e^{-\eta^2} \, d\eta = \frac{\sqrt{\pi}}{2}.$$

The velocity profile then is given as

$$u = u_0(1 - erf \ \eta) = u_0 \left(1 - erf \frac{y}{2\sqrt{vt}} \right). \tag{7.15}$$

The same profile is plotted in Figure 7.12.

7.8   Non-Isothermal Poiseuille Flow

Consider steady non-isothermal laminar ($Re_H < \sim 1000$) Poiseuille flow of an incompressible Newtonian fluid in a long two-dimensional channel with height H as shown in Figure 7.13. Again, the flow is established and maintained by an applied pressure gradient across the channel. However, in this flow the thermal effects are also considered.

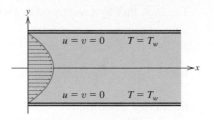

FIGURE 7.13 Non-isothermal flow fully developed flow through a two-dimensional channel with height H.

Governing Equations

In addition to the usual governing equations of continuity and momentum

Continuity

$$\frac{\partial u}{\partial x} + \frac{\partial v}{\partial y} = 0.$$

x-momentum

$$\rho \left(u\frac{\partial u}{\partial x} + v\frac{\partial u}{\partial y} \right) = \rho g_x - \frac{\partial P}{\partial x} + \frac{\partial \tau_{xx}}{\partial x} + \frac{\partial \tau_{xy}}{\partial y}.$$

y-momentum

$$\rho \left(u\frac{\partial v}{\partial x} + v\frac{\partial v}{\partial y} \right) = \rho g_y - \frac{\partial P}{\partial y} + \frac{\partial \tau_{yx}}{\partial x} + \frac{\partial \tau_{yy}}{\partial y}.$$

the two-dimensional energy equation must be included

Energy equation

$$\rho C \underbrace{\left(u\frac{\partial T}{\partial x} + v\frac{\partial T}{\partial y} \right)}_{\text{thermal convection}} = -\underbrace{\left(\frac{\partial q_x}{\partial x} + \frac{\partial q_y}{y} \right)}_{\text{heat conduction}} + \Phi_v,$$

where Φ_v is the viscous dissipation defined as

$$\Phi_v = \underbrace{\tau_{xx}\left(\frac{\partial u}{\partial x} \right) + \tau_{yy}\left(\frac{\partial v}{\partial y} \right) + \tau_{xy}\left(\frac{\partial u}{\partial y} + \frac{\partial v}{\partial x} \right)}_{\text{viscous heating (viscous dissipation)}}.$$

Constitutive Relations

The constitutive behavior of the Newtonian fluid in laminar flow is given in terms Newton's law of viscosity.

$$\underbrace{\begin{bmatrix} \tau_{xx} & \tau_{xy} \\ \tau_{xy} & \tau_{yy} \end{bmatrix}}_{\text{Viscous stress } (\tau)} = \mu \underbrace{\begin{bmatrix} \left(2\dfrac{\partial u}{\partial x}\right) & \left(\dfrac{\partial u}{\partial y} + \dfrac{\partial v}{\partial x}\right) \\ \left(\dfrac{\partial u}{\partial y} + \dfrac{\partial v}{\partial x}\right) & \left(2\dfrac{\partial v}{\partial y}\right) \end{bmatrix}}_{\text{Rate of deformation } (\Lambda)}.$$

The heat flux by conduction is given by "Fourier's Law" of conduction

$$q_x \mathbf{i} + q_y \mathbf{j} = -\kappa \left(\frac{\partial T}{\partial x} \mathbf{i} + \frac{\partial T}{\partial y} \mathbf{j} \right),$$

where κ is the thermal conductivity of the fluid.

Boundary Conditions

The usual no-slip boundary condition is used at the channel walls (i.e., $u(x, y = \pm H/2) = 0$ and $v(x, y = \pm H/2) = 0$). The temperature of the fluid is assumed to be constant at the walls $T = T_w$.

Assumptions

The fluid flow and temperature are assumed to be fully developed. This implies that x-derivatives of the velocity and temperature fields are zero. The body forces due to gravity are again neglected (i.e., $g_x = g_y = 0$). The properties of the fluid (e.g., density, viscosity, thermal conductivity) are assumed to be constant and independent of the temperature.

Solution

By assuming that the fluid properties do not dependent on the temperature, the continuity and momentum equations are also independent of the temperature (i.e., the temperature does not appear in those equations). This implies that we can first solve for the fluid flow, and then for the temperature field.

Since the governing equations, boundary conditions, and assumptions are identical to Poiseuille flow, the velocity field is

$$u = \frac{1}{2\mu} \frac{dP}{dx} \left[y^2 - \left(\frac{H}{2} \right)^2 \right].$$

Using the assumptions of steady, fully developed conditions, the energy equation reduces to

$$-\frac{\partial q_y}{\partial y} + \tau_{xy} \frac{\partial u}{\partial y} = 0,$$

but $\tau_{xy} = \mu \dfrac{\partial u}{\partial y}$ and $q_y = -\kappa \dfrac{\partial T}{\partial y}$. Hence,

$$\frac{d^2 T}{dy^2} = -\frac{\mu}{\kappa} \left(\frac{\partial u}{\partial y} \right)^2,$$

and since $\dfrac{\partial u}{\partial y} = \dfrac{1}{\mu} \dfrac{dP}{dx} y$,

$$\frac{\partial^2 T}{\partial y^2} = -\frac{\mu}{\kappa} \left(\frac{1}{\mu} \frac{dP}{dx} \right)^2 y^2.$$

Using the Poiseuille flow solution, the foregoing is expressed in terms of the volumetric flow rate $\dot{Q}$ as

$$\frac{d^2 T}{dy^2} = -\frac{\mu}{\kappa} \left(-12 \frac{\dot{Q}}{H^3} \right)^2 y^2,$$

Integrating twice, and by applying the boundary conditions, we get

$$T = T_w + \frac{3}{4} \frac{\mu}{\kappa} \frac{\dot{Q}^2}{H^2} \left(1 - \left(\frac{2y}{H} \right)^4 \right). \tag{7.16}$$

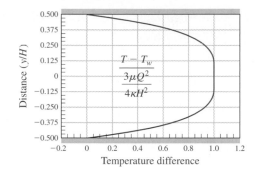

FIGURE 7.14 Temperature distribution in non-isothermal, fully developed flow through a two-dimensional channel.

The final result shows that because of to viscous action, the fluid heats up to a maximum temperature T_{max}

$$T_{max} = T_w + \frac{3}{4}\frac{\mu}{\kappa}\frac{\dot{Q}^2}{H^2}, \qquad (7.17)$$

that occurs at the centerline as shown in Figure 7.14. As expected, the temperature rise is proportional to the viscosity (i.e., the higher the friction between fluid layers, the higher is the heat generated). It also increases with the volumetric flow rate, since this implies increased rate of friction. However, the temperature rise decreases with increasing conductivity because heat conduction increases. The heat generation decreases with increasing channel height because the velocity gradients are reduced.

REFERENCES

R.B. BIRD, R.C. ARMSTRONG, and O.HASSAGER, *Dynamics of Polymeric Liquids*, 2nd ed., New York, Wiley-Interscience, 1987.

G.K. BATCHELOR, F.R.S., *An Introduction to Fluid Dynamics*. Cambridge, England, Cambridge University Press, 1970.

R.B. BIRD, W.E. STEWARD, and E.N. LIGHTFOOT, *Transport Phenomena*. New York, Wiley, 1960.

R.L. PANTON, *Incompressible Flow*, New York, Wiley, 1984.

PROBLEMS

1. Find the maximum velocity and volumetric flow rate in a smooth 5 *cm*-diameter pipe so that the flow remains laminar.

2. Find the pressure gradient required to sustain constant flow of 0.0016 m^2/s though a two dimensional channel with height 0.02 *m*, by assuming fully developed flow conditions and the fluid to be (a) air, (b) water, and (c) glycerin.

3. A constant pressure gradient of 0.20 kPa/m is applied across a two dimensional channel with height of 0.2 *m*. By assuming fully developed conditions, find the volumetric flow rate of glycerin per unit width. What is the maximum velocity in the channel.

4. The maximum velocity in a fully developed water flow through a channel (height 0.04 *m*) is 0.05 *m/s*. Find the volumetric flow rate through the channel and the required pressure gradient to sustain the flow. What is the maximum shear in the flow?

5. Modify the classical Poiseuille problem described in Section 7.2 and consider now that the upper and lower walls of the channel move with constant velocity, U_1 and U_2, respectively. By solving the equations of motion (mass and momentum), find the velocity distribution in the channel.

6. For the modified classical plane Poiseuille flow described in the previous problem use $U_1 = 0.3$ *m/s*, and $U_2 = 0$. Find the pressure gradient required to sustain constant volumetric rate of 0.26 m^2/s of glycerin. Assume the height of the channel is 0.15 *m*. What is the maximum shear in the flow? Repeat the calculations for the case when the plate is moving in the opposite direction with velocity $U_1 = -0.3$ *m/s*.

7. Find the velocity profile for steady, fully developed laminar flow through a circular annulus (i.e., the flow is confined in an annulus) with an out radius R and an inner radius kR, where $0 < k < 1$.

8. Consider the same annulus flow described in the problem above and now consider that: (a) the outer wall moves

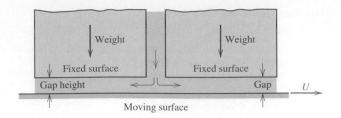

FIGURE 7.15   Schematic for Problem 9.

with velocity u_0, (b) the inner wall moves with velocity u_0. What happens when both walls move with the same velocity?

9. A common application in fluid mechanics is flow in a narrow gap between two moving surfaces; or equivalently, between two surfaces where one is moving relative to the other. The purpose on the fluid (visosity μ) is to keep the two surfaces separate (i.e., for lubrication purposes). Therefore, fluid is pumped between the surfaces and flows through the gap as shown in Figure 7.15. The flow then is a combination of Poiseuille and Couette flow. Show that the Navier-Stokes equations for this problem reduce to

$$-\frac{dP}{dx} + \mu \frac{\partial^2 u}{\partial y^2} = 0,$$

provided that the gap height is constant. Find an expression for the load bearing capacity of the fluid, i.e., the relation between the weight of the solid body supported by the fluid and the volumetric flow rate.

10. Consider flow of a thin, uniform film on an incompressible Newtonian fluid over a long inclined plate with an angle θ with respect to the horizontal. Find the velocity distribution across the thickness δ of the film when the plate is moving against the flow direction with velocity V. Find the critical value of V for which the volumetric flow rate vanishes.

11. Two immiscible incompressible liquids A and B with densities ρ_A and ρ_B ($\rho_A > \rho_B$) and viscosities μ_A and μ_B flow between two long parallel plates. The flow is induced by moving the lower plate with speed V while keeping the upper plate fixed. Find the velocity distribution as a function of the distance in the direction normal to the flow direction. Consider the thickness of each layer to be H_A and H_B.

12. Consider fully developed Poiseuille flow between two plates $2H$ apart where the fluid is allowed to slip along the solid surface with a slip velocity u_w given by $-\tau_{xy} = \beta u_w$ where β is a slip coefficient. Find the velocity distribution and the volumetric flow rate through the channel.

13. Consider fully developed Poiseuille flow in a pipe with radius R where the fluid is allowed to slip along the solid surface with a slip velocity u_w given by $-\tau_{xy} = \beta u_w$ where β is a slip coefficient. Find the velocity distribution and the volumetric flow rate through the channel.

14. Find the velocity profile for steady flow between two rotating cylinders; the inner cylinder rotates with velocity $\omega_1 R_1$ and the outer cylinder with velocity $\omega_2 R_2$, where ω_1 and ω_2 are angular velocities. What happens when both cylinders move with the same velocity?

15. Find the velocity profile for fully developed flow down a pipe, which is also rotating with velocity ωR. [Note that the flow is a linear combination of axial pressure driven flow and radial flow due to the roatation of the pipe.]

16. Consider two vertical plates (in the direction of gravity) at a distance h apart where flow of a liquid is induced by gravity and by one of the plates which is moving with a velocity V in the same direction as gravity. Using the conservation of laws, find the velocity and stress distribution in the flow.

17. Find the maximum velocity and stress in the flow of glycerine in a vertical two dimensional channel with height of $0.07\ m$ when one of the walls is moving in the same direction as gravity with velocity $0.1\ m/s$.

18. Consider a vertical pipe with fully developed flow established by gravity. Using the conservation of mass and momentum determine the velocity distribution and mass flow rate as a function of the diameter of the pipe and the fluid properties.

19. Find the volumetric flow rate of glycerine induced by gravity in vertical a $0.08\ m$ diameter pipe.

20. A horizontal $0.05\ m$ diameter pipe delivers water at a rate of $8.1 \times 10^{-5}\ m^3/s$. Find the required pressure gradient to sustain this flow.

21. A horizontal $0.01\ m$ pipe is induced by a constant pressure gradient of $0.034\ Pa/m$. Find the volumetric flow rate and the maximum velocity.

22. Consider a vertical pipe with diameter D where the flow is induced by both gravity and a pressure gradient. Using the conservation of mass and momentum determine the velocity distribution and mass flow rate s a function of the diameter of the pipe and the fluid properties. Under what conditions the effect of gravity can be neglected?

23. Fully developed flow of low-density polyethelene ($\rho = 914.2\ kg/m^3$, $\mu = 64.0\ Pa \cdot s$) at the rate of $0.1\ m^3/s$ is induced in a vertical pipe with diameter of $0.4\ m$, due to an applied pressure gradient and gravity. What is the required pressure gradient?

24. A $0.01\ m$ thin film of glycerin flows down an inclined surface with angle $30°$ with respect to the horizontal direction. Find the maximum velocity and the volumetric flow rate per unit width.

25. Consider a modified problem with flow down an inclined surface as discussed in Section 7.4 where the surface is moving with velocity U in the same direction as gravity. Find the velocity distribution and the volumetric flow rate per unit width.

26. A $0.065\ m$ polymer film ($\rho = 914.2\ kg/m^3$, $\mu = 64.0\ Pa \cdot s$) flows down an inclined surface at an angle of $40°$ degrees

with respect to the horizontal direction. If the surface is moving in the same direction as the film at 0.1 m/s. Find the volumetric flow rate.

27. Find the pressure gradient required to sustain flow of a power-law Non-Newtonian fluid through a two dimensional channel relative to the pressure gradient required for a Newtonian fluid and the same volumetric flow rate. Consider $n = 0.6$, $n = 0.8$, $n = 1.2$ and $n = 1.4$, $\mu_0 = 1$ $Pa \cdot s$ and $H = 0.1$ m.

28. Find the pressure gradient required to sustain a volumetric flow rate per unit width of 0.76 m^2/s of a power-law Non-Newtonian fluid ($\rho = 750$ kg/m^3, $\mu = 4.6 \times 10^{-1}$ $Pa \cdot s$) through a two dimensional channel with height of 0.16 m. Use $n = 0.6$, $n = 0.8$, $n = 1.2$, and $n = 1.4$.

29. How long will it take a fluid particle 10 m below the surface of a deep lake to move when the free surface is suddenly moved with velocity of 4.2 m/s?

30. How long will it take for the shear induced at the free surface of a lake by a 20 $miles/hr$ wind to reach at a depth of 40 m?

31. Calculate and plot the shear at a 2-m depth, induced by a 35 $miles/hr$ wind at the surface of a lake.

32. Calculate and plot the vorticity at a depth of 5 m due to a 25 $miles/hr$ wind at the surface of a lake.

33. Find the velocity as a function of time at a depth of 0.3 m when the surface of a deep lake is suddenly moved with velocity 3 m/s.

34. Reconsider the classical Couette flow of a Newtonian fluid between two long plates at a distance H apart, with the to plate moving with velocity U. Find the velocity when the plate is suddenly stopped. (Hint: define a new variable as $w = u - \dfrac{yU}{H}$, where u is the horizontal velocity.)

35. A polymeric film ($\rho = 914.2$ kg/m^3, $\mu = 64.0$ $Pa \cdot s$) 0.12 m long, is drawn from a die as discussed in Section 7.5. Find the force required to pull the film when the velocity at the exit of the die is 0.1 m/s and the height is 0.015 m.

36. A polymeric film (($\rho = 914.2$ kg/m^3, $\mu = 64.0$ $Pa \cdot s$) 0.16 m long, is drawn from a die as discussed in Section 7.5. Find the force required to pull the film when the velocity at the exit of the die is 0.1 m/s and the height is 0.015 m.

37. Glycerin ($\kappa = 0.286$ $W/m°C$) is flowing in a two-dimensional channel with height of 0.02 m. The flow is sustained by a 0.24 Pa/m pressure gradient. If the channel wall temperature is $30°$ degrees, find the temperature distribution in the channel. What is the maximum temperature in the channel?

38. Find the pressure gradient that will produce a temperature increase of $1.5°$ degree in a 0.03 m channel due to flow of a polymeric melt ($\rho = 1120$ kg/m^3, $\mu = 1.5$ $Pa \cdot s$, $\kappa = 0.1$ $W/m°C$).

39. Reconsider the drawing of a thin film as discussed in Section 7.5 and assume now that the film is drawn by a force aided by gravity. Using the equations of motion find the velocity and film thickness as a function of the length of the film.

8 Combined Analytic and Experimental Solutions

> *... neither theory nor experiments alone will ever solve complex problems of fluid flows — a combination of both is needed.*[1]
>
> —Learned Committee, Academie des Paris, 1775.

As discussed in earlier chapters, fluid flow phenomena can become intractable to exact analysis because of turbulence, complex constitutive fluid behavior, and flow geometry. Therefore, the study of fluid flow often involves the combined use of theory, experimentation, and numerical modeling. To understand the behavior of flows that are more complex than those encountered in earlier chapters as well as flows associated with problems that are more typical of common applications, we must first discuss the behavior of the flow near solid surfaces. Because of friction at the fluid-solid interface, a special fluid layer develops adjacent to the surface known as the *boundary layer*. The effects of this layer play a critical role in the dynamics of fluid flow and contribute significantly to many of the complexities encountered in common applications.

Subsequently, following a discussion of the dynamics of the boundary layer, we discuss in detail two important classes of problems: (a) *external* flows past bluff bodies, and (b) *internal* flows through channels and pipes, under both turbulent and laminar conditions. These complex problems are solved by a combination of experimental data and theoretical considerations. Since the advent of faster and larger computers and significant developments in numerical analysis, many complex problems for which a theoretical description is possible can now be solved numerically.

8.1 Boundary Layer Concept

Historically, flow phenomena have been studied by the most famous thinkers of antiquity and, more recently, by the most notable mathematicians and experimentalists. Over time, though, a significant gap of understanding developed between theory and practice, to the extent that while theory could predict with reasonable accuracy fluid flow behavior away from solid surfaces, it could not predict such commonly observed phenomena as fluid resistance or the vortical flow behind bluff bodies. Interestingly, though, early experimentalists could not provide a sufficient theoretical framework to explain them.

The discrepancies between theory and practice were resolved in 1904 by Ludwig Prandtl, a German fluid dynamicist, with his *boundary layer* theory. The theory concentrates on a thin fluid layer near solid surfaces where the viscous effects provide the missing link in explaining the earlier shortcomings of both theory and practice. Boundary layer theory remains one of the most important concepts in fluid dynamics, one that has been pivotal in shaping the field of fluid dynamics.

According to the theory, the flow past solid surfaces may be divided into two regions: an "inner" region confined to a thin layer adjacent to solid surfaces, and an "outer" region away from the solid surface (Figure 8.1). For instance, in uniform flow of air past a flat

---

[1]Members: Jean d' Alembert, Antoine Condorcet, Abbe Charles Bossut, Adrian Marie Legendre, Gaspard Monge.

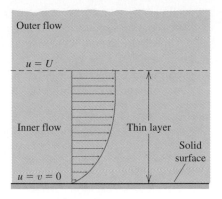

FIGURE 8.1 Schematic of the boundary layer.

plate, the thickness of the boundary layer 0.5 *m* downstream from the leading edge for a velocity of 10 *m/s* is about 0.5 *cm*. Inside this thin layer, the velocity changes from zero at the solid surface (no-slip condition) to the free-stream velocity $u = U$. Because of this large velocity gradient, *fluid deformation, and the associated viscous forces are large and cannot be neglected.*

8.1.1 Boundary Layer Theory

Figure 8.2 shows uniform flow past a flat plate where, resulting from the action of viscosity, a thin layer develops, within which the velocity varies from the no-slip value $u = 0$ at the wall to the uniform flow velocity $u = U$ at the edge of the layer. As shown in the figure, the thickness of the layer increases in the streamwise direction x (the direction along the main flow). At a critical local Reynolds number based on the distance from the upstream edge of the plate L, $Re_L \approx 3 \sim 5 \times 10^5$, the flow undergoes a transition from laminar to turbulent flow. Eventually, for $Re \geq 10^6$ the flow becomes fully turbulent.

The small thickness of the boundary layer compared with the streamwise length of the plate, suggests that the governing equations (momentum and mass conservation) could be simplified from their original form. Indeed, as shown subsequently, by using widely accepted arguments, the general theory can be simplified to what are known as the *boundary layer equations.*

The boundary layer equations are normally derived through an order-of-magnitude analysis of the laminar equations of motion, after they are scaled, using as characteristic

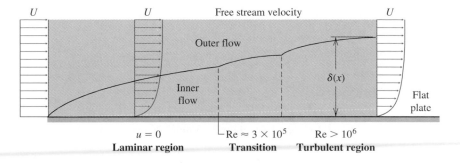

FIGURE 8.2 Transition regions for the boundary layer along a flat plate.

scaling parameters the freestream velocity U and the length of the plate L. Unless otherwise stated, the discussion immediately following refers to scaled variables and quantities.

Key to scaling arguments is the fact that the boundary layer is thin. Therefore, the nondimensional thickness of the layer δ is small, giving $\delta \ll 1$. By definition, the scaled streamwise component of the velocity u and the scaled distance x in the streamwise direction from the leading edge of the plate cannot exceed unity. Therefore, we can write $u \sim 1$ and $x \sim 1$. For this reason, the magnitude of the derivative $\dfrac{\partial u}{\partial x}$ is of order ~ 1.

Now, since the continuity equation requires that

$$\frac{\partial u}{\partial x} + \frac{\partial v}{\partial y} = 0, \quad \text{or} \quad \frac{\partial u}{\partial x} \sim \frac{\partial v}{\partial y},$$

it implies that $\dfrac{\partial v}{\partial y} \sim 1$. However, since the scaled transverse direction y is of the same order of magnitude as the boundary layer thickness δ, we must then have $v \sim \delta$. Proceeding with the same order-of-magnitude analysis for a sample of velocity terms, we find the following estimates:

$$\frac{\partial^2 u}{\partial x^2} \sim 1, \quad \frac{\partial v}{\partial x} \sim \delta, \quad \frac{\partial^2 v}{\partial x^2} \sim \delta, \quad \frac{\partial u}{\partial y} \sim \frac{1}{\delta}, \quad \frac{\partial^2 u}{\partial y^2} \sim \frac{1}{\delta^2}.$$

The nondimensional momentum equations, and the relative size of all scaled velocity terms in them, with the Reynolds number defined as $Re \equiv UL/\nu$, are listed below.

x-momentum

$$\frac{\partial u}{\partial t} + u\frac{\partial u}{\partial x} + v\frac{\partial u}{\partial y} = -\frac{\partial P}{\partial x} + \frac{1}{Re}\left(\frac{\partial^2 u}{\partial x^2} + \frac{\partial^2 u}{\partial y^2}\right).$$

$$1 \;+\; 1\;1 \;+\; \delta\frac{1}{\delta} \;=\; -\frac{\partial P}{\partial x} + \frac{1}{Re}\left(1 + \frac{1}{\delta^2}\right).$$

y-momentum

$$\frac{\partial v}{\partial t} + u\frac{\partial v}{\partial x} + v\frac{\partial v}{\partial y} = -\frac{\partial P}{\partial y} + \frac{1}{Re}\left(\frac{\partial^2 v}{\partial x^2} + \frac{\partial^2 v}{\partial y^2}\right).$$

$$\delta \;+\; 1\frac{\delta}{1} \;+\; \delta\frac{\delta}{\delta} \;=\; -\frac{\partial P}{\partial y} + \frac{1}{Re} \quad (\delta + 1/\delta).$$

The scaling reveals that most terms in the x-momentum equation are of order ~ 1. For the viscous terms we have

$$\frac{\partial^2 u}{\partial x^2} \sim 1, \quad \text{and} \quad \frac{\partial^2 u}{\partial y^2} \sim \frac{1}{\delta^2},$$

which imply that

$$\frac{\partial^2 u}{\partial y^2} \gg \frac{\partial^2 u}{\partial x^2}.$$

Therefore, the term $\dfrac{\partial^2 u}{\partial x^2}$, being smaller than $\dfrac{\partial^2 u}{\partial y^2}$, can be dropped from the equation. Now, in order for the remaining viscous term to be of order ~ 1 we must have

$$\frac{1}{Re}\frac{1}{\delta^2} \sim 1 \rightarrow \frac{1}{Re} \sim \delta^2, \quad \text{or} \quad \delta \sim \sqrt{\frac{1}{Re}}.$$

The nondimensional boundary layer thickness, then, is inversely proportional to the square root of Re (i.e., the thickness decreases as Re increases). Therefore, for meaningful boundary layer analysis, Re must be sufficiently large in order to satisfy the basic assumption of a thin layer used in the derivation of the boundary layer equations.

If we now define a local Reynolds number $Re = \dfrac{U x_a}{\nu}$, where x_a is the actual (dimensional) distance from the front edge of the plate and δ_a the actual thickness of the boundary layer, we can write

$$\delta \equiv \frac{\delta_a}{x} \sim \sqrt{\frac{\nu}{Ux}}, \quad \text{or} \quad \delta_a \sim \sqrt{\frac{\nu x}{U}}.$$

This result verifies the schematic in Figure 8.2, where the boundary layer thickness is shown to increase in the downstream direction.

Because of the small thickness of the boundary layer, most of the terms in the y-momentum equation are of order δ or smaller. Moreover, since $1/Re \sim \delta^2$, the viscous terms are at least of order δ. By eliminating such small terms, the y-momentum equation then reduces to

$$\frac{\partial P}{\partial y} \sim 0 \Rightarrow P = P(x).$$

This result shows that the *pressure across the thickness of the boundary layer is constant and varies only in the direction of the flow.* Of course, given the small thickness of the layer, this is not a surprising result.

By neglecting terms of order δ (or smaller) and by returning to the actual dimensional form of the equations, the *complete boundary layer equations* are given as

Continuity

$$\frac{\partial u}{\partial x} + \frac{\partial v}{\partial y} = 0. \tag{8.1}$$

x-momentum

$$\frac{\partial u}{\partial t} + u\frac{\partial u}{\partial x} + v\frac{\partial u}{\partial y} = -\frac{1}{\rho}\frac{\partial P}{\partial x} + \frac{1}{\rho}\frac{\partial \tau}{\partial y}. \tag{8.2}$$

For laminar flow of Newtonian fluids, the viscous stress τ can be replaced by the constitutive relation $\tau = \mu\dfrac{\partial u}{\partial y}$.

In the preceding equations, the unknowns are the two components of the velocity vector u and v. The pressure gradient $\dfrac{dP}{dx}$ is determined by writing the momentum equation at the edge of the boundary layer, where the viscous stresses vanish

$$\frac{\partial U}{\partial t} + U\frac{dU}{dx} = -\frac{1}{\rho}\frac{dP}{dx}. \tag{8.3}$$

In steady flow, then, the pressure gradient is

$$-\frac{1}{\rho}\frac{\partial P}{\partial x} = U\frac{\partial U}{\partial x},$$

which reduces to

$$\frac{\partial}{\partial x}\left(\rho\frac{U^2}{2} + P\right) = 0.$$

By integration, the pressure as a function of x is

$$\rho\frac{U(x)^2}{2} + P(x) = \rho\frac{U_0^2}{2} + P_0 = Constant,$$

or

$$P(x) = P_0 + \rho \left(\frac{U_0^2}{2} - \frac{U(x)^2}{2} \right),$$ (8.4)

which is the well-known steady Bernoulli equation. Subscript 0 refers to conditions at a point along the streamline at the edge of the boundary layer where both the velocity and pressure are known. Typically, these conditions are taken at the front end of the plate ($x = 0$). For unsteady flows then, the pressure gradient is given by Equation (8.3), and in steady flows it is given by Equation (8.4). Note also that in steady flow past a flat plate, since U is constant, $dP/dx = 0$.

The velocity field within the boundary layer is, in principle, determined from Equations (8.1)–(8.2). An exact analytic solution, however, is not possible because of the nonlinear form of the equations. General computational solutions are possible, however, for flows that can be modeled.

8.1.2   Approximate Momentum Integral Theory

The boundary layer can be modeled approximately using an integral approach in which the governing equations are integrated across the thickness of the layer (Figure 8.3). As shown subsequently, this integration leads to a simple, ordinary differential equation to be integrated along the main flow direction x. To account for the development of the boundary layer and to ensure that its thickness is confined within the limits of integration, the upper limit of ξ is chosen to be larger than the expected thickness of the layer (Figure 8.3). By integration,

$$\int_0^\xi \left(u \frac{\partial u}{\partial x} + v \frac{\partial u}{\partial y} \right) dy = \int_0^\xi -\frac{1}{\rho} \frac{\partial P}{\partial x} \, dy + \int_0^\xi \frac{1}{\rho} \frac{\partial \tau}{\partial y} \, dy.$$ (8.5)

According to the continuity equation,

$$v = -\int_0^y \frac{\partial u}{\partial x} \, dy,$$

and for steady flow, the pressure gradient is

$$\frac{1}{\rho} \frac{dP}{dx} = -U \frac{dU}{dx}.$$

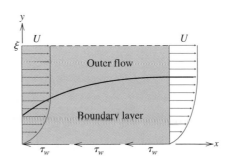

FIGURE 8.3   Integral approach to model the growth of the boundary layer.

By substitution into Equation (8.5), we get

$$\int_0^{\xi} u \frac{\partial u}{\partial x} dy - \int_0^{\xi} \underbrace{\left(\int_0^y \frac{\partial u}{\partial x} dy \right)}_{v} \frac{\partial u}{\partial y} dy = \int_0^{\xi} \underbrace{U \frac{dU}{dx}}_{1/\rho(dP/dx)} dy + \int_0^{\xi} \frac{1}{\rho} \frac{\partial \tau}{\partial y} dy. \qquad (8.6)$$

Defining

$$dB = \frac{\partial u}{\partial y} dy \Rightarrow B = u,$$

and

$$A = \int_0^y \frac{\partial u}{\partial x} dy \Rightarrow dA = \frac{\partial u}{\partial x} dy,$$

by integration by parts, the second term in Equation (8.6), becomes

$$\int_0^{\xi} A \, dB = AB|_0^{\xi} - \int_0^{\xi} B \, dA = U \int_0^{\xi} \frac{\partial u}{\partial x} dy - \int_0^{\xi} u \frac{\partial u}{\partial x} dy.$$

Finally, since ξ is independent of x, by integrating once again by parts and by rearranging the various terms, Equation (8.5) reduces to

$$\frac{\partial}{\partial x} \left\{ U^2 \underbrace{\int_0^{\xi} \left[\frac{u}{U} \left(1 - \frac{u}{U} \right) \right] dy}_{\delta_2(x)} \right\} + U \frac{\partial U}{\partial x} \underbrace{\int_0^{\xi} \left(1 - \frac{u}{U} \right) dy}_{\delta_1(x)} = \frac{\tau_w}{\rho}, \qquad (8.7)$$

where τ_w is the shear stress at the wall. Note that in the range $y \geq \delta$, where δ is now the actual dimensional thickness of the boundary layer, the velocity ratio $u/U = 1$. Therefore, since in this range the term $\left(1 - \frac{u}{U} \right)$ in Equation (8.7) is zero, the upper limit of the integrals ξ can be changed to δ.

Equation (8.7) can be simplified further to

$$\frac{\partial}{\partial x} [U^2 \delta_2(x)] + \delta_1(x) U \frac{\partial dU}{\partial x} = \frac{\tau_w}{\rho}, \qquad (8.8)$$

where the variable $\delta_1(x)$ is the *displacement thickness* and $\delta_2(x)$ is the *momentum thickness*. Equation (8.8) is known as the *momentum-integral equation*, or as the *von Karman's integral equation*. For laminar flow of a Newtonian fluid $\tau_w = \mu \frac{\partial u}{\partial y} \Big|_{y=0}$.

$\delta_1(x)$ is related to the reduction in the mass flow rate per unit depth $\dot{m}_d$ through the boundary layer ($0 \leq y \leq \delta$) as a result of the viscosity induced velocity slowdown ($U - u$) (Figure 8.4),

$$\dot{m}_d = \rho \int_0^{\delta} (U - u) \, dy.$$

As demonstrated in Example 3.3, the rate of mass flow defect $\dot{m}_d$ is due to the displacement of fluid from the near-wall region (i.e., the boundary layer) to the outer region. When $\dot{m}_d$ is expressed in terms of an equivalent thickness δ_1 of uniform flow with the same mass flow rate — that is, $\dot{m}_d = \rho U \delta_1(x)$, we have

$$\dot{m}_d = \rho \delta_1(x) U = \rho \int_0^{\delta} (U - u) \, dy,$$

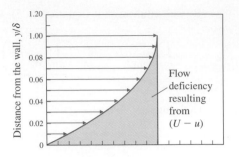

FIGURE 8.4 Definition of the displacement thickness $\delta_1(x)$.

which leads to the definition of $\delta_1(x)$ as

$$\delta_1(x) = \int_0^\delta \left(1 - \frac{u}{U}\right) dy. \tag{8.9}$$

The *momentum thickness* $\delta_2(x)$ represents the defect in the x-momentum per unit depth $\dot{\mathbf{M}}_d$ associated with the slowdown of the velocity within the boundary layer (Figure 8.5),

$$\dot{\mathbf{M}}_d = \rho \int_0^\delta u(U - u)\, dy.$$

As demonstrated in Example 3.11, this momentum deficiency produces a net force in the streamwise direction. $\delta_2(x)$ is defined as the thickness of a uniform flow that carries the same momentum (i.e., $\dot{\mathbf{M}}_d = \rho\delta_2(x)U^2$). Therefore,

$$\rho\delta_2(x)U^2 = \rho \int_0^\delta u(U - u)\, dy,$$

or

$$\delta_2(x) = \int_o^\delta \left[\frac{u}{U}\left(1 - \frac{u}{U}\right)\right] dy. \tag{8.10}$$

Since the boundary layer thickness increases in the streamwise direction, both $\delta_1(x)$ and $\delta_2(x)$ are functions of x.

Note that since Equation (8.8) was derived without any assumption concerning the nature of the flow, it can be applied to both laminar and turbulent flows.

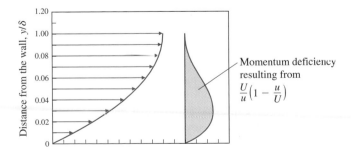

FIGURE 8.5 Definition of the momentum thickness $\delta_2(x)$.

8.2 Laminar Boundary Layer Flow

8.2.1 "Exact" Laminar Similarity Solutions

The boundary layer equations Equations (8.1)–(8.2) represent a set of nonlinear parabolic partial differential equations for which there is no known exact analytic solution. For laminar flow, however, they can be solved using numerical methods. See, for instance, the finite difference solution of the two-dimensional, laminar boundary layer equations discussed in Section 13.5.1.

The closest to an "exact" solution to Equations (8.1)–(8.2) is the Blasius solution for steady flow past a flat plate. The solution is based on a special transformation of the variables in the boundary layer equations. The basis of the transformation is the order of magnitude result

$$\frac{\delta}{x} \sim \sqrt{\frac{v}{Ux}} \quad \text{or} \quad \delta \sim \sqrt{\frac{vx}{U}}.$$

Blasius used this fact to introduce a new nondimensional variable η defined as

$$\eta = \frac{y}{\delta} = y\sqrt{\frac{U}{vx}}, \tag{8.11}$$

and a nondimensional streamfunction $f(\eta)$ given by

$$f(\eta) = \frac{\psi}{\sqrt{vxU}} \Rightarrow \psi = f(\eta)\sqrt{vxU}. \tag{8.12}$$

The variable η combines into a single independent variable the two independent coordinate variables x and y. This transformation reduces the two-dimensional boundary layer equations into what appears to be an one-dimensional problem as a function of the new independent variable η.

Because of the introduction of the streamfunction, the continuity equation is satisfied automatically. The momentum equation can be rearranged using the streamfunction, and subsequent manipulation based on the definitions of η and $f(\eta)$;

$$u = \frac{\partial \psi}{\partial y} = \frac{\partial \psi}{\partial \eta}\frac{\partial \eta}{\partial y} = f'\sqrt{vxU}\sqrt{\frac{U}{vx}} = f'U,$$

where f' indicates the derivative with respect to η. Also,

$$\frac{\partial u}{\partial x} = \frac{\partial}{\partial x}(f'U) = U\frac{df'}{d\eta}\frac{\partial \eta}{\partial x} = Uf''\frac{\partial \eta}{\partial x},$$

but

$$\frac{\partial \eta}{\partial x} = -\frac{\eta}{2x}.$$

Therefore,

$$\frac{\partial u}{\partial x} = -\frac{\eta}{2x}Uf''.$$

Similarly,

$$\frac{\partial u}{\partial y} = \frac{\partial}{\partial y}(f'U) = U\frac{df'}{d\eta}\frac{\partial \eta}{\partial y} = U\sqrt{\frac{U}{vx}}f'',$$

and

$$\frac{\partial}{\partial y}\left(\frac{\partial u}{\partial y}\right) = \frac{\partial}{\partial \eta}\left(\sqrt{\frac{U}{vx}}Uf'\right)\frac{\partial \eta}{\partial y} = \frac{U^2}{vx}f'''.$$

The transverse velocity v is expressed as

$$v = -\frac{\partial \psi}{\partial x} = -\left(\sqrt{\nu x U}\frac{\partial f}{\partial x} + \frac{f}{2}\sqrt{\frac{\nu U}{x}}\right)$$

$$= -\left(\sqrt{\nu x U}\frac{df}{d\eta}\frac{\partial \eta}{\partial x} + \frac{f}{2}\sqrt{\frac{\nu U}{x}}\right) = \frac{1}{2}\sqrt{\frac{\nu U}{x}}(\eta f' - f).$$

The final form of the momentum equation for flow past a flat plate where $\frac{dp}{dx} = 0$ is then

$$2f''' + ff'' = 0. \tag{8.13}$$

This is a third-order, non-linear ordinary differential equation.

The boundary conditions in terms of the new variables are as follows: along the surface $y = 0$ and, hence, $\eta = 0$, the velocity is zero (i.e., $u = 0 = Uf'$). Therefore, $f' = 0$. In addition, the streamfunction along the surface is set to zero, $f = 0$. At a distance away from the boundary layer $\eta = $ large, $u = U = Uf'$, which yields $f' = 1$.

Equation (8.13) is a nonlinear ordinary differential equation whose solution can be obtained fairly easily using numerical methods. For example, using a fourth-order Runge-Kutta integration scheme, the solution to Equation (8.13) is given in Table 8.1. The same solution is plotted in Figure 8.6. Since the numerical integration can be performed easily with a high degree of accuracy, for all practical purposes, the solution can be considered to be an "exact" solution to boundary layer flow past a flat plate. Henceforth, this solution will be referred to as an exact solution.

This solution represents the flow for an infinite number of boundary layer problems; irrespective of the coordinate location (x and y) and fluid conditions (U and ν), the solution in terms of f, f', f'' for any combination of these parameters is identical, provided that the parameters yield the same numerical value for η. Such a solution is called a *similarity solution*, and the variable η a *similarity variable*.

By defining the thickness of the boundary layer as the distance from the wall where the velocity is 99% of the freestream velocity (i.e., $f' = 0.99$), the numerical solution shows that the nondimensional thickness of the boundary layer is $\eta = 5$. In dimensional coordinates, the thickness of the boundary layer, then, is

$$\delta = \frac{5x}{\sqrt{Re_x}} = \frac{5x}{\sqrt{Ux/\nu}}. \tag{8.14}$$

Here, the Reynolds number Re_x, is based on the distance x from the front edge of the plate.

The numerically exact solution can now be used to determine other characteristic flow quantities. By definition, the local wall shear stress τ_w is given as

$$\tau_w = \mu\frac{\partial u}{\partial y}\bigg|_{y=0} = \mu U\sqrt{\frac{U}{\nu x}}f''\bigg|_{\eta=0}.$$

Using the numerical solution $f''(0) = 0.332$, τ_w becomes

$$\tau_w = 0.332\frac{\rho U^2}{\sqrt{Re_x}}. \tag{8.15}$$

Note that the shear stress at the surface decreases with the distance along the surface according to $\sim x^{-0.5}$.

TABLE 8.1   Blasius Solution.

η	f	f'	f''
0.0000000E+00	0.0000000E+00	0.0000000E+00	0.3320600
0.2500000	1.0376875E-02	8.3006032E-02	0.3319165
0.5000000	4.1494049E-02	0.1658866	0.3309136
0.7500000	9.3284436E-02	0.2483208	0.3282084
1.000000	0.1655748	0.3297825	0.3230098
1.250000	0.2580366	0.4095603	0.3146356
1.500000	0.3701439	0.4867927	0.3025829
1.750000	0.5011419	0.5605230	0.2866015
2.000000	0.6500322	0.6297698	0.2667536
2.250000	0.8155764	0.6936100	0.2434452
2.500000	0.9963216	0.7512640	0.2174131
2.750000	1.190646	0.8021722	0.1896628
3.000000	1.396821	0.8460487	0.1613615
3.250000	1.613085	0.8829061	0.1337038
3.500000	1.837715	0.9130443	0.1077739
3.750000	2.069094	0.9370083	8.4430709E-02
4.000000	2.305766	0.9555216	6.4236179E-02
4.250000	2.546470	0.9694083	4.7435224E-02
4.500000	2.790157	0.9795170	3.3984236E-02
4.750000	3.035983	0.9866555	2.3614302E-02
5.000000	3.283299	0.9915444	1.5911143E-02
5.250000	3.531620	0.9947910	1.0394325E-02
5.500000	3.780600	0.9968815	6.5830108E-03
5.750000	4.029997	0.9981864	4.0417481E-03
6.000000	4.279651	0.9989761	2.4056220E-03
6.250000	4.529459	0.9994394	1.3880555E-03
6.500000	4.779355	0.9997029	7.7646959E-04
6.750000	5.029300	0.9998482	4.2111927E-04
7.000000	5.279273	0.9999259	2.2145297E-04
7.250000	5.529260	0.9999662	1.1292604E-04
7.500000	5.779254	0.9999865	5.5846038E-05
7.750000	6.029253	0.9999964	2.6787688E-05
8.000000	6.279252	1.000001	1.2465006E-05

It is customary to express the local shear stress in nondimensional form using the *skin friction coefficient* C_f defined as

$$C_f = \frac{\tau_w}{\frac{1}{2}\rho U^2}. \tag{8.16}$$

For a flat plate, then, C_f is obtained as

$$C_f = \frac{0.664}{\sqrt{Re_x}}.$$

The total resistance force F_v due to viscosity, or *skin friction*, along the plate surface A is obtained by integration

$$F_v = \int_A \tau_w \, dA = \int_0^L \tau_w W \, dx,$$

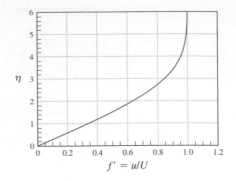

FIGURE 8.6   Blasius's solution for the nondimensional velocity as a function of the similarity variable η.

as

$$F_v = \int_0^L 0.332 \frac{\rho U^2}{\sqrt{Re_x}} W \, dx = 0.332 \rho U^2 \sqrt{\frac{\nu}{U}} W \int_0^L \frac{1}{\sqrt{x}} \, dx = 0.664 \rho U^2 W \sqrt{\frac{\nu L}{U}},$$

where L and W are the length and width of the flat plate, respectively. The result can be rearranged into

$$F_v = 0.664 \; \rho U^2 \frac{LW}{\sqrt{Re_L}}, \qquad (8.17)$$

where the Reynolds number now is defined based on the length L of the plate, $Re_L = \frac{UL}{\nu}$.

EXAMPLE 8.1

Problem Statement   Find the maximum boundary layer thickness along the surface of the vertical tail of the glider shown in Figure 8.7, flying at 2 m/s. Assume that the tail is flat, with a streamwise length of 0.5 m and height 1 m. The glider is flying at an altitude of 1500 m, where the average temperature is $6°C$. For the conditions given, find the flow-induced wall stress at the location of maximum boundary layer thickness.

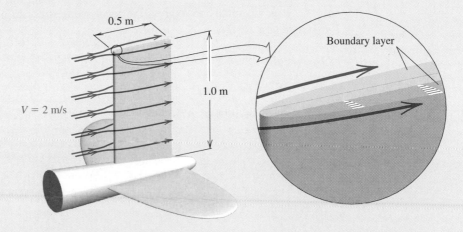

FIGURE 8.7   Schematic of Example 8.1.

Governing Equations For laminar conditions, the thickness of the boundary layer is obtained from the Blasius solution as

$$\delta = \frac{5x}{\sqrt{Re_x}}.$$

x is the streamwise distance from the front edge of the tail, and Re_x is the local Reynolds number based on x. The wall shear stress τ_w is given as

$$\tau_w = 0.332\frac{\rho U^2}{\sqrt{Re_x}}.$$

The flow past a flat tail remains laminar for $Re_x < 3 \times 10^5$.

Assumptions It is assumed that the flow is steady, and that effects from the interaction between the tail and its supporting surface and those due to end-tail effects are negligible.

SOLUTION Since the thickness of the boundary layer increases with x, the maximum thickness occurs at the most downstream point of the tail (i.e., at the back end). For the conditions given, the viscosity $v = 1.66 \times 10^{-5} \ m^2/s$, and the density of air $\rho = 1.050 \ kg/m^3$. Therefore,

$$Re_L = \frac{UL}{v} = \frac{2 \ m/s \times 0.5 \ m}{1.66 \times 10^{-5} \ m^2/s} = 6.02 \times 10^4.$$

Under these conditions, the flow is laminar. The thickness of the boundary layer then is

$$\delta = \frac{5x}{\sqrt{Re_x}} = \frac{5 \times 0.5 \ m}{\sqrt{6.02 \times 10^4}} = 0.01 \ m.$$

The wall shear stress τ_w at the same location is

$$\tau_w = 0.332 \times \frac{\rho U^2}{\sqrt{Re_x}} = 0.332 \times \frac{1.050 \ kg/m^3 \times 2^2 \ m^2/s^2}{\sqrt{6.02 \times 10^4}} = 5.68 \times 10^{-3} \ Pa.$$

EXAMPLE 8.2

Problem Statement Find the variation of the total skin friction force F_v as a function of the velocity of the glider described in Example 8.1. Consider velocities in the range of 2 m/s to 4 m/s.

Governing Equations For laminar flow, the total force due to the viscous effects on a flat tail is given by

$$F_v = 0.664 \ \rho U^2 \frac{LW}{\sqrt{Re_L}}.$$

Re_L is the Reynolds number based on the length L of the tail in the direction of the flow, and W the width of the tail transverse to the flow direction.

Assumptions The flow is assumed to be steady, with negligible end-tail effects.

SOLUTION   At the maximum glider speed of $4\ m/s$, the Reynolds number based on the entire lenght of the tail is $Re = 1.2 \times 10^5$, which indicates that the flow remains laminar along the entire length of the tail. The laminar viscous force

$$F_v = 0.664 \rho U^2 \frac{LW}{\sqrt{Re_L}},$$

can be rearranged as

$$F_v = 0.664 \rho W \sqrt{\nu L U^3}.$$

By substituting the numerical values

$$F_v = 0.664 \times 1.127\ kg/m^3 \times 1\ m \sqrt{1.66 \times 10^{-5}\ m^2/s \times 0.5\ m \times U^3}$$

$$\Rightarrow F_v = 2.16 \times 10^{-3} U^{1.5}.$$

The result shows that the viscous force increases with $U^{1.5}$. When the glider accelerates from $U = 2\ m/s$ to $U = 4\ m/s$, the force varies from $F_v = 6.1 \times 10^{-3}\ N$, to $F_v = 1.72 \times 10^{-2}\ N$ (Figure 8.8).

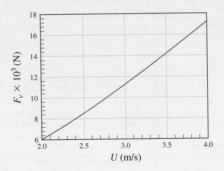

FIGURE 8.8   Example 8.2: the skin friction force F_v as a function of the glider velocity U.

EXAMPLE 8.3

Problem Statement   *Vorticity generated within the boundary layer.* The velocity gradient within the boundary layer is also responsible for the generation of vorticity ω (Figure 8.9). At a glider speed of $2\ m/s$, find and plot the vorticity distribution within the boundary at a distances $x = 0.1\ m$, $x = 0.3\ m$, and $x = 0.5\ m$ from the leading edge of the tail.

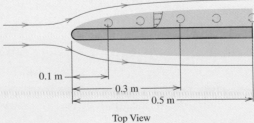

Top View

FIGURE 8.9   Schematic of Example 8.3.

Governing Equations The vorticity is defined (Chapter 5) as

$$\omega = \frac{\partial v}{\partial x} - \frac{\partial u}{\partial y}.$$

Assumptions It is assumed that the flow is steady and that end-plate effects are negligible.

SOLUTION By definition,

$$\omega = \frac{\partial v}{\partial x} - \frac{\partial u}{\partial y}.$$

However, because of the small thickness of the boundary layer, as shown in Section 8.1, the term $\frac{\partial v}{\partial x}$ is small and can be neglected. Therefore,

$$\omega \approx \frac{\partial u}{\partial y}.$$

According to the coordinate and variable transformations given by Equations (8.11) and (8.12) we have

$$\frac{\partial u}{\partial y} = U\sqrt{\frac{U}{\nu x}}\, f''.$$

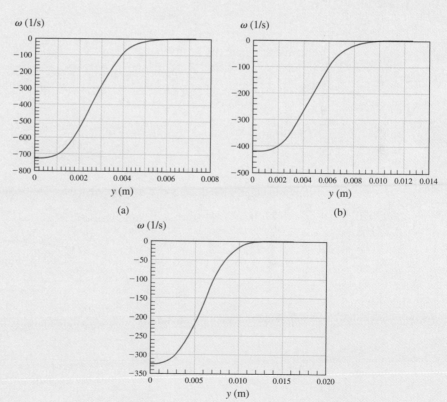

FIGURE 8.10   Example 8.3: Vorticity ω at $x = 0.1, 0.3, 0.8$ m from the forward edge of the tail.

For the conditions of the problem,

$$\omega(\eta) \approx -2 \ m/s \sqrt{\frac{2 \ m/s}{1.66 \times 10^{-5} \times x}} f''.$$

The negative sign of the vorticity indicates that fluid particles within the layer rotate in the clockwise direction. Using the tabulated results from the exact Blasius solution (Table 8.1) and by replacing η with $y\sqrt{\dfrac{U}{\nu x}}$, the distribution of $\omega(y)$ as a function of y within the boundary layer at the three selected positions is shown in Figure 8.10.

The results show that the magnitude of the vorticity decreases with x. However, by integrating across the thickness of the layer, the total vorticity ω_{total} is

$$\omega_{total} = \int_0^\delta \omega \, dy = -\int_0^\delta \frac{\partial u}{\partial y} \, dy = -\int_0^\delta du = -U.$$

Since the freestream velocity is constant, the total vorticity is also constant! For this reason, while the thickness of the layer increases with x, the average vorticity ω_a decreases with x, according to

$$\omega_a \equiv \frac{\omega_{total}}{\delta(x)} = -\frac{U}{\delta(x)}.$$

Substituting,

$$\delta = \frac{5x}{\sqrt{Re_x}},$$

the average vorticity is given as

$$\omega_a = -U \frac{\sqrt{Re_x}}{5x} = -U\sqrt{\frac{U}{25\nu x}} = -2 \ m/s \times \sqrt{\frac{2 \ m/s}{25 \times 1.66 \times 10^{-5} \ m^2/s \times x}}.$$

Therefore,

$$\omega_a = -\frac{138.8}{\sqrt{x}} \ s^{-1},$$

showing clearly the decrease of ω_a with distance. This result is also plotted in Figure 8.11.

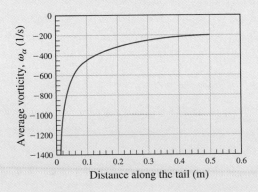

FIGURE 8.11   Example 8.3: Average vorticity ω_a along the tail.

8.2.2   Approximate Laminar Boundary Layer Flow

As shown in Figure 8.12, the exact Blasius solution can be approximated with reasonable accuracy by a number of simple functions. Therefore, these functions can be used in Equation (8.8) to analytically determine flow details within the boundary layer. This approach is demonstrated using two simple velocity distributions: (a) a quadratic function, and (b) a sinusoidal function. The same procedure can be repeated for a number of other functions that approximate the general velocity profile within the boundary layer.

Case I

Using a quadratic approximation, the velocity profile within the boundary layer is given as

$$\frac{u}{U} = 2\left(\frac{y}{\delta(x)}\right) - \left(\frac{y}{\delta(x)}\right)^2,$$

where $\delta(x)$ is the thickness of the boundary layer as a function of the streamwise distance x from the front end of the plate. Consistent with the theory, this velocity profile satisfies the boundary conditions — that is, at the plate surface $y = 0$, $u = v = 0$, and at the edge of the boundary layer $y = \delta(x)$ $u = U$. Additionally, since at $y = \delta(x)$, $\frac{\partial u}{\partial y} = 0$, the shear τ vanishes there.

For flow past a flat plate, the velocity U is constant; hence, $U\dfrac{dU}{dx} = 0$. Equation (8.8) then reduces to

$$\tau_w = \rho U^2 \frac{d}{dx}(\delta_2),$$

where δ_2 and τ_w are defined as

$$\delta_2 = \int_0^\delta \frac{u}{U}\left(1 - \frac{u}{U}\right) dy,$$

and

$$\tau_w = \mu \left.\frac{\partial u}{\partial y}\right|_{y=0}.$$

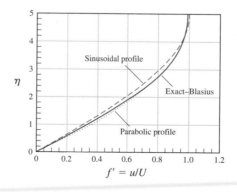

FIGURE 8.12   Approximate velocity distributions within the boundary layer by using simple functions.

To facilitate the calculations, we introduce the nondimensional variable $\eta = \frac{y}{\delta}$. Therefore, $\frac{u}{U} = 2\eta - \eta^2$ and $\delta\, d\eta = dy$. By substitution, the momentum displacement thickness is

$$\delta_2 = \int_0^1 (2\eta - \eta^2)(1 - 2\eta + \eta^2)\delta\, d\eta = \int_0^1 (2\eta - 5\eta^2 + 4\eta^3 - \eta^4)\delta\, d\eta = \frac{2}{15}\delta.$$

The shear stress is given as

$$\tau_w = \mu \frac{\partial u}{\partial y}\bigg|_{y=0} = \mu \frac{\partial u}{\partial \eta}\frac{\partial \eta}{\partial y}\bigg|_{\eta=0} = \frac{\mu}{\delta}\frac{\partial u}{\partial \eta}\bigg|_{\eta=0} = \frac{\mu}{\delta}(2 - 2\eta)\bigg|_{\eta=0} U = 2\frac{\mu}{\delta}U.$$

Equation (8.8) then becomes

$$2\frac{\mu}{\delta} = \rho U^2 \frac{d\delta}{dx}.$$

By rearranging

$$\delta \frac{d\delta}{dx} = \frac{1}{2}\frac{d\delta^2}{dx} = \frac{15\mu U}{\rho U^2},$$

and integrating, we get

$$\delta = \sqrt{30\frac{\nu x}{U}}.$$

Finally,

$$\delta = \frac{5.48x}{\sqrt{Re_x}}.$$

The assumed simple quadratic velocity profile predicts a boundary thickness that is remarkably very close to the exact solution $\delta = 5x/\sqrt{Re_x}$. Actually, the result is within 10% of the exact solution.

The distribution of the shear stress then along the wall is

$$\tau_w = 2\frac{\mu}{\delta}U = 0.365\frac{\rho U^2}{\sqrt{Re_x}},$$

and the skin friction coefficient

$$C_f = \frac{\tau_w}{\frac{1}{2}\rho U^2} = \frac{0.730}{\sqrt{Re_x}}.$$

The total resistance force F_v at the plate surface is obtained by integrating τ_w along the surface of the plate as,

$$F_v = \int_0^L \tau_w W\, dx,$$

where L and W are the length and width of the flat plate. By integration,

$$F_v = \int_0^L 0.365\frac{\rho U^2}{\sqrt{Re_x}}W\, dx = 0.365\rho U^2 \sqrt{\frac{\nu}{U}}W \int_0^L \frac{1}{\sqrt{x}}\, dx = 0.730\rho U^2 W \sqrt{\frac{\nu L}{U}},$$

and after rearrangement, the force is obtained as

$$F_v = 0.730\rho U^2 \frac{LW}{\sqrt{Re_L}}.$$

The approximate results for τ_w, C_f, and F_v are also within 10% of the exact solution.

Case II

Now, by assuming the sinusoidal velocity profile $\frac{u}{U} = \sin\left(\frac{\pi}{2}\frac{y}{\delta}\right)$, we follow the same steps as before to evaluate the thickness of the boundary layer. Again, along the plate surace $y = 0$, $u = v = 0$, and at the edge of the boundary layer $y = \delta(x)$, $u = U$ (which explains the factor $\pi/2$), and the shear $\tau = 0$. Introducing the same nondimensional variable $\eta = \frac{y}{\delta}$, the shear stress at the wall is

$$\tau_w = \mu\frac{\partial u}{\partial y}\bigg|_{y=0} = \mu\frac{\partial u}{\partial \eta}\frac{\partial \eta}{\partial y}\bigg|_{\eta=0} = \mu\frac{\pi}{2}\frac{U}{\delta}.$$

The momentum displacement thickness δ_2 is evaluated as

$$\delta_2 = \int_0^\delta \frac{u}{U}\left(1 - \frac{u}{U}\right)dy = \delta\int_0^1 \sin\left(\frac{\pi}{2}\eta\right)\left[1 - \sin\left(\frac{\pi}{2}\eta\right)\right]d\eta.$$

Upon integration,

$$\delta_2 = \delta\left[-\frac{2}{\pi}\cos\left(\frac{\pi}{2}\eta\right) - \frac{\eta}{2} + \frac{1}{2\pi}\sin(\pi\eta)\right]_0^1 = 0.137\delta.$$

Combining the above into the momentum equation we get,

$$\mu\frac{\pi U}{2\delta} = 0.137\rho U^2\frac{d\delta}{dx}.$$

By rearranging the terms and integrating for the boundary layer thickness, we get

$$\delta = \frac{4.79x}{\sqrt{Re_x}}.$$

Again, this is remarkably close to the exact solution (to within an error of 5%). The shear stress is obtained as

$$\tau_w = \mu\frac{\pi U}{2}\frac{\sqrt{Re_x}}{4.79x} = 0.328\frac{\rho U^2}{\sqrt{Re_x}},$$

and the skin friction coefficient as

$$C_f = \frac{0.656}{\sqrt{Re_x}}.$$

The total force along a flat plate with length L and width W is now

$$F_v = 0.656\rho U^2 W\sqrt{\frac{\nu L}{U}} = 0.656\rho U^2\frac{LW}{\sqrt{Re_L}}.$$

The approximate results for τ_w, C_f, and F_v are also within 1.2% of the exact solution! Figure 8.13 compares the variation in the boundary layer thickness as predicted by the approximate profiles with the exact Blasius solution. The figure shows that the simple velocity profiles are quite accurate in representing $\delta(x)$. Table 8.2 summarizes the results for laminar flow past a flat plate.

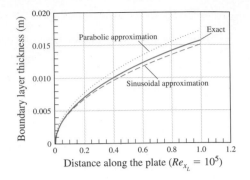

FIGURE 8.13 The growth of the boundary layer thickness using parabolic and sinusoidal approximations, compared with the exact solution.

TABLE 8.2 Summary of Flow Parameters for Laminar Flow Past a Flat Plate.

Parameter	Exact	Quadratic	Sinusoidal
$\delta = \dfrac{C_1 x}{\sqrt{Re_x}}$	$C_1 = 5$	$C_1 = 5.48$	$C_1 = 4.79$
$\tau_w = C_2 \dfrac{\rho U^2}{\sqrt{Re_x}}$	$C_2 = 0.332$	$C_2 = 0.365$	$C_2 = 0.328$
$C_f = \dfrac{C_3}{\sqrt{Re_x}}$	$C_3 = 0.664$	$C_3 = 0.730$	$C_3 = 0.656$
$F_v = C_4 \rho U^2 \dfrac{LW}{\sqrt{Re_L}}$	$C_4 = 0.664$	$C_4 = 0.730$	$C_4 = 0.656$

EXAMPLE 8.4

Problem Statement   Using the quadratic and sinusoidal approximations for the velocity distribution within the boundary layer, calculate the total skin force on the vertical tail discussed in Examples 8.1–8.3. Compare the results with those predicted by the Blasius solution.

Governing Equations   The total viscous force on a flat plate (here the vertical tail) is given by the general expression

$$F_v = C_i \rho U^2 \frac{LW}{\sqrt{Re_L}}.$$

L is the length of the tail in the direction of flow, and W is its width transverse to the flow direction. Re_L is the length Reynolds number, defined with respect to L. The constant C_i depends on the particular approximation used. For the exact solution, $C_e = 0.664$; for the quadratic approximation, $C_q = 0.730$; and for the sinusoidal approximation, $C_s = 0.656$.

Assumptions   The flow is assumed to be steady and with negligible end-plate effects.

SOLUTION   The total force can be rearranged as

$$F_v = C_i \rho W \sqrt{\nu L U^3}.$$

Therefore,

$$\frac{F_{v_q}}{F_{v_{exact}}} = \frac{C_q}{C_e} = \frac{0.730}{0.664} = 1.0994,$$

where F_{v_q} is the force predicted by the quadratic approximation. The exact viscous force obtained in Example 8.2 is $F_{v_{exact}} = 6.1 \times 10^{-3} N$. Therefore,

$$F_{v_q} = 1.0994 \times 6.1 \times 10^{-3} N = 6.71 \times 10^{-3} N.$$

This corresponds to a relative error of 9.94%. Similarly, the estimate of the force using the sinusoidal approximation F_{v_s} is obtained from the ratio

$$\frac{F_{v_s}}{F_{v_{exact}}} = \frac{C_s}{C_e} = \frac{0.656}{0.664} = 0.988.$$

Therefore,

$$F_{v_s} = 0.988 \times 6.1 \times 10^{-3} N = 6.03 \times 10^{-3} N.$$

This corresponds to a relative error of only 1.2%!

EXAMPLE 8.5

Problem Statement *Rate of mass flow ejected due to the development of the boundary layer.* While the boundary layer is growing because of the viscosity-induced slowdown of the velocity within the layer, excess flow is ejected from the boundary layer (Figure 8.14). By approximating the velocity within the boundary with a quadratic function, calculate the total rate of mass ejected in a uniform flow past a flat plate, and compare the result with the exact Blasius solution. Consider L to be the length of the plate in the direction of the flow, W the width of the plate transverse to the flow direction, and U the velocity outside of the boundary layer.

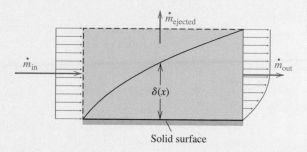

FIGURE 8.14 Schematic of Example 8.5.

Governing Equations Using mass conservation for the open system shown in Figure 8.14, the ejected mass flow rate $\dot{m}_e$ is given as

$$\dot{m}_e + \int_{inlet} \rho(\mathbf{V} \cdot \mathbf{n}) \, dA + \int_{outlet} \rho(\mathbf{V} \cdot \mathbf{n}) \, dA = 0.$$

Assumptions The flow is assumed to be laminar and steady, with negligible end-plate effects.

SOLUTION   At the inlet of the control volume, $\mathbf{V} = U\mathbf{i}$, and $\mathbf{n} = -\mathbf{i}$. At the exit, $\mathbf{V} = u\mathbf{i}$, and $\mathbf{n} = \mathbf{i}$, where $u = u(y)$ is the velocity distribution within the boundary layer.

The conservation of mass simplifies to

$$\dot{m}_e - \int_0^\delta \rho U W \, dy + \int_0^\delta \rho u W \, dy = 0.$$

By rearranging,

$$\dot{m}_e = \rho W \int_0^\delta (U - u) \, dy,$$

which can be finally written as

$$\dot{m}_e = \rho U \delta_1 W.$$

The displacement thickness δ_1 is defined as

$$\delta_1 = \int_0^\delta \left(1 - \frac{u}{U}\right) dy.$$

Using a quadratic approximation for u/U,

$$\delta_1 = \int_0^\delta \left(1 - 2\frac{y}{\delta} + \frac{y^2}{\delta^2}\right) dy = \frac{1}{3}\delta.$$

The ejected mass flow rate then from the approximate solution is

$$\dot{m}_e = \frac{1}{3}\rho U \delta W.$$

Recall from Chapter 4 that the difference in the numerical value of two streamlines $(\psi_e - \psi_s)$ represents the volumetric flow rate per unit depth (m^2/s) between the two streamlines. Therefore, the rate of mass leaving the control volume at $x = L$ is

$$\dot{m}_{out} = \rho(\psi_e - \psi_s)W,$$

where ψ_s and ψ_e are the streamfunction values at the two ends of the exit plane $y = 0$ and $y = \delta(L)$, respectively.

The amount of mass entering through the same height δ is

$$\dot{m}_{in} = \rho U \delta W.$$

Therefore, the rejected mass is

$$\dot{m}_e = \dot{m}_{in} - \dot{m}_{out} = \rho W[U\delta - (\psi_e - \psi_s)].$$

Using the definition,

$$\psi = f(\eta)\sqrt{\nu U x},$$

the numerical value of the streamline along the surface is $\psi_s = 0$. The value of the streamline ψ_e at the edge of the boundary layer is calculated, using the Blasius solution, as

$$\psi_e = 3.283299\sqrt{\nu U L},$$

which can be rearranged as

$$\psi_e = 3.283299\frac{U L}{\sqrt{Re_x}}.$$

But according to the exact theory,

$$\delta = \frac{5L}{\sqrt{Re_x}}.$$

Therefore,

$$\psi_e = \frac{3.283299}{5}U\delta = 0.657U\delta.$$

Finally,

$$\dot{m}_e = \rho W[U\delta - 0.657U\delta] = 0.343\rho U \delta W.$$

The quadratic approximation predicts the ejected mass flow rate within an error of about 2.92%.

8.3   Turbulent Boundary Layer Flow

Since the integral-momentum boundary layer equation is general, it applies to turbulent flow as well. Turbulent flow, however, is significantly more complex than laminar flow, and depends on such factors as geometry, surface roughness, and the like. Fortunately, experimental data are available to assist in the analysis of turbulent boundary layer flow.

For instance, the behavior of the flow near surfaces is described quite accurately from experimental results obtained for turbulent flow in pipes. According to these data, a thin *viscous sublayer* develops near the solid surface, where the flow is dominated by viscous effects. Away from the surface, the flow becomes fully turbulent. In an *overlap region*, or *transition layer*, the flow is influenced by both viscous and turbulent effects (Figure 8.15). Despite the viscous sublayer, the flow is also affected by the roughness of the surface. At smooth walls, the surface irregularities are entirely within the viscous layer,

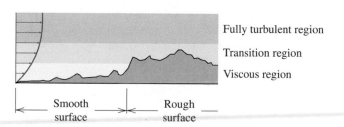

Fully turbulent region

Transition region

Viscous region

Smooth surface

Rough surface

FIGURE 8.15   Structure of the turbulent boundary layer.

and their effect is minimum. However, at rough surfaces, they penetrate the viscous layer, and interfere with the main flow field.

According to experimental evidence, across the viscous sublayer the shear stress τ_w appears to be constant. Therefore,

$$\tau_w = \mu \frac{\partial u}{\partial y} \Rightarrow \frac{\tau_w}{\rho} \approx \nu \frac{\Delta u}{\Delta y} = \nu \frac{u}{y},$$

so that $u = (\tau_w/\rho\nu)y$, from which it follows that in the viscous sublayer u varies linearly with y.

Since $\dfrac{\tau_w}{\rho}$ has units of velocity square (m^2/s^2), we can define an equivalent velocity as

$$U^* = \sqrt{\frac{\tau_w}{\rho}}.$$

The above is known as the *friction velocity*. Within the viscous sublayer, then, the velocity varies according to

$$\frac{u}{U^*} = \frac{U^* y}{\nu}. \tag{8.18}$$

This equation is known also as the *law of the wall*. The thickness of the viscous sublayer γ_s and the overlap region γ_o (measured from the wall) are found experimentally to be

$$\gamma_s \approx \frac{5\nu}{U^*} \quad \text{and} \quad \gamma_o \approx \frac{20\nu}{U^*}.$$

The influence of surface roughness e — defined as the average height of the surface irregularities — on turbulent flow is usually quantified by the ratio $\dfrac{eU^*}{\nu}$. Depending on the value of $\dfrac{eU^*}{\nu}$, surfaces can be categorized as

(a) Smooth, when $\dfrac{eU^*}{\nu} < 5$. This condition defines the *smooth-pipe zone of flow*,

(b) Rough, when $\dfrac{eU^*}{\nu} > 70$. This condition defines the *rough-pipe zone of flow*.

In the range $5 < \dfrac{eU^*}{\nu} < 70$, surface roughness affects the flow as it undegoes a transition from the viscous sublayer flow to fully turbulent flow. Equivalently, this range defines the *frictional transition flow zone*.

Smooth Surfaces

In flow past smooth surfaces, the velocity distribution is found experimentally to vary according to the logarithmic law

$$\frac{u}{U^*} = \alpha \ln \frac{yU^*}{\nu} + \beta, \tag{8.19}$$

where the constants are $\alpha = 2.5$ and $\beta = 5.0$. Equation (8.19) is valid for $\ln(yU^*/\nu) > 3$.

Therefore, as shown in Figure 8.16, for $\ln(yU^*/\nu) < 3$, the velocity for both the viscous sublayer and the overlap region is described by Equation (8.18), and for

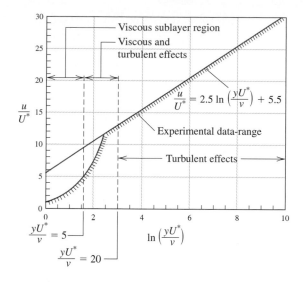

FIGURE 8.16 Velocity distribution in a turbulent boundary layer.

$\ln(yU^*/\nu) > 3$, by Equation (8.19). Away from the wall the velocity is expressed using

$$\frac{u}{U} = \left(\frac{y}{\delta}\right)^{\frac{1}{n}},\tag{8.20}$$

which is valid in the region $4{,}000 < Re < 3.2 \times 10^6$. Here, Re is based on the diameter of the pipe. When the same results are used for flat surfaces, the Reynolds number is based on the streamwise distance from the leading edge of the surface. The constant n is in the range of 6–10. Since in practice the value $n = 7$ is used quite frequently, this expression is known as the *seventh-root law*. For smooth surfaces, Blasius established experimentally that for Re less than 3×10^6, the shear stress along the wall could be expressed as

$$\tau_w = 0.0233\rho U^2 \left(\frac{\nu}{U\delta}\right)^{\frac{1}{4}}.$$

Rough Surfaces

At rough surfaces, the velocity profile for the frictional transition zone depends on the roughness e, and it appears to also follow the logarithmic relation

$$\frac{u}{U^*} = \alpha \ln \frac{yU^*}{\nu} + \beta(e),\tag{8.21}$$

where β now depends on the roughness itself (Figure 8.17).

In the fully rough zone, the velocity profile is found to vary according to

$$\frac{u}{U^*} = \alpha \ln \frac{y}{e} + \beta,\tag{8.22}$$

which is independent of ν and, hence, Re. The constants are found to be $\alpha = 2.5$ and $\beta = 8.5$

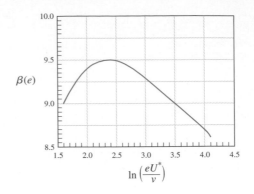

FIGURE 8.17   Variable $\beta(e)$ as a function of the roughness in transitional turbulent flow.

8.3.1   Approximate Turbulent Boundary Layer Flow

The seventh-root law and the wall shear expression established by Blasius, can be used in the same procedure as in the previous section to determine the growth of turbulent boundary layers as well as other useful flow parameters such as the shear stress and the skin friction. Note that the seventh-root law is actually valid away from the solid surface. However, since the distance from the wall at which the law is valid is not known with certainty, for simplicity we assume that the expression is valid everywhere within the boundary layer.

Again, for constant flow past a flat plate, $\dfrac{dp}{dx} = -\rho U \dfrac{dU}{dx} = 0$, Equation (8.8) reduces to

$$\tau_w = \rho U^2 \frac{d}{dx}(\delta_2).$$

Using the nondimensional variable $\eta = \dfrac{y}{\delta}$, the velocity profile becomes $\dfrac{u}{U} = \eta^{\frac{1}{7}}$. The momentum displacement thickness δ_2 is evaluated as

$$\delta_2 = \int_0^\delta \frac{u}{U}\left(1 - \frac{u}{U}\right) dy = \int_0^1 \eta^{\frac{1}{7}}\left(1 - \eta^{\frac{1}{7}}\right)\delta d\eta = \left|\frac{7}{8}\eta^{1+\frac{1}{7}} - \frac{7}{9}\eta^{1+\frac{2}{7}}\right|_0^1 \delta = \frac{7}{72}\delta.$$

According to the experimental results by Blasius, the shear stress at the wall is given by

$$\tau_w = 0.0233\rho U^2 \left(\frac{\nu}{U\delta}\right)^{\frac{1}{4}}.$$

Substituting these expressions into the integral momentum equation

$$0.0233\rho U^2 \left(\frac{\nu}{U\delta}\right)^{\frac{1}{4}} = \rho U^2 \frac{7}{72}\frac{d\delta}{dx} \Rightarrow \int_0^x 0.239\left(\frac{\nu}{U}\right)^{\frac{1}{4}} dx = \int_0^\delta \delta^{\frac{1}{4}} d\delta,$$

and, upon integration, we get,

$$\delta = 0.382\left(\frac{\nu}{U}\right)^{\frac{1}{5}} x^{\frac{4}{5}},$$

or

$$\frac{\delta}{x} = \frac{0.382}{Re_x^{\frac{1}{5}}}. \tag{8.23}$$

By substitution, the shear stress along the wall and the skin friction coefficient are obtained as

$$\tau_w = 0.0297 \frac{\rho U^2}{Re_x^{\frac{1}{5}}}, \tag{8.24}$$

and

$$C_f = \frac{0.0594}{Re_x^{\frac{1}{5}}}. \tag{8.25}$$

The total skin friction force F_v along the plate is evaluated by integration,

$$F_v = \int_0^L 0.0297 \frac{\rho U^2}{Re_x^{\frac{1}{5}}} W \, dx = 0.0297 \rho U^2 W \int_0^L \frac{1}{Re_x^{\frac{1}{5}}} dx$$

$$F_v = 0.0297 \rho U^2 W \int_0^L \left(\frac{\nu}{U}\right)^{\frac{1}{5}} \left(\frac{1}{x}\right)^{\frac{1}{5}} dx = 0.0297 \rho U^2 W \left(\frac{\nu}{U}\right)^{\frac{1}{5}} \int_0^L \left(\frac{1}{x}\right)^{\frac{1}{5}} dx$$

$$= 0.0297 \frac{5}{4} \rho U^2 W \left(\frac{\nu}{U}\right)^{\frac{1}{5}} L^{\frac{4}{5}} = 0.0297 \frac{5}{4} \rho U^2 W \frac{L}{Re_L^{\frac{1}{5}}},$$

which is finally rearranged as

$$F_v = 0.037 \rho U^2 \frac{LW}{Re_L^{\frac{1}{5}}}. \tag{8.26}$$

EXAMPLE 8.6

Problem Statement Find the total friction force on a square plate with dimensions of $0.5 \, m$ in each side, due to uniform air flow with velocity $100 \, m/s$ (density $\rho = 1.165 \, kg/m^3$, and viscosity $\nu = 1.60 \times 10^{-5} \, m^2/s$).

Governing Equations For laminar conditions $Re_x < 3 \times 10^5$,

$$\tau_w = 0.332 \frac{\rho U^2}{\sqrt{Re_x}}, \quad \text{and} \quad F_v = 0.664 \rho U^2 \frac{LW}{\sqrt{Re_L}},$$

and for turbulent conditions $Re_x > 3 \times 10^5$,

$$\tau_w = 0.0297 \frac{\rho U^2}{Re_x^{\frac{1}{5}}}, \quad \text{and} \quad F_v = 0.037 \rho U^2 \frac{LW}{Re_L^{\frac{1}{5}}}.$$

Assumptions The flow is assumed to be steady with negligible end-plate effects. Also, we assume that for transition flow we can use the same expressions as for fully turbulent flow.

SOLUTION Since the flow remains laminar for $Re < 3 \times 10^5$, the transition to turbulent flow occurs at

$$x \leq \frac{3 \times 10^5 \times 1.6 \times 10^{-5} \, m^2/s}{100 \, m/s} = 0.048 \, m.$$

Above this limit, the flow is fully turbulent. For laminar flow, the skin friction force is

$$F_{v_1} = 0.664 \rho U^2 \frac{LW}{\sqrt{Re_L}} = 0.664 \times 1.165 \ kg/m^3 \times (100 \ m/s)^2 \times \frac{0.048 \ m \times 0.5 \ m}{\sqrt{3 \times 10^5}}$$

$$\Rightarrow F_{v_1} = 0.0619 \ N.$$

For the section of the plate where the flow is turbulent,

$$Re_L = \frac{UL}{\nu} = \frac{100 \ m/s \times (0.5 - 0.048) \ m}{1.60 \times 10^{-5} \ m^2/s} = 2.825 \times 10^6.$$

The skin friction force then for this section is

$$F_{v_2} = 0.037 \rho U^2 \frac{LW}{Re_L^{\frac{1}{5}}} = 0.037 \times 1.165 \ kg/m^3 \times (100 \ m/s)^2$$

$$\times \frac{(0.5 - 0.048) \ m \times 0.5 \ m}{(2.825 \times 10^6)^{\frac{1}{5}}} \Rightarrow F_{v_2} = 4.994 \ N,$$

and the total force is

$$F_v = F_{v_1} + F_{v_2} = 0.339 \ N + 4.994 \ N = 5.333 \ N.$$

By assuming turbulent flow along the entire length of the plate, the force turns out to be 5.41 N, which is close to the more exact calculation given previously.

EXAMPLE 8.7

Problem Statement   Find the thickness of the boundary layer along a square plate with dimensions of 1 m for each side due to uniform air flow at $30°C$. The velocity away from the plate is 150 m/s.

Governing Equations   The thickness of the boundary layer under laminar conditions $(Re_x < Re_{cr})$ is,

$$\delta = \frac{5x}{\sqrt{Re_x}} = 5\sqrt{\frac{x\nu}{U}},$$

and for turbulent flow $(Re_x > Re_{cr})$ is

$$\delta = \frac{0.382x}{Re_x^{\frac{1}{5}}} = 0.382x^{\frac{4}{5}} \left(\frac{\nu}{U}\right)^{\frac{1}{5}},$$

where Re_{cr} is the critical Reynolds number beyond which laminar flow changes into turbulent flow.

Assumptions   It is assumed that the flow is steady and that end-plate effects are negligible. Following the approach in earlier problems, we assume that the critical Reynolds number is $Re_{cr} = 3 \times 10^5$.

SOLUTION   For air at $30°C$ the density is $\rho = 1.165 \ kg/m^3$, and the viscosity is $\nu = 1.6 \times 10^{-5}$. The flow remains laminar for

$$Re_x = \frac{Ux}{\nu} < 3 \times 10^5,$$

which occurs over a distance x from the front end of the plate

$$x \le 3 \times \frac{10^5 \times 1.6 \times 10^{-5} \ m^2/s}{150 \ m/s} = 0.032 \ m.$$

Beyond x, the flow begins to turn turbulent. By considering the different flow regions, the thickness δ is plotted in Figure 8.18. As expected, the maximum thickness of the boundary layer occurs at the downstream end of the plate. Note also the sudden change in the boundary thickness when the transition to turbulent flow starts. In reality, the change from laminar to turbulent flow occurs gradually over the transition region, and the variation in the thickness is not as abrupt as shown in Figure (8.18).

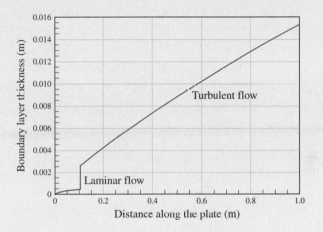

FIGURE 8.18 Example 8.7: Boundary layer thickness as a function of distance along a flat plate.

EXAMPLE 8.8

Problem Statement Determine the average vorticity generated along a flat, square plate with dimensions of $1.m$ for each side, due to uniform air flow at $30°C$. The uniform velocity is $120 \ m/s$.

Governing Equations Using the boundary layer approximation $\frac{\partial v}{\partial x} \approx 0$, the average vorticity within the boundary layer is found in Example 8.3 to be

$$\omega_a \approx -\frac{U}{\delta(x)}.$$

The thickness of the boundary layer under laminar conditions ($Re_x < 3 \times 10^5$) is

$$\delta = \frac{5x}{\sqrt{Re_x}},$$

and for turbulent flow ($Re_x > 3 \times 10^5$) is

$$\delta = \frac{0.382x}{Re_x^{\frac{1}{5}}}.$$

Assumptions   It is assumed that the flow is steady and that end-plate effects are negligible. The flow is assumed to become fully turbulent at $Re_x > 3 \times 10^5$.

SOLUTION   For air at $30°C$, the density is $\rho = 1.165 \ kg/m^3$ and the viscosity is $\nu = 1.6 \times 10^{-5} \ m^2/s$. The flow remains laminar for $Re \leq 3 \times 10^5$,

$$Re = \frac{Ux}{\nu} \leq 3 \times 10^5.$$

Laminar flow occurs over a distance x from the leading edge given by

$$x \leq \frac{3 \times 10^5 \times 1.6 \times 10^{-5} \ m^2/s}{120 \ m/s} = 0.04 \ m.$$

Above this limit, the flow is fully turbulent.
    For laminar flow

$$\omega_a = -\frac{U}{5x}\sqrt{\frac{Ux}{\nu}} = -\frac{120 \ m/s}{5x}\sqrt{\frac{120 \ m/sx}{1.6 \times 10^{-5} \ m^2/s}} = -\frac{65727}{\sqrt{x}},$$

and for turbulent flow,

$$\omega_a = -\frac{U}{0.382x}\left(\frac{Ux}{\nu}\right)^{\frac{1}{5}} = -\frac{120 \ m/s}{0.382x}\left(\frac{120 \ m/sx}{1.6 \times 10^{-5} \ m^2/s}\right)^{\frac{1}{5}}$$
$$= -\frac{7450}{x^{\frac{4}{5}}}.$$

By considering the different flow regimes, the average vorticity ω_a is plotted in Figure 8.19. As expected, the average vorticity decreases along the length of the plate.

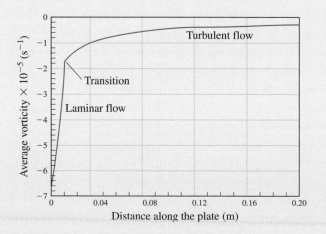

FIGURE 8.19   Example 8.8: Average vorticity along the flat plate.

8.4   Dynamics of External Flows

8.4.1   Flow Separation

The discussion so far has focused on boundary layer flow bounded by a solid surface, and an outer flow region where the flow is considered as inviscid. The flow within the boundary layer is coupled to the inviscid flow by means of the externally imposed pressure gradient. In all cases considered, the outer flow is shown to follow smoothly the solid surface. Under certain conditions, however, this does not happen, and the fluid moves away from solid surfaces. This phenomenon is commonly referred to as *flow separation*. Obviously, for this to happen the direction of the shear at the surface $\tau = \mu \dfrac{\partial u}{\partial y}\Big|_{y=0}$ must change sign: as shown in Figure 8.20(a) for forward flow $\mu \dfrac{\partial u}{\partial y}\Big|_{y=0} > 0$, and for reversed (back) flow $\mu \dfrac{\partial u}{\partial y}\Big|_{y=0} < 0$ (Figure 8.20(c)). The onset of separation is expressed by the limiting condition, where the shear at the wall vanishes and $\mu \dfrac{\partial u}{\partial y}\Big|_{y=0} = 0$ (Figure 8.20(b)).

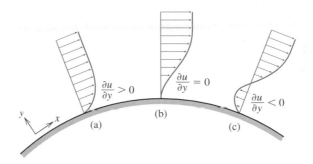

FIGURE 8.20   Schematic showing conditions for attached and separated flow.

The velocity within the boundary layer increases from a zero value at the solid surface to its freestream value at the edge of the layer, $y = \delta$. Therefore, as $y \to \delta$, $\dfrac{\partial u}{\partial y}\Big|_{y \to \delta} > 0$. In reversed flow, then, the slope changes sign from a negative value at the solid surface to a positive value at the edge of the boundary layer. Consequently, as $\dfrac{\partial u}{\partial y}$ must go through zero, separation is associated with a velocity profile that has an inflection point (a point at which the slope changes sign).

The shear stress is maximum at the solid surface and then reduces to zero at the edge of the boundary layer. Consequently, for $y \to \delta$ and irrespective of whether there is separation or not, we always have

$$\frac{\partial \tau}{\partial y}\Big|_{y \to \delta} < 0 \quad \text{or} \quad \mu \frac{\partial^2 u}{\partial y^2}\Big|_{y \to \delta} < 0. \tag{8.27}$$

The remaining question now is to establish the external flow conditions under which separation can possibly occur. If we apply Equation (8.2) along the solid surface (where $u = v = 0$), we get

$$\mu \frac{\partial^2 u}{\partial y^2}\Big|_{y=0} = \frac{dp}{dx}.$$

However, according to the Bernoulli equation,

$$\frac{dp}{dx} = -\rho U \frac{dU}{dx}.$$

By combining the momentum and Bernoulli equations, we get

$$\mu \frac{\partial^2 u}{\partial y^2}\bigg|_{y=0} = -\rho U \frac{dU}{dx}. \tag{8.28}$$

At this point, we can identify two types of external flow: (a) accelerating flows and (b) decelerating flows. When the flow accelerates, $\left(\dfrac{dU}{dx} > 0\right)$, according to Equation (8.28) the curvature of the velocity within the boundary layer is negative $\left(\dfrac{\partial^2 u}{\partial y^2} < 0\right)$. The velocity then increases monotonically to the outer free-stream velocity. Therefore, since there is no inflection point, the flow will not separate.

However, when the flow decelerates $\left(\dfrac{dU}{dx} < 0\right)$, the curvature of the velocity at the solid surface is positive $\left(\dfrac{\partial^2 u}{\partial y^2}\bigg|_{y=0} > 0\right)$, while at the edge of the boundary layer (as shown by Equation (8.27)), it is negative $\left(\dfrac{\partial^2 u}{\partial y^2}\bigg|_{y\to\delta} < 0\right)$. Therefore, in decelerating flow, the velocity profile may develop an inflection point. Consequently, *separation may possibly occur only in decelerating flows.*

To summarize, then: flow around a solid surface remains attached to the surface as long as the flow accelerates, or equivalently, when the pressure gradient is "favorable" $\left(\dfrac{dp}{dx} < 0\right)$. When the flow decelerates, the resulting "adverse" pressure gradient $\left(\dfrac{dp}{dx} > 0\right)$ causes the flow near the surface to slow down to the extent that the flow *may* be forced to reverse direction and thus to separate.

Consider, for instance, the incompressible flow through the converging-diverging nozzle shown in Figure 8.21. Because of the decrease in area in the converging part of the nozzle, the flow accelerates, whereas in the diverging part, the flow decelerates. Separation, then, can develop only in the diverging section of the nozzle. In the converging part of

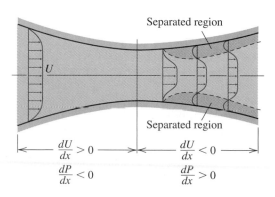

FIGURE 8.21   Flow in a converging diverging nozzle: separation may develop in the diverging part of the nozzle.

the nozzle, the flow remains attached to the wall. Note that the flow will not always separate when the flow decelerates, but if separation is to take place, the flow must be decelerating.

8.4.2 Wake Dynamics

Another consequence of the boundary layer is that the interaction of the flowing fluid with the walls produces vorticity ω (i.e., fluid particles within the layer develop the tendency to rotate). The average vorticity generated within the boundary layer is

$$\omega_a = \frac{1}{\delta} \int_0^\delta \omega \, dy = \frac{1}{\delta} \int_0^\delta \left(\frac{\partial v}{\partial x} - \frac{\partial u}{\partial y} \right) dy,$$

and, since the term $\dfrac{\partial v}{\partial x}$ within the boundary layer is small,

$$\omega_a \approx -\frac{1}{\delta} \int_0^\delta \frac{\partial u}{\partial y} \, dy = -\frac{U}{\delta}.$$

The rotation of the fluid can also be expressed using the concept of circulation according to the definition

$$\frac{d\Gamma}{dt} = \int_0^\delta u\omega \, dy - \int_0^\delta u \frac{\partial u}{\partial y} \, dy = -\int_0^\delta u \, du = -\frac{U^2}{2}. \tag{8.29}$$

Therefore, for constant bulk flow the rate of circulation produced within the boundary layer is constant.

As discussed in Section 8.4.1, when the pressure gradient is positive, the flow can separate by releasing vorticity into the main flow. Vorticity is also released into the main flow at sharp corners or sudden changes in the geometry where the flow cannot follow the solid surface smoothly (Figure 8.22). Whenever vorticity is released into the main flow, the rotating fluid accumulates behind the body into large, distinct vortices. When the vortices reach a critical size, they shed from the body and are convected downstream by the main flow, thus forming the *wake* (Figure 8.22). This vortex shedding is known as the *von Karman vortex street*, named for *Theodore von Karman*, the famous German aerodynamicist. The resulting shedding of large periodic vortices in the wake is a major characteristic of many real flows. In flow past bluff bodies or bodies with complicated geometry, separation can occur at more than one location. Consequently, wake dynamics in bluff bodies can be quite complicated. For instance, in flow past a circular cylinder, separation occurs at two locations — one at each side of the cylinder, roughly 90° from the forward stagnation point. In this case, the wake is sustained by an antisymmetric shedding of vortices from either side of the cylinder.

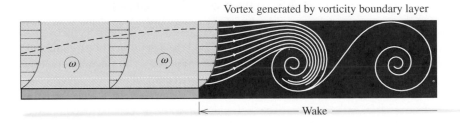

FIGURE 8.22 Wake development as a result of vorticity shed from the boundary layer.

EXAMPLE 8.9

Problem Statement *Rate of circulation shed into the wake.* Calculate the average vorticity and the rate of circulation shed from one side of a keel attached to a boat traveling at 6 *knots* (1 *knot* ≈ 0.514 *m/s*). The keel is 2 *m* long along the direction of the flow, and is 3 *m* below the hull of the boat. For water at 20°C, $\rho = 998.2 \ kg/m^3$ and $\nu = 1.004 \times 10^{-6} \ m^2/s$.

Governing Equations At separation, the average vorticity released into the main flow is

$$\omega_a \approx -\frac{U}{\delta(x)},$$

where $\delta(x)$ is the thickness of the boundary layer. The rate at which circulation is released is

$$\frac{d\Gamma}{dt} = \int_0^\delta u\omega \, dy = -\frac{U^2}{2}.$$

Assumptions It is assumed that the flow is laminar and steady and that end-plate effects are negligible.

SOLUTION Under the given conditions, Re_L based on the streamwise length of the keel is

$$Re_L = \frac{UL}{\nu} = \frac{6 \ knots \times 0.514 \ m/s \ knot \times 2 \ m}{1.004 \times 10^{-6} \ m^2/s} = 6.143 \times 10^6.$$

The flow, then, is fully turbulent. The thickness of the boundary layer is

$$\delta(L) = \frac{0.382L}{Re_L^{\frac{1}{5}}} = \frac{0.382 \times 2 \ m}{(6.143 \times 10^6)^{\frac{1}{5}}} = 3.35 \times 10^{-2} \ m.$$

The average vorticity, then, is

$$\omega_a = -\frac{U}{\delta(L)} = -\frac{6 \ knots \times 0.514 \ m/s/knot}{3.35 \times 10^{-2} \ m} = -92.06 \ s^{-1}.$$

The rate at which circulation Γ is released in the main flow is

$$\frac{d\Gamma}{dt} = -\frac{U^2}{2} = \frac{(6 \ knots \times 0.514 \ m/s/knot)^2}{2} = -4.76 \ m^2/s^2.$$

8.4.3 Drag and Lift Forces

As expected, antisymmetric shedding of vortices produces unsteady pressure and velocity fields around bluff bodies. Therefore, bluff bodies in crossflow experience a net unsteady force $\mathbf{F} = \mathbf{F}(t)$. The frequency, and magnitude with which $\mathbf{F}$ fluctuates is directly related to the frequency and strength of the shedding vortices. Therefore, the dynamics of the wake and of the induced forces are related to each other.

The net induced force $\mathbf{F}$ can be decomposed into a force component $\mathbf{F}_D$ along the direction of flow, and another component $\mathbf{F}_L$ in the transverse direction as shown in Figure 8.23. $\mathbf{F}_D$ is known as the *drag force*, and $\mathbf{F}_L$ as the *lift force*. By definition, the drag

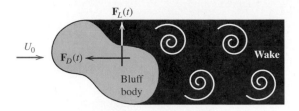

FIGURE 8.23 Induced drag and lift forces.

force represents the resistance encountered by bodies traveling through fluids. This force explains the need to supply power (i.e., fuel) to keep cars, airplanes, and other velocities in motion. The lift force is responsible for the ability of heavier-than-air bodies to fly.

A major objective in separated flows is to evaluate the drag and lift forces on bodies subjected to crossflow. To achieve this, of course, we need the exact pressure and force distribution along the surface of the body. However, because of the complexity of the flow geometry and because for most practical applications the flow is turbulent, these forces are usually evaluated experimentally and presented in empirical or tabular form.

As shown in Chapter 6, nondimensional analysis implies that the induced force on a cylinder (or other bluff bodies) is expressed by the general functional relationship

$$\frac{\mathbf{F}}{\rho U^2 D^2} = f(Re_D, e),$$

where e is the surface roughness. In the foregoing expression, D is a characteristic length scale and Re_D is the Reynolds number based on D. For a circular cylinder, D is the diameter of the cylinder.

Consistent with the foregoing expression, induced forces are typically expressed in terms of the *drag* C_D and *lift* coefficients C_L, defined as

$$C_D = \frac{F_D}{\frac{1}{2}\rho U^2 A_D} = f(Re, e) \tag{8.30}$$

and

$$C_L = \frac{F_L}{\frac{1}{2}\rho U^2 A_L} = f(Re, e). \tag{8.31}$$

Unless defined otherwise, A_D and A_L are the projected areas normal to the direction of the drag and lift forces, respectively. The factor $1/2$ is included so that the $\frac{1}{2}\rho U^2$ represents an equivalent dynamic pressure. The Re is always based on a length scale that is characteristic of the problem, (e.g., the diameter D of a circular cylinder or the height of a channel). Because of their importance, these forces and their behaviors are well documented for a large number of configurations and flow conditions.

8.4.4 Flow Past a Circular Cylinder

To better understand the development of the wake and its dynamics, below we will discuss in detail the flow past a two dimensional circular cylinder. The classical pictures by Prandtl and Tietjens in Figures 8.24 and 8.25 show clearly the events that lead to the development of the wake. Figure 8.24 shows details of the flow near the surface of the cylinder, and Figure 8.25 shows the accumulation of the shed vorticity and the early stages of the wake development. Initially, the flow separates at a single point at the rear stagnation point. This rear stagnation point divides into two separation points, which move forward until

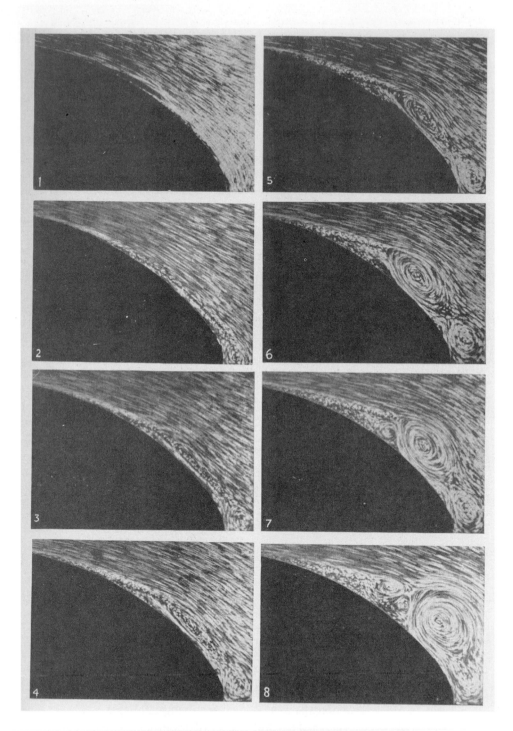

FIGURE 8.24   Development of the boundary layer in flow past a circular cylinder (Prandtl and Tietjens, 1934. Taken from: Introduction to Fluid Dynamics, G.K. Batchelor, Cambridge University Press, 1970, by permission).

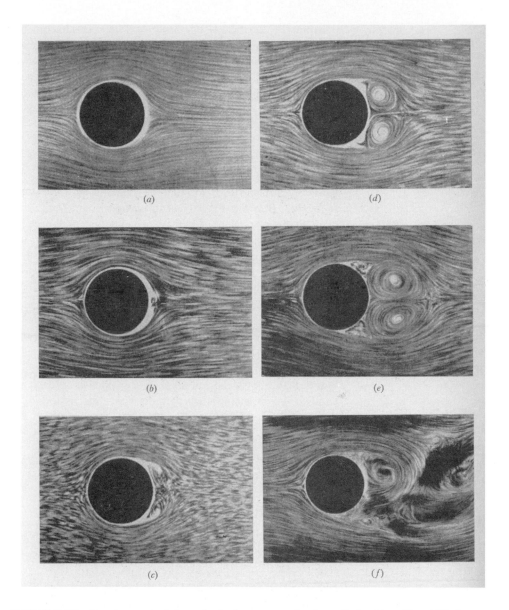

FIGURE 8.25 Development of the wake in flow past a circular cylinder (Prandtl and Tietjens, 1934. Taken from: Introduction to Fluid Dynamics, G.K. Batchelor, Cambridge University Press, 1970, by permission).

they reach a place of near equilibrium at about 78°–82° from the forward stagnation point. At separation, the boundary layer is disrupted, and vorticity is introduced into the region directly behind the body. This vorticity accumulates behind the body and forms two large vortices. As the vortices draw circulation from the vorticity shed at separation, they grow stronger. As a result, the vortices induce velocities that cause the flow close to the solid boundary to decelerate, affecting the location of the separation points. The flow remains symmetric until small disturbances destroy the balance between the two vortices (Figure 8.25(f)).

The wake dynamics are complicated further by inertia effects as expressed by Re. Depending on the range of Re, the flow can be (a) subcritical ($Re < 10^5$), where the flow is laminar; (b) critical ($Re \approx 3 \times 10^5$), where a reattachment bubble forms; (c) transcritical ($3 \times 10^5 \leq Re \leq 5 \times 10^5$), when the reattachment bubble disappears and the flow becomes turbulent; (d) supercritical ($Re > 5 \times 10^5$), where the flow is fully turbulent.

In the subcritical region, flow separation, as defined by a vanishing shear stress, develops about 78°–82° from the forward stagnation point. The drag coefficient in this region is about $C_D = 1.14$. Traditionally, the shedding frequency is measured by the non-dimensional *Strouhal number*, defined as

$$St = \frac{f_s D}{U},$$

where $f_s = 1/T_s$, T_s is the time of shedding between two consecutive vortices from the same side, D is the diameter of the cylinder, and U the freestream velocity.

In the critical region, the flow experiences laminar separation around 94°, but it reattaches downstream as turbulent flow. Experimental measurements downstream from the laminar separation point show an abrupt increase in the skin friction. Final separation is observed around 147°. In the supercritical region, the boundary layer flow is fully turbulent. In this regime the separation points vary between 65°–145°. This is due to the fact that the turbulent flow penetrates the adverse pressure gradient further downstream than the limit observed in laminar flow. In this regime, the drag coefficient drops abruptly, because the separation points move away from the forward stagnation points, resulting is a smaller wake behind the cylinder. For this reason, golf balls have dimples in order to ensure that the flow around the ball is turbulent so that the drag force is minimized. Figure 8.26 shows the variation of the drag coefficient as a function of the Reynolds number, which is consistent with the above discussions.

Summary of Flow Regimes in Flow Past a Circular Cylinder

After many years of careful experimentation, the flow behavior and the wake dynamics for flow past a circular cylinder over the whole range of Re have been documented fairly well:

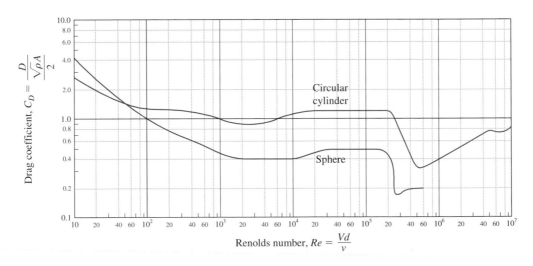

FIGURE 8.26 Drag coefficient for flow past a smooth cylinder and flow past a sphere as a function of the Reynolds number. Data from Report 253 by A.F. Zahm and NACA, and from A. Roshko, *J. of Fluid Mechanics*, **10**, (1961), p. 345.

1. For $Re \ll 1$. In this flow region, the flow maintains its symmetry about both axis as the flow does not separate. The drag force on the cylinder is due to the "deformation resistance," which is attributed to the disturbance of the flow due to the presence of the body.

2. $4 \leq Re \leq 40$. In this region, the flow separates, resulting in two large vortices behind the cylinder. These vortices remain attached to the cylinder up to $Re = 40$. Therefore, the flow maintains its symmetry around the x-axis. The drag force increases with Re, while the lift force is zero.

3. $40 \leq Re \leq 150$. In this range, the regular shedding begins and the symmetry of the flow is entirely lost. The drag and lift forces assume a periodic variation, which is related to the frequency of shedding.

4. $200 \leq Re \leq 400$. The flow inside the vortices undergoes a change from laminar to turbulent. As a consequence, a loss of regularity in the flow is observed. This transition is known as the Tritton transition region.

5. $400 \leq Re \leq 3 \times 10^5$. The antisymmetric von Karman vortex street reappears, and the flow becomes regular.

6. $3 \times 10^5 \leq Re \leq 5 \times 10^5$. The flow undergoes rapid changes. As discussed previously, the flow could be subcritical, transcritical, or supercritical. Associated with these changes also are changes in the drag and lift forces.

7. $5 \times 10^5 \leq Re \leq 3 \times 10^6$. The flow is completely supercritical (i.e., the flow is fully turbulent), and the separation points move towards the rear stagnation point. The width of the wake narrows and the drag coefficient drops abruptly. This drop in the drag force is known as the "drag crisis."

8. $Re \geq 3 \times 10^6$. The drag force increases from its low during the drag crisis but remains smaller than the drag observed at lower Re. The flow remains fully turbulent.

The Strouhal number S_t is defined only for $Re \geq 40$, since shedding does not start until $Re = 40$. At low Re, the S_t is about 0.12 and then increases to about 0.2 in the critical region. When the flow loses regularity it is difficult to identify a single characteristic frequency. In the supercritical region, S_t increases to about 0.35–0.45.

8.4.5   Flow Past Bodies with Arbitrary Shapes

The fluid dynamics in flows past bodies of arbitrary shapes are similar to the flow phenomena and transition regions discussed in the flow past a circular cylinder. Because of different shapes, however, flow separation, the development of the wake, the frequency of shedding and the variation in the magnitude and frequency of the induced forces are quantitatively different from those of flow past a cylinder. Because of the variability of the flow behavior and the fact that in most practical applications the flow is turbulent, these flows are mostly determined experimentally. Figures 8.27–8.29 show the drag and lift coefficients for a selected number of geometries.

Shape	Reference area A	Drag coefficient C_D
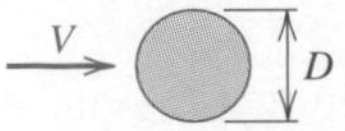	Db	1.17
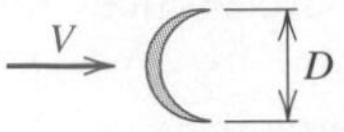	Db	1.20
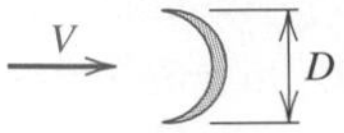	Db	2.30
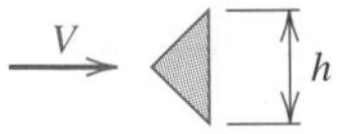	hb	1.55
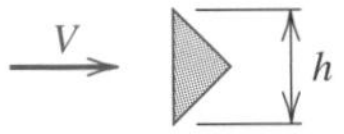	hb	2.00
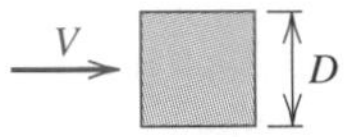	Db	2.05
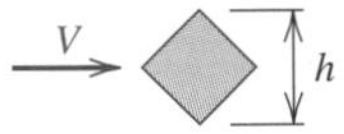	hb	1.55

FIGURE 8.27 Drag coefficient of various two-dimensional geometries for $10^4 \geq Re \leq 10^6$. Reference area is the frontal area of the bodies. Data taken from Hoerner (1958).

Shape	Reference area A	Drag coefficient C_D
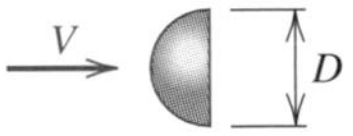	$\dfrac{\pi D^2}{4}$	0.42
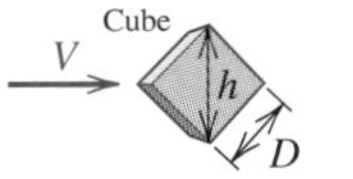 Cube	Dh	0.80
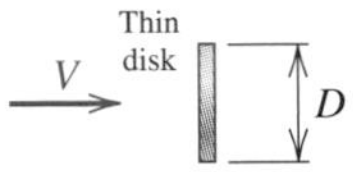 Thin disk	$\dfrac{\pi D^2}{4}$	1.10
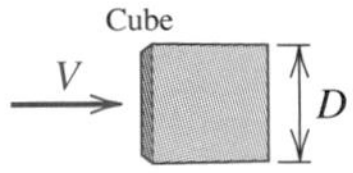 Cube	D^2	1.05
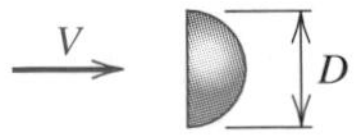	$\dfrac{\pi D^2}{4}$	1.17

FIGURE 8.28 Drag coefficient of various three-dimensional geometries for $10^4 \geq Re \leq 10^6$ configurations. Reference area is the frontal area of the bodies. Data taken from Hoerner (1958).

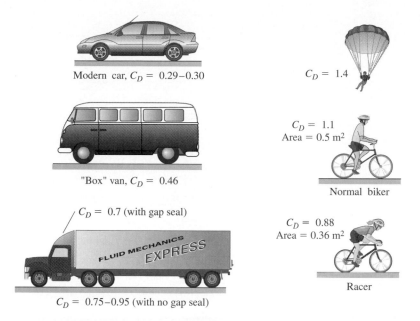

Modern car, C_D = 0.29–0.30

"Box" van, C_D = 0.46

C_D = 0.7 (with gap seal)

C_D = 0.75–0.95 (with no gap seal)

C_D = 1.4

C_D = 1.1
Area = 0.5 m²

Normal biker

C_D = 0.88
Area = 0.36 m²

Racer

FIGURE 8.29 Drag coefficient of various configurations. Data taken from Hoerner (1958).

EXAMPLE 8.10

Problem Statement In skydiving, upon jumping parachuters accelerate and then reach a constant, terminal velocity (Figure 8.30). If the total mass of the parachuter and the gear is 80 kg, find the terminal velocity. Consider that the fully deployed parachute has a diameter of 4 m.

FIGURE 8.30 Schematic of Example 8.10

Governing Equations By selecting both the parachute and the parachutist as a control volume, when the control volume drops with constant terminal velocity, the acceleration vanishes $\left(\text{i.e., } m\dfrac{dU}{dt} = 0\right)$. The conservation of linear momentum then reduces to

$$m\frac{dU}{dt} = \sum \mathbf{F} = 0. \tag{8.32}$$

Assumptions Assume that density variations are negligible.

SOLUTION The forces acting on the selected control volume are the weight $\mathbf{W} = -mg\mathbf{j}$ and the drag force $\mathbf{F}_D = F_D\mathbf{j}$. By definition,

$$F_D = C_D\frac{1}{2}\rho U^2 A,$$

where A is the frontal area of the parachute $A = \dfrac{\pi D^2}{4}$. Using experimental results, the drag coefficient is $C_D = 1.4$. Upon substitution into Equation (8.32), we get

$$C_D\frac{1}{2}\rho U^2 A = mg,$$

or,

$$U = \sqrt{2\frac{mg}{C_D\rho A}}.$$

In terms of numerical values, the terminal velocity U is

$$U = \sqrt{\frac{2 \times 80 \ kg \times 9.81 \ m/s^2 \times 4}{1.4 \times 1.126 \ kg/m^3 \times \pi \times 4^2 \ m^2}} = 8.901 \ m/s.$$

EXAMPLE 8.11

Problem Statement Consider a spherical metal projectile ($\rho_c = 10^4 \ kg/m^3$) with 0.01-m diameter, being fired with an initial velocity of 100 m/s. How long will the projectile take to travel 500 m?

Governing Equations By considering the projectile as a control volume, the conservation of momentum in the horizontal direction yields

$$m\frac{du}{dt} = \sum_{i}^{N} F_{x_i},$$

where F_{x_i} are the forces acting on the projectile in the direction of motion. In this case, while the drag is acting to retard the projectile, its weight (gravity) acts in the vertical direction and, therefore, does not affect the forward motion of the projectile.

Assumptions Assume that the drag coefficient is constant at a value $C_D = 0.1$ and that air is at $40°C$, where $\rho = 1.127 \ kg/m^3$.

SOLUTION The drag force is defined as

$$F_d = -C_D\frac{1}{2}\rho_a u^2 A_d.$$

$A_d = \pi r^2$ is the frontal area of the projectile (the negative sign shows that the force is slowing down the projectile). By substitution, the momentum equation becomes

$$m\frac{du}{dt} = -C_D\frac{1}{2}\rho_a u^2 A_d,$$

or

$$\frac{du}{dt} = -C^2,$$

where $C = C_D\frac{1}{2}\frac{\rho_a}{m}A_d = C_D\frac{1}{2}\frac{\rho_a}{\rho_c}\frac{A_d}{\mathcal{V}}$ and $\mathcal{V}$ is the volume of the projectile.

By integration,

$$-\int_{u_0}^{u}\frac{du}{u^2} = \int_0^t C\, dt,$$

we get

$$\frac{1}{u} - \frac{1}{u_0} = Ct,$$

which simplifies to

$$u = \frac{u_0}{Cu_0 t + 1}.$$

Now, the distance traveled L, is found by integration

$$\frac{dx}{dt} = u = \frac{u_0}{Cu_0 t + 1},$$

as

$$\int_0^L dx = \int_0^t \frac{u_0}{Cu_0 t + 1}\, dt.$$

The integration gives

$$L = \frac{1}{C}\ln(Cu_0 t + 1).$$

By rearranging,

$$t = \frac{e^{LC} - 1}{Cu_0}.$$

In terms of the numerical values,

$$C = C_D\frac{1}{2}\frac{\rho_a}{\rho_c}\frac{A_d}{\mathcal{V}} = 0.1 \times \frac{1}{2} \times \frac{1.127\ kg/m^3}{10000\ kg/m^3} \times \frac{6}{0.01^3\ m^3} \times \frac{0.01^2}{4}\ m^2,$$

$$C = 0.845 \times 10^{-3}\ m^{-1}.$$

The time the projectile needs to travel 500 m is

$$t = \frac{e^{LC} - 1}{Cu_0} = \frac{e^{500\ m \times 0.845 \times 10^{-3}\ m^{-1}} - 1}{0.845 \times 10^{-3}\ m^{-1} \times 100\ m/s} = 6.22\ s.$$

Therefore, if we exclude the effects of the drag force, the travel time will be underestimated by about 20%.

8.4.6   Fundamentals of Aerodynamic Applications

The preceding discussion clearly shows that drag and lift forces are important to aerodynamic applications, of which the most obvious example is aircraft flight. As shown in the schematic in Figure 8.31, aircraft are made up of several structural parts such as the main body or *fuselage*, the wings, the control surfaces, the engines, and the landing gear, Among these parts, we can identify *lifting surfaces* and *non-lifting surfaces*, depending on whether they produce lift forces or not. However, as the flow past an aircraft is significantly more complex than the flow past a circular cylinder, the design depends heavily not only on experimental data but also on the vast knowledge base accumulated over many years in designing aircraft. Given the enormous scope of the topic, we introduce the basic vocabulary and discuss only fundamental concepts in aerodynamics.

The lift force required for flight is produced mainly by the wing and, to a lesser degree, by the horizontal tail. Lift develops as a result of the asymmetric shape of the cross-sectional area of the wing known as an *airfoil* section (Figure 8.32). Because of the shape, the velocity accelerates faster at the upper surface of the airfoil than at the lower surface. Roughly speaking, according to the Bernoulli effect, the pressure at the lower surface is higher than that at the upper surface, resulting in a net vertical force.

The wing is often at an *angle of attack* α, defined as the angle the airfoil makes with respect to the horizontal direction. Following established procedure, the drag force is expressed in terms of an overall drag coefficient C_D. This coefficient represents the cumulative effect of all drag-producing components N,

$$C_D = \sum_{j=1}^{j=N} C_{D_j}.$$

Therefore,

$$C_D = \frac{F_D}{\frac{1}{2}\rho U^2 A} \Rightarrow F_D = C_D \frac{1}{2}\rho U^2 A.$$

Similarly, the lift force is determined in terms of the total lift coefficient C_L, representing the sum of all lift-producing components

$$C_L = \sum_{j=1}^{j=M} C_{L_j}.$$

Therefore,

$$C_L = \frac{F_L}{\frac{1}{2}\rho U^2 A} \Rightarrow F_L = C_L \frac{1}{2}\rho U^2 A.$$

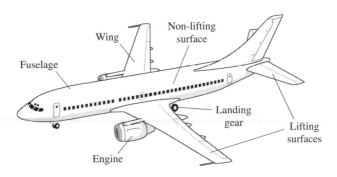

FIGURE 8.31   Schematic of an aircraft.

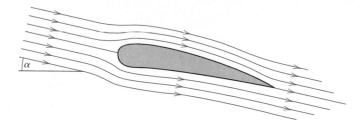

FIGURE 8.32 Flow past an airfoil at an "angle of attack" α.

Here, the area A is the *platform area* defined as the maximum projected area of the lifting surfaces. For an aircraft, this is the maximum projected area of the wings.

The C_{L_j} and C_{D_j} coefficients include the effects of Re, the surface condition, and the shape of the components. However, as a result of the complex flow field and geometry, a number of other effects must be considered. In can be shown both experimentally and theoretically that for a thin airfoil section, the lift increases linearly with increasing angle of attack

$$C_L = C_{Lo} + \alpha \frac{\partial C_L}{\partial \alpha},$$

where $\dfrac{\partial C_L}{\partial \alpha}$ is the slope of the experimental curve relating the lift coefficient to α as shown in Figure 8.33, and C_{Lo} the lift coefficient at zero angle of attack. When the lift is measured relative to the *zero lift line*, defined as the angle at which the airfoil is producing no lift, the lift coefficient is expressed as

$$C_L = \alpha \frac{\partial C_L}{\partial \alpha} = \alpha C_{L,\alpha}.$$

Lift increases with α until a limiting angle of attack α_s, known as the *stall angle*. At this angle, because of the large inclination, the flow at the upper part of the airfoil separates completely, leading to complete loss of lift and a sharp increase in the drag coefficient (Figure 8.33). Therefore, at α_s the lifting surface becomes completely ineffective. Stalling follows shortly after the point of maximum lift C_{Lmax}. The speed at C_{Lmax} is defined as the *stall speed* and represents the speed below which the aircraft looses its ability to produce sufficient lift to maintain flight.

Typically, experimental data for airfoil sections are obtained under controlled laboratory conditions and do not account for the fact that wing sections have finite dimensions that can modify the experimental data. As shown in Figure 8.34, when vorticity is shed

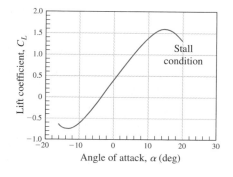

FIGURE 8.33 Lift coefficient as a function of the angle of attack α for a NACA 4412 airfoil section.

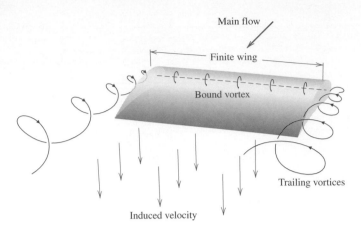

FIGURE 8.34   (a) Schematic showing the effects of a finite wing.

from the side of the wing, it accumulates into a system of vortices just behind the wing. These *trailing vortices* can be quite strong and often extend for several miles behind the aircraft. Consequently, an aircraft leaves behind a flow disturbance that can be dangerous to smaller planes when they are caught in it. Flight controllers account for this effect by controlling take-off time intervals and separation distances between aircraft.

The trailing vortices affect also the flow field around the wing by inducing a net downward velocity (*downwash*). The net effect of this downwash is to change the direction of the freestream velocity as seen by the wing. Effectively, then, the trailing vortices induce a negative angle of attack α_i

$$\alpha_i = -\frac{C_L}{\pi A_r},$$

where A_r is the geometric aspect ratio of the lifting surface defined as the span area b^2 (the area formed by a square using the length of the airfoil section b as one side) divided by the platform area A $\left(\text{i.e., } A_r = \frac{b^2}{A}\right)$. The reduction in the effective angle of attack α_i reduces the produced lift

$$C_L = \alpha C_{L,\alpha} - \frac{C_L}{\pi A_r}.$$

This means that in order to offset the finite aspect ratio effects, the wing must be at a higher angle of attack. The formation of the trailing vortices can be inhibited by placing short extensions perpendicular to the wings, known as *winglets*.

The induced angle of attack also produces an additional drag force C_{D_i} that is found to increase parabolically with lift

$$C_{D_i} = \frac{C_L^2}{\pi A e_r}$$

e_r is the *airplane efficiency factor* that depends on the geometry of the plane and corrects for deviations in the ideal pressure distribution along the wing. Under ideal conditions, the pressure distribution is elliptic. Typical values of e_r vary from 0.95 to 1.0. Also, because of the condition of the surface, a *parasitic drag* C_{D_p} develops, which appears to vary according to

$$C_{D_p} = \kappa C_L^2.$$

The constant κ is usually in order of 10^{-3}.

Consequently, at the very least the overall drag coefficient $C_{D_{total}}$ for an aircraft includes the obvious drag due to the shape C_D, flow and surface conditions C_{D_p}, and those contributions induced by the lift force C_{D_i}. Therefore,

$$C_{D_{total}} = C_D + C_{D_p} + C_{D_i} = C_D + \kappa C_L^2 + \frac{C_L^2}{\pi A_r e} = \frac{F_D}{\frac{1}{2}\rho U^2 A}.$$

However, real aircraft analysis includes many more contributions that are beyond the scope of the introductory nature of the material.

EXAMPLE 8.12

Problem Statement   Calculate the minimum flight speed of an aircraft flying at constant altitude. Consider an aircraft with maximum take-off mass of 30,740 kg, $C_L = 1.2$, and a cruising altitude of 10,000 m. The total lifting surface is 140 m^2.

Governing Equations   When the aircraft is flying at cruising altitude, the velocity is constant. Therefore, the conservation of momentum gives

$$\sum \mathbf{F} = 0,$$

where $\mathbf{F}$ are the forces acting on the aircraft.

Assumptions   Assume constant speed at cruising altitude.

SOLUTION   By selecting the aircraft as a control volume, the forces in the direction of flight are the drag force F_D and the thrust force T generated by the engines. Hence,

$$\sum F_x = F_D - T = 0, \quad \text{or} \quad T = F_D.$$

For horizontal flight, the balance of forces in the transverse direction gives

$$\sum F_y = F_L - W = 0 \quad \text{or} \quad C_L \frac{1}{2}\rho U^2 A = W = mg.$$

By rearranging,

$$U = \sqrt{\frac{2mg}{C_L \rho A}}.$$

At the flying altitude, $\rho = 0.414 \ kg/m^3$ and the acceleration due to gravity is 9.776 m/s^2. Therefore,

$$U = \sqrt{\frac{2mg}{C_L \rho A}} = \sqrt{\frac{2 \times 30,740 \ kg \times 9.776 \ m/s^2}{1.2 \times 0.414 \ kg/m^3 \times 140 \ m^2}} = 92.96 \ m/s.$$

The cruising speed is minimum when the overall lift coefficient is maximum $C_{L_{max}}$. Of course, at all times the minimum speed must be kept above the stall speed of the aircraft so that the lifting surfaces remain effective in producing lift. The speed can be reduced by increasing the platform area A of the aircraft. During landing, for instance, where lower approach speeds are required, the speed is reduced by deploying flaps. However, the increased platform area increases also the effective lift force.

Consequently, since $(C_D \propto C_L^2)$, the total drag force increases suddenly. In such a case, the engines of the aircraft must be able to produce enough thrust to sustain flight.

EXAMPLE 8.13

Problem Statement   Calculate the effect of the altitude on the minimum cruising speed of an aircraft. Consider an aircraft with maximum take-off mass of $30{,}740\ kg$, $C_{L_{max}} = 1.2$, and the cruising altitude varying from $1000\ m$ to $10{,}000\ m$. The lifting area is $140\ m^2$.

Governing Equations   According to the result obtained in Example 8.12, the minimum speed is

$$U = \sqrt{\frac{2W}{C_L \rho A}}.$$

Assumptions   Assume constant speed at cruising altitude and constant C_L.

SOLUTION   Use

$$U = \sqrt{\frac{2W}{C_L \rho A}}.$$

Therefore,

$$U = \sqrt{\frac{2 \times 30{,}740\ kg \times g}{1.2 \times \rho \times 140\ m^2}} = 19.13 \times \sqrt{\frac{g}{\rho}}\ m/s.$$

Using the information in Appendix A for standard atmospheric conditions and by repeating the same calculations as previously, the minimum speed is calculated and plotted in Figure 8.35. As shown in the figure, the minimum cruising speed increases with altitude. This fact, along with reduced weather disturbances, explains the fact that aircraft fly at high altitude.

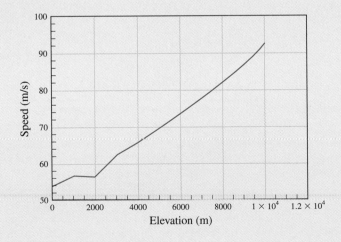

FIGURE 8.35   Example 8.13. Minimum speed as a function of the elevation

EXAMPLE 8.14

Problem Statement Determine the thrust T of an aircraft as function of level-flight speed at an altitude of 10,000 m. Consider again an aircraft with maximum take-off mass of 30,740 kg, $C_{L_{max}} = 1.2$, and cruising speed varying from the minimum speed U_{min} at $C_{L_{max}}$ to three times U_{min}. The overall drag coefficient C_D independent of C_L is equal to 0.015. The lifting surface is 140 m^2, aspect ratio $A_r = 8$, and $e_r = 0.98$.

Governing Equations For cruising at constant velocity and elevation, conservation of momentum requires that

$$\sum \mathbf{F} = 0,$$

where $\mathbf{F}$ are the forces acting on the aircraft.

Assumptions Assume constant speed at cruising altitude.

SOLUTION By selecting the aircraft as a control volume, the balance of forces in the vertical direction gives

$$\sum F_y = F_W - F_L = 0,$$

or

$$W = F_L,$$

where

$$F_L = C_L \frac{1}{2} \rho U^2 A.$$

At any speed, then,

$$C_L = \frac{2W}{\rho U^2 A}.$$

In the direction of flight we have the drag force F_D and the thrust force T generated by the engines. Hence,

$$\sum F_x = F_D - T = 0,$$

or

$$T = F_D,$$

where

$$F_D = C_{D_{total}} \frac{1}{2} \rho U^2 A.$$

The total drag coefficient is given as

$$C_{D_{total}} = C_D + C_{D_i} + C_{D_p} = C_D + \kappa C_L^2 + \frac{C_L^2}{\pi A_r e}.$$

By definition, for the same platform area at any speed we have

$$T = \frac{C_D}{C_L} W.$$

At minimum speed, $C_L = C_{L_{max}} = 1.2$. Using the results in Example 8.13,

$$U_{min} = \sqrt{\frac{2W}{\rho C_L A}} = \sqrt{\frac{2 \times 30{,}740 \ kg \times 9.776 \ m/s^2}{1.2 \times 0.4135 \ kg/m^3 \times 140 \ m^2}} = 93.02 \ m/s.$$

At minimum speed, conditions, $C_L = C_{L_{max}} = 1.2$, the drag coefficient is

$$C_D = 0.015 + 10^{-3} \times 1.2^2 + \frac{1.2^2}{\pi \times 8 \times 0.98} = 0.0749.$$

The thrust force is

$$T = \frac{C_D}{C_L} W = \frac{0.0749}{1.2} \times 30740 \ kg/m^3 \times 9.779 \ m/s^2 /1000 \ N/kN = 18.76 \ kN.$$

At the maximum speed, $U = 3 \times U_{min} = 3 \times 93.02 \ m/s = 279.06 \ m/s$. The lift coefficient is

$$C_L = \frac{2W}{\rho U^2 A} = \frac{2 \times 30740 \ kg \times 9.776 \ m/s^2}{0.4135 \ kg/m^3 \times (279.06 \ m/s)^2 \times 140 \ m^2} = 0.1333.$$

The drag coefficient at this speed is

$$C_{D_{total}} = C_D + C_{D_i} + C_{D_p} = C_D + \kappa C_L^2 + \frac{C_L^2}{\pi A_r e},$$

$$C_{D_{total}} = 0.015 + 10^{-3} \times 0.1333^2 + \frac{0.1333^2}{\pi \times 8 \times 0.98} = 0.01574.$$

The thrust force then is $T = 35.48 \ kN$.

Again using Appendix A, the thrust as a function of level flight velocity is shown in Figure 8.36. According to the figure, as the velocity increases, the thrust initially decreases and then starts to increase. Therefore, the optimum operating speed is at the point of minimum thrust. In this case, the speed is about 133 m/s, which corresponds roughly to about 300 $miles/hr$.

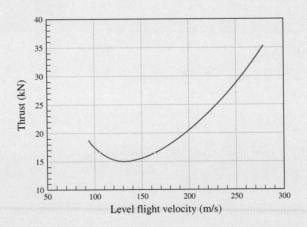

FIGURE 8.36 Example 8.14: Thrust as a function of the level flight velocity.

8.5   Internal Flows

8.5.1   Energy Considerations in Pipe Flow

In internal flows such as flow through pipes, channels, etc., the flow is established and sustained by an applied pressure that must be sufficient to overcome the resistance to flow. Figure 8.37 shows a typical case, on which uniform flow enters a circular pipe. Because of the boundary layer effects, a velocity gradient develops along the internal wall surface while the velocity towards the center remains uniform. The thickness of the boundary layer grows in the downstream direction until it reaches a critical distance L_e, where the flow becomes fully developed. In circular pipes the flow changes from laminar to turbulent in the range of $2500 < Re_D < 3000$ with Re_D based on the diameter D. When the flow is laminar, the entrance region shown previously is found to be

$$L_e = 0.058 D Re_D,$$

When the flow is turbulent,

$$L_e = 4.4 D Re_D^{\frac{1}{6}}.$$

In internal flows, a major objective is to determine the relationship between the pressure gradient $\dfrac{dP}{dx}$ and the induced flow rate $\dot{Q}$. As shown in Chapter 7, for fully developed laminar pipe flow of a Newtonian fluid, the volumetric flow rate is found to be

$$\dot{Q} = \frac{\pi}{8\mu}\frac{dP}{dx}R^4 = \frac{\pi}{128\mu}\frac{dP}{dx}D^4,$$

where μ is the viscosity of the fluid. R and D are respectively, the radius and diameter of the pipe. However, under general flow conditions, this relationship is not so simple, it is complicated by such factors as the geometry, the nature of the flow, and the roughness of the surface, Contrary to external flows, the boundary layer effects are now confined within the geometry of the flow. Therefore, effects such as separation, vortex formation, and unsteady effects are not only confined, but they are now affecting the entire flow domain. In general, we can write

$$\Delta p = f\left(V, \mu, \rho, D, L, e\right).$$

V is the average velocity through the pipe and ρ and μ are are the density and viscosity of the fluid. L and D are the length and diameter of the pipe, and e is the roughness of the surface. By organizing the variables in nondimensional groups, we find that the nondimensional pressure is given as

$$\frac{\Delta p}{\rho V^2} = f\left(Re, \frac{L}{D}, \frac{e}{D}\right). \tag{8.33}$$

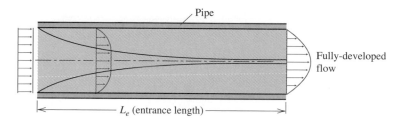

FIGURE 8.37   Flow development at the entrance of a pipe.

Exact solutions to internal flows are possible only for a few special cases, and numerical solutions are possible for laminar flows and for a limited number of turbulent flows. Hence, since in most practical applications the flow is turbulent, as in external flows, internal flows are described using experimental data.

To account for all the complexities of internal flows, problems are approached from a finite control volume point of view by using the energy equation. For instance, in pipe flow as shown in Figure 8.38, we can write

$$\frac{\partial E}{\partial t} + \int \rho \left(\frac{V^2}{2} + gz + \frac{P}{\rho} + u \right) (\mathbf{V} \cdot \mathbf{n}) \, dA = \dot{Q} - \dot{W}, \tag{8.34}$$

where V is the average velocity through the pipe defined as

$$V = \frac{1}{A} \int (\mathbf{V} \cdot \mathbf{n}) \, dA.$$

For steady flow, Equation (8.34) simplifies to

$$\int \rho \left(\frac{V^2}{2} + gz + \frac{P}{\rho} + u \right) (\mathbf{V} \cdot \mathbf{n}) \, dA = \dot{Q} - \dot{W}.$$

Assuming further that the conditions across the inlet and exit are uniform and by rearranging, we get

$$\underbrace{\left(\frac{V^2}{2} + gz + \frac{P}{\rho} \right)_1}_{E_1} - \underbrace{\left(\frac{V^2}{2} + gz + \frac{P}{\rho} \right)_2}_{E_2} - \frac{\dot{W}}{\dot{m}} = \underbrace{-\frac{\dot{Q}}{\dot{m}} + (u_2 - u_1)}_{Losses} = h_{losses}. \tag{8.35}$$

The right-hand side of Equation (8.35) represents the losses h_{losses} occurring between the inlet and the exit. These losses include the irreversible heat generation due to viscous dissipation as well as other energy losses associated with factors identified earlier. By definition, h_{losses} is always positive, and it represents *energy leaving* the control volume. Recall also that $-\frac{\dot{W}}{\dot{m}}$ is *energy gained* by the control volume—for instance, when a pump is used to pump liquid. Since pumps appear often in piping systems, it is more convenient to define $\frac{\dot{W}_{in}}{\dot{m}} = -\frac{\dot{W}}{\dot{m}}$ and rewrite the steady energy equation as

$$\underbrace{\left(\frac{V^2}{2} + gz + \frac{P}{\rho} \right)_1}_{E_1} - \underbrace{\left(\frac{V^2}{2} + gz + \frac{P}{\rho} \right)_2}_{E_2} + \frac{\dot{W}_{in}}{\dot{m}} = h_{losses}.$$

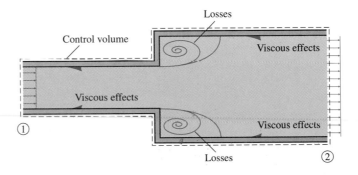

FIGURE 8.38   Finite control volume in a general pipe flow.

However, if energy is removed (e.g., when a turbine is part of the system) W_{in} is negative. For general unsteady problems, the energy equation may be written as

$$\frac{\partial E}{\partial t} + \dot{m}\left(\frac{V^2}{2} + gz + \frac{P}{\rho}\right)_1 - \dot{m}\left(\frac{V^2}{2} + gz + \frac{P}{\rho}\right)_2 + \dot{W}_{in} = \dot{m}h_{losses}. \qquad (8.36)$$

In conclusion, then, by lumping all possible losses into experimentally determined loss terms, the solution to flows complicated by turbulence and geometry become possible. However, the lumped analysis gives results for average quantities in the flow, such as the velocity and energy loss. In the limit, of no losses, the energy equation reduces to the familiar Bernoulli Equation. Despite the fact that losses are expressed in units of energy/mass flow rate, historically they are known as *head losses*. As shown subsequently, the net losses can be classified as major losses and minor losses.

8.5.2   Major Losses

Major losses are those associated with viscous dissipation that occurs along the fluid/solid interface. According to the energy equation in steady flow through a horizontal pipe, these losses result in a pressure loss between the inlet and the exit

$$\frac{P_1 - P_2}{\rho} = h_{major}.$$

Returning to the nondimensional arrangement of the flow variables given by Equation (8.33), the losses are given as (and by including the factor $1/2$)

$$\frac{\Delta p}{\frac{1}{2}\rho V^2} = \frac{h_{major}}{\frac{1}{2}V^2} = f\left(Re, \frac{L}{D}, \frac{e}{D}\right).$$

Again, this relationship must be established experimentally. In practice, the dependence of h_{major} on Re and e is expressed by an experimentally determined *friction coefficient* $f\left(Re, \frac{e}{D}\right)$, known as the *Darcy friction factor* in the form

$$h_{major} = f\frac{L}{D}\frac{V^2}{2}. \qquad (8.37)$$

In laminar, fully developed flow in a circular smooth ($e = 0$) pipe, we have already found that

$$\frac{\Delta P}{L} = \frac{128\mu}{\pi D^4}Q = \frac{128\mu}{\pi D^4}V\frac{\pi D^2}{4}.$$

By simplifying and substituting the definition for the major losses, we get

$$h_{major} = \frac{64}{Re}\frac{L}{D}\frac{V^2}{2}.$$

Comparing this result with the definition of f, the friction factor in laminar flow is

$$f = \frac{64}{Re}.$$

The friction factor for flow through a pipe is described by the well-known *Moody diagram* shown in Figure 8.39. As predicted by dimensional analysis, the diagram shows the dependence of f on Re and the relative roughness $\frac{e}{D}$. Values of $\frac{e}{D}$ for typical applications are given in Figure 8.40.

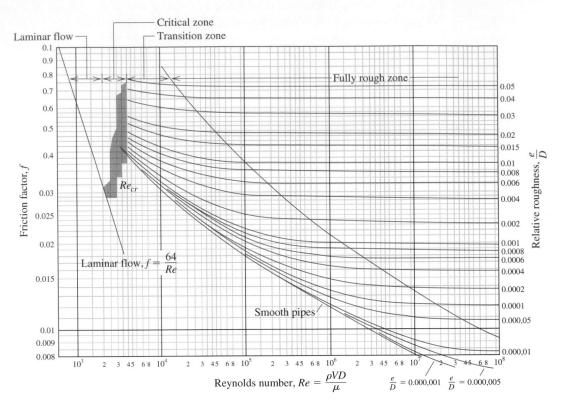

FIGURE 8.39 Moody diagram: friction factor in pipe flow. From Moody, L.F. "Friction Factors for Pipe Flow," *Transactions of the ASME*, 66, 8, November 1944, pp. 671–684, used by permission.

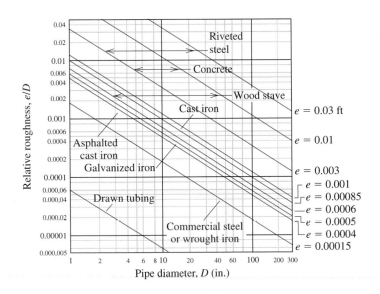

FIGURE 8.40   Relative surface roughness for various pipes. (From Moody, L.F. "Friction Factors for Pipe Flow," *Transactions of the ASME*, 66, 8, November 1944, pp. 671–684, used by permission.)

For smooth pipes, the friction factor in the range $Re < 10^5$ is approximated using the Blasius solution

$$f = \frac{0.3164}{Re^{0.25}}.$$

In the transition region, the friction depends on both the roughness and Re. For fully turbulent flow in rough pipes, the friction factor is independent of the Re and is only a function of the relative roughness.

The graphical results shown in the Moody diagram have been correlated by Colebrook (1938) and represented in mathematical form by the *Colebrook formula*

$$\frac{1}{\sqrt{f}} = -0.869 \ln \left(\frac{e/D}{3.7} + \frac{2.51}{Re\sqrt{f}} \right). \qquad (8.38)$$

Unfortunately, since f appears in both sides of Equation (8.38), it cannot be solved explicitly for the friction factor. The solution then must be obtained iteratively. According to Miller (1981), the initial guess for the friction factor f_o obtained by

$$f_o = 0.25 \left[0.434 \ln \left(\frac{e/D}{3.7} + \frac{5.74}{Re^{0.9}} \right) \right]^{-2},$$

when used in Equation (8.38), yields the correct f in one iteration.

Given the various experimental inaccuracies and correlation errors, the results given here must be considered to be accurate to within 10%. Also, since the experimental results are usually obtained using new pipes, the flow and surface roughness in old pipes are different, leading to higher losses.

Noncircular Pipes and Conduits

The results obtained for circular pipes or conduits can be applied also to flows through noncircular sections by performing calculations using an equivalent diameter $\bar{D}$ defined as

$$\bar{D} = \frac{4 \times \text{Flow Area}}{\text{Wetted Perimeter}}$$

When this definition is applied to a circular cross-sectional area with radius r, it predicts correctly that

$$\bar{D} = \frac{4\pi r^2}{2\pi r} = 2r = D,$$

where D is the diameter of the pipe. If instead the cross-sectional area is a square with each side h, we have

$$\bar{D} = \frac{4h^2}{4h} = h.$$

EXAMPLE 8.15

Problem Statement   Determine the major losses and the pressure drop in a 1.2-m horizontal section of a smooth pipe with 0.05 m diameter when water at $20°\,C$ is flowing steadily with velocity 0.02 m/s (Figure 8.41). At these conditions, $\rho = 998.2\ kg/m^3$ and $\nu = 1.004 \times 10^{-6}\ m^2/s$.

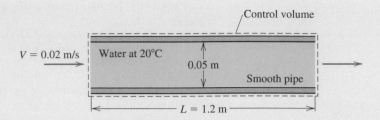

FIGURE 8.41   Schematic of Example 8.15.

Governing Equations   The steady energy equation applied between the two ends of the pipe denoted by 1 and 2, as shown in Figure 8.41 is given as

$$\left(\frac{V^2}{2} + gz + \frac{P}{\rho}\right)_1 - \left(\frac{V^2}{2} + gz + \frac{P}{\rho}\right)_2 = h_{major},$$

where the major losses are calculated using the definition,

$$h_{major} = f\frac{L}{D}\frac{V^2}{2}.$$

Assumptions   Assume steady, fully developed flow with constant properties.

SOLUTION   For the conditions given

$$Re = \frac{VD}{\nu} = \frac{0.02 \; m/s \times 0.05 \; m}{1.004 \times 10^{-6} \; m^2/s} = 996.$$

At this Re, the flow is laminar. Since the flow is fully developed, f is obtained from the analytic result for smooth pipes as

$$f = \frac{64}{Re} = \frac{64}{996} = 0.0643.$$

The major losses are

$$h_{major} = f\frac{L}{D}\frac{V^2}{2} = 0.0643 \times \frac{1.2 \; m}{0.05 \; m} \times \frac{0.02^2}{2} \; m^2/s^2 = 0.31 \times 10^{-3} \; m^2/s^2.$$

For a horizontal pipe, $z_1 = z_2$. Mass conservation for a pipe with constant diameter yields $V_1 = V_2$. Therefore, the energy equation reduces to

$$\frac{P_1 - P_2}{\rho} = h_{major} = 0.31 \times 10^{-3} \; m^2/s^2.$$

The pressure drop is

$$\Delta P = \rho h_{major} = 998.2 \; kg/m^3 \times 0.31 \times 10^{-3} \; m^2/s^2 = 0.31 \; Pa.$$

EXAMPLE 8.16

Problem Statement   Determine the head loss in a 2.0-m horizontal section of a smooth pipe ($e/D = 0$) with diameter of 0.1 m, when water at 20° C flows steadily with velocity ranging from 0.0633 m/s to 2.48 m/s. At these conditions, $\rho = 998.2 \ kg/m^3$ and $\nu = 1.004 \times 10^{-6} \ m^2/s$.

Governing Equations   The head loss is evaluated as

$$h_{major} = f \frac{L}{D} \frac{V^2}{2},$$

where the friction factor $f = f\,(e/d, Re)$ is found using the Moody diagram.

Assumptions   Assume steady, fully developed flow with constant properties.

SOLUTION   For the lower limit of the velocity,

$$Re = \frac{VD}{\nu} = \frac{0.0633 \ m/s \times 0.1 \ m}{1.004 \times 10^{-6} \ m^2/s} = 6305.$$

For smooth pipes ($e/D = 0$), at $Re \leq 10^5$, we can use the experimental results by Blasius

$$f = \frac{0.3164}{Re^{0.25}}.$$

In this case we get $f = 0.0355$. The major loss is

$$h_{major} = f \frac{L}{D} \frac{V^2}{2} = 0.0355 \times \frac{2 \ m}{0.1 \ m} \times \frac{0.0633^2}{2} \ m^2/s^2 = 0.00142 \ m^2/s^2.$$

In units of power,

$$Loss = \dot{m} h_{major} = (\rho V A)\, h_{major}$$

$$Loss = 998.2 \ kg/m^3 \times \pi \times \frac{0.1^2 \ m^2}{4} \times 0.0633 \ m/s \times 0.00142 \ m^2/s^2 = 7.05 \times 10^{-4} \ W.$$

In the upper limit of the velocity,

$$Re = \frac{VD}{\nu} = \frac{2.48 \ m/s \times 0.1 \ m}{1.004 \times 10^{-6} \ m^2/s} = 0.247 \times 10^6.$$

Since $Re > 10^5$, we will use the Moody diagram. For $e/D = 0$, the diagram (or Colebrook formula) gives for the friction factor $f = 0.015$. The major loss then is

$$h_{major} = f \frac{L}{D} \frac{V^2}{2} = 0.015 \times \frac{2 \ m}{0.1 \ m} \times \frac{2.48^2}{2} m^2/s^2 = 0.923 \ m^2/s^2.$$

In units of power,

$$Loss = \dot{m} h_{major} = (\rho V A)\, h_{major}$$

$$Loss = 998.2 \ kg/m^3 \times \pi \times \frac{0.1^2 \ m^2}{4} \times 2.48 \ m/s \times 0.923 \ m^2/s^2 = 17.95 \ Watts.$$

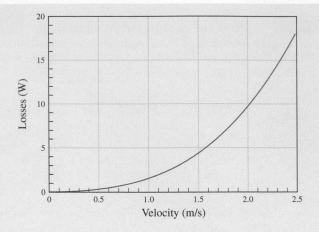

FIGURE 8.42   Example 8.16: Energy losses as a function of the velocity in the pipe.

Figure 8.42 shows the variation of the total losses as function of the velocity of the water.

EXAMPLE 8.17

Problem Statement   Calculate the level of water in a reservoir required to maintain constant steady flow of 1.5 m/s in a 50-m pipe with diameter of 0.2 m (Figure 8.43). The inner surface of the pipe has a roughness ratio $\dfrac{e}{D} = 0.0004$.

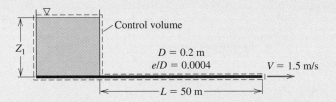

FIGURE 8.43   Schematic of Example 8.17.

Governing Equations   The energy equation applied between the free surface and the exit of the pipe is

$$\left(\frac{V^2}{2} + gz + \frac{P}{\rho}\right)_1 - \left(\frac{V^2}{2} + gz + \frac{P}{\rho}\right)_2 = h_{major},$$

where

$$h_{major} = f\frac{L}{D}\frac{V^2}{2}.$$

Assumptions   Assume steady flow of water at $20°\,C$, with constant properties.

SOLUTION   At the free surface, $V_1 \approx 0$ and $P_1 = P_{atm}$, where P_{atm} is the atmospheric pressure. By selecting the level of the pipe and the reference elevation, $z_2 = 0$. Also, since the pipe is releasing water to the atmosphere, $P_2 = P_{atm}$. For these conditions, the energy equation reduces to

$$gz_1 - \frac{V_2^2}{2} = h_{major} = f\frac{L}{D}\frac{V_2^2}{2},$$

which, by rearranging the terms simplifies to

$$z_1 = \frac{V_2^2}{2g}\left[1 + f\frac{L}{D}\right].$$

For the given flow conditions,

$$Re = \frac{VD}{\nu} = \frac{1.5\ m/s \times 0.2\ m}{1.004 \times 10^{-6}\ m^2/s} = 298,805.$$

Substituting Re, and $\dfrac{e}{D} = 0.0004$ into the Colebrook formula (or the Moody diagram), we get $f = 0.0176$.

The reservoir elevation then is obtained as

$$z_1 = \frac{V_2^2}{2g}\left[1 + f\frac{L}{D}\right] = \frac{1.5^2\ m^2/s^2}{2 \times 9.81\ m/s^2}\left[1 + 0.0176 \times \frac{50\ m}{0.2\ m}\right] = 0.619\ m.$$

8.5.3 Minor Losses

Minor losses represent all losses associated with the geometry of the flow, such as losses due to a sudden expansion or contraction, turns in the flow, various fittings, or inlet and exit effects. In general, minor losses are expressed in terms of an experimentally determined *loss coefficient K*, defined as

$$h_{minor} = K\frac{V^2}{2}.$$

In addition, minor losses are often expressed as an equivalent length L_e of the same pipe and flow conditions that produces the same loss in the form of major loss

$$h_{minor} = f\frac{L_e}{D}\frac{V^2}{2}.$$

Figure 8.44 shows the effect of the inlet on energy loss. As expected, when the flow encounters sharp corners, it can separate, thus resulting in significant energy losses ($K = 0.5$). Clearly, the losses can be minimized by carefully designing the inlet profile. For example, when the inlet is slightly rounded, $K = 0.2$. For a well-rounded inlet, the loss coefficient is reduced to $K = 0.04$! Similar losses can occur when flow goes though sudden expansions, contractions, or turns. (Figure 8.45). Figure 8.46 shows the loss coefficients for sudden changes in the cross-sectional area.

In summary, then: general pipe-flow problems are analyzed using the energy equation in the form

$$\frac{\partial E}{\partial t} + \dot{E}_1 - \dot{E}_2 + \dot{W}_{in} = \dot{m}\underbrace{\sum h_{major} + \sum h_{minor}}_{Total\ Losses}, \qquad (8.39)$$

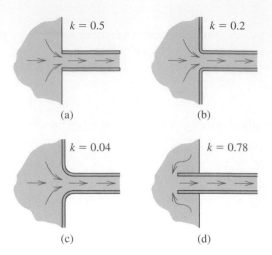

FIGURE 8.44   Effect of inlet conditions on minor losses: (a) sharp corner; (b) rounded corner; (c) smooth gradually varying inlet; (d) pipe extended in the reservoir.

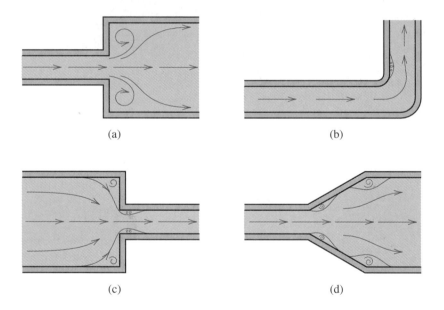

FIGURE 8.45   The effects of geometry on minor losses: (a) sudden expansion; (b) 90° bend; (c) sudden contraction; (d) gradually varying expansion.

where $\dot{E}$ is the energy flux at either end of the pipe defined as

$$\dot{E} = \dot{m}\left(\frac{V^2}{2} + gz + \frac{P}{\rho}\right).$$

The summation $\sum h_{major\ losses}$ is over all segments of pipes where viscous losses occur. $\sum h_{minor\ losses}$ accounts for all the minor losses due to the geometry. Table 8.3 shows the minor losses occurring in a variety of pipe geometries and fittings.

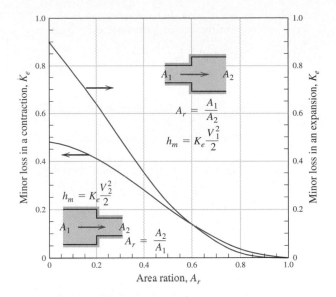

FIGURE 8.46 Losses in sudden changes in the cross-sectional area.

TABLE 8.3 Minor Losses*

Fitting	Equivalent Length L_e/D**
Valves (fully open)	
Ball valve	3
Gate valve	8
Angle valve	150
Globe valve	340
Pipe Patterns	
Standard elbow: 90°	30
Standard elbow: 45°	16

* From "Flow of Fluids through Valves, Fittings, and Pipe," New York: Crane Co. Technical Paper No. 410, 1982.

** Based on $h_{ml} = f \dfrac{L_e}{D} \dfrac{V^2}{2}$.

EXAMPLE 8.18

Problem Statement Reconsider Example 8.17 and calculate the level of water in the reservoir under the same flow and system conditions by including now the minor losses as well ($V = 1.5 \ m/s$, $L = 50 \ m$, $D = 0.2 \ m$ and $\dfrac{e}{D} = 0.0004$).

Governing Equations The energy equation applied between the free surface and the exit of the pipe is

$$\frac{\partial E}{\partial t} + \dot{m}\left(\frac{V^2}{2} + gz + \frac{P}{\rho}\right)_1 - \dot{m}\left(\frac{V^2}{2} + gz + \frac{P}{\rho}\right)_2 + \dot{W}_{in}$$

$$= \dot{m}\left(h_{major} + \sum h_{minor}\right),$$

where the major losses are calculated as

$$h_{major} = f \frac{L}{D} \frac{V^2}{2}.$$

The $\sum h_{minor}$ represents the sum of all minor losses due to the geometry of the system.

Assumptions   Assume steady flow of water at $20° C$ with constant properties.

SOLUTION   Using the same conditions as in Example 8.17 — that is, $\dot{W} = 0$, $V_1 \approx 0$, $P_1 = P_2 = P_{atm}$, $z_2 = 0$, for steady flow $\frac{\partial E}{\partial t} = 0$ — the energy equation reduces to

$$g z_1 - \frac{V_2^2}{2} = \underbrace{h_{major} + \sum h_{minor}}_{Total\ Losses}.$$

The losses now include both, major and minor losses. The major losses are defined again as

$$h_{major} = f \frac{L}{D} \frac{V_2^2}{2}.$$

In this problem, minor losses occur at the inlet to the straight section. For a square-edge inlet, the minor loss is

$$h_{minor} = K \frac{V_2^2}{2},$$

where $K = 0.5$.
  By rearranging the terms, the energy equation now simplifies to

$$z_1 = \frac{V_2^2}{2g} \left[1 + f \frac{L}{D} + K \right].$$

For the given flow conditions,

$$Re = \frac{VD}{\nu} = \frac{1.5\ m/s \times 0.2\ m}{1.004 \times 10^{-6}\ m^2/s^2} = 298,805.$$

Substituting Re and $\frac{e}{D} = 0.0004$ into the Colebrook formula (or the Moody diagram) we get $f = 0.0176$.
  The reservoir elevation is

$$z_1 = \frac{V_2^2}{2\ g} \left[1 + f \frac{L}{D} + K \right] = \frac{1.5^2\ m^2/s^2}{2 \times 9.81\ m/s^2} \left[1 + 0.0176 \times \frac{50\ m}{0.2\ m} + 2 \right] = 0.849\ m.$$

Because of the additional minor losses in the system, the required reservoir level is now 37% higher. This indicates the importance of identifying and including all relevant losses in a system.

EXAMPLE 8.19

Problem Statement Find the power required to pump water at a rate of $0.06\ m^3/s$ through a 0.2-m diameter pipe made of commercial steel, as shown in Figure 8.47. Assume that the pump is 87% efficient.

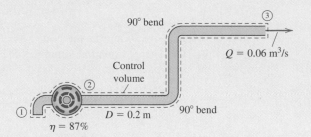

FIGURE 8.47 Schematic of Example 8.19.

Governing Equations The energy equation applied between any two points 1 and 2 in steady flow is

$$\frac{\partial E}{\partial t} + \dot{m}\left(\frac{V^2}{2} + gz + \frac{P}{\rho}\right)_1 - \dot{m}\left(\frac{V^2}{2} + gz + \frac{P}{\rho}\right)_2 + \dot{W}_{in} = Losses.$$

The losses include those due to viscous dissipation (major), the geometry of the piping system (minor), and those in the pump.

Assumptions Assume steady flow of water at $20°\,C$. Also, consider that the pump receives and delivers water at atmospheric pressure conditions.

SOLUTION Consider first a control volume around the pump, as shown in Figure 8.47. By neglecting changes in the kinetic and potential energies, the energy equation is given as

$$\dot{m}\left(\frac{P_1 - P_2}{\rho}\right) + \dot{W}_a = Losses. \tag{8.40}$$

$\dot{W}_a$ is the actual power supplied by the pump. The mass flow rate is $\dot{m} = \rho Q$.

Because of substantial losses occurring at the pump, the required power is higher than what is required under ideal conditions, when there are no losses. For this reason, the efficiency of the pump η defined as the ratio of the theoretical power $\dot{W}_t$ to the actual $\dot{W}_a$ is always less than 100%

$$\eta = \frac{\dot{W}_t}{\dot{W}_a} < 1.$$

Energy losses at the pump are equal to $\left(\dot{W}_a - \dot{W}_t\right)$, or

$$Losses = \dot{W}_a - \eta \dot{W}_a = (1 - \eta)\,\dot{W}_a.$$

By substituting the losses into Equation (8.40) and by simplifying,

$$\dot{W}_a = \frac{\rho Q}{\eta}\left(\frac{P_2 - P_1}{\rho}\right).$$

The pressure drop $(P_2 - P_1)$ or, since $P_1 = P_3$, $(P_2 - P_3)$ is related to the losses in the pipe. Using the energy equation for steady flow for points 2 and 3,

$$\left(\frac{V^2}{2} + gz + \frac{P}{\rho}\right)_2 - \left(\frac{V^2}{2} + gz + \frac{P}{\rho}\right)_3 = Losses_{2\to3}.$$

At the inlet, $z_2 = 0$, and at the exit, $z_3 = 50\ m$. Also, because of mass conservation, $V_2 = V_3$. The energy equation simplifies to

$$\left(\frac{P_2 - P_3}{\rho}\right) = \underbrace{h_{major} + h_{minor}}_{Total\ Losses} + gz_3.$$

The total losses include the major losses due to viscous dissipation in the 300-m long pipe h_{major}, and the minor losses due to the geometry h_{minor}. Therefore,

$$h_{major} = f\frac{L}{D}\frac{V^2}{2}.$$

Re is calculated using

$$Re = \frac{VD}{\nu} = \frac{4Q}{\pi D\nu} = \frac{4 \times 0.06\ m^3/s}{\pi \times 0.2\ m \times 1.004 \times 10^{-6}\ m^2/s} = 380,450.$$

For a commercial steel pipe, $\frac{e}{D} = 0.0002$. From the Moody diagram (or the Colebrook formula), $f = 0.0158$. The major losses are

$$h_{major} = f\frac{L}{D}\frac{V^2}{2} = 0.0158 \times \frac{300\ m}{0.2\ m} \times \frac{8 \times 0.06^2\ m^6/s^2}{\pi^2 \times 0.2^4\ m^4} = 43.22\ m^2/s^2.$$

The minor losses are those due to the geometry of the system. In this case, minor losses occur at the two 90° bends. For a standard 90° elbow, the loss is given in terms of an equivalent length of the same pipe L_e. For each elbow, $\frac{L_e}{D} = 30$. Therefore,

$$h_{minor} = 2f\frac{L_e}{D}\frac{V^2}{2}.$$

By substitution,

$$h_{minor} = 2 \times 0.0158 \times 30 \times \frac{8 \times 0.06^2\ m^6/s^2}{\pi^2 \times 0.2^4\ m^2} = 1.729\ m^2/s^2.$$

The pressure drop then is

$$\left(\frac{P_2 - P_3}{\rho}\right) = 43.22\ m^2/s^2 + 1.729\ m^2/s^2 + 9.81\ m/s^2 \times 50\ m = 535.45\ m^2/s^2.$$

Finally, the power input to the pump is

$$\dot{W}_a = \frac{\rho \dot{Q}}{\eta}\left(\frac{P_2 - P_1}{\rho}\right) = \frac{998.2\ kg/m^3}{0.87} \times 0.06\ m^3/s \times 535.45\ m^2/s^2 = 36.86\ kW.$$

8.5.4   Solution Procedure for Single-Pipe Problems

The velocity V through pipes or conduits depends on the applied pressure ΔP, the diameter D, the length L, the roughness e, and the geometry of the system. Additionally, the flow depends on the density ρ and the viscosity ν of the fluid. Often, the flow is expressed in terms of the volumetric flow rate Q. In such a case, the solution procedure is modified slightly by replacing V with $\dfrac{Q}{A}$, where A is the cross-sectional area.

In hydraulic design, for a fixed system (given geometry) and fluid (given ρ and ν), the objective is to determine one of the remaining flow parameters. Design considerations, for instance, may require determining the diameter of a pipe D in order to minimize losses while delivering a fixed flow rate (Q), using a given pump (ΔP). Since the variables depend on each other in a nonlinear manner, the solution to such problems often involves iterative procedures. The following is an outline of various solution procedures for the most frequently encountered types of problems.

Type I: Given V, D, L, find ΔP

The solution to such problems is straightforward: the energy equation, rearranged as

$$\frac{\Delta P}{\rho} = \left[\frac{V_2^2}{2} - \frac{V_1^2}{2} + g\,(z_2 - z_1)\right] + \sum h_{major\ losses} + \sum h_{minor\ losses},$$

is solved directly for ΔP. With V, D, L known, the major losses for every section in the system can be calculated using the Moody diagram (or the Colebrook formula), with the proper value of $\left(\dfrac{e}{D}\right)$ obtained from Figure 8.40 or other sources. The minor losses are similarly obtained from appropriate tables.

Type II: Given ΔP, V, D, find L

The solution to this type of problems is also straightforward: the length of the pipe can be obtained explicitly by rearranging the energy equation as

$$f\frac{L}{D}\frac{V^2}{2} = E_1 - E_2 - \sum h_{minor\ losses} - \sum h_{major\ losses}.$$

The major losses here include the losses in additional pipe sections that may be part of the system. The friction factor $f = f\left(Re, \dfrac{e}{D}\right)$ for the unknown length of pipe with the proper $\dfrac{e}{D}$ is determined using the Moody diagram (or the Colebrook formula). Again, the minor losses are obtained from appropriate tables.

Type III: Given ΔP, L, D, find V or Q

For this type of problem, the velocity V cannot be expressed in an explicit form as previously, because the major and minor losses are functions of the velocity itself. Therefore, these problems must be solved iteratively. A typical solution procedure for solving such problems is the following:

1. Assume a value for V (or $Q = VA$).
2. Determine the major and minor losses (whenever possible) in terms of f.
3. Solve the energy equation with the current estimate of V (or Q) for the friction factor f.

4. Calculate $Re = \dfrac{VD}{\nu} = \dfrac{4Q}{\pi D \nu}$, and, with the proper $\left(\dfrac{e}{D}\right)$, determine f from the Moody diagram.

5. If the calculated value of f from the energy equation is the same as that obtained from the Moody diagram, the solution is complete. Otherwise, use the current value of the friction factor to get a new guess for V (or Q) and repeat steps (2)–(5).

In the above procedure, several steps can be taken to accelerate convergence to the final solution. For instance, an upper-limit estimate for the velocity can be obtained from the energy equation by neglecting major losses. The entire iteration procedure can also be programmed using a computer program utilizing the Colebrook formula.

EXAMPLE 8.20

Problem Statement Consider the simple fountain shown in Figure 8.48. If the power input to the pump is $0.483\ kW$, find the minimum velocity and volumetric flow rate such that water can be pumped to a height of $3\ m$ as shown in the figure. The pipe with diameter of $0.25\ m$ is made of commercial steel with relative roughness $e/D = 0.0001$. The converging nozzle at the exit of the pipe introduces a 50% reduction in the flow area. Consider that the nozzle angle is $60°$ ($K = 0.06$).

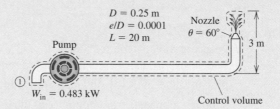

FIGURE 8.48 Schematic of Example 8.20.

Governing Equations By selecting the control volume as shown in the figure, the energy equation is

$$\frac{\partial E}{\partial t} + \dot{m}\left(\frac{V^2}{2} + gz + \frac{P}{\rho}\right)_1 - \dot{m}\left(\frac{V^2}{2} + gz + \frac{P}{\rho}\right)_2 + \dot{W}_{in} = Losses.$$

The losses are due to viscous dissipation h_{major} (major), and because of the geometry of the system (minor) h_{minor},

$$Losses = \sum h_{major} + \sum h_{minor}.$$

Assumptions Assume steady flow, with uniform, constant flow conditions.

SOLUTION Relative to the horizontal pipe, at the inlet to the pump $z_1 = 0$. Also $P = P_{atm}$ and $V = V_1$. At the highest point of the jet (point 2), $z_2 = 3\ m$, $P = P_{atm}$, and $V_2 = 0$. Using the assumption of steady flow $\frac{\partial E}{\partial t} = 0$ and constant flow conditions

$$\dot{m}\left(\frac{V_1^2}{2} - gz_2\right) + \dot{W}_{in} = \dot{m}\left(\sum h_{major} + \sum h_{minor}\right).$$

By definition, $Q = V_1 A$ and $\dot{m} = \rho Q = \rho V_1 A$. The cross-sectional area A is $A = \dfrac{\pi D^2}{4} = 1.96 \times 10^{-3}\ m^2$. The major losses are those due to viscous effects in the pipe — that is,

$$h_{major} = f\frac{L}{D}\frac{V_1^2}{2} = f\frac{20m}{0.25m}\frac{V_1^2}{2} = 80f\frac{V_1^2}{2}.$$

The minor losses occur at the 90° turn and at the 60° converging nozzle where the area reduction is 50%. According to Table 8.3, the minor loss due to the bend in terms of an equivalent length is $L_e/D = 30$, of the same pipe and flow conditions — that is,

$$h_{bend} = f\frac{L_e}{D}\frac{V_1^2}{2} = 30f\frac{V_1^2}{2}.$$

The loss at the nozzle is obtained given as $K = 0.06$. Hence,

$$h_{nozzle} = K\frac{V_1^2}{2} = 0.06\frac{V_1^2}{2}.$$

The energy equation then becomes

$$(\rho V_1 A)\left[\frac{V_1^2}{2} - gz_2\right] + \dot{W}_{in} = (\rho V_1 A)\left[80f\frac{V_1^2}{2} + 30\ f\frac{V_1^2}{2} + 0.06\frac{V_1^2}{2}\right],$$

which simplifies to

$$\frac{V_1^3}{2}(-1 + 110f + 0.06) + V_1 gz_2 = \frac{\dot{W}_{in}}{\rho A}.$$

In terms of the known numerical values,

$$V_1^3(55f - 0.47) + 29.43V_1 = \frac{483\ W}{998.2\ kg/m^3 \times 1.96 \times 10^{-3}\ m^2} = 246.87\ m^3/s^3.$$

As expected, this equation requires an iterative solution procedure since $f = f(Re, e/D)$. A good initial guess for V_1 is the velocity when viscous losses are neglected $f = 0$ — that is,

$$29.43V_{1_{max}} - 0.47V_{1_{max}}^3 = 246.87\ m^3/s^3.$$

The preceding equation gives an initial velocity of about $V_{1_{max}} = 8.52\ m/s$. However, when major losses are accounted for, the final velocity must be less than $V_{1_{max}}$. Therefore, we will assume an initial guess of $V_{1_{guess}} = 0.78V_{1_{max}} = 6.67\ m/s$. By solving,

$$f_{guess} = \frac{0.47}{55} + \frac{246.87 - 29.43V_1}{55V_1^3} = 0.016. \tag{8.41}$$

In terms of the current assumed velocity,

$$Re = \frac{V_1 D}{\nu} = \frac{6.67\ m/s \times 0.25\ m}{1.004 \times 10^{-6} m^2/s} = 1660856.$$

For the given pipe $\frac{e}{D} = 0.0001$, the Moody diagram gives $f_{estim} = 0.0129$. However, since f_{guess} is not the same as f_{estim}, the solution procedure must be repeated.

A new reasonable guess for the velocity is to use the current $f_{estim} = 0.0129$ in Equation (8.41), — that is,

$$0.0129 = \frac{0.47}{55} + \frac{246.87 - 29.43V_1}{55V_1^3}.$$

In this case, a new guess for the velocity is $V_1 = 6.35 \ m/s$. The Reynolds number is

$$Re = \frac{V_1 D}{\nu} = \frac{6.35 \ m/s \times 0.25 \ m}{1.004 \times 10^{-6} m^2/s} = 1,581,175.$$

The new estimate for the friction factor is $f_{estim} = 0.01297$. This is very close to f_{guess}. One more iteration and we get the final velocity of about $V_1 = 6.3 \ m/s$. The volumetric flow rate is then

$$Q = VA = 6.3 \ m/s \times 1.96 \times 10^{-3} \ m^2 = 0.0123 \ m^3/s.$$

Type IV: Given ΔP, L, V, find D

Type IV problems also require an iterative solution procedure similar to the one described for Type-III problems. Here, though, the iteration is performed for D. Unlike the previous procedure, since D does not appear in the energy equation, an initial estimate cannot be obtained using a simplified form of the energy equation.

Since D depends on both the Re and the roughness ratio, in order to initiate the iteration procedure it is often advantageous to start with the assumption that the flow is in the fully turbulent region, where f is only a function of the roughness ratio. With this assumption, Re and the roughness ratio are temporarily decoupled. After one or two iterations, the solution procedure will clearly indicate the region where the final solution lies.

EXAMPLE 8.21

Problem Statement   Consider the system shown in Figure 8.49. Find the diameter of the pipe so that when the valve is fully opened the pipe delivers water at the rate of $4.24 \times 10^{-4} \ m^3/s$. The loss coefficient for the fully opened valve is $K_v = 3$. Consider also that the pipe is smooth.

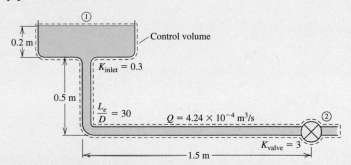

FIGURE 8.49   Schematic of Example 8.21.

Governing Equations   By selecting the control volume as shown in the figure, the energy equation is

$$\frac{\partial E}{\partial t} + \dot{m}\left(\frac{V^2}{2} + gz + \frac{P}{\rho}\right)_1 - \dot{m}\left(\frac{V^2}{2} + gz + \frac{P}{\rho}\right)_2 + \dot{W}_{in} = Losses.$$

The losses are due to viscous dissipation h_{major} (major) and to the geometry of the system (minor) h_{minor} — that is,

$$Losses = \sum h_{major} + \sum h_{minor}.$$

Assumptions   Assume steady flow, with uniform constant flow conditions.

SOLUTION   For the selected control volume as shown in Figure 8.49, for steady flow $\frac{\partial E}{\partial t} = 0$ and constant flow conditions, the energy equation applied between the free surface and the exit of the pipe becomes

$$\left(\frac{V^2}{2} + gz + \frac{P}{\rho}\right)_1 - \left(\frac{V^2}{2} + gz + \frac{P}{\rho}\right)_2 = Losses.$$

At the free surface we have $V_1 \approx 0$, $P = P_{atm}$, and $z_1 = 0.7$ m. At the exit of the pipe, $z_2 = 0$ and $P_2 = P_{atm}$. The energy equation then reduces to

$$gz_1 - \frac{V_2^2}{2} = \sum h_{major} + \sum h_{minor}.$$

The major losses are

$$h_{major} = f\frac{L}{D}\frac{V^2}{2} = f\frac{L}{D}\frac{Q^2}{2A^2} = \frac{8Q^2}{\pi^2}f\frac{L}{D}\frac{1}{D^4}.$$

The minor losses are due to the rounded inlet $K_i = 0.3$, the 90° elbow $\frac{L_e}{D} = 30$, and the valve $K_v = 3$. Therefore,

$$h_{minor} = K_i\frac{V^2}{2} + K_v\frac{V^2}{2} + f\frac{L_e}{D}\frac{V^2}{2} = 0.3\frac{V^2}{2} + 3\frac{V^2}{2} + 30f\frac{V^2}{2}.$$

In terms of Q and D, the minor losses can be written as

$$h_{minor} = \frac{8Q^2}{\pi^2}\frac{1}{D^4}(3.3 + 30f).$$

By substitution, the energy simplifies to

$$\frac{\pi^2 gz_1}{8Q^2} = \frac{1}{D^4}\left(4.3 + 30f + f\frac{L}{D}\right).$$

Since f is a function of D, the preceding nonlinear equation must be solved iteratively for D; by assuming a value for D, we solve for f,

$$f = \frac{\pi^2 gz_1 D^5 -- 34.4Q^2 D}{8Q^2(30D + L)}.$$

In terms of the known numerical values,

$$f = \frac{67.77 \times D^5 - 6.184 \times 10^{-6} \times D}{0.144 \times 10^{-5}(30 \times D + 2)}.$$

The calculated value of f is compared with the friction factor predicted by the smooth pipe equation,

$$f = \frac{0.3164}{Re^{0.25}}.$$

If the two values agree, the current value of D is the final solution. Otherwise, the solution procedure is repeated.

By assuming $D = 0.03$, the friction factor is calculated to be $f = 0.35$. The corresponding value of the velocity is obtained as

$$V = \frac{4Q}{\pi D^2} = \frac{4 \times 4.24 \times 10^{-4} \ m^3/s}{\pi \times 0.03^2 \ m^2} = 0.6 \ m/s.$$

The Reynolds number is

$$Re = \frac{0.6 \ m/s \times 0.03 \ m}{1.004 \times 10^{-6} \ m^2/s} = 17{,}928.$$

For a smooth surface

$$f = \frac{0.3164}{Re^{0.25}} = \frac{0.3164}{17{,}928^{0.25}} = 0.0273.$$

Since the two values of f do not agree, we need a new estimate for D.

By assuming $D = 0.02$, the friction factor is calculated from the energy equation to be $f = 0.025$. The corresponding value of the velocity is obtained as

$$V = \frac{4Q}{\pi D^2} = \frac{4 \times 4.24 \times 10^{-4} \ m^3/s}{\pi \times 0.02^2 \ m^2} = 1.35 \ m/s.$$

The Reynolds number is

$$Re = \frac{1.35 \ m/s \times 0.02 \ m}{1.004 \times 10^{-6} \ m^2/s} = 26{,}892.$$

For a smooth surface,

$$f = \frac{0.3164}{Re^{0.25}} = \frac{0.3164}{26892^{0.25}} = 0.0247,$$

which is almost identical to that calculated by the energy equation. Therefore, the unknown diameter is $D = 0.02$.

8.5.5   Flow in Piping Networks

In the previous sections we considered single-pipe systems or multiple pipes connected in series. In reality, piping systems can be very complicated. For instance, the delivery system for water in a building involves pipes of various diameters connected in a complex network. The solution to such complex systems follows the same solution procedures as in single pipes but must be modified to account for the fact that pipes are connected not only in series but in parallel.

Consider the piping system connected as shown in Figure 8.50. The flow in each pipe is governed by the energy equation applied at the inlet and exit of each pipe. Obviously, then, pipes with common junctions have the same pressure drop across them (i.e., the

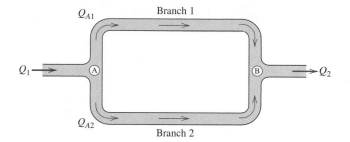

FIGURE 8.50 Schematic of a piping network.

same head losses). Note also that because of mass conservation, the net inflow/outflow in each junction is zero.

According to the foregoing observations, the pressure difference $\Delta P = P_A - P_B$ in Figure 8.50 for branch 1 is the same for branch 2. Also, for junction A, we have $Q_1 = Q_{A_1} + Q_{A_2}$, and for junction B, $Q_2 = Q_{B_1} + Q_{B_2}$. Since network problems require iterative solutions procedures that can be very lengthy, they are typically solved using specialized computer programs.

EXAMPLE 8.22

Problem Statement   Consider the simple network shown in Figure 8.51. Water at the rate of 2.2 m^3/s, is divided between two pipes before merging at a common junction. The first pipe with diameter 0.3 m and length 20 m, is made with cast iron ($e = 0.2159 \times 10^{-4}$ m). The second pipe with diameter 0.18 m and length 15 m is made with stainless steel ($e = 0.0381 \times 10^{-4}$ m). For this system, calculate the pressure drop across the two pipes, and the volumetric flow rate in each pipe.

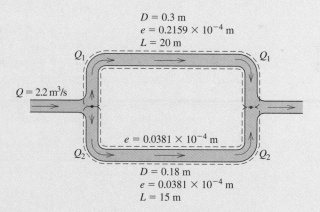

FIGURE 8.51   Schematic of Example 8.22.

Governing Equations   For piping networks the following is true:

- The net inflow/outflow in each junction is zero. This implies that there is no accumulation of mass at the junctions.
- Pipes or conduits with common end points experience the same pressure drop.

For the problem here, mass conservation at junction A requires that

$$Q_0 = Q_1 + Q_2,$$

where Q_0 is the volumetric flow rate entering the junction.

The pressure drop between junctions A and B is the same for both pipes (i.e., the major losses in the two pipes are the same).

$$h_{major_1} = h_{major_2} \Rightarrow f_1 \frac{L_1}{D_1} \frac{V_1^2}{2} = f_2 \frac{L_2}{D_2} \frac{V_2^2}{2}.$$

Assumptions   Assume no minor losses at the two junctions and elsewhere in the system.

SOLUTION   Using mass conservation,

$$Q_0 = 2.2 \ m^3/s = Q_1 + Q_2.$$

The energy equation expressed in terms of the volumetric flow rate is

$$f_1 \frac{L_1}{D_1} \frac{Q_1^2}{2D_1^4} = f_2 \frac{L_2}{D_2} \frac{Q_2^2}{2D_2^4}.$$

In term of numerical values,

$$f_1 \times \frac{20 \ m}{0.3 \ m} \times \frac{Q_1^2}{0.3^4 \ m^4} = f_2 \times \frac{15 \ m}{0.18 \ m} \times \frac{Q_2^2}{0.18^4 \ m^4},$$

or,

$$0.104 f_1 Q_1^2 = f_2 Q_2^2.$$

Because f in each pipe depends on the unknown Q, the solution requires an iterative procedure following the steps given below:

1. Determine the proper roughness ratio for each pipe. Here

$$\frac{e_1}{D_1} = \frac{0.2159 \times 10^{-4} \ m}{0.3 \ m} = 7.2 \times 10^{-5} \text{ and}$$

$$\frac{e_2}{D_2} = \frac{0.0381 \times 10^{-4} \ m}{0.18 \ m} = 2.12 \times 10^{-5}.$$

2. Assume a value for Q_1, and get an estimate for Q_2 from

$$Q_2 = 2.2 m^3/s - Q_1.$$

3. Calculate Re in each pipe, using

$$Re = \frac{VD}{\nu} = \frac{4Q}{\pi D \nu}.$$

4. Using the Moody diagram (or the Colebrook formula), obtain the friction factor for each pipe f_1 and f_2, using the current values of Q_1 and Q_2.

5. If the friction factors for each pipe satisfy

$$0.104 f_1 \left(\frac{Q_1}{Q_2} \right)^2 = f_2, \tag{8.42}$$

then the solution is complete. Otherwise, continue the iteration by repeating steps (2–5).

For this problem, we obtain an initial guess for Q_1 by assuming that the two friction factors are identical. Therefore, $0.104 Q_1^2 = Q_2^2$, which, combined with the conservation of mass, gives $Q_1 = 0.21 \ m^3/s$ and $Q_2 = 1.99 \ m^3/s$. With these values, we get

$$Re_1 = \frac{4 \times 0.21 \ m^3/s}{\pi \times 0.3 \ m \times 1.004 \times 10^{-6}} = 887717,$$

and

$$Re_2 = \frac{4 \times 1.99 \ m^3/s}{\pi \times 0.18 \ m \times 1.004 \times 10^{-6}} = 14{,}020{,}289.$$

According to the Moody diagram, $f_1 = 0.01315$ and $f_2 = 0.00952$. Checking Equation (8.42)

$$0.104 \times 0.01315 \times \left(\frac{0.21}{1.99} \right)^2 = 1.523 \times 10^{-5} = f_{2_{est}}.$$

This estimate does not agree with the calculated f_2. Assuming a new value for $Q_1 = 1.2 \ m^3/s$, according to mass conservation $Q_2 = 1.0 \ m^3/s$. With these values, we get $Re_1 = 5{,}072{,}667$, and $Re_2 = 7{,}045{,}371$, for which $f_1 = 0.01169$ and $f_2 = 9.9 \times 10^{-3}$. Checking again Equation (8.42),

$$0.104 \times 0.01169 \times \left(\frac{1.2}{1.0} \right)^2 = 1.75 \times 10^{-3} = f_{2_{est}},$$

which again does not agree with f_2. Eventually, the value of $Q_1 = 1.65 \ m^3/s$ and $Q_2 = 0.55 \ m^3/s$, give $Re_1 = 6{,}974{,}918$ and $Re_2 = 3{,}874{,}954$. The Moody diagram gives $f_1 = 0.01157$ and $f_2 = 0.0104$. A final check of Equation (8.42)

$$0.104 \times 0.01157 \times \left(\frac{1.65}{0.55} \right)^2 = 0.0108 = f_{2_{est}},$$

which is close enough to the calculated value of f_2. The pressure drop ΔP across the pipes is given as

$$\Delta P = \rho h_{major} = f_1 \frac{L_1}{D_1} \frac{16 Q_1^2}{2\pi^2 D_1^4} = 0.01157 \times \frac{20}{0.3} \times \frac{16 \times 1.65^2}{2 \times \pi^2 \times 0.3^4}$$

$$= 210.14 \ Pa.$$

REFERENCES

G.A. TOKATY, *A History and Philosophy of Fluid Mechanics*, New York, Dover, 1994.

E.A. AVALONE and T. BAUMEISTER, *Marks' Standard Handbook for Mechanical Engineers*, 9th ed., New York, McGraw Hill, 1987.

R.B. BIRD, W.E. STEWART, and E.N. LIGHTFOOT, *Transport Phenomena*, New York, Wiley, 1960.

C.F. COLEBROOK, "Turbulent Flow in Pipes with Particular Reference to the Transition Region Between the Smooth and Rough Pipe Laws," *Journal of the Institute of Civil Engineers*, London, 11, 1938–1939, pp. 133–156.

S.F. HOERNER, *Fluid-Dynamic Drag*, Published by the author. 148 Busteed Drive, Midland Park, NJ, (1958).

A.F. ZAHM, Report 253, National Aeronautics and Space Administration.

A. ROSHKO, "Experiments on Flow Past a Circular Cylinder at Very High Reynolds Number," *Journal of Fluid Mechanics*, Vol. 10, (1961).

R.W. MILLER, *Flow Measurement Engineering Handbook*, 2nd ed., New York, McGraw-Hill, 1981.

L.M. MILNE-THOMSON, *Theoretical Aerodynamics*, 4th ed., New York, Dover, 1973.

R.S. SHEVELL, *Fundamentals of Flight*, 2nd ed., Englewood Cliffs, N.J., Prentice Hall, 1989.

J.J. BERTIN and M.L. SMITH, *Aerodynamics for Engineers*, 2nd ed., Englewood Cliffs, N.J., Prentice Hall, 1989.

W.A. MAIR, and D.L. BIRDSALL, *Aircraft Performance*, Aerospace Series 5, New York, Cambridge University Press, 1992.

PROBLEMS

1. Uniform air flow at 18 m/s is flowing past a flat plate. Determine the thickness of the boundary layer at a distance 2.8 m from the leading edge. Find the average vorticity at the same location.

2. A wind tunnel is designed to operate at 25 m/s. Due to the development of the boundary layer, the effective size of the test section is reduced. If the physical size of the test section is a square with a 0.8-m side, and a length of 3 m, find the effective test section area. State clearly your assumptions.

3. Compare the maximum thickness of the boundary layer that develops due to uniform flow past a flat plate when the fluid is: (a) air, (b) water, (c) glycerine. Consider the plate as a square with 1 m side, and fluid velocity of 10 m/s.

4. Find the total viscous force on a rectangular plate due to uniform flow of air at 50 m/s. The length of the plate in the direction of the flow is 4.2 m, and the width is 6.4 m.

5. Find the total viscous force on a rectangular plate due to uniform air flow as a function of the freestream velocity in the range from 0.1 m/s to 60 m/s. The length of the plate in the direction of the flow is 8 m, and the width is 6 m.

6. Using Blasius' solution, show that the component of the velocity within the boundary layer in the direction normal to the main flow relative to the free-stream velocity is small. What is an estimate of the magnitude of the derivative $\dfrac{\partial v}{\partial y}$? Use water flowing at 2.23 m/s past a flat surface, 2.8 m long.

7. Using the linear function $u = U\dfrac{y}{\delta}$, as an approximation for the velocity distribution across the boundary layer, determine the displacement, and momentum thickness.

8. Using the linear function $u = U\dfrac{y}{\delta}$, as an approximation for the velocity distribution across the boundary layer,

determine the growth of the layer, and the skin friction coefficient, as a function of the distance along a flat plate.

9. Determine the growth of the layer, and the skin friction coefficient, as a function of the distance along a plate when the external flow is given by $U = ax$. Approximate the velocity distribution across the boundary layer with the linear function $u = U\dfrac{y}{\delta}$.

10. Determine the displacement, and momentum thickness by approximating the velocity distribution within the boundary layer using $u = U\left(\dfrac{3}{2}\dfrac{y}{\delta} - \dfrac{1}{2}\dfrac{y^3}{\delta^3}\right)$.

11. Determine the boundary layer thickness and the skin friction coefficient as a function of the distance along a flat plate using the velocity approximation $u = U\left(\dfrac{3}{2}\dfrac{y}{\delta} - \dfrac{1}{2}\dfrac{y^3}{\delta^3}\right)$.

12. Determine the displacement, and momentum thickness by approximating the velocity distribution within the boundary layer using $u = U\left(\dfrac{y}{\delta}\right)^{0.5}$.

13. Determine the displacement, and momentum thickness by approximating the velocity distribution within the boundary layer using $u = U(1 - e^{-ay})$. Note, that for this profile the boundary layer thickness extends to infinity.

14. In lubrication flows through a narrow converging gap of local height $h(x)$ and length L. For $h/L \ll 1$, $(h/L)^2 Re_L \ll 1$ and $\Delta h/L \ll 1$, where Δh is the change in the height from the inlet and exit, show that the two-dimensional Navier-Stoke equation reduces to $\partial u/\partial x + \partial v/\partial y = 0$, and $\mu \partial^2 u/\partial y^2 = dP/dx$.

15. *Flow Past a Wedge.* A coordinate transformation similar to the one introduced by Blasius was used by Falkner and Skan in 1934 for the flow past a wedge with a total angle $\theta = 2\pi \dfrac{m}{m+1}$ as shown in Figure 8.52. In this case, the

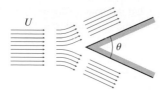

FIGURE 8.52 Schematic for Problem 15.

external flow is given by $U = U_o x^m$, and the new variables are given as

$$\eta = y\sqrt{\frac{m+1}{2}\frac{U}{\nu x}} \quad \text{and} \quad \psi = \sqrt{\frac{2}{m+1}}\sqrt{\nu U_o x}^{\frac{m+1}{2}} f(\eta).$$

Show that according to these variables the boundary layer equations reduce to

$$f''' + ff'' + \beta(1 - f'^2) = 0,$$

where $\beta = \dfrac{2m}{m+1} = \dfrac{\theta}{\pi}$. What are the proper boundary conditions?

16. Determine the total ejected mass flow rate in uniform flow past a flat plate by approximating the velocity within the boundary layer using the linear function $u = U\dfrac{y}{\delta}$. Compare the result to the exact Blasius solution. Consider L the length of the plate in the direction of the flow and W the width of the plate normal to the flow direction.

17. Calculate the total ejected mass flow rate in uniform flow past a flat plate by approximating the velocity within the boundary layer using a simusoidal approximation. Compare the result to the exact Blasius solution. Consider L the length of the plate in the direction of the flow and W the width of the plate normal to the flow direction.

18. Calculate the total ejected mass flow rate in uniform flow past a flat plate by approximating the velocity within the boundary layer using the function $u = U\left(\dfrac{3}{2}\dfrac{y}{\delta} - \dfrac{1}{2}\dfrac{y^3}{\delta^3}\right)$. Compare the result to the exact Blasius solution. Consider L the length of the plate in the direction of the flow and W the width of the plate normal to the flow direction.

19. The streamfunction

$$\psi = V_0\left[y + \frac{(\delta(x) - y)^3}{3\delta^2(x)}\right] \quad 0 \le y \le \delta(x)$$

approximates the streamfunction in a laminar boundary layer for a flat plate, where V_0 is the velocity away from the plate, y is the coordinate direction normal to the main flow and $\delta(x)$ is the thickness boundary layer. Find the stress distribution and the total force on a plate of length L.

20. Determine the length of a two dimensional channel required before uniform flow at the entrance of the pipe reaches

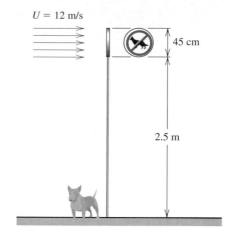

FIGURE 8.53 Schematic for Problem 24.

fully develop conditions. Use glycerine with velocity of $0.6\ m/s$ and a channel height of $0.01\ m$. (Hint: due to mass conservation, the freestream velocity changes.)

21. In the previous problem *estimate* the pressure drop between the inlet of the channel and the point where the flow becomes fully developed. State clearly your assumptions.

22. Calculate the length of a two dimensional channel required before uniform flow at the entrance of the pipe reaches fully develop conditions. Use water with velocity of $0.6\ m/s$ and a channel height of $0.04\ m$. Hint: due to mass conservation, the freestream velocity changes.

23. In the previous problem *estimate* the pressure drop between the inlet of the channel and the point where the flow becomes fully developed. State clearly your assumptions.

24. Calculate the maximum moment induced by a $12\ m/s$ wind on the support of the sign shown in Figure 8.53.

25. Consider a tall, $180\ m$ high building with a $40\ m$ square base. Find the total force induced by a $35\ miles/hr$ wind normal to one of the faces of the building. What is the maximum moment induced on the structure?

26. Calculate the power required by a $1000\ kg$-car to overcome the drag force as a function of the velocity. Consider velocities in the range of 15–$65\ miles/hrs$. Use a drag coefficient of $C_d = 0.25$, and a $4\ m^2$ frontal area.

27. Typical truck designs have an almost square frontal area. Calculate the power required to overcome the drag force, as a function of the velocity. Consider velocities in the range of 10–$55\ mile/hr$, a total frontal area of $10\ m^2$ and a drag coefficient of $C_d = 1.05$.

28. Determine the terminal velocity of a $110\ kg$ sky diver. How long will it take to reach the terminal velocity. Use a frontal area (normal to the direction of the fall) of $0.15\ m^2$ and a $C_d = 0.1$.

29. Find the annual cost associated with outfitting taxis with advertisement panels. Assume that the frontal area of the

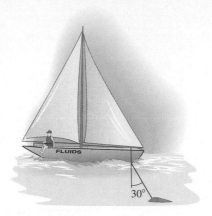

FIGURE 8.54   Schematic for Problem 32.

panel is a triangle with a 40 *cm* base and 30 *cm* height. The length of the panel is 1.5 *m*. Consider that the taxi operates 6 days a week with two 10-hour shifts per day, with an average speed of 25 *miles/hr*. To evaluate the cost consider that the heating value of gas is about 39000 *Joules/liter*, and that the price is 50 *cents/liter*.

30. Find the time it takes a 1 *cm* spherical object with density of 1660 kg/m^3 and initial velocity of 120 *m/s* to travel 300 *m* in air. Use $C_d = 0.5$.

31. Consider a 20 *cm* diameter balloon filled with helium and attached to a pole by a cord. Due to a wind with velocity V, the balloon is deflected. Find the angle the balloon makes with respect to the horizontal direction as a function of the wind velocity.

32. Find the speed of the boat when the device attached to the boat is deflected 30° degrees with respect to the vertical direction as shown in Figure 8.54. The mass of the attachment is 0.120 *kg* and its specific gravity $SG = 0.85$.

33. Crystals are often produced in solutions where due to chemical reactions small, dust-like crystals develop and slowly settle to the bottom of the reactor. For particles with diameter $d = 100 \times 10^{-6}$ *m* find the settling rate (speed) as a function of time. Consider that the drag force per unit mass is given by $F = \dfrac{u_p}{\tau}(1 + 0.15\,Re^{0.67})$, where Re is the Reynolds number based on the fluid properties (ρ_f, μ_f, etc.) and the particle velocity u_p. τ is a characteristic time term defined as $\tau = \dfrac{\rho_f d^2}{18\mu_f}$. The density of the particles is $\rho_p = 1560$ kg/m^3 and the specific gravity of the solution is $SG - 1.2$ with absolute viscosity μ_f equal to that of water.

34. Find the percentage reduction in power needed to drive a car that has been redesigned so that from an initial drag coefficient of 0.45 it dropped to 0.3. Consider a speed of 60 *mph* and a frontal area of 2.5 m^2.

35. Find the force and moment a chimney 1 *m* in diameter and 10 *m* high will experience due to a 25 *mph* wind.

36. An aircraft with velocity of 280 *km/hr* at level flight has a lift coefficient of 0.42 and a drag coefficient of 0.008. If the total mass is 1000 *kg*, what is the effective lifting area?

37. Find the time it takes a tennis ball with original velocity of 110 *miles/hr* (typical velocity in professional games) to travel 20 *m*. The mass of the ball is 0.56 *g* and its diameter 6.4 *cm*.

38. A steel sphere (specific gravity 7.82) with 5 *cm* diameter is released in a large tank filled with liquid three times as dense as water and six times as viscous. Find the terminal velocity that the sphere will attain.

39. A solid sphere with density ρ_s that settles slowly (creeping motion) in a very viscous fluid with viscosity μ and density ρ, in addition to the bouyancy force, it experiences a drag force. According to Stokes the force is equal to $F_d = 6\pi\mu Rv$, where R and v are the diameter and velocity of the sphere. Show how this problem can form the basis for an instrument to measure the viscosity of an unknown fluid.

40. The Stokes drag force experienced by a sphere (previous problem) settling slowly is valid when the Reynolds number Re_D based on the diameter of the sphere D is less than 0.1. Find the maximum diameter of an aluminum sphere ($\rho_s = 2700$ kg/m^3) settling in a bath of glycerin that will fulfill the assumptions of creeping flow. What is the corresponding velocity.

41. Using the last two problems as background information, find the time it will take an aluminum sphere with the maximum possible radius (so that the creeping flow limit is satisfied) to reach the terminal velocity when the liquid is water. Compare the results to the case of the liquid being glycerin.

42. A spherical balloon contains helium and ascends through a standard (assumed to be constant) atmosphere. The mass and payload of the balloon is 160 *kg*. What is the diameter of the balloon so that it will ascend at a constant rate of 2 *m/s*?

43. Calculate the minimum flight speed of an aircraft flying at constant altitude. Consider, an aircraft with maximum take-off mass of 42,000 *kg*, $C_L = 1.16$ and the cruising altitude of 10,000 *m*. The total lifting surface is 160 m^2.

44. Determine the thrust T of an aircraft as function of level-flight speed at an altitude of 10,000 *m*. Consider, again an aircraft with maximum take-off mass of 42,000 *kg*, $C_{L_{max}} = 1.16$ and cruising speed varying from the minimum speed U_{min} at $C_{L_{max}}$ to twice the minimum velocity U_{min}. The overall drag coefficient C_D independent of C_L is equal to 0.018. The lifting surface is 160 m^2, aspect ratio $A_r = 10$, and $e_r = 0.985$.

45. A 1200 kg aircraft is at level flight at 226 km/hr. The lift coefficient is 0.38 and the drag coefficient independent of the lift is 0.05. Find the platform area of the aircraft.

46. Show that an non-powered aircraft glides at an angle α given by $\alpha = \tan^{-1}\dfrac{C_D}{C_L}$, where C_D and C_L are the total drag and lift coefficients. At what conditions will the slope be minimum.

47. Find the approach velocity when the aircraft described in Example 8.4.6 is gliding at an angle of $3°$ with respect to the horizontal direction.

48. Find the approach velocity when the aircraft described in Problem 44 is gliding at an angle of $3.5°$ with respect to the horizontal direction.

49. What is the friction factor for (a) a smooth pipe and $Re = 10^5$, (b) transition flow, $Re = 5 \times 10^4$, and $e/D = 0.002$, (c) fully turbulent flow, $Re = 5 \times 10^6$, and $e/D = 0.03$.

50. What is the friction factor for (a) laminar flow and $Re = 5 \times 10^2$, (b) smooth pipe, $Re = 2 \times 10^5$, (c) completely rough pipe, $Re = 10^7$, and $e/D = 0.025$.

51. Determine the variation of the pressure, kinetic and potential, and total energy as a function of the length of the pipe in a 0.42 m diameter steel pipe 1000 m long. Assume that the pipe is at an angle of $45°$ degrees with respect to the horizontal and that the inlet pressure is 450 kPa and the exit pressure is atmospheric. Use water as working fluid.

52. Oil with density of 910 kg/m^3 and viscosity of 0.0091 Ns/m^2 flows in a 400 m commercial steel pipe 20 cm in diameter at a rate of 0.3 m^3/s. Determine and plot the pressure drop as a function of the distance along the pipe.

53. Glycerin flows in a 10 cm horizontal pipe 2 cm in diameter. What is the maximum flow rate through the pipe so that the flow remains laminar? What is the corresponding pressure drop across the pipe?

54. Water flows in a smooth pipe at $Re = 10^6$. After years of use the surface inside the pipe deteriorates to such extent so that, for the same pressure drop, only 60% of the original flow rate flows through. What is the roughness of the pipe?

55. Find the appropriate height H so that water at a rate of 0.01 m/s flows through the pipe (Figure 8.55).

56. The hypodermic needle shown in Figure 8.56 is used to supply a drug with density 920 kg/m^3 and viscosity $0.870 \times 10^{-3} Ns/m^2$. If the piston moves with constant velocity of 1.cm/s, find the force needed to move the piston. Assume a friction factor of 0.0001.

57. The medical device shown in Figure 8.57 is used to supply a drug with density 940 kg/m^3 and viscosity $0.9 \times 10^{-3} Ns/m^2$. If the piston moves with constant velocity of 1 cm/s, find the force needed to

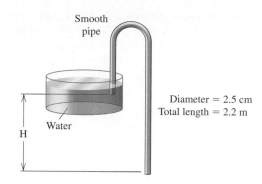

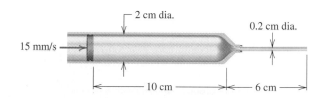

FIGURE 8.55 Schematic for Problem 55.

FIGURE 8.56 Schematic for Problem 56.

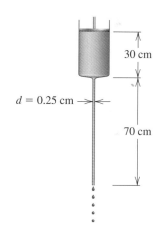

FIGURE 8.57 Schematic for Problem 57.

move the piston. Assume a friction factor of 0.0001. Find the number of drops per minute delivered by the device, by assuming spherical drops with diameter 12% larger than the diameter of the delivery tube.

58. What is the diameter of a galvanized iron pipe that has the same friction factor as a 25 cm diameter cast-iron pipe for flow at $Re = 10^5$.

59. Find the power required to pump water at a rate of 0.1 m^3/s, through a 0.25 m diameter pipe made of commercial steel as shown in Figure 8.5.3. Assume that the pump is 92% efficient.

60. Consider the simple fountain shown in Example 8.5.3. If the power input to the pump is $\mathcal{P} = 0.5\ kW$, find the minimum velocity and volumetric flow rate that water can be pumped to a height of 8 m as shown in the figure. The pipe with diameter of 0.22 m is made of commercial steel with relative roughness $e/D = 0.0002$. The converging nozzle at the exit of the pipe introduces a 50% reduction in the flow area. Consider that the nozzle angle is $60°$ ($K = 0.06$).

61. A centrifugal pump with 80% efficiency delivers water at $20°\ C$ from one tank to another. The total length of the steel pipe is 120 m and its diameter 6 cm. If the power required is 80 kW, calculate the volumetric flow rate. Assume that the difference in elevation between the two tanks is 40, and that the piping has two 90-degree elbows.

62. Two open tanks filled with water has their free surfaces 30 m apart. The tanks are connected through two smooth circular pipes each 80 m long, but with different diameters $D_1 = 20\ cm$ and $D_2 = 30\ cm$. If the joint between the two pipes is an abrupt expansion, find the volumetric flow rate through the piping?

63. A 200 m long pipe delivers 0.6 m^3 of water. Find the power needed to pump the water. The diameter of the pipe is 20 cm and the roughness-to-diameter ratio is 2×110^{-4}.

64. Determine the head loss in a 12.0 m, horizontal section of a pipe ($e/D = 0.0001$) with diameter of 0.16 m, when water at $20°\ C$ flows steadily with velocity of 0.12 m/s.

65. Find the power required to pump water in a 300 m long smooth pipe with 0.1 m diameter as a function of the volumetric flow rate. Consider velocities in the range 0.1 m/s–10 m/s.

66. Find the power required to pump water in a 300 m long pipe with 0.1 m diameter as a function of the roughness of the pipe for 0.5 m/s velocity. Consider roughness ratios in the range $1 \times 10^{-5} - 1 \times 10^{-3}$.

67. Find the diameter of 18 m smooth pipe through glycerin will flow at a rate of 0.5 m^3/s when the pressure drop is 120 Pa/m. Neglect minor losses.

68. Consider a 20 m long smooth pipe with diameter 6 cm is attached to a tank through a square entrance to a tank filled with water. The exit of the pipe is H m below the free surface. Find and plot the volumetric flow rate as a function of the height H.

69. A horizontal smooth pipe 15 cm in diameter exits from a tank filled with water, 2 m below the free surface in the tank. If the pipe is 10 m long, find the volumetric flow rate. Consider a square-edge entrance.

70. Calculate the level of water in a reservoir required to maintain constant steady flow of 0.8 m/s, in a 25 m, long pipe with diameter of 0.16 m (similar to Figure 8.5.2). The inner surface of the pipe has a roughness ratio $\dfrac{e}{D} = 0.00034$.

71. Two reservoirs are connected in series by two pipes made of wrought iron. The pipes have lengths of $L_1 = 300\ m$ and $L_2 = 250\ m$, and diameters of $D_1 = 20\ cm$ and $D_2 = 40\ cm$. If the water flow rate is 0.1 m^3/s find the elevation difference between the two reservoirs. Does it make a difference if the order with which the pipes are connected together is reversed?

72. What is the discharge between two water reservoirs connected by a 1 m concrete pipe that is 2590 m long. The difference in elevation of the free surface between the two reservoirs is 30 m.

73. A reservoir with water surface elevation of 60 m discharges into the atmosphere by way of a 0.62 m diameter pipe. If the pipe length is 590 m, the elevation of the discharge point is 30 m, the friction factor is 0.012, and the entrance loss coefficient is $K = 0.5$, find the volumetric flow rate. If an 85% efficient turbine is attached to the discharge-end, find the net power produced.

74. A circular tank 1 m in diameter and height of 2 m filled with water drains through a 2 cm diameter hole that it is connected to a short smooth pipe 80 cm long via a square entrance. Find the time it will take for the tank to drain.

75. Two pipes are connected in parallel between two water reservoirs with 10 m difference in elevation. The first pipe is made of cast iron and has a length of 1000 m and diameter of 50 cm. The second pipe has the same dimensions as the first pipe, but being newer, its roughness is 15% less than the roughness of the first pipe. Find the volumetric flow rate in each pipe.

76. A volumetric flow rate of 0.7 m^3/s of water is being discharged from a water reservoir through a turbine with 80% efficiency. What is the level of the reservoir so that a net of 74.6 kW is produced by the turbine. The water is brought to the turbine using a 60 cm diameter pipe that is 50 m long pipe with average roughness of 0.005 cm. The system also includes two 90-degree elbows and an entrance loss coefficient of $K = 0.45$.

77. Consider a system similar to the one shown in Figure 8.5.4. Find the diameter of the pipe so that when the valve is fully opened, the pipe delivers water at the rate of 0.02 m^3/s. The loss coefficient for the fully opened valve is $K_v = 3$. Consider also that the pipe is smooth.

78. A 15.24 cm diameter pipe leads from a water reservoir to a point 488 m away and 11 m below the reservoir level, where it branches into a 7.62 cm diameter pipe 305 m long with an open end, and a 5 cm diameter 122 m long pipe with a nozzle (loss coefficient $K = 0.06$) attached to its end. Both pipes discharge into atmosphere at a point 18 m below the reservoir level. Assume a friction coefficient of 0.01. Find: (a) the flow rate in each pipe, (b) the pressure at the junction.

79. Consider a straight 100 m length of steel pipe and 60 cm diameter attached to a water reservoir with its free surface 3 m above the pipe inlet. The flow is controlled by a valve

placed at the exit of the pipe. If the valve is fully closed and then opened completely, find the time it will take for the flow to established steady flow conditions. Assume a constant friction factor of 0.006.

80. Two water reservoirs with 6 m difference in their elevation, are connected with an 80 m steel pipe. What must the pipe diameter be so that the pipe could deliver 80 liters/s? Is this volumetric flow rate possible?

81. What is the volumetric flow rate in a 40 cm diameter steel pipe, when the head loss is 8 m per 500 m of pipe?

82. Determine the diameter of a wrought-iron pipe to deliver 0.8 m^3/s of water when the total available head is 10 m.

9 Ideal Inviscid Flow

Early theoretical studies of fluid flow have neglected the effects of fluid friction (i.e., the effects of viscosity). This assumption was based on the fact that for most fluids the numerical value of the viscosity is small. For instance, the viscosity of water is $\mu \approx 10^{-4}\ kg/ms$. Using this assumption, the fluid is considered to be *inviscid* or *ideal*, and the theory developed around it is known as *inviscid theory*, or *potential theory*. However, despite their elegance, inviscid theories fail to represent commonly observed phenomena that are associated with viscosity, such as flow separation, generation of vorticity, and formation of the wake.

As already discussed in Chapter 8, important viscous effects are confined within the boundary layer whose thickness is proportional to $1/\sqrt{Re}$. Despite the small thickness of the boundary layer, however, its effects cannot be readily neglected. In decelerating flows, for instance, with adverse pressure gradient ($dP/dx > 0$), viscous effects within the boundary layer can lead to separation, vortex shedding, and the development of the wake. Irrespective then of the magnitude of Re or the numerical value of the viscosity, in separated flows the viscous effects cannot be neglected.

However, when the flow accelerates ($dP/dx < 0$), because of the favorable pressure gradient the flow does not separate and remains attached to the surface. In this case, the viscous effects remain confined within the boundary layer and the bulk flow is unaffected by them. Therefore, potential theory is reasonable to use in attached flows at high Re.

Despite the apparent shortcomings of the inviscid assumption, the analysis of inviscid flows is quite revealing, and it is worth exploring it further. Moreover, it turns out that the theory is useful in many applications, especially in aerodynamics, where the assumptions of the theory are nearly satisfied. Subsequently, we first discuss the implications of the ideal fluid assumption and then proceed with the development of the fundamental theory and its applications.

9.1 Inviscid Theory: Euler Equations

In the absence of viscous effects (i.e., viscous forces), the momentum equation for incompressible fluids in three dimensions simplifies to

$$\rho\left(\frac{\partial u}{\partial t} + u\frac{\partial u}{\partial x} + v\frac{\partial u}{\partial y} + w\frac{\partial u}{\partial z}\right) = \rho g_x - \frac{\partial P}{\partial x}$$

---

[1] G.H. Hardy, *A Mathematician's Apology.*

$$\rho\left(\frac{\partial v}{\partial t}+u\frac{\partial v}{\partial x}+v\frac{\partial v}{\partial y}+w\frac{\partial v}{\partial z}\right)=\rho g_y-\frac{\partial P}{\partial y} \tag{9.1}$$

$$\rho\left(\frac{\partial w}{\partial t}+u\frac{\partial w}{\partial x}+v\frac{\partial w}{\partial y}+w\frac{\partial w}{\partial z}\right)=\rho g_z-\frac{\partial P}{\partial z}.$$

This equation is commonly referred to as the *Euler equation*, in recognition of Euler for his contribution to the development of the theory. The equations show that flow is induced by pressure and gravity forces but not by viscous forces. Note that Equation 9.1 has been derived in Chapter 6 with the use of order-of-magnitude arguments for inertia-dominated flows (large *Re*).

For ideal flow, the continuity equation retains its familiar form

$$\nabla\cdot\mathbf{V}=\frac{\partial u}{\partial x}+\frac{\partial v}{\partial y}+\frac{\partial w}{\partial z}=0. \tag{9.2}$$

The behavior of ideal fluids represents the mathematical limit when the *viscosity and all associated effects are ignored*. Therefore, the solution to the Euler equation does not predict anything associated with viscosity. For instance, even at large *Re*, the Euler equation cannot predict turbulent phenomena.

Because of lack of friction, particles rotating at a rate $\boldsymbol{\omega}$, will rotate indefinitely at the same rate. Therefore,

$$\frac{D\boldsymbol{\omega}}{Dt}=0, \tag{9.3}$$

which implies that $\boldsymbol{\omega}=$ constant. Consequently, an initially *irrotational flow* $\boldsymbol{\omega}=0$ will remain irrotational. Using the definition of *circulation* Γ,

$$\Gamma=\oint_s\mathbf{V}\cdot\mathbf{t}\,ds=\int_A\boldsymbol{\omega}\cdot\mathbf{n}\,dA,$$

it follows then that

$$\frac{D\Gamma}{Dt}=0\Rightarrow\Gamma=\text{constant}. \tag{9.4}$$

Therefore, *in inviscid flow the total circulation is conserved*. Moreover, by neglecting the viscosity, we also neglect viscous dissipation, implying that the total mechanical energy in the flow is conserved.

Another consequence of the inviscid assumption is that, because of the absence of friction, fluid particles move past solid surfaces without resistance. Therefore, instead of the usual no-slip boundary condition, we now have *perfect slip*; unimpeded by the solid surface, fluid particles flow tangentially to it (Figure 9.1). Moreover, since fluid cannot penetrate solid surfaces, we also have $\mathbf{V}\cdot\mathbf{n}=0$. Clearly, these boundary conditions are identical to those defining streamlines. Therefore, in ideal flows *solid boundaries coincide with flow streamlines*. As shown in subsequent sections, this property is exploited in many flow situations.

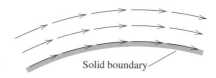

Solid boundary

FIGURE 9.1 Inviscid fluid-solid boundary conditions.

9.1.1   Euler Equations in Streamline Coordinates

Since, in inviscid flow streamlines are key parameters, it is useful to express the governing laws in terms of a coordinate system that follows naturally the local streamlines. An appropriate such system is one defined along the tangential **s**, and normal directions **n**, of local streamlines. This coordinate system is also known as a "natural coordinate system".

Consider then an infinitesimal control volume around a point N in inviscid flow, with velocity V, and pressure P as shown in Figure 9.2. A natural coordinate system then with origin at point N, is one drawn along the tangential **s**, and normal **n**, directions of the local streamline. Using a Taylor's expansion series, and by retaining only the linear terms, the flow conditions along the boundary of the control volume relative to those at point N are as shown in Figure 9.3.

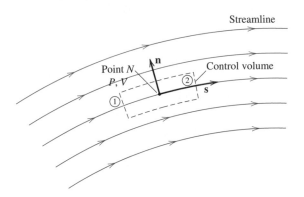

FIGURE 9.2   Flow along a streamline.

Conservation of momentum requires

$$\frac{\partial}{\partial t}\int \rho V\, d\mathcal{V} + \int \rho \mathbf{V}\,(\mathbf{V}\cdot\mathbf{n})\, dS = \sum \mathbf{F}. \tag{9.5}$$

Using Figure 9.3 and by assuming constant conditions at each face of the control volume, the **s** component of Equation (9.5) can be written as

$$\frac{\partial}{\partial t}\left(\rho V\, ds\, dn\, d\tau\right) + \rho\left(V + \frac{\partial V}{\partial s}\frac{ds}{2}\right)\left[\left(V + \frac{\partial V}{\partial s}\frac{ds}{2}\right) - \left(V - \frac{\partial V}{\partial s}\frac{ds}{2}\right)\right] dn\, d\tau =$$

$$\left[-\left(P + \frac{\partial P}{\partial s}\frac{ds}{2}\right) + \left(P - \frac{\partial P}{\partial s}\frac{ds}{2}\right)\right] dn\, d\tau + \rho g_s,$$

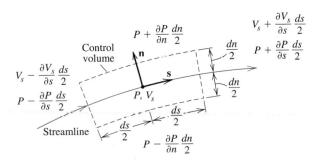

FIGURE 9.3   Differential fluid element around a streamline.

where $d\tau$ is the thickness of the control volume. With respect to the chosen coordinate system, the component of the gravity along **s** is $g_s = -g\dfrac{\partial z}{\partial s}$. Finally, by combining each term and by neglecting small differential terms, the equation simplifies to

$$\frac{\partial V}{\partial t} + V\frac{\partial V}{\partial s} = -\frac{1}{\rho}\frac{\partial P}{\partial s} - g\frac{\partial z}{\partial s}. \tag{9.6}$$

The velocity in the direction normal to the streamline is, by definition, zero (i.e., $V_n = 0$). For this direction, Equation (9.5) is written as

$$0 = \left[-\left(P + \frac{\partial P}{\partial n}\frac{dn}{2}\right) + \left(P - \frac{\partial P}{\partial n}\frac{dn}{2}\right)\right] ds\,d\tau + \rho g_n\, ds\, dn\, d\tau - F_c,$$

where the component of the gravitational acceleration is given by $g_n = -g\dfrac{\partial z}{\partial n}$. However, fluid particles in a curved trajectory experience a centripetal force F_c. For the trajectory shown in the figure

$$F_c = -\rho\frac{V^2}{R}\, ds\, dn\, d\tau,$$

where R is the local radius of curvature. The negative sign $(-)$ is included to indicate that the force is always pointing toward the origin of the subscribed circle.

Finally, by combining the various terms and by neglecting small quantities, the normal **n** component of the momentum equation is obtained as

$$\frac{V^2}{R} = \frac{1}{\rho}\frac{\partial P}{\partial n} + g\frac{\partial z}{\partial n}. \tag{9.7}$$

This result shows that the curved streamlines induce a pressure variation in the direction normal to them.

EXAMPLE 9.1

Problem Statement *Pressure variation in a bend.* Consider uniform flow V in a circular bend, as shown in Figure 9.4. Find the pressure variation across the bend as a function of the local radius. If $V = 10\ m/s$ and the inner and outer radii are $r_1 = 2\ m$ and $r_2 = 5\ m$, respectively, what is the pressure difference between the two locations?

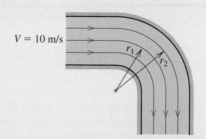

FIGURE 9.4 Pressure variation across a bend.

Governing Equations The pressure across the bend is determined by Equation (9.7)

$$\frac{V^2}{R} = \frac{1}{\rho}\frac{\partial P}{\partial n} + g\frac{\partial z}{\partial n}.$$

For a circular arc, the normal direction coincides with the radial r direction ($\mathbf{n} \equiv \mathbf{r}$).

Assumptions   It is assumed that the flow is inviscid and two-dimensional and that the plane of the flow is at constant elevation, $z = $ constant.

SOLUTION   In the absence of gravity in the normal direction (here, $\mathbf{r}$ direction), Equation (9.7) reduces to

$$\frac{V^2}{r} = \frac{1}{\rho}\frac{\partial P}{\partial r} \Rightarrow dP = \rho V^2 \frac{dr}{r}.$$

Therefore,

$$\int_{P_1}^{P_2} dP = \int_{r_1}^{r_2} \rho V^2 \frac{dr}{r} \Rightarrow P_2 - P_1 = \rho V^2 \ln \frac{r_2}{r_1}.$$

In terms of numerical values,

$$P_2 - P_1 = 1000 \ kg/m^3 \times 10^2 \ m^2/s^2 \ln \frac{5}{2} = 91.6 \ kPa.$$

9.1.2   The Bernoulli Equation

The momentum equation along a streamline (Equation (9.6)) can be manipulated further by integrating it between two arbitrary points (1) and (2),

$$\int_1^2 \frac{\partial V}{\partial t} ds + \int_1^2 V\frac{\partial V}{\partial s} ds + \int_1^2 \frac{1}{\rho}\frac{\partial P}{\partial s} ds + \int_1^2 g\frac{\partial z}{\partial s} ds = 0.$$

For incompressible flow, $\rho = $ constant, the equation simplifies to

$$\int_1^2 \frac{\partial V}{\partial t} ds + \frac{1}{2}\left(V_2^2 - V_1^2\right) + \frac{1}{\rho}(P_2 - P_1) + g(z_2 - z_1) = 0. \tag{9.8}$$

Equation (9.8) is known as the *unsteady Bernoulli equation*. For steady flow ($\partial/\partial t = 0$), it reduces to the well-known *steady Bernoulli equation*

$$\frac{1}{2}\left(V_2^2 - V_1^2\right) + \frac{1}{\rho}(P_2 - P_1) + g(z_2 - z_1) = 0. \tag{9.9}$$

Each term in Equation (9.9) represents changes in the kinetic energy, flow work, and potential energy per unit mass. However, since the two points along the streamline are arbitrary, the Bernoulli equation implies that

$$\underbrace{\frac{1}{2}V^2}_{Kinetic\ Energy} + \underbrace{\frac{P}{\rho}}_{Flow\ Work} + \underbrace{gz}_{Potential\ Energy} = \text{constant}. \tag{9.10}$$

The steady Bernoulli equation then states that *for isothermal, incompressible fluids, in the absence of heat transfer and viscous dissipation, the energy along a streamline is conserved.* Using equivalent assumptions, the Bernoulli equation was derived in Chapter 3 from the energy equation for a control volume drawn around a streamline.

EXAMPLE 9.2

Problem Statement *Bernoulli effect.* For inviscid flow, the Bernoulli equation indicates that when the velocity along a streamline increases, the pressure decreases. This trend is generally true in viscous flows as well (although the Bernoulli equation is not valid anymore). In nature, animals take advantage of this phenomenon to ventilate their burrows. If a mount is built at one of the entrances, the wind velocity increases, thus creating a pressure differential that induces air flow through the burrow (Figure 9.5).

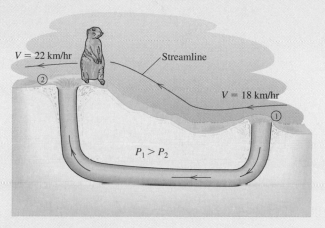

FIGURE 9.5 Bernoulli effect in nature.

For the mount shown in Figure 9.5, find the pressure difference due to wind blowing at 18 km/hr. The effect of the mount is to increase the wind velocity to 22 km/hr (consider $\rho = 1.225\ kg/m^3$).

Governing Equations The Bernoulli equation applied between a point at the inlet (1) and another at the exit of the burrow (2), where both points lie along the same streamline, as shown in Figure 9.5, is given by

$$\int_1^2 \frac{\partial V}{\partial t}\,ds + \frac{1}{2}\left(V_2^2 - V_1^2\right) + \frac{1}{\rho}\left(P_2 - P_1\right) + g\left(z_2 - z_1\right) = 0.$$

Assumptions The flow is assumed to be steady, with negligible viscous effects. Also, it is assumed that elevation changes are not significant.

SOLUTION Using the assumptions for steady flow $\frac{\partial V}{\partial t}$ and negligible elevation changes $z_1 - z_2 \approx 0$, the Bernoulli equation reduces to

$$\frac{1}{2}(V_2^2 - V_1^2) + \frac{1}{\rho}(P_2 - P_1) = 0.$$

The pressure ΔP difference is

$$P_1 - P_2 = \frac{\rho}{2}\left(V_2^2 - V_1^2\right).$$

Substituting the numerical values $V_1 = 18000\ m/3600,\ s = 5\ m/s$, and $V_2 = 22000\ m/3600\ s = 6.11\ m/s$, in terms of numerical values we get

$$\Delta P = P_1 - P_2 = \frac{1.225\ kg/m^3}{2}\left(6.11^2\ m^2/s^2 - 5^2\ m^2/s^2\right) = 7.55\ Pa.$$

EXAMPLE 9.3

Problem Statement The arrangement shown in Figure 9.6 is used in boats as a passive ventilation system for the area below deck. For the conditions shown in the figure ($L = 3.0\ m$, $D = 0.8\ m$), find the induced air velocity as a function of the boat's velocity. Consider speeds between 2 and 10 *knots* ($1\ knot \approx 0.514\ m/s$).

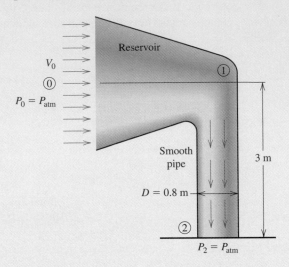

FIGURE 9.6   Ventilation induced by the Bernoulli effect.

Governing Equations   For inviscid flow, the Bernoulli equation applied between two points (0) and (1) along the same streamline is given by

$$\int_0^1 \frac{\partial V}{\partial t}\,ds + \frac{1}{2}(V_1^2 - V_0^2) + \frac{1}{\rho}(P_1 - P_0) + g(z_1 - z_0) = 0.$$

The counterpart of the Bernoulli equation in viscous flow is the energy equation which accounts for losses in the flow. Between points (1) and (2), the energy equation is expressed as

$$\left(\frac{P}{\rho} + \frac{V^2}{2} + gz\right)_1 - \left(\frac{P}{\rho} + \frac{V^2}{2} + gz\right)_2 = Losses.$$

Losses may be those due to viscous dissipation (major losses) or those due to the flow geometry (minor losses).

Assumptions   In this problem, it is assumed that the flow from far upstream (freestream conditions) to point (1) inside the converging section of the pipe is inviscid, but the flow in the straight section is fully viscous. Below deck, the pressure is assumed to be hydrostatic. Consider the volume inside the converging nozzle as a reservoir at constant pressure.

SOLUTION   Applying the Bernoulli equation between points denoted as (0) and (1), for steady flow and no changes in the elevation, we have

$$\frac{1}{2}\left(V_1^2 - V_0^2\right) + \frac{1}{\rho}(P_1 - P_{atm}) = 0,$$

where V_0 is the speed of the boat. Because the $90°$ turn the velocity V_1 is zero. Therefore,

$$P_1 = P_{atm} + \frac{\rho}{2}V_0^2,$$

by applying the energy equation between the two ends of the straight section,

$$\left(\frac{P}{\rho} + \frac{V^2}{2} + gz\right)_1 - \left(\frac{P}{\rho} + \frac{V^2}{2} + gz\right)_2 = h_{major},$$

where the major losses, h_{major}, occur between points (1) and (2). Since the diameter is constant, $V_1 = V_2 = V$. Also, at the exit of the pipe $P_2 = P_{atm} + \rho g z_1$, with $z_1 = 3\ m$ and $z_2 = 0$. The major losses in the pipe are

$$h_{major} = f\frac{L}{D}\frac{V^2}{2}.$$

By substitution,

$$\frac{V_o^2}{2} = \left(f\frac{L}{D} + 1\right)\frac{V^2}{2}.$$

In fully developed laminar flow, $f = \dfrac{64}{Re}$, and in turbulent flow at $3000 < Re < 10^5$, $f = \dfrac{0.3164}{Re^{0.25}}$, where Re is evaluated based on the diameter and the air velocity of the straight section. For $Re > 10^5$, f is evaluated using the Moody diagram. The kinematic viscosity of air at standard conditions is assumed to be $1.47 \times 10^{-5}\ m^2/s$.

At the lower boat speed of $2\ knots$, $V_o = 2 \times 0.514\ m/s = 1.028\ m/s$. Hence,

$$\left(f\frac{3}{0.8} + 1\right)V^2 = 1.028^2\ m^2/s^2,$$

$$(3.75f + 1)V^2 = 1.0568\ m^2/s^2.$$

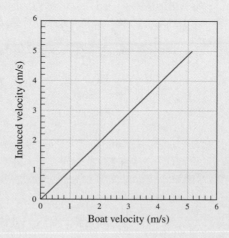

FIGURE 9.7 Example 9.3. Rate of ventilation as a function of the boat speed.

Figure 9.7 shows the induced volumetric air flow rate calculated for a range of boat speeds. As expected, the flow rate increases with the speed of the boat. The induced velocity, though, is slightly less than the boat velocity because of the effect of the losses. The results also indicate that one could have assumed ideal conditions everywhere in the flow, as the effect of the viscosity is rather small.

EXAMPLE 9.4

Problem Statement   Consider a pipe with a variable cross-sectional area connected to a pump (Figure 9.8). Find the time it takes for the water in the pipe to accelerate from rest to its final steady state velocity. Assume that the cross-sectional area varies linearly along the length of the pipe ($L = 50\ m$) from an initial area of $0.3\ m^2$ to a final area of $0.15\ m^2$. Consider that the pump pressurizes the water to $250\ kPa$.

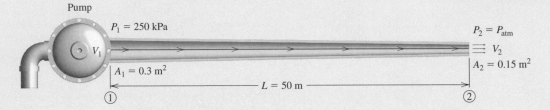

FIGURE 9.8   Example 9.4. Application of the unsteady Bernoulli equation.

Governing Equations   This is an unsteady problem for which we need to use the unsteady Bernoulli equation (Equation 9.8)

$$\int_{s=0}^{s=L} \frac{\partial V}{\partial t}\, ds + \frac{V_2^2 - V_1^2}{2} + \frac{P_2 - P_1}{\rho} + g\,(z_2 - z_1) = 0,$$

where subscripts 1 and 2 denote, respectively, points at the inlet and exit of the pipe.

Mass conservation in the pipe requires that $VA = $ constant, where A is the cross-sectional area of the pipe.

Assumptions   It is assumed that the flow is inviscid and that the pressure at the exit of the pipe is at standard conditions $P_2 = 101.3\ kPa$.

SOLUTION   Using mass conservation,

$$V_1 = V_2 \frac{A_2}{A_1} = V_2 \frac{0.15}{0.33} = 0.5 V_2.$$

For a horizontal pipe, $z_1 = z_2 = 0$. Therefore, the Bernoulli equation reduces to

$$\int_{s=0}^{s=L} \frac{\partial V}{\partial t}\, ds + \frac{V_2^2 - 0.25 V_2^2}{2} + \frac{101300\ Pa - 250000\ Pa}{998.2\ kg/m^3} = 0,$$

or

$$\int_{s=0}^{s=L} \frac{\partial V}{\partial t}\, ds + 0.375\, V_2^2 - 148.97\ m^2/s^2 = 0. \qquad (9.11)$$

Again, because of mass conservation, the velocity V at any point in the pipe is given by

$$V = V_2 \frac{A_2}{A},$$

where

$$A = A_1 - \frac{A_1 - A_2}{L}\, s.$$

The unsteady term is evaluated below as

$$\int_{s=0}^{s=L} \frac{\partial V}{\partial t}\, ds = \int_{s=0}^{s=L} \frac{\partial}{\partial t}\left(V_2 \frac{A_2}{A}\right) ds.$$

Since the limits of integration are independent of time, the time derivative can be brought outside the integral as

$$\frac{\partial}{\partial t}\int_{s=0}^{s=L} V_2 \frac{A_2}{A_1 - \dfrac{A_1 - A_2}{L}s}\, ds = \frac{\partial V_2}{\partial t}\int_{s=0}^{s=L} \frac{A_2}{A_1 - \dfrac{A_1 - A_2}{L}s}\, ds$$

$$= -\frac{A_2 L}{A_1 - A_2}\left[\ln\left(A_1 - \frac{A_1 - A_2}{L}s\right)\right]_{s=0}^{s=L} \frac{\partial V_2}{\partial t}$$

$$= \left(\frac{A_2 L}{A_1 - A_2}\right)\left(\ln \frac{A_1}{A_2}\right)\frac{\partial V_2}{\partial t}.$$

In terms of numerical values we get,

$$\int_{s=0}^{s=L} \frac{\partial V}{\partial t}\, ds = \left(\frac{0.15 \times 50}{0.3 - 0.15}\right) \times \ln \frac{0.3}{0.15} \times \frac{\partial V_2}{\partial t} = 34.66\,\frac{\partial V_2}{\partial t}\; m^2/s^2.$$

Equation (9.10) then reduces to

$$34.66\,\frac{\partial V_2}{\partial t} + 0.375\, V_2^2 - 148.97 = 0.$$

Separating variables we get

$$\int_{V_2=0}^{V_{2s}} \frac{34.66}{148.97 - 0.375\, V_2^2}\, dV_2 = \int_{t=0}^{t=t_s} dt,$$

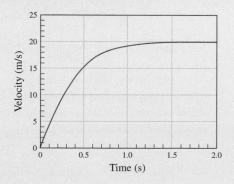

FIGURE 9.9   Example 9.4. Velocity as a function of time.

where subscript s refers to steady state conditions. Since V_2 is only a function of time, the partial derivative $(\partial/\partial t)$ was changed to a total derivative (d/dt). By integration and applying the boundary condition that at $t = 0$ $V_2 = 0$, we get

$$V_2 = 19.93 \left(\frac{e^{4.16t} - 1}{e^{4.16t} + 1} \right) \ m/s.$$

Figure 9.9 shows the velocity as a function of time. At sufficiently large time (here $t \approx 2 \ sec$), the velocity reaches a steady state value of $V_2 = 19.93 \ m/s$.

9.2 Exact Solutions for Irrotational Inviscid Flows

As shown in Figure 9.10, a solid body in inviscid flow acts to deflect the local streamlines in order to accommodate the geometry of the body. This property is directly related to the fact that because of lack of fluid friction, solid boundaries are the same as flow streamlines. This implies that the streamline distribution (i.e., streamfunction $\psi = \psi(x, y)$), is sufficient to completely construct the velocity field.

Now, recall that the definition of the streamfuction leads to

$$\frac{\partial^2 \psi}{\partial x} + \frac{\partial^2 \psi}{\partial y} = -\omega, \qquad (9.12)$$

where ω is the vorticity. For irrotational flows $\omega = 0$ and for more general rotational inviscid flows, $\omega = $ constant. Two-dimensional inviscid problems then reduce to solving either the Poisson equation $\nabla^2 \psi = -\omega$ for rotational flows, or the Laplace equation $\nabla^2 \psi = 0$ for irrotational flows (the most common application), for $\psi = \psi(x, y)$, subject to the proper boundary conditions.

The velocity field $\mathbf{V}(x, y)$ then is determined fully using the definitions

$$u = \frac{\partial \psi}{\partial y} \quad \text{and} \quad v = -\frac{\partial \psi}{\partial x}.$$

Note that because of the use of ψ, the conservation of mass for two-dimensional flows is automatically satisfied.

Finally, the momentum equation in the form given by the Bernoulli equation,

$$\int_1^2 \frac{\partial V}{\partial t} ds + \frac{1}{2} \left(V_2^2 - V_1^2 \right) + \frac{1}{\rho} \left(P_2 - P_1 \right) + g \left(y_2 - y_1 \right) = 0,$$

integrated along any streamline determined from Equation (9.11), yields the pressure distribution.

Obviously, the combination of $\nabla^2 \psi = 0$ and the Bernoulli equation is equivalent to the two-dimensional Euler equation for irrotational flow; a solution to these equations satisfies two-dimensional equivalent of the continuity and the inviscid momentum equation (Equations (9.1) and (9.2)).

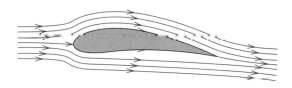

FIGURE 9.10   Inviscid flow past solid bodies.

Irrotational invisid flows can be alternatively defined using the velocity potential ϕ. Recall that, by definition, ϕ automatically satisfies the condition $\omega = 0$. Mass conservation is satisfied by solving the simple Laplace equation $\nabla^2\phi = 0$, for $\phi = \phi(x, y)$. Very much as in the case with the streamfunction ψ, the velocity potential ϕ is sufficient to determine the velocity field by using

$$u = \frac{\partial \phi}{\partial x}, \quad \text{and} \quad v = \frac{\partial \phi}{\partial y}.$$

The proper boundary conditions for either $\nabla^2\psi = 0$ or $\nabla^2\phi = 0$ along solid boundaries can be derived by simultaneously solving $\mathbf{u} \cdot \mathbf{n} = 0$ and $\mathbf{u} \cdot \mathbf{t} = V_s$, where V_s is the velocity along the surface. These conditions are

$$\frac{\partial \phi}{\partial n} = 0, \quad \frac{\partial \psi}{\partial s} = 0, \quad \text{and} \quad \frac{\partial \psi}{\partial n} = V_s, \quad \frac{\partial \phi}{\partial s} = V_s,$$

where s and n are, respectively, the tangential and normal directions to the boundary.

Again, the pressure distribution is derived from the Bernoulli equation. By subsituting $V_s ds = d\phi$ into Equation (9.4), the Bernoulli equation in terms of ϕ is now given as

$$\frac{\partial \phi}{\partial t} + \frac{V^2}{2} + \frac{P}{\rho} + gy = \text{constant.} \tag{9.13}$$

9.2.1 Elementary Flows

In mathematics, a large number of known elementary functions are exact analytic solutions to the Laplace equation. Remarkably, some of these functions, when plotted graphically, form patterns that resemble fluid flows. For this reason, these functions form the basis for analyzing inviscid flows. This is demonstrated as follows for some key inviscid flows.

Uniform Flow

The functions,

$$\psi = (V_o \cos\alpha)\, y - (V_o \sin\alpha)\, x \tag{9.14}$$

and

$$\phi = (V_o \sin\alpha)\, y + (V_o \cos\alpha)\, x, \tag{9.15}$$

which respectively satisfy $\nabla^2\psi = 0$ and $\nabla^2\phi = 0$, when represented in graphical form yield patterns that resemble those of uniform flow V_o at an angle α with respect to the horizontal (Figure 9.11).

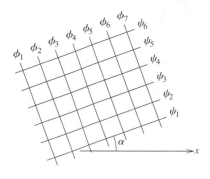

FIGURE 9.11   Uniform flow.

As shown in the figure, both the streamlines and the velocity potential are represented as straight lines that are normal to each other. For horizontal uniform flow, $\alpha = 0$,

$$\psi = V_o y, \quad \text{and} \quad \phi = V_o x.$$

By definition, we get

$$u = \frac{\partial \psi}{\partial y} = \frac{\partial \phi}{\partial x} = V_o \quad \text{and} \quad v = -\frac{\partial \psi}{\partial x} = \frac{\partial \phi}{\partial x} = 0,$$

which is, indeed, the velocity field for uniform flow.

Using polar coordinates ($x = r \cos \theta$, and $y = r \sin \theta$), the radial v_r and azimuthal v_θ velocities are given by

$$v_r = \frac{1}{r}\frac{\partial \psi}{\partial \theta} = \frac{\partial \phi}{\partial r} \quad \text{and} \quad v_\theta = -\frac{\partial \psi}{\partial r} = \frac{1}{r}\frac{\partial \phi}{\partial \theta}.$$

Uniform flow, then, in r, θ, coordinates is expressed as

$$\psi = V_o r \left(\cos \alpha \sin \theta - \sin \alpha \cos \theta \right)$$

and

$$\phi = V_o r \left(\sin \alpha \sin \theta + \cos \alpha \cos \theta \right).$$

The corresponding velocity components are

$$v_r = \frac{1}{r}\frac{\partial \psi}{\partial \theta} = \frac{\partial \phi}{\partial r} = V_o \left(\sin \alpha \sin \theta + \cos \alpha \cos \theta \right)$$

and

$$v_\theta = -\frac{\partial \psi}{\partial r} = \frac{1}{r}\frac{\partial \phi}{\partial \theta} = V_o \left(\sin \alpha \cos \theta - \cos \alpha \sin \theta \right).$$

EXAMPLE 9.5

Problem Statement   Find the streamfunction, and velocity potential for uniform flow of 40 m/s, at an angle $60°$ with respect to the horizontal.

Governing Equations   In Cartesian coordinates, the streamfunction ψ and velocity potential ϕ are given by

$$\psi = (V_o \cos \alpha)\, y - (V_o \sin \alpha)\, x,$$

and

$$\phi = (V_o \sin \alpha)\, y + (V_o \cos \alpha)\, x.$$

Assumptions   The flow is assumed to be two-dimensional and inviscid.

SOLUTION   For uniform flow at $60°$, $\theta = 60$. Also $V_o = 40$ m/s. By definition,

$$\psi = (V_o \cos \alpha)\, y - (V_o \sin \alpha)\, x.$$

Therefore,

$$\psi = 40 \cos 60 \; y - 40 \sin 60 \; x = 20 \; y - 34.64 \; x. \tag{9.16}$$

Using the velocity potential ϕ, we have,

$$\phi = (V_o \sin \alpha)\, y + (V_o \cos \alpha)\, x,$$

which yields

$$\phi = 40 \sin 60 \; y + 40 \cos 60 \; x = 34.64 \; y + 20 \; x. \tag{9.17}$$

It is obvious that Equations (9.15), and (9.16) represent straight lines that are perpendicular to each other. The actual patterns can be obtained by plotting lines for a selected number of ψ_k, and ϕ_k values.

Line Source/Sink Flow

An ideal *source* or *sink* in two dimensions is represented as a mathematical point that releases or absorbs fluid at a constant volumetric flow rate per unit depth $Q\; m^2/s$ (Figure 9.12). Often, the source is identified as *line* because the flow extends to infinity in the direction normal to the plane of the flow. As expected, because of symmetry, the velocity field is purely radial. For constant density ρ, in order to conserve mass the same flow rate must cross any circular boundary with radius r and radial velocity v_r according to

$$\rho Q = \pm \rho 2\pi r v_r \Rightarrow v_r = \pm \frac{Q}{2\pi r} = \frac{m}{r}.$$

The constant $m = \pm \dfrac{Q}{2\pi}$, measured in units of m^2/s, is included for convenience. The positive sign is associated with a source indicating release of fluid, and the negative sign is associated with a sink, where fluid is being absorbed.

The streamfunction ψ for this flow is given as

$$\psi = m \tan^{-1} \frac{y}{x}. \tag{9.18}$$

The streamfunction increases in the counterclockwise direction. By definition, the volumetric flow rate per unit depth Q between two streamlines ψ_1 and ψ_2, is given as the difference $Q = \psi_2 - \psi_1$.

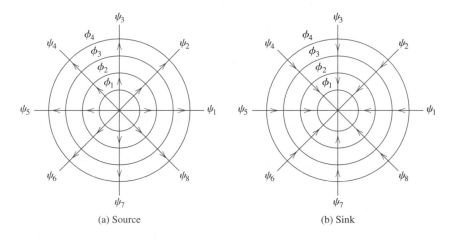

(a) Source

(b) Sink

FIGURE 9.12 Source/sink flow.

The velocity potential function ϕ is given by

$$\phi = m \ln \left(x^2 + y^2\right)^{\frac{1}{2}}.$$ (9.19)

Figure 9.12 shows that iso-velocity potential lines are concentric circles and that stream-lines are straight radial lines emanating from the origin.

The velocity field is obtained by differentiation,

$$u = \frac{\partial \psi}{\partial y} = m \frac{1}{1 + \left(\frac{y}{x}\right)^2} \frac{1}{x} = \frac{mx}{x^2 + y^2}$$ (9.20)

and

$$v = -\frac{\partial \psi}{\partial x} = -m \frac{1}{1 + \left(\frac{y}{x}\right)^2} \left(-\frac{y}{x}\right) = \frac{my}{x^2 + y^2}.$$ (9.21)

Alternatively, we could use the velocity potential to get the same result — for example,

$$u = \frac{\partial \phi}{\partial x} = \frac{m}{2} \frac{2x}{x^2 + y^2} = \frac{mx}{x^2 + y^2}$$

and

$$v = \frac{\partial \phi}{\partial y} = \frac{m}{2} \frac{2y}{x^2 + y^2} = \frac{my}{x^2 + y^2}.$$

Note that by expressing ψ, and ϕ, in terms of polar coordinates,

$$\psi = m\,\theta, \text{ and } \phi = m \ln r,$$

(where θ is expressed in radians), these functions are easier to manipulate:

$$v_r = \frac{1}{r}\frac{\partial \psi}{\partial \theta} = \frac{\partial \phi}{\partial r} = \frac{m}{r}, \text{ and } v_\theta = -\frac{\partial \psi}{\partial r} = \frac{1}{r}\frac{\partial \phi}{\partial \theta} = 0.$$

As expected, to conserve mass the velocity field is purely radial (i.e., $v_\theta = 0$). It can be shown easily that the velocity components in polar coordinates, when transformed into Cartesian coordinates, yield the same expressions as those given by Equations (9.19) and (9.20).

EXAMPLE 9.6

Problem Statement   Consider an ideal two-dimensional source delivering water at a steady rate of 2 m^2/s. Find (a) the pressure distribution as a function of the distance from the source; (b) the time it takes for a fluid particle, originally at the origin, to travel 1 m.

Governing Equations   The Bernoulli equation applied between points (1) and (2) along the same streamline is

$$\int_{s=0}^{s=L} \frac{\partial V}{\partial t}\, ds + \frac{V_2^2 - V_1^2}{2} + \frac{P_2 - P_1}{\rho} + g\,(z_2 - z_1) = 0.$$

The time t_s a particle takes to travel a distance r_s is determined using the particle velocity V_p

$$V_p = \frac{dr}{dt}, \quad \text{or} \quad t_s = \int_{r=0}^{r=r_s} \frac{1}{V_p}\, dr.$$

Assumptions   It is assumed that the flow is steady, two-dimensional, and inviscid and that the plane of the flow is assumed to be at the same elevation.

SOLUTION   The streamlines in a two-dimensional source flow are radial lines emanating from the origin (Figure 9.12). The steady Bernoulli equation applied between two points along an arbitrary streamline on a horizontal plane ($z_1 = z_2$) simplifies to

$$\frac{V_2^2 - V_1^2}{2} + \frac{P_2 - P_1}{\rho} = 0.$$

The fluid velocity along this streamline is

$$V = \frac{Q}{2\pi r}.$$

Since the velocity V at the origin is singular ($V = \infty$), the pressure distribution is evaluated starting from a point ϵ, at a radial distance $\epsilon = 1.0 \ m$ away from the origin as

$$P - P_\epsilon = \frac{\rho}{2}\left(V_\epsilon^2 - V^2\right),$$

or

$$P - P_\epsilon = \frac{\rho}{2}\left(\frac{Q}{2\pi}\right)^2\left(\frac{1}{\epsilon^2} - \frac{1}{r^2}\right).$$

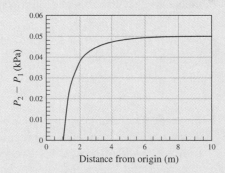

FIGURE 9.13   Pressure variation along a streamline in source flow.

The final pressure distribution is

$$P - P_\epsilon = \frac{998.2 \ kg/m^3}{2000}\left(\frac{2 \ m^2/s}{2\pi}\right)^2\left(\frac{1}{1.0^2 \ m^2} - \frac{1}{r^2 \ m^2}\right)$$

$$= 0.0506\left(\frac{1}{1.0^2} - \frac{1}{r^2}\right) kPa.$$

The result is plotted in Figure 9.13. As expected, since the velocity away from the source decreases, the pressure at a large distance away reaches asymptotically a constant value.

The time t_s a particle takes to travel 1 m is

$$t_s = t_s = \int_{r=0}^{r=r_s}\frac{1}{V_p}\,dr = \int_{r=0}^{r=1m}\frac{2\pi}{Q}r\,dr = \frac{\pi}{2} = 1.57 \ s.$$

Line Vortex Flow

An ideal vortex flow is obtained by reversing the role of the streamfunction and velocity potential for source flow as

$$\psi = -m \ln \left(x^2 + y^2 \right)^{\frac{1}{2}} = -m \ln r, \tag{9.22}$$

$$\phi = m \tan^{-1} \frac{y}{x} = m\theta. \tag{9.23}$$

Therefore, streamlines are now concentric circles, and iso-velocity potential lines are radial straight lines emanating from the origin (Figure 9.14). The negative sign in the streamfunction is included in order to fix the azimuthal velocity as positive in the counterclockwise direction. By differentiation, the velocity field is

$$v_r = 0, \text{ and } v_\theta = \frac{m}{r}.$$

The circulation Γ around a closed circle, which includes the origin, is determined by integration

$$\Gamma = \oint \underbrace{\mathbf{V} \cdot \mathbf{t}}_{v_\theta} \, ds = \oint_0^{2\pi} \frac{m}{r} r \, d\theta = 2\pi m.$$

Therefore, the constant m is a measure of the total circulation, otherwise known as the *strength of the vortex*. Both the circulation and the strength of the vortex are expressed in units of m^2/s. In terms of Γ, the vortex is more conveniently expressed as

$$\psi = -\frac{\Gamma}{2\pi} \ln \left(x^2 + y^2 \right)^{\frac{1}{2}} = -\frac{\Gamma}{2\pi} \ln r, \tag{9.24}$$

and

$$\phi = \frac{\Gamma}{2\pi} \tan^{-1} \frac{y}{x} = \frac{\Gamma}{2\pi} \theta. \tag{9.25}$$

In vortex flow, fluid particles rotate around the origin with total circulation Γ. This may at first appear as a contradiction, because the flow is assumed to be irrotational. A calculation of $\boldsymbol{\omega} = \nabla \times \mathbf{V}$ shows that $\boldsymbol{\omega} = 0$ everywhere in the flow, except at the origin. The circulation Γ is finite only when the integration path includes the origin. Any integration path that does not include the origin yields $\Gamma = 0$.

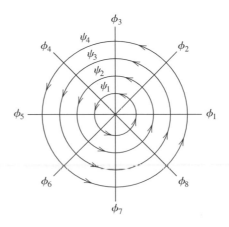

FIGURE 9.14   Ideal line vortex.

EXAMPLE 9.7

Problem Statement   Find the pressure distribution as a function of r for a line vortex with circulation $\Gamma = 20 \ m^2/s$. Consider the fluid to be water at $20°C$. How much time does it take for a fluid particle to travel one complete revolution at a distance $0.8 \ m$ from the origin?

Governing Equations   In vortex flow, streamlines are concentric circles with constant velocity. Therefore, according to the Bernoulli equation, the pressure along streamlines is constant. Consequently, the pressure varies only in the direction normal to the local streamlines. According to Equation (9.7), for circular streamlines

$$\frac{V^2}{r} = \frac{1}{\rho}\frac{\partial P}{\partial r} + g\frac{\partial z}{\partial r}.$$

The time a fluid particle takes to complete an entire revolution along a streamline with radius r is evaluated as

$$\int_0^{t_s} dt = t_s = \int_{s=0}^{s=2\pi r} \frac{1}{v_\theta} ds = \int_{\theta=0}^{\theta=2\pi} \frac{1}{v_\theta} r \, d\theta.$$

Assumptions   The flow is assumed to be steady, inviscid, and two-dimensional.

SOLUTION   For two-dimensional flow on a horizontal plane, $\partial z/\partial r = 0$. Therefore, Equation (9.7) simplifies to

$$\frac{1}{\rho}\frac{dP}{dr} = \frac{V^2}{r}.$$

For a vortex flow, $V = \dfrac{\Gamma}{2\pi r}$. Therefore,

$$\frac{1}{\rho}\frac{dP}{dr} = \left(\frac{\Gamma}{2\pi r}\right)^2 \frac{1}{r}.$$

Since the velocity is singular at the origin, the pressure distribution is evaluated starting from a point ϵ, at a distance $\epsilon = 1.0 \ m$ away from the origin. Hence,

$$\int_{P_\epsilon}^{P} dP = \rho\left(\frac{\Gamma}{2\pi}\right)^2 \int_\epsilon^r \frac{1}{r^3} dr = \rho\left(\frac{\Gamma}{2\pi}\right)^2 \left(-\frac{1}{2}r^{-2}\right)\Big|_\epsilon^r$$

$$= \frac{\rho}{2}\left(\frac{\Gamma}{2\pi}\right)^2 \left(\frac{1}{\epsilon^2} - \frac{1}{r^2}\right).$$

In terms of numerical values,

$$\Delta P = \frac{998.2 \ kg/m^3}{2 \times 1000 \ Pa/kPa}\left(\frac{20 \ m^2/s}{2\pi}\right)^2 \left(\frac{1}{1.2} - \frac{1}{r^2}\right) = 5.06\left(\frac{1}{1.2} - \frac{1}{r^2}\right) \ kPa.$$

Figure 9.15 shows the pressure variation as a function of the distance from the origin. The time a particles takes for a complete revolution is

$$\int_0^{t_s} dt = \int_0^{2\pi r} \frac{1}{v_\theta} ds = \int_0^{2\pi} \frac{1}{v_\theta} r \, d\theta = \frac{2\pi r^2}{\Gamma} \int_0^{2\pi} d\theta.$$

Finally,

$$t_s = \frac{(2\pi r)^2}{\Gamma} = \frac{(2\pi \times 0.8\ m)^2}{20\ m^2/s} = 1.26\ s.$$

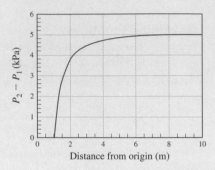

FIGURE 9.15　Pressure variation as a function of the distance from the origin for a line vortex.

EXAMPLE 9.8

Problem Statement　*Irrotationality of line vortex flow*. Calculate the circulation around a closed loop that does not include the origin, and show that the flow in an ideal line vortex is irrotational.

Governing Equations　The circulation along a closed loop within the flow is evaluated as

$$\Gamma = \oint \mathbf{V} \cdot \mathbf{t}\, ds,$$

where $\mathbf{V}$ is the velocity vector and $\mathbf{t}$ is the unit tangent vector along the arclength s.

Assumptions　The flow is assumed to be ideal and two-dimensional.

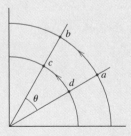

FIGURE 9.16　Schematic of Example 9.8.

SOLUTION　Consider the closed arbitrary loop shown in Figure 9.16. The term inside the integral is equal to the azimuthal velocity $\mathbf{V} \cdot \mathbf{t} = v_\theta$, and $ds = r\,d\theta$. The circulation, then, is

$$\Gamma = \oint v_\theta r\, d\theta = \int_a^b v_\theta r\, d\theta + \int_b^c v_\theta r\, d\theta + \int_c^d v_\theta r\, d\theta + \int_d^a v_\theta r\, d\theta$$

$$= \Gamma = \int_a^b m\, d\theta + 0 - \int_d^c m\, d\theta + 0.$$

Since the angle between *ab* and *cd* is the same, the two non-zero terms cancel out, giving $\Gamma = 0$.

9.2.2   Superposition of Elementary Flows

The Laplace equation is a linear homogeneous elliptic equation. Therefore, for its solution we can use the *superposition principle*: if ψ_1 and ψ_2 are both solutions to the Laplace equation, then $\psi = \psi_1 + \psi_2$ is also a solution to $\nabla^2 \psi = 0$. As shown subsequently, this property allows the use of elementary flow solutions as building blocks to construct more complicated inviscid flows.

Source and a Sink

A source and a sink, at a distance $2a$ apart, with equal strength m as shown in Figure 9.17, can be combined to yield

$$\psi = \underbrace{m \tan^{-1} \frac{y}{x + a}}_{Source} - \underbrace{m \tan^{-1} \frac{y}{x - a}}_{Sink},$$

and

$$\phi = \underbrace{m \ln \left[(x + a)^2 + y^2 \right]^{\frac{1}{2}}}_{Source} - \underbrace{m \ln \left((x - a)^2 + y^2 \right)^{\frac{1}{2}}}_{Sink}.$$

According to the superposition principle, the resulting flow is also inviscid and irrotational. After considerable manipulation, the above equations simplify to

$$x^2 + \left(y + a \cot \frac{\psi}{m} \right)^2 = a^2 \csc^2 \frac{\psi}{m},$$

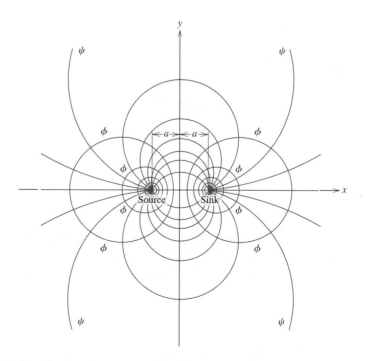

FIGURE 9.17   Combination of a source and a sink.

and

$$y^2 + \left(x - a \coth \frac{\phi}{m} \right)^2 = a^2 \operatorname{csch}^2 \frac{\phi}{m}.$$

The streamline pattern is a set of circles with the origin along the y-axis at $y_0 = \left(-a \cot \frac{\psi}{m} \right)$, and radius $r = \left(a \csc \frac{\psi}{m} \right)$. The velocity potential lines are also circles, but with their origin along the x-axis at $x_0 = \left(a \coth \frac{\phi}{m} \right)$, and with radius $r = \left(a \operatorname{csch} \frac{\phi}{m} \right)$ (Figure 9.17).

The velocity components are

$$u = \frac{m(x+a)}{(x+a)^2 + y^2} - \frac{m(x-a)}{(x-a)^2 + y^2},$$

and

$$v = \frac{my}{(x+a)^2 + y^2} - \frac{my}{(x-a)^2 + y^2}.$$

Doublet

The doublet is obtained at the limit when the distance a between the source and the sink is zero (Figure 9.18). Normally, at this limit the two flows offset the effect of each other. Therefore, the streamfunction and velocity potential functions here are obtained by assuming that the product $2am = \lambda$, remains constant and finite. Under these conditions,

$$\psi = -\frac{\lambda y}{x^2 + y^2} = -\frac{\lambda \sin \theta}{r}, \quad \text{and} \quad \phi = \frac{\lambda x}{x^2 + y^2} = \frac{\lambda \cos \theta}{r}. \tag{9.26}$$

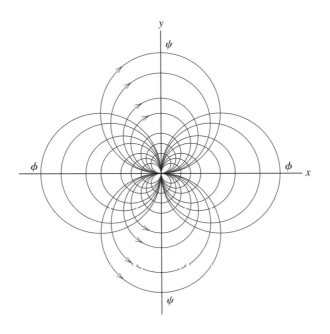

FIGURE 9.18   Ideal doublet.

After rearranging, ψ and ϕ are given as

$$x^2 + \left(y + \frac{\lambda}{2\psi}\right)^2 = \left(\frac{\lambda}{2\psi}\right)^2, \quad \text{and} \quad y^2 + \left(x - \frac{\lambda}{2\psi}\right)^2 = \left(\frac{\lambda}{2\phi}\right)^2.$$

The streamfunction represents a set of circles with their origin along the y-axis at $y_0 = \left(-\frac{\lambda}{2\psi}\right)$, and radius $r = \frac{\lambda}{2\psi}$. Obviously, the circles are tangent at the origin. The iso-velocity potential lines are circles along the x-axis. The velocity components in Cartesian coordinates are

$$u = \frac{\lambda\left(y^2 - x^2\right)}{\left(x^2 + y^2\right)^2}, \quad \text{and} \quad v = -\frac{2\lambda xy}{\left(x^2 + y^2\right)^2}, \tag{9.27}$$

and in polar coordinates as

$$v_r = -\frac{\lambda\cos\theta}{r^2}, \quad \text{and} \quad v_\theta = -\frac{\lambda\sin\theta}{r^2}. \tag{9.28}$$

Uniform Flow and a Source

The combination of a uniform flow and a source is given by

$$\psi = V_o r \sin\theta + m\theta, \tag{9.29}$$

and

$$\phi = V_o r \cos\theta + m\ln r. \tag{9.30}$$

The graphical representation of this combination produces the flow pattern shown in Figure 9.19. Because of the competition of the incoming uniform flow and the flow released by the source, a *stagnation point* where the velocity is zero develops along the negative x-axis. As shown in the figure, at the stagnation point the incoming streamline splits into two branches. Away from the stagnation point, the streamlines progressively bend. Eventually, at the back of the body, they align with the uniform flow. Since the location of the stagnation point is along $\theta = \pi$, the value of this streamline is $\psi_s = m\pi$.

By sketching the streamline pattern, as shown in Figure 9.19, an elliptical shape defined by the streamline emanating from the stagnation point develops. As noted earlier,

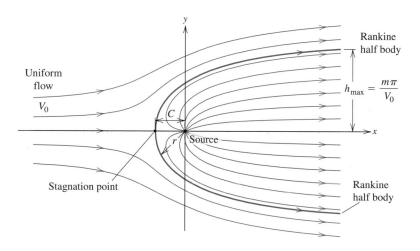

FIGURE 9.19 Half Rankine body: Uniform flow past a source.

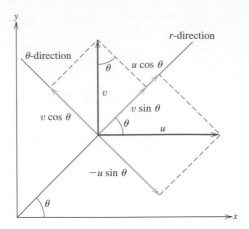

FIGURE 9.20   Velocity vector relationships between polar and Cartesian coordinates.

in inviscid flow any streamline can be replaced by a solid boundary without affecting the flow pattern. Therefore, the flow is identical to that past an elliptic shape with geometry given by the condition $\psi_s = m\pi$ as

$$\psi_s = m\pi = V_o r \sin\theta + m\theta \Rightarrow r = \frac{m(\pi - \theta)}{V_o \sin\theta}. \qquad (9.31)$$

This elliptic body is known as a *half Rankine body*. Characteristic geometric features of this shape are the location of the stagnation point and its maximum height.

The velocity distribution around the body is evaluated by differentiation

$$v_r = V_o \cos\theta + \frac{m}{r}, \qquad (9.32)$$

and

$$v_\theta = -V_o \sin\theta. \qquad (9.33)$$

The Cartesian velocity components u, v, can be determined from the components in polar coordinates using the geometry in Figure 9.20

$$v_r = u\cos\theta + v\sin\theta, \text{ and } v_\theta = -u\sin\theta + v\cos\theta,$$

which gives

$$u = v_r \cos\theta + v_\theta \sin\theta, \text{ and } v = v_\theta \cos\theta + v_r \sin\theta.$$

Therefore,

$$u = V_o + \frac{m}{r}\cos\theta = V_o + \frac{mx}{x^2 + y^2}, \qquad (9.34)$$

and

$$v = \frac{m}{r}\sin\theta = \frac{my}{x^2 + y^2}. \qquad (9.35)$$

As expected, the vertical component of the velocity does not depend on the magnitude of the uniform flow; because of superposition, the uniform flow, being horizontal, contributes only to the u component of the velocity.

The x-coordinate of the stagnation point c, along the x-axis is obtained by requiring that $u = 0$. Hence,

$$x_0 = -\frac{m}{V_o} = -c, \qquad (9.36)$$

where $c = \dfrac{m}{V_o}$. The maximum thickness of the body h_{max} is obtained at the limit $\theta \to 0$,

$$r \sin \theta = h_{max} = \frac{m\pi}{V_o}. \tag{9.37}$$

This result can be confirmed also by considering conservation of the volumetric flow rate Q: the total volumetric flow rate released by the source $Q = 2m\pi$ changes into uniform flow with velocity V_o with total height

$$h_{total} = 2h_{max} = \frac{Q}{V_o} = \frac{2m\pi}{V_o} \Rightarrow h_{max} = \frac{m\pi}{V_o}.$$

The magnitude of the velocity at any point in the flow field is

$$V^2 = V_o^2 \left(1 + \frac{c^2}{r^2} + 2\frac{c}{r} \cos \theta \right). \tag{9.38}$$

Using the Bernoulli equation, between a point in the uniform flow far away from the body (subscript 0) and any point along the same streamline

$$\frac{P_o}{\rho} + \frac{V_o^2}{2} = \frac{P}{\rho} + \frac{V^2}{2},$$

we find the variation of the pressure as

$$P = P_o - \frac{\rho V_o^2}{2} \left(\frac{c^2}{r^2} + 2\frac{c}{r} \cos \theta \right). \tag{9.39}$$

The pressure coefficient C_p is expressed as

$$C_p = \frac{(P - P_o)}{\frac{1}{2}\rho V_o^2} = -\left(\frac{c^2}{r^2} + 2\frac{c}{r} \cos \theta \right). \tag{9.40}$$

Figure 9.21 shows the variation of the velocity ratio $\dfrac{V}{V_o}$ and pressure coefficient along the surface of the body — that is, for $\dfrac{c}{r} = \dfrac{\sin \theta}{(\pi - \theta)}$.

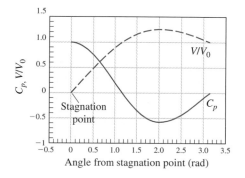

FIGURE 9.21   Speed and pressure coefficient variation along a half Rankine body.

EXAMPLE 9.9

Problem Statement   Consider an oil spill in which a ship capsized in a lake releases oil at the rate of $150\ m^3/s$. If the speed of the current at the location of the spill is $0.5\ m/s$ and the average depth is $20\ m$, determine the area that is affected by the oil.

Assumptions   This problem is modeled by assuming that the oil release can be represented by a point source and that the water and oil behave as ideal fluids. It is also assumed that the flow can be approximated as a two-dimensional flow.

Governing Equations   Using the basic assumptions, the flow is represented as a combination of uniform flow (effect of the current) and a two-dimensional point source for which the streamfunction is

$$\psi = V_o r \sin\theta + m\theta.$$

SOLUTION   The combination of a uniform flow, and a source yield a characteristic half Rankine body described by

$$r = \frac{m\,(\pi - \theta)}{V_o \sin\theta},$$

which qualitatively is the same as that described in Figure 9.19. In this problem, the oil is confined within this Rankine body. The maximum thickness of the body is a

$$h_{total} = \frac{Q}{V_o}.$$

Here, Q is the rate of released per unit depth (i.e., $Q = \dfrac{150\ m^3/s}{20\ m} = 7.5\ m^2/s$). Then,

$$h_{total} = \frac{Q}{V_o} = \frac{7.5\ m^2/s}{0.5\ m/s} = 15.0\ m.$$

By definition, the volumetric flow rate per unit depth is $Q = 2m\pi$. Hence,

$$m = \frac{Q}{2\pi} = \frac{7.5\ m^2/s}{2\pi} = 1.194\ m^2/s.$$

The oil penetrates a distance x_s ahead of the point of release given by

$$x_s = -\frac{m}{V_o} = -\frac{1.194\ m^2/s}{0.5\ m/s} = -2.387\ m.$$

Full Rankine Body: Uniform Flow Combined with a Source and a Sink

The natural extension of the flow past a half Rankine body is to include a sink with strength m equal to that of the source. If the source and sink are placed at distance a from the origin, as shown in Figure 9.22, for $\psi(x, y)$ and $\phi(x, y)$ we have

$$\psi = V_o y + m \tan^{-1} \frac{y}{x + a} - m \tan^{-1} \frac{y}{x - a}, \tag{9.41}$$

and

$$\phi = V_o x + m \ln\left((x + a)^2 + y^2\right)^{\frac{1}{2}} - m \ln\left((x - a)^2 + y^2\right)^{\frac{1}{2}}. \tag{9.42}$$

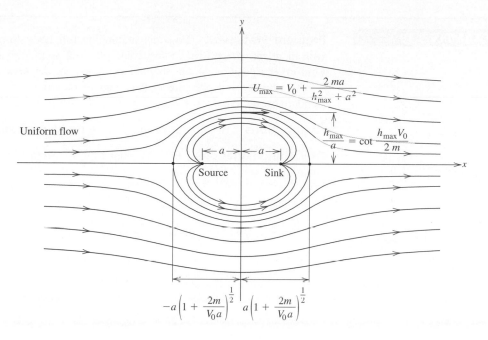

FIGURE 9.22 Full Rankine body: uniform flow past a source and a sink.

As expected, since flow released by the source is absorbed by the sink, we now have flow past a closed oval-shaped body, known as a *Rankine body* (Figure 9.22).

The velocity field in Cartesian coordinates is given as

$$u = V_o + \frac{m(x+a)}{y^2 + (x+a)^2} - \frac{m(x-a)}{y^2 + (x-a)^2}, \tag{9.43}$$

$$v = \frac{my}{y^2 + (x+a)^2} - \frac{my}{y^2 + (x-a)^2}. \tag{9.44}$$

The streamline coinciding with the body $\psi = 0$ defines the actual shape of the body. The coordinates of the two stagnation points, P and S, are obtained by requiring that $u = 0$ along $y = 0$—that is,

$$u = V_o + \frac{m}{x+a} - \frac{m}{x-a} = 0 \Rightarrow x_s = \pm a \left(1 + \frac{2m}{V_o a}\right)^{\frac{1}{2}}. \tag{9.45}$$

The maximum height of the body h_{max} is obtained by using the expression for $\psi = 0$, and $x = 0$—that is,

$$0 = V_o y + m \tan^{-1} \frac{y}{a} - m \tan^{-1} -\frac{y}{a},$$

or in simpler form,

$$\frac{h_{max}}{a} = \cot \frac{h_{max} V_o}{2m}. \tag{9.46}$$

Unfortunately, this equation must be solved numerically for h_{max}. The maximum velocity u_{max}, occurs at h_{max},

$$u_{max} = V_o + \frac{2ma}{h_{max}^2 + a^2}. \tag{9.47}$$

EXAMPLE 9.10

Problem Statement   Consider a Rankine full body formed by a source and a sink with identical strengths of 25 m^2/s and placed 2 m apart, and a uniform flow of 15 m/s. Find (a) the position of the stagnation points; (b) the maximum thickness of the body; (c) the velocity and pressure at the point of maximum thickness. Consider air flow at standard conditions ($P_1 = P_{atm} = 101.3 kPa$ and $\rho = 1.225\ kg/m^3$).

Assumptions   The flow is assumed to be two-dimensional and inviscid.

Governing Equations   For a combination of a source, a sink, and a uniform flow, the streamlines are

$$\psi = V_o y + m \tan^{-1} \frac{y}{x+a} - m \tan^{-1} \frac{y}{x-a}.$$

The velocity field in Cartesian coordinates is given as

$$u = V_o + \frac{m(x+a)}{y^2 + (x+a)^2} - \frac{m(x-a)}{y^2 + (x-a)^2},$$

$$v = \frac{my}{y^2 + (x+a)^2} - \frac{my}{y^2 + (x-a)^2}.$$

The pressure is obtained using the Bernoulli equation applied along the streamline that emanates from a point away from the body and ends at the point of maximum thickness,

$$\int_{s=0}^{s=L} \frac{\partial V}{\partial t} ds + \frac{V_2^2 - V_1^2}{2} + \frac{P_2 - P_1}{\rho} + g(z_2 - z_1) = 0.$$

SOLUTION   The stagnation points are defined as the points where $u = v = 0$. Because of symmetry, these points lie on the x-axis. For this problem $a = 1\ m$, $V_o = 15\ m/s$, and $m = 25\ m^2/s$. According to Equation (9.44)

$$x_s = \pm a \left(1 + \frac{2m}{V_o a} \right)^{\frac{1}{2}} = \pm 1\ m \times \left(1 + \frac{2 \times 25\ m^2/s}{15\ m/s \times 1\ m} \right)^{\frac{1}{2}} = \pm 2.082\ m.$$

The maximum height of the body h_{max} is obtained by following the streamline $\psi = 0$ and by setting $x = 0$,

$$\frac{h_{max}}{a} = \cot \frac{h_{max} V_o}{2m} = \cot \frac{h_{max} \times 15\ m/s}{2 \times 25\ m^2/s} = \cot(0.3 \times h_{max}).$$

By trial and error, we get $h_{max} = 1.74\ m$. The velocity at $x = 0$ and $y = h_{max} = 1.74\ m$, is $v = 0$, and

$$u = V_o + \frac{m(x+a)}{y^2 + (x+a)^2} - \frac{m(x-a)}{y^2 + (x-a)^2} = V_o + \frac{2ma}{y^2 + a^2},$$

$$u_{max} = 15\ m/s + \frac{2 \times 25\ m^2/s \times 1\ m}{1.74^2\ m^2 + 1^2\ m^2} = 27.41\ m/s.$$

The Bernoulli equation for steady flow $\dfrac{\partial V}{\partial t} = 0$; with no gravity effects (planar flow), it reduces to

$$\frac{V_2^2 - V_1^2}{2} + \frac{P_2 - P_1}{\rho} = 0,$$

or

$$P_2 = P_1 + \frac{\rho}{2}\left(V_1^2 - V_2^2\right) = 101.3 \; kPa + \frac{1.225}{2000}\left(15^2 - 27.41^2\right) = 100.98 \; kPa.$$

Ideal Flow Past a Cylinder

A more practical problem than the flow past a Rankine body is the flow past a circular cylinder. This can be obtained by combining a uniform flow and a doublet with the appropriate strength (Figure 9.23). This is not surprising, because, as the source and sink are brought closer together, the oval-shaped Rankine body becomes progresively more circular; in the limit when the two overlap, of course, we get a doublet.

$$\psi = V_o r \sin\theta - \frac{\lambda \sin\theta}{r}. \tag{9.48}$$

By selecting the strength of the doublet as $\lambda = a^2 V_o$, we can write

$$\psi = V_o \sin\theta \left(r - \frac{a^2}{r}\right). \tag{9.49}$$

In this form, we observe that the circle $a = r$ defines the streamline $\psi = 0$. Therefore, the preceding streamfunction describes flow past a cylinder (Figure 9.24(a)). The velocity field in polar coordinates is given as

$$v_r = \frac{1}{r}\frac{\partial \psi}{\partial \theta} = V_o\left(1 - \frac{a^2}{r^2}\right)\cos\theta, \tag{9.50}$$

and

$$v_\theta = -\frac{\partial \psi}{\partial r} = -V_o\left(1 + \frac{a^2}{r^2}\right)\sin\theta. \tag{9.51}$$

Along the cylinder, $r = a$,

$$v_\theta = -2V_o \sin\theta \quad \text{and} \quad v_r = 0, \tag{9.52}$$

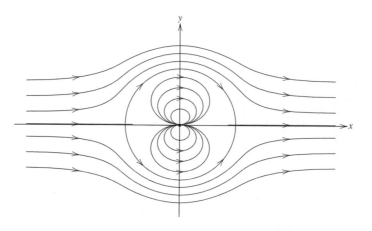

FIGURE 9.23   Flow past a cylinder: Uniform flow past a doublet.

which shows that the flow is tangent to the cylinder. The maximum velocity occurs at the top of the cylinder ($\theta = 90°$), $v_{max} = 2V_o$. Since the flow field is completely symmetric with respect to both the x and y axes, according to the Bernoulli equation, the pressure distribution is also symmetric. Therefore, the net force experienced by the cylinder is zero! Of course, given the assumption of inviscid flow, this is not surprising.

The flow symmetry can be changed by superimposing a vortex with strength $\left(\dfrac{\Gamma}{2\pi}\right)$ as

$$\psi = V_o \sin\theta \left(r - \frac{a^2}{r}\right) - \frac{\Gamma}{2\pi}\ln\frac{r}{a}. \tag{9.53}$$

The velocity field is now

$$v_r = V_o \left(1 - \frac{a^2}{r^2}\right)\cos\theta, \tag{9.54}$$

and

$$v_\theta = -V_o\left(1 + \frac{a^2}{r^2}\right)\sin\theta + \frac{\Gamma}{2\pi r}. \tag{9.55}$$

Because of superposition, the added circulation contributes only to the azimuthal velocity. Again, the radial velocity vanishes along the cylinder, and the azimuthal velocity is

$$v_\theta = -2V_o\sin\theta + \frac{\Gamma}{2\pi a}. \tag{9.56}$$

The stagnation points for this flow are defined by

$$\sin\theta_s = \frac{\Gamma}{4\pi a V_o}. \tag{9.57}$$

Because of the circulation, the stagnation points are not along the $x\text{-}axis$, but as shown in Figure 9.23. In the limit of no circulation $\Gamma = 0$, the stagnation points at $\theta_s = 0$ and $\theta_s = 180$ of the earlier flow are recovered. When $\dfrac{\Gamma}{4\pi a V_o} = 1$, there is only a single stagnation point at the top of the cylinder (Figure 9.24(c)). For values of $\dfrac{\Gamma}{4\pi a V_o} > 1$, a single stagnation point flow develops away from the cylinder (Figure 9.24(d)).

Because of symmetry with respect to the y-axis, the drag force is zero. For real flows this statement of course is not true; it is well known that in viscous flows, the drag force is finite. This point demonstrates the shortcomings of potential theory to account for viscous-related phenomena.

Since the addition of a vortex makes the flow asymmetric with respect to the $x\text{-}axis$, we expect a finite lift force per unit depth L. By definition,

$$L = \oint (-P\mathbf{n})\cdot\mathbf{j}\,ds.$$

Substituting the outward pointing unit normal vector $\mathbf{n} = \dfrac{dy}{ds}\mathbf{i} - \dfrac{dx}{ds}\mathbf{j}$ (s is measured in the counterclockwise direction) and by performing the scalar product, the lift force is calculated using

$$L = \oint P\frac{dx}{ds}ds = \oint P\,dx = \int_0^{2\pi} P\frac{dx}{d\theta}d\theta = -\int_0^{2\pi} Pa\sin\theta\,d\theta, \tag{9.58}$$

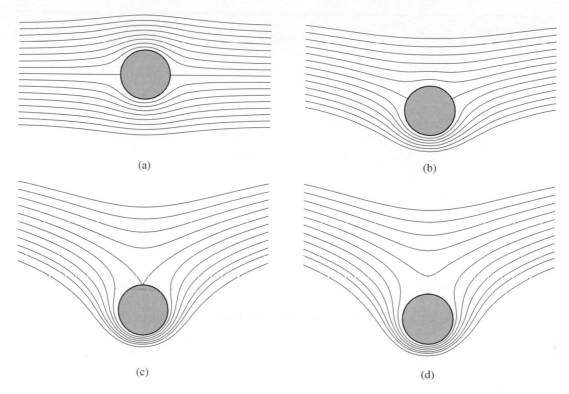

(a) (b)

(c) (d)

FIGURE 9.24 Flow past a cylinder: Uniform flow past a doublet: (a) $\Gamma = 0$; (b) $\dfrac{\Gamma}{4\pi a V_o} < 1$; (c) $\dfrac{\Gamma}{4\pi a V_o} = 1$; (d) $\dfrac{\Gamma}{4\pi a V_o} > 1$.

where we used the fact that $x = a\cos\theta$. According to the Bernoulli equation, the pressure is

$$P = P_o + \frac{\rho V_o^2}{2} - \frac{\rho}{2}\left(-2V_o\sin\theta + \frac{\Gamma}{2\pi a}\right)^2.\qquad (9.59)$$

Substituting Equation (9.58) into Equation (9.57), by integration, we get the lift force as

$$L = -\rho V_o \Gamma !\qquad (9.60)$$

This is a remarkable result and points to the fact that the net lift force is proportional to the circulation Γ and V_o, and is independent of the geometry of the cylinder. This phenomenon is known as the *Magnus Effect*. This result has been generalized by W.M. Kutta (1902) and N. Joukowski (1906) as follows:

> *The lift force experienced by a body in inviscid flow is proportional to the total circulation around the body. The direction of the lift is 90° from the direction of the flow, rotated in the opposite direction as the circulation.*

As expected, this result is quite useful in theoretical aerodynamics, where the calculation of the lift force is important. In baseball, pitchers use the same phenomenon in the *screwball*, where the ball, under the influence of an initial rotation (spin), curves away from the batter.

EXAMPLE 9.11

Problem Statement: Consider a semicircular structure with radius R and length L, subjected to a cross-flow as shown in Figure 9.25. Find the optimum location for a window so that the force on the structure induced by the flow is minimum.

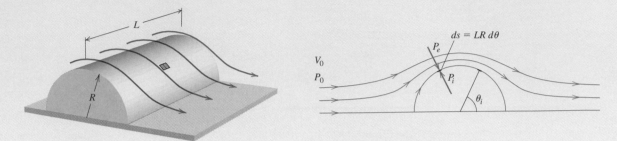

FIGURE 9.25   Proper placing of a window to eliminate flow-induced forces.

Governing Equations   The net force on the structure is obtained by

$$\mathbf{F} = \int_S d\mathbf{F},$$

where $d\mathbf{F}$ is the force on an infinitesimal segment of the boundary ds due to the difference between the internal pressure P_i and the external pressure P_e,

$$d\mathbf{F} = (P_i - P_e)\, \mathbf{n}\, ds.$$

The outward unit normal vector $\mathbf{n}$ is $\mathbf{n} = \cos\theta\, \mathbf{i} + \sin\theta\, \mathbf{j}$.

By selecting a streamline that follows the surface of the structure, according to the Bernoulli equation the pressure at any point along the surface is given as

$$P = P_o + \rho\frac{V_o^2}{2} - \rho\frac{V^2}{2},$$

where P_o, and V_o, are the pressure and velocity far upstream from the cylinder.

Basic Assumptions   The flow is assumed to be steady and ideal. It is also assumed that the finite length of the structure does not alter the given velocity distribution.

SOLUTION   The external pressure along the boundary of the structure P_e is obtained by substituting the external velocity $v_\theta = -2V_o \sin\theta$ into the Bernoulli equation,

$$P_e = P_o + \rho\frac{V_o^2}{2}\left(1 - 4\sin^2\theta\right).$$

The internal pressure P_i is determined by the local pressure at the location of the window θ_i

$$P_i = P_o + \rho\frac{V_o^2}{2}\left(1 - 4\sin^2\theta_i\right).$$

The total force is

$$\mathbf{F} = \int_0^\pi (P_i - P_e)(\cos\theta\mathbf{i} + \sin\theta\mathbf{j})\, LR\, d\theta.$$

Substituting the expressions for the pressure and by simplifying, we get

$$\mathbf{F} = 2\rho V_o^2 LR \int_0^\pi \left(\sin^2\theta - \sin^2\theta_i \right) (\cos\theta \mathbf{i} + \sin\theta \mathbf{j}) \, d\theta.$$

The force F_x in the x direction is

$$F_x = 2\rho V_o^2 LR \int_0^\pi \left(\sin^2\theta - \sin^2\theta_i \right) \cos\theta \, d\theta,$$

which, upon integration, gives

$$F_x = 2\rho V_o^2 LR \left[\frac{\sin^3\theta}{3} - \sin^2\theta_i \sin\theta \right]_0^\pi = 0.$$

So, independent of the location of the window, there is no force acting in this direction. This is not surprising, because both the external pressure distribution and the pressure inside the structure, are symmetric with respect to the y-axis; hence, they cannot induce a net force in the x-direction.

The force F_y in the y-direction is

$$F_y = 2\rho V_o^2 LR \int_0^\pi \left(\sin^2\theta - \sin^2\theta_i \right) \sin\theta \, d\theta,$$

which, upon integration gives

$$F_y = 2\rho V_o^2 LR \left[\frac{1}{3}\cos^3\theta - \cos\theta + \sin^2\theta_i \cos\theta \right]_0^\pi = 2\rho V_o^2 LR \left(-2\sin^2\theta_i + \frac{4}{3} \right).$$

The vertical force vanishes $F_y = 0$, when the window is placed at an angle θ_i with respect to the horizontal direction so that

$$-2\sin^2\theta_i + \frac{4}{3} = 0,$$

or

$$\theta_i = \sin^{-1}\sqrt{\frac{2}{3}} = 54.74°.$$

Therefore, when the window is placed at $\theta_i = 54.74°$, the structure will not experience a vertical force!

EXAMPLE 9.12

Problem Statement   *Magnus effect: Flettner boat.* In the 1920s Flettner, a German engineer, applied the same idea to the design of a boat powered by the lift force produced by two rotating cylinders.

Consider a ship operating on the Magnus principle, with two rotors each $8\ m$ high, with diameter of $1.5\ m$, and rotating at $800\ rev/min$ (Figure 9.26). Find the net propulsion force the ship experiences when the ship travels at $8\ km/hr$, with a $25\ km/hr$ wind blowing in a direction normal to the direction of the boat.

Assumptions   The flow is assumed to be steady and inviscid.

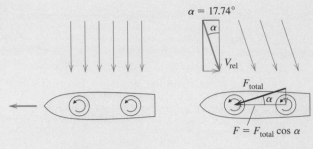

(a) Absolute velocity diagram   (b) Relative velocity diagram

FIGURE 9.26   Magnus effect used for propulsion.

Governing Equations   The force on a rotating cylinder per unit height of the cylinder is given by

$$\frac{F}{H} = \rho V_{rel} \Gamma.$$

By accounting for the velocity of the boat, V_{rel} is the relative velocity of the air past the rotor. The circulation of the rotor is evaluated using the definition,

$$\Gamma = \oint \mathbf{V} \cdot \mathbf{n}\, ds.$$

SOLUTION   The force on each rotating cylinder is given by

$$F = H\rho V_{rel} \Gamma.$$

H is the height of the rotor. V_{rel} is the relative velocity of the air past the rotor. According to Figure 9.26, the relative velocity vector is

$$\mathbf{V}_{rel} = -V_{air}\mathbf{j} + V_{boat}\mathbf{i}.$$

In this problem, $V_{air} = 25000\ m/3600\ s = 6.94\ m/s$, and $V_{boat} = 8000\ m/3600\ s = 2.22\ m/s$. Therefore,

$$\mathbf{V}_{rel} = -6.94\ \mathbf{j} + 2.22\ \mathbf{i},$$

and

$$V_{rel} = |\mathbf{V}_{rel}| = \sqrt{6.94^2 + 2.22^2}\ m/s = 7.29\ m/s.$$

The direction of the force is 90° from the direction of the flow, rotated in the opposite direction as the circulation as shown in Figure 9.26.

The circulation is evaluated as

$$\Gamma = \oint \mathbf{V} \cdot \mathbf{n}\, ds = \int_0^{2\pi} \omega r\, r\, d\theta = \omega r\, 2\pi r$$

$$\Gamma = 800\ rev/min \times \frac{2\pi}{60\ s/min} \times \frac{1.5\ m}{2} \times 2\pi \times \frac{1.5\ m}{2} = 296.09\ m^2/s.$$

The net force on each rotor is

$$F = H\rho V_{rel}\Gamma = 8 \times 1.225\ kg/m^3 \times 7.29\ m/s \times 296.09\ m^2/s = 21.153\ kN.$$

The total force, then, on both rotors is $F_{total} = 2 \times 21.153 \, kN = 42.306 \, kN$. However, this force is acting in the direction shown in the figure. The force responsible for propulsion is in the direction of the boat's motion — that is,

$$F_x = F_{total} \cos\left(\tan^{-1}\frac{8}{25}\right) = 42.306 \, kN \cos 17.74° = 40.29 \, kN.$$

Using the same principle, the environmentalist group of Jean Cousteau operates an experimental ship named *Alcyon*.

REFERENCES

L.M. MILNE-THOMSON, *Theoretical Aerodynamics*, New York, Dover, 1966.

W. LI and S. LAM, *Principles of Fluid Mechanics*, Reading, MA, Addison-Wesley, 1976.

R. VON MISES, *Theory of Flight*, New York, Dover, 1959.

P.G. SAFFMAN, *Vortex Dynamics*, Cambridge, England, Cambridge University Press, 1993.

S. GOLDSTEIN, (ed.), *Modern Developments in Fluid Mechanics*, New York, Dover, 1965.

B. THWAITES (ed.), *Incompressible Aerodynamics*, New York, Dover, 1960.

A.M. KUETHE and C. CHOW, *Foundations of Aerodynamics, Bases of Aerodynamic Design*, 4th ed., New York, Wiley, 1986.

R.L. PANTON, *Incompressible Flow*, New York, Wiley, 1984.

I.G. CURRIE, *Fundamental Mechanics of Fluids*, 2nd ed., New York, McGraw-Hill, 1993.

H. LAMB, *Hydrodynamics*, 6th ed., New York, Dover, 1945.

PROBLEMS

1. Determine the velocity potential and streamfunction for a combination of a source and uniform flow located at (a) (1,2) and, (b) (−1,2) of a Cartesian coordinate system.

2. Find the range L each jet coming out of the reservoir shown in Figure 9.27 will reach.

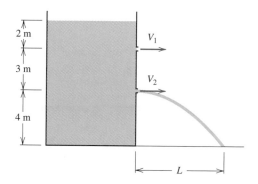

FIGURE 9.27 Schematic for Problem 2.

3. The velocity potential for unsteady flow $U(t)$ past a circular cylinder with radius R is given by

$$\phi = U(t)\left[r + \frac{R^2}{r}\right]\cos\theta.$$

Using the unsteady Bernoulli equation, find the pressure distribution along the surface of the cylinder.

4. A 12 *cm* fire hose with a 5 *cm* diameter nozzle discharges water at a rate of 1.4 $m^3/minute$. By assuming that the nozzle discharges to atmospheric conditions, and by neglecting fluid friction, find the force on the hose.

5. The velocity potential for unsteady flow $U(t)$ past a circular cylinder with radius R is given by

$$\phi = U(t)\left[r + \frac{R^2}{r}\right]\cos\theta.$$

Using the unsteady Bernoulli equation, find the total force that the cylinder experiences. How do you interpret the results.

6. Consider a pipe with variable cross-sectional area connected to a pump, similar to the one shown in Figure 9.1.1. Find the time it takes for the water inside the pipe to accelerate from rest to its final steady state velocity. Assume that the cross-sectional area varies linearly along the length of the pipe ($L = 68 \, m$), from an initial area of 0.4 m^2 to a final area of 0.25 m^2. Consider that the pump pressurizes the water to 310 kPa.

7. Show that the definition of the streamfunction and velocity potential in polar coordinates, when applied to uniform flow at 45 degrees with respect to the horizontal, leads to the result obtained using Cartesian coordinates.

8. The streamfunction $\psi = Ar^4\sin(4\theta)$, represents ideal flow past a wedge. Find the u_r and u_θ components of the velocity. What is the velocity potential function for this flow.

9. The streamfunction $\psi = Ar^n \sin(n\theta)$, represents ideal flow past a wedge. Find the u_r and u_θ components of the velocity. What is the velocity potential function for this flow.

10. Water assumed to be an ideal, irrotational flows with velocity 4 m/s in a channel with constant width of 1.5 m. Following a straight section, the channel has a $90°$ degree circular bend with inner radius of 2 m and outer radius of 3.5 m. If the velocity V in this part varies according to $Vr = constant$, where r is the radial distance, find the velocity along the inside and outside of the channel. What is the pressure difference in the direction normal to the flow direction?

11. Show that the functions for the velocity potential and streamfunction for a source (or sink) satisfy the Laplace equations $\nabla^2\phi = 0$, $\nabla^2\psi = 0$.

12. For the combination of a source and a sink with strength of 6 m^2/s, respectively located at $(-1.2,0)$ and $(2,0)$, find the velocity vector at $(1,1)$, $(3,1)$ and $(0,4)$.

13. *Method of images.* In inviscid flow streamlines can be replaced by solid surfaces without affecting the flow. An *image* source then placed as shown in Figure 9.28 produces a symmetric flow pattern with respect to the streamline along the x-axis. If this streamline is replaced by a solid surface, a source and its image model the flow due to a source near a solid surface. By using the method of images find the pressure distribution relative to the pressure at the stagnation point, along a solid surface due to a source with strength $m = 20$ m^2/s, placed at a distance $H = 0.1$ m away from the surface.

14. Using the method of images, find the velocity distribution in a $90°$-degree corner due to a source with strength $m = 12$ m^2/s, placed at a distance $H = 0.1$ m from each side of the corner.

15. Using the method of images find the velocity potential, streamfunction and velocity components for flow due to a source at the center of a two-dimensional channel with height $2H$. Hint: since the method of images requires complete symmetry we need an infinite number of image sources.

16. Consider ideal uniform flow past a flat plate. At a point on the plate there is a sink with strength Q. By locating the sink at the origin of a Cartesian coordinate system with the x-axis aligned with the plate. Determine the streamfunction and velocity potential functions that describe this flow. What is the pressure variation along the plate? Where is the maximum pressure?

17. Consider ideal uniform flow past a flat plate. At a point on the plate there is a source with strength Q. By locating the source at the origin of a Cartesian coordinate system with the x-axis aligned with the plate. Determine the streamfunction and velocity potential functions that describe this flow. What is the pressure variation along the plate? Where is the maximum pressure?

18. Air at 1.5 m/s is flowing parallel to a plate and encounters air being released from a jet normal to the plate direction. The jet releases air at 0.6 m^3/s per unit depth. By considering the jet as a source, (a) find the pressure distribution along the plate relative to the pressure at the source, (b) find the stagnation points, and (c) how far will the air from the jet penetrate in the main flow?

19. Find the circulation around a square area with side of 1 m due to source flow with strength $m = 10$ m^2/s, as shown in Figure 9.29.

20. A source with strength of 3.2 m/s is combined with a uniform flow of 11 m/s. Find and sketch, the shape of the half Rankine body that develops. What is the thickness of the body far away from the origin of the source? Where is the stagnation point? Find also the point where the pressure is minimum.

21. The stagnation point in the flow due to a source and uniform flow of 5 m/s is 2 m away from the location of the source and in the direction of the incoming uniform flow. For the half Rankine body that develops, find the maximum height, maximum velocity and its location and the minimum pressure.

22. Consider a full Rankine body formed by a source and a sink with identical strengths of 16 m^2/s placed 1.8 m apart, and

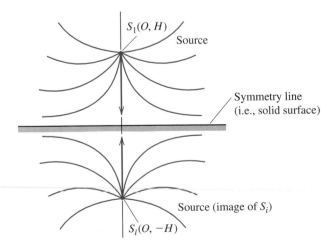

FIGURE 9.28   Schematic for Problem 13.

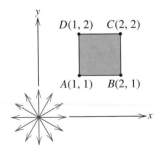

FIGURE 9.29   Schematic for Problem 19.

a uniform flow of 12 m/s. Find: (a) the position of the stagnation points, (b) the maximum thickness of the body, and (c) the velocity and pressure at the point of maximum thickness. Consider air flow at standard conditions ($P_1 = P_{atm} = 101.3\ kPa$ and $\rho = 1.225\ kg/m^3$).

23. Flow past a complete Rankine body is modeled using a combination of a uniform flow V_0, a source and a sink with equal strengths (m) at a distance a apart. For $V_0 = 18\ m/s$, $a = 0.4\ m$ and $m = 12\ m^2/s$, find the position of the stagnation points, the maximum thickness of the body and the maximum velocity.

24. Consider uniform air flow at 12 m/s past a Rankine body 1.5 m long and 0.75 m wide. Find (a) the maximum velocity, and (b) the position of the stagnation points.

25. Show that the functions for the velocity potential and streamfunction for a vortex satisfy the Laplace equations $\nabla^2 \phi = 0$, $\nabla^2 \psi = 0$.

26. Flow in a drain or flow induced by a tornado can be modeled as a combination of a sink and a vortex with a common origin. For such a combination find the velocity potential and streamfunction and evaluate the velocity field.

27. The wind 4 *miles* from the center of a tornado is measured to be 40 *mph*, and the pressure 98 kPa. Find the velocity and pressure 1 *miles* away from the center of the tornado. Assume the tornado as a combination of a sink and a vortex with a common origin.

28. The velocity potential function for an ideal spiral vortex is

$$\phi = \frac{\Gamma}{2\pi}\theta - \frac{m}{2\pi}\ln r$$

where Γ and m are constants, and θ is the angle the radial direction r makes with respect to the horizontal. (a) Select numerical values for the constants and plot selected values of constant potential lines, (b) find the velocity field, (c) show that the angle β the velocity vector makes with respect to the local radial line is constant everywhere in the flow.

29. For a combination of a sink ($m = 10\ m^2/s$) and a vortex ($\Gamma = 4\ m^2/s$) find the velocity at (3,2), (0,1) and (2,0).

30. Typhoons and hurricanes can be modeled as *circular vortices* where the vorticity is assumed to be constant in a circular core with radius a and zero outside this core. Find the velocity and pressure distribution as a function of the distance from the origin of the vortex.

31. Find the pressure and velocity distribution as a function of the distance from the origin of a circular vortex when the maximum velocity is 20 m/s at a distance 32 m from the center. Assume standard atmospheric conditions far away from the origin.

32. Determine the velocity distribution along a flat surface due to a vortex with circulation $\Gamma = 14\ m^2/s$ in the clockwise direction placed at a distance of 0.6 m away. What is the

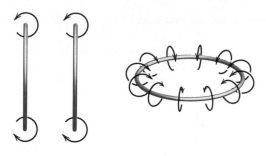

FIGURE 9.30   Schematic for Problem 34.

pressure variation along the surface (relative to the pressure far away from the vicinity of the vortex)?

33. What happens to a vortex with circulation Γ, next to a long horizontal wall? Will the direction of the circulation change the situation?

34. The geometry of a circular vortex filament (ring) is shown in Figure 9.30. Discuss the interaction and motion of two such filaments placed next to each other.

35. Show that the functions for the velocity potential and streamfunction for a doublet satisfy the Laplace equations $\nabla^2 \phi = 0$, $\nabla^2 \psi = 0$.

36. What is the required doublet strength to represent ideal inviscid flow past a circular cylinder in a 40 m/s uniform cross-flow?

37. A circular cylinder 20 cm in diameter, rotates at 100 rpm in a uniform air flow of 60 m/s. What is the lift that the cylinder will experience?

38. Sketch the streamlines and identify the location of the stagnation points for the combination of a uniform flow (V_0) and a doublet ($\lambda = a^2 V_0$) and a vortex (Γ) when:

1. $a = 0.1$, $V_0 = 1\ m/s$, $\Gamma = 1.2566\ m^2/s$
2. $a = 0.1$, $V_0 = 2\ m/s$, $\Gamma = 2.22\ m^2/s$
3. $a = 0.1$, $V_0 = 2\ m/s$, $\Gamma = -2.22\ m^2/s$
4. $a = 0.1$, $V_0 = 3\ m/s$, $\Gamma = 4.15\ m^2/s$
5. $a = 0.1$, $V_0 = 3\ m/s$, $\Gamma = -4.15\ m^2/s$

39. Find the circulation required to shift the stagnation points in flow past a circular cylinder with radius of 0.25 m to points defined by $\theta_1 = 80°$ and $\theta_2 = 100°$. The uniform velocity is 12 m/s. What is the magnitude and direction of the resulting lift force?

40. Find the circulation required to shift the stagnation points in flow past a circular cylinder with radius of 0.18 m to points defined by $\theta_1 = 240°$ and $\theta_2 = 300°$. The uniform velocity is 6 m/s. What is the magnitude and direction of the resulting lift force?

41. Find the circulation required to produce a single stagnation point at the top of the cylinder in flow past a circular cylinder with radius of 0.32 m. The uniform velocity is 14 m/s.

What is the magnitude and direction of the resulting lift force?

42. Find the magnitude and direction of the force on a vortex with circulation $\Gamma = 16 \ m^2/s$ due to a uniform flow of $8 \ m/s$.

43. Find the stagnation points in a $0.35 \ m$ radius cylinder with circulation of $\Gamma = -8.5 \ m^2/s$.

44. A cylinder with $0.15 \ m$ radius is in a cross-flow of an ideal fluid (density ρ) with velocity of $V_0 = 18 \ m/s$ and pressure p_o. If $2(p - p_o)/\rho V_0^2 = -8$ at the top with pressure p, find the circulation and velocity there.

45. In a two-dimensional flow past a circular cylinder with radius R, the maximum velocity is 6 times that of the incoming fluid velocity. Find the circulation and distance of the stagnation point for the center of the cylinder.

46. Consider two vortices with circulation Γ_1 and Γ_2 at a distance H apart. What will happen if they are allowed to moved? Consider two cases: (a) when the two vortices have the same sign and (b) have different sign.

47. Consider a ship operating on the Magnus principle, with two rotors each $6.6 \ m$ high with diameter of $1.2 \ m$ and rotating at $660 \ rev/min$. Find the net propulsion force the ship experiences when the ship travels at $6.4 \ km/hr$ with a $18 \ km/hr$ wind blowing in a direction normal to the direction of the boat.

48. A baseball pitcher throws the ball at $95 \ mph$. What spin (angular rotation) he needs to impose on the ball to cause $0.5 \ m$ downward deflection (consider using a computer program to get a precise answer). Consider that the ball has a mass of $0.145 \ kg$ and radius of $3.7 \ cm$.

10 Dynamics of Rotating Fluids: Turbomachinery

Many devices that form the backbone of the industrial and technological bases are *fluid machines*, devices whose operation is based on the use of fluids. For instance, fluid machines are used to produce power (turbines), to achieve flight (jet engines), to provide cooling (fans), or to distribute fluid across a pipe network (pumps). Obviously, there is hardly any industrial activity that it is not directly or indirectly dependent upon these devices.

Depending on the intended usage, fluid machines may be distinguished either as *energy absorbing devices*, (e.g., pumps) or as *energy producing devices*, (e.g., turbines). Furthermore, these machines may be distinguished by the basis of their operation (1) as *positive displacement machines*, when they operate by displacing fluids through changes in the volume (such as a reciprocating piston) or (2) as *turbomachines* when their operation depends on the dynamics of rotating fluids (pumps, turbines).

The flow in fluid machines is quite complicated, consisting almost always of unsteady turbulent flow in very complicated geometries. Therefore, detailed analysis of such flows is very difficult. In practice, the performance of fluid machines is typically determined through a combination of theoretical considerations and extensive experimentation. However, simple analyses, such as those covered in this chapter, that are based on idealized flow situations can be quite useful. Despite their simplicity, simple analyses provide insight into the flow behavior, and reveal physical trends that are similar to those found in the actual flows.

This chapter, following a discussion of the dynamics of rotating fluids, focuses on turbomachines by analyzing both energy-absorbing and energy-producing devices.

10.1 Conservation of Angular Momentum Concept

The principle for the conservation of linear momentum leads naturally to the conservation of angular momentum **H** by using the definition

$$\mathbf{H} = m(\mathbf{r} \times \mathbf{V}),$$

where **r** is the position vector and **V** the velocity vector. Note that the conservation of angular momentum principle is not a separate governing law but a special application of the conservation of linear momentum.

---

[1] Aristotle, *De Caelo*. Taken from Sir Thomas Heath, *A History of Greek Mathematics*.

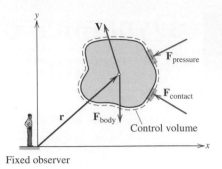

FIGURE 10.1   Schematic for conservation of angular momentum.

10.1.1   Conservation of Angular Momentum for a Closed System

For the closed system shown in Figure 10.1, the conservation of angular momentum with respect to a fixed coordinate system is obtained by taking the vector product of the conservation of linear momentum as

$$\mathbf{r} \times \frac{D}{Dt}(m\mathbf{V}) \equiv \mathbf{r} \times \sum \mathbf{F}.$$

Since the mass is constant, differentiation by parts gives

$$\frac{D}{Dt}[\mathbf{r} \times (m\mathbf{V})] = \frac{D\mathbf{r}}{Dt} \times (m\mathbf{V}) + \mathbf{r} \times \frac{D}{Dt}(m\mathbf{V}).$$

However, since

$$\frac{D\mathbf{r}}{Dt} = \mathbf{V} \Rightarrow \frac{D\mathbf{r}}{Dt} \times (m\mathbf{V}) = m(\mathbf{V} \times \mathbf{V}) = 0,$$

it simplifies to

$$\frac{D}{Dt}[m(\mathbf{r} \times \mathbf{V})] \equiv \mathbf{r} \times \sum \mathbf{F}.$$

In the general case, when the velocity varies within the control volume, the conservation of angular momentum is expressed in integral form as

$$\frac{D}{Dt}\int \rho(\mathbf{r} \times \mathbf{V})\,d\mathcal{V} \equiv \sum(\mathbf{r} \times \mathbf{F}) = \mathbf{T}_{net}. \tag{10.1}$$

The term $\sum(\mathbf{r} \times \mathbf{F})$ is the net torque $\mathbf{T}_{net}$ produced by the forces acting on the closed system. These are the usual body forces (e.g., due to gravity) $\mathbf{F}_b$, the surface forces $\mathbf{F_s}$, and those forces due to contact. However, since fluid rotation is induced by rotating shafts (or, conversely, fluid flow induces shaft rotation) the moment due to contact forces is customarily represented as an equivalent shaft torque $\mathbf{T}_{shaft}$. Therefore, the net torque on the control volume is given as

$$\mathbf{T}_{net} = \underbrace{\int \mathbf{r} \times d\mathbf{F_s}}_{surface} + \underbrace{\int \rho(\mathbf{r} \times \mathbf{g})\,d\mathcal{V}}_{gravity} + \underbrace{\mathbf{T}_{shaft}}_{contact}. \tag{10.2}$$

The final form of the conservation of angular momentum principle for a closed system, relative to a fixed coordinate system, is

$$\frac{D}{Dt}\int \rho(\mathbf{r} \times \mathbf{V})\,d\mathcal{V} \equiv \int \mathbf{r} \times d\mathbf{F_s} + \int \rho(\mathbf{r} \times \mathbf{g})\,d\mathcal{V} + \mathbf{T}_{shaft}. \tag{10.3}$$

EXAMPLE 10.1

Problem Statement Find the torque required to sustain constant rotation of 4800 rev/hr of the disk shown in Figure 10.2. The gap between the rotating, and the stationary disk $H = 0.01\ m$, is filled with oil having viscosity $\mu = 0.1\ Pa \cdot s$. The diameter of the rotating disk is $0.4\ m$.

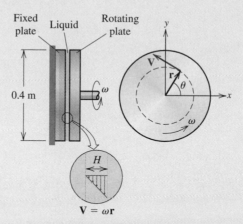

FIGURE 10.2   Torque required to sustain constant rotation.

Governing Equations By selecting the rotating disk as a control volume, the conservation of angular momentum is given by

$$\frac{d}{dt} \int \rho (\mathbf{r} \times \mathbf{V})\, d\mathcal{V} \equiv \int \mathbf{r} \times d\mathbf{F_s} + \int \rho (\mathbf{r} \times \mathbf{g})\, d\mathcal{V} + \mathbf{T}_{shaft}.$$

$\mathbf{V}$ is the linear velocity of the disk, $\mathbf{F}_s$ the surface forces, $\mathbf{g}$ the gravity, and $\mathbf{T}_{shaft}$ the torque on the control volume due to the contact forces. All terms are expressed relative to a fixed (inertial) coordinate system.

Assumptions We assume that the rotation of the disk is steady. Also, since the gap between the two disks is small, by neglecting any end effects, we can assume that the flow between the disks can be approximated by Couette flow.

SOLUTION For steady rotation, $\dfrac{d}{dt} \displaystyle\int \rho (\mathbf{r} \times \mathbf{V})\, d\mathcal{V} = 0$. Using the fixed Cartesian coordinate system shown in Figure 10.2, $\mathbf{r} = (r \cos\theta\, \mathbf{i} + r \sin\theta\, \mathbf{j})$. The net moment due to gravity ($g\mathbf{k}$) then vanishes according to

$$\int \rho (\mathbf{r} \times \mathbf{g})\, d\mathcal{V} = \int_0^R \int_0^{2\pi} (rg \sin\theta\, \mathbf{i} - rg \cos\theta\, \mathbf{j})\, r d\theta\, dr = 0.$$

Therefore, the conservation of angular momentum reduces to

$$\mathbf{T}_{shaft} = -\int \mathbf{r} \times d\mathbf{F_s}.$$

The surface force is due to the viscous shear stress τ at the disk-oil interface. Hence,

$$d\mathbf{F}_s = \tau_s\, dA = \tau_s\, 2\pi r\, dr.$$

Using the assumption of Couette flow, the shear stress is $\tau_s = \mu\dfrac{V}{H}$. This stress is acting in a direction opposite to that of the velocity. At a radial distance $\mathbf{r}$ from the axis of rotation, the velocity is given by

$$\mathbf{V} = \omega \times \mathbf{r} = \omega\mathbf{k} \times (r\cos\theta\,\mathbf{i} + r\sin\theta\,\mathbf{j}) = -r\omega\sin\theta\,\mathbf{i} + r\omega\cos\theta\,\mathbf{j}.$$

The surface force then is expressed as

$$d\mathbf{F}_s = -\underbrace{\frac{\mu}{H}(-r\omega\sin\theta\,\mathbf{i} + r\omega\cos\theta\,\mathbf{j})}_{\tau_s}\,\underbrace{2\pi r\,dr}_{dA}.$$

Therefore,

$$\mathbf{T}_{shaft} = -\int_0^R (r\cos\theta\,\mathbf{i} + r\sin\theta\,\mathbf{j}) \times \left(\mu\frac{r\omega}{H}\sin\theta\,\mathbf{i} - \mu\frac{r\omega}{H}\cos\theta\,\mathbf{j}\right)2\pi r\,dr,$$

or

$$\mathbf{T}_{shaft} = 2\pi\mu\frac{\omega}{H}\int_0^R r^3\,dr\,\mathbf{k} = \pi\mu\frac{\omega}{H}\frac{R^4}{2}\mathbf{k}.$$

In terms of the numerical values,

$$T_{shaft} = \pi \times 0.1\ Ns/m^2 \times \frac{4800\ rev/hr \times 2\pi}{3600\ s/hr} \times \frac{1}{0.01\ m} \times \frac{0.2^4\ m^4}{2} = 0.21055\ Nm.$$

EXAMPLE 10.2

Problem Statement   Consider the rotating disk in Example 10.1. Find the steady angular velocity and the time it takes for the disk to reach steady state conditions when a constant torque of $0.21055\ N \cdot m$ is applied in the vertical direction. The disk is made from aluminum ($\rho = 2700\ kg/m^3$) and has a thickness of $2\ cm$.

Governing Equations   By selecting the same control volume as in Figure 10.2, the conservation of angular momentum is given by

$$\frac{d}{dt}\int \rho(\mathbf{r} \times \mathbf{V})\,d\mathcal{V} \equiv \int \mathbf{r} \times d\mathbf{F_s} + \int \rho(\mathbf{r} \times \mathbf{g})\,d\mathcal{V} + \mathbf{T}_s.$$

$\mathbf{V}$ is the linear velocity of the disk, $\mathbf{F}_s$ are the surface forces, $\mathbf{g}$ is the gravity, and $\mathbf{T}_s$ is the torque on the control volume. All terms are expressed relative to a fixed (inertial) coordinate system.

Assumptions   We assume that the flow between the disks can be approximated by Couette flow and that edge effects are negligible.

SOLUTION   Again, similar to Example 10.1, the moment due to gravity cancels out. However, since this is a time-dependent, the unsteady term must be retained here. Hence,

$$\frac{d}{dt}\int \rho(\mathbf{r} \times \mathbf{V})\,d\mathcal{V} \equiv \int \mathbf{r} \times d\mathbf{F_s} + \mathbf{T}_s.$$

In the previous example,

$$dF_s = \left(\mu \frac{r\omega}{H} \sin\theta \, \mathbf{i} - \mu \frac{r\omega}{H} \cos\theta \, \mathbf{j} \right) 2\pi r \, dr,$$

which finally gives,

$$\int \mathbf{r} \times d\mathbf{F_s} = -\pi\mu \frac{\omega}{H} \frac{R^4}{2} \mathbf{k}.$$

The unsteady term $\dfrac{d}{dt} \displaystyle\int \rho(\mathbf{r} \times \mathbf{V}) \, d\mathcal{V}$ can be evaluated using the results from Example 10.1:

$$\mathbf{r} = (r\cos\theta \, \mathbf{i} + r\sin\theta \, \mathbf{j}),$$

and

$$\mathbf{V} = -r\omega\sin\theta \, \mathbf{i} + r\omega\cos\theta \, \mathbf{j}.$$

Therefore,

$$\mathbf{r} \times \mathbf{V} = (r\cos\theta \, \mathbf{i} + r\sin\theta \, \mathbf{j}) \times (-r\omega\sin\theta \, \mathbf{i} + r\omega\cos\theta \, \mathbf{j}) = r^2\omega \, \mathbf{k}.$$

By substitution we get,

$$\frac{d}{dt} \int \rho(\mathbf{r} \times \mathbf{V}) \, d\mathcal{V} = \frac{d}{dt} \int \rho r^2 \omega \, 2\pi h r \, dr = \pi\rho h \frac{R^4}{2} \frac{d\omega}{dt} \mathbf{k},$$

where h is the thickness of the disk. Combining the individual terms,

$$\pi\rho h \frac{R^4}{2} \frac{d\omega}{dt} = T_s - \pi\mu \frac{R^4}{2H}\omega,$$

or

$$\alpha \frac{d\omega}{dt} = T_s - \beta\omega,$$

where

$$\alpha \equiv \pi\rho h \frac{R^4}{2} \quad \text{and} \quad \beta \equiv \pi\mu \frac{R^4}{2H}.$$

Separating variables,

$$\int_{\omega=0}^{\omega} \left(\frac{\alpha}{\beta\omega - T_s} \right) d\omega = -\int_0^t dt,$$

and integrating, we get

$$\frac{\alpha}{\beta} \ln \frac{T_s - \beta\omega}{T_s} = -t.$$

Finally,

$$\omega = \frac{T_s}{\beta} \left(1 - e^{-\frac{\beta}{\alpha}t} \right) = \frac{2T_s H}{\pi\mu R^4} \left(1 - e^{-\frac{\mu}{\rho h H}t} \right) = 8.3775(1 - e^{-0.1852\,t}) \; rad/s.$$

At large t, the steady state angular velocity is obtained as

$$\omega_{steady} = \frac{2T_s H}{\pi \mu R^4}.$$

This is the same result obtained in Example 10.1. The numerical values give $\omega_s = 4800\ rev/hr$. Figure 10.3 shows the angular velocity as a function of time. The disk reaches steady state conditions at about $t = 26\ s$.

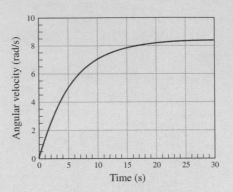

FIGURE 10.3   Angular speed as a function of time for Example 10.2.

Often, however, it is convenient to express the conservation of angular momentum relative to a *rotating coordinate system* as shown in Figure 10.4. In this case, it is expressed as

$$\frac{d}{dt} \int \rho(\mathbf{r} \times \mathbf{V}_r)\, d\mathcal{V} \equiv \mathbf{T}_{net} - \int \mathbf{r} \times \left[\underbrace{2\omega \times \mathbf{V}_r}_{Coriolis} + \underbrace{\omega \times (\omega \times \mathbf{r})}_{Centripetal} + \dot{\omega} \times \mathbf{r} \right] \rho\, d\mathcal{V} \quad (10.4)$$

In the preceding expression, ω is the rate of rotation of the coordinate system and $\mathbf{V}_r$ is the fluid velocity *expressed relative to the rotating coordinate system*. In addition,

$$\bullet\ \mathbf{T}_{net} = \underbrace{\int \mathbf{r} \times d\mathbf{F_s}}_{surface} + \underbrace{\int \rho(\mathbf{r} \times \mathbf{g})\, d\mathcal{V}}_{gravity} + \underbrace{\mathbf{T}_{shaft}}_{contact}$$

represents the net torque produced by the forces acting on the control volume.

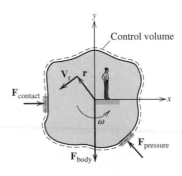

FIGURE 10.4   Conservation of angular momentum with respect to a rotating coordinate system.

- $2\omega \times \mathbf{V}_r$ is the *Coriolis* acceleration. This is the acceleration due to the motion of the fluid within the moving control volume.
- $\omega \times (\omega \times \mathbf{r})$ is the *centripetal acceleration*. This is the acceleration due to the angular velocity.
- $\dot{\omega} \times \mathbf{r}$ is a tangential acceleration due to rotating coordinate system.

EXAMPLE 10.3

Problem Statement   Rework Example 10.2 by selecting a coordinate system that it is attached to and rotates with the disk.

Governing Equations   For a non-inertial coordinate system, the conservation of angular momentum is written as

$$\frac{d}{dt} \int \rho(\mathbf{r} \times \mathbf{V}_r) \, d\mathcal{V} \equiv \mathbf{T}_{net} - \int \mathbf{r} \times [2\omega \times \mathbf{V}_r + \omega \times (\omega \times \mathbf{r}) + \dot{\omega} \times \mathbf{r}]\rho \, d\mathcal{V},$$

where

$$\mathbf{T}_{net} = \int \mathbf{r} \times d\mathbf{F}_s + \int \rho(\mathbf{r} \times \mathbf{g}) \, d\mathcal{V} + \mathbf{T}_{shaft}.$$

Assumptions   We use the same assumptions as in Example 10.2 (i.e., Couette flow between the disks with negligible edge effects).

SOLUTION   For a coordinate system rotating with the disk, we have $\mathbf{V}_r = 0$. Also, the moment due to gravity $\int \rho(\mathbf{r} \times \mathbf{g}) \, d\mathcal{V} = 0$. The conservation of angular momentum simplifies to

$$\int \mathbf{r} \times d\mathbf{F}_s + \mathbf{T}_{shaft} = \int \mathbf{r} \times [\omega \times (\omega \times \mathbf{r}) + \dot{\omega} \times \mathbf{r}]\rho \, d\mathcal{V}.$$

Now, with respect to the attached coordinate system, $\omega = \omega \mathbf{k}$, and $\mathbf{r} = r\,\mathbf{e}_r$. The terms in the right-hand side of the equation are evaluated as follows:

- $\omega \times \mathbf{r} = \omega \mathbf{k} \times r\,\mathbf{e}_r = -r\omega\,\mathbf{e}_\theta$
- $\omega \times (\omega \times \mathbf{r}) = \omega \mathbf{k} \times (-r\omega\,\mathbf{e}_\theta) = r\omega^2\,\mathbf{e}_r$
- $\mathbf{r} \times (\omega \times \omega \times \mathbf{r}) = r\,\mathbf{e}_r \times r\omega^2\,\mathbf{e}_r = 0$
- $\dot{\omega} \times \mathbf{r} = \dot{\omega} \mathbf{k} \times r\,\mathbf{e}_r = r\dot{\omega}\,\mathbf{e}_\theta$
- $\mathbf{r} \times (\dot{\omega} \times \mathbf{r}) = r\mathbf{e}_r \times r\dot{\omega}\,\mathbf{e}_\theta = r^2\dot{\omega}\,\mathbf{k}$

By substitution we get,

$$\int \mathbf{r} \times d\mathbf{F}_s + \mathbf{T}_{shaft} = \int \rho r^2 \dot{\omega} \, d\mathcal{V}.$$

This result is identical to that obtained in Example 10.2. The solution from this point on is also identical to Example 10.2. This indicates that the final solution is independent of the choice of the coordinate system (as it should be, since nature does not function on the basis of coordinate systems!).

10.1.2 Conservation of Angular Momentum for an Open System

The conservation of angular momentum for an open system relative to a fixed coordinate system is expressed using Reynold's transport theorem developed in Chapter 3 as

$$\frac{\partial}{\partial t} \int \rho(\mathbf{r} \times \mathbf{V})\, d\mathcal{V} + \underbrace{\int \rho(\mathbf{r} \times \mathbf{V})\mathbf{V} \cdot \mathbf{n}\, dA}_{\text{flux term}} \equiv \int \mathbf{r} \times d\mathbf{F_s} + \int \rho(\mathbf{r} \times \mathbf{g})\, d\mathcal{V} + \mathbf{T}_{shaft} \quad (10.5)$$

In terms of a rotating coordinate system, it is expressed as

$$\frac{\partial}{\partial t} \int \rho(\mathbf{r} \times \mathbf{V}_r)\, d\mathcal{V} + \int \rho(\mathbf{r} \times \mathbf{V}_r)\mathbf{V}_r \cdot \mathbf{n}\, dA \equiv \int \mathbf{r} \times d\mathbf{F_s} + \int \rho(\mathbf{r} \times \mathbf{g})\, d\mathcal{V}$$

$$+ \mathbf{T}_{shaft} - \int \mathbf{r} \times [2\omega \times \mathbf{V}_r + \omega \times (\omega \times \mathbf{r}) + \dot{\omega} \times \mathbf{r}]\rho\, d\mathcal{V}. \quad (10.6)$$

As with a closed system, ω is the rate of rotation of the coordinate system. The fluid velocity $\mathbf{V}_r$ is *expressed relative to the rotating coordinate system*. The other terms are identical to those for a closed system.

EXAMPLE 10.4

Problem Statement   Consider the sprinkler system with two arms rotating at a constant angular velocity of $40\ rev/min$ as shown in Figure 10.5. The sprinkler delivers water at a rate of $12\ liters/min$, the radius of rotation extents $20\ cm$ and the arm diameter is $0.5\ cm$. Find the frictional torque that develops during the operation of the sprinkler.

Governing Equations   A control volume is selected to include the volume formed by the revolution of the arms as shown in Figure 10.5. For this selection of the control volume, the conservation of mass is expressed as

$$\frac{\partial}{\partial t} \int \rho\, d\mathcal{V} + \int_{\mathcal{V}} \rho(\mathbf{V} \cdot \mathbf{n})\, dA = 0,$$

where $\mathbf{V}$ is expressed relative to the control volume and $\mathbf{n}$ is the outward pointing unit normal vector at locations where mass crosses the boundary of the control volume.

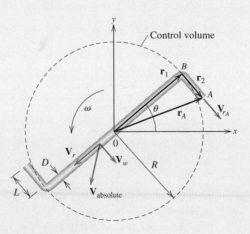

FIGURE 10.5   Operation of a sprinkler.

For the same control volume, the conservation of angular momentum is expressed as

$$\frac{\partial}{\partial t}\int \rho(\mathbf{r}\times\mathbf{V})\,d\mathcal{V}+\int \rho(\mathbf{r}\times\mathbf{V})\mathbf{V}\cdot\mathbf{n}\,dA \equiv \int \mathbf{r}\times d\mathbf{F_s}+\int \rho(\mathbf{r}\times\mathbf{g})\,d\mathcal{V}+\mathbf{T}_{shaft}.$$

Again, all the terms are expressed relative to a fixed coordinate system. Since the torque develops in reaction to the $\mathbf{T}_{shaft}$,

$$\mathbf{T}_{friction} = -\mathbf{T}_{shaft}.$$

Assumptions   We are assuming that the flow is steady and uniform.

SOLUTION   For steady rotation, the unsteady term vanishes (i.e., $\frac{\partial}{\partial t}\int \rho(\mathbf{r}\times\mathbf{V})\,d\mathcal{V}=0$).

For this problem, we also have $\int \mathbf{r}\times d\mathbf{F_s}=0$ and $\int \rho(\mathbf{r}\times\mathbf{g})\,d\mathcal{V}=0$. The conservation of angular momentum then reduces to

$$\int \rho(\mathbf{r}\times\mathbf{V})\mathbf{V}\cdot\mathbf{n}\,dA \equiv \mathbf{T}_{shaft}.$$

Since both arms are identical, we evaluate the contribution of one of them and multiply the result by two.

With respect to a fixed coordinate system, the absolute fluid velocity $\mathbf{V}$ in the preceding expression is the sum of the velocity relative to the arm $\mathbf{V}_r$ and the velocity due to the rotation of the arm $\mathbf{V}_\omega$. Therefore,

$$\mathbf{V} = \mathbf{V}_r + \mathbf{V}_\omega.$$

The magnitude of the velocity relative to the arm is found by using conservation of mass,

$$V_r = \frac{\dot{Q}}{2\,A},$$

where $\dot{Q}$ is the total volumetric flow rate (for both arms) and A the cross-sectional area of the arm. Therefore,

$$V_r = \frac{\dot{Q}}{2\,A} = \frac{12\ liter/min \times 10^{-3}\ m^3/liter \times 4}{2 \times 60\ s/min \times \pi \times 0.005^2\ m^2} = 5.09\ m/s.$$

According to Figure 10.5, at the exit of the arm we have,

$$\mathbf{V}_{r_A} = V_r \sin\theta\,\mathbf{i} - V_r \cos\theta\,\mathbf{j}.$$

The velocity induced by rotation at point A is

$$\mathbf{V}_{\omega_A} = \omega \times \mathbf{r}_A,$$

where $\omega = \omega\,\mathbf{k}$. Also,

$$\mathbf{r}_A = \mathbf{r}_1 + \mathbf{r}_2.$$

Using the geometry,

$$\mathbf{r}_1 = R\cos\theta\,\mathbf{i} + R\sin\theta\,\mathbf{j},$$

and

$$\mathbf{r}_2 = L\sin\theta\,\mathbf{i} - L\cos\theta\,\mathbf{j},$$

where L is the length of the arm that is turned $90°$. Hence,

$$\mathbf{r}_A = (R\cos\theta + L\sin\theta)\,\mathbf{i} + (R\sin\theta - L\cos\theta)\,\mathbf{j}.$$

The velocity induced by the rotation is

$$\mathbf{V}_{\omega_A} = \omega \times \mathbf{r}_A = \omega\mathbf{k} \times [(R\cos\theta + L\sin\theta)\,\mathbf{i} + (R\sin\theta - L\cos\theta)\,\mathbf{j}]$$

$$\mathbf{V}_{\omega_A} = (-\omega R\sin\theta + \omega L\cos\theta)\,\mathbf{i} + (\omega R\cos\theta + \omega L\sin\theta)\,\mathbf{j}.$$

The absolute velocity

$$\mathbf{V}_A = \mathbf{V}_{r_A} + \mathbf{V}_{\omega_A},$$

is expressed as

$$\mathbf{V}_A = (V_r\sin\theta - \omega R\sin\theta + \omega L\cos\theta)\mathbf{i} + (-V_r\cos\theta + \omega R\cos\theta + \omega L\sin\theta)\mathbf{j}.$$

Therefore,

$$\mathbf{r}_A \times \mathbf{V}_A = ((R\cos\theta + L\sin\theta)\,\mathbf{i} + (R\sin\theta - L\cos\theta)\,\mathbf{j})(V_r\sin\theta - \omega R\sin\theta$$
$$+ \omega L\cos\theta)\,\mathbf{i} + (-V_r\cos\theta + \omega R\cos\theta + \omega L\sin\theta)\,\mathbf{j}$$

$$\mathbf{r}_A \times \mathbf{V}_A = -(RV_r - \omega R^2 - \omega L^2).$$

Since in most sprinklers $L \ll R$,

$$\mathbf{r}_A \times \mathbf{V}_A = -R(V_r - \omega R).$$

The conservation of angular momentum then for a single arm gives

$$\mathbf{T}_{shaft} = \int \rho(-R(V_r - \omega R))\mathbf{V}\cdot\mathbf{n}\,dA = -\rho\frac{\dot{Q}}{2}R(V_r - \omega R).$$

For both arms and by using $\mathbf{T}_{shaft} = -\mathbf{T}_{friction}$,

$$\mathbf{T}_{friction} = \rho\dot{Q}R(V_r - \omega R).$$

In terms of numerical values,

$$\mathbf{T}_{friction} = \rho\dot{Q}R(V_r - \omega R) = 998.2\ kg/m^3 \times 12\ liter/min \times 10^{-3}\ m^3/liter$$

$$\times \frac{1}{60\ s/min} \times 0.2\ m \times \left(5.09\ m/s - 40 \times 2 \times \pi \times \frac{1}{60\ s/min} \times 0.2\ m\right)$$

$$= 0.17\ N\cdot m.$$

EXAMPLE 10.5

Problem Statement   Reconsider the sprinkler in Example 10.4 by selecting instead a control volume that is attached to and rotates with the sprinkler.

Governing Equations   Selecting a control volume that rotates with the arms of the sprinkle as shown in Figure 10.6, the conservation of mass is expressed as

$$\frac{\partial}{\partial t}\int \rho\,d\mathcal{V} + \int_{\mathcal{V}} \rho(\mathbf{V}\cdot\mathbf{n})\,dA = 0,$$

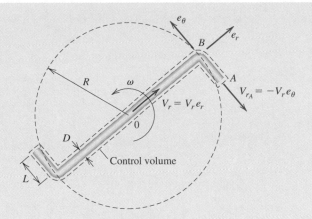

FIGURE 10.6 Schematic of Example 10.5.

where $\mathbf{V}$ is expressed relative to the control volume and $\mathbf{n}$ is the outward pointing unit normal vector at locations where mass crosses the boundary of the control volume.

For the same control volume the conservation of angular momentum is expressed as

$$\frac{\partial}{\partial t}\int \rho(\mathbf{r}\times\mathbf{V}_r)\,d\mathcal{V}+\int \rho(\mathbf{r}\times\mathbf{V}_r)\mathbf{V}_r\cdot\mathbf{n}\,dA \equiv \int \mathbf{r}\times d\mathbf{F_s}+\int \rho(\mathbf{r}\times\mathbf{g})\,d\mathcal{V}$$

$$+\,\mathbf{T}_{shaft}-\int \mathbf{r}\times[2\omega\times\mathbf{V}_r+\omega\times(\omega\times\mathbf{r})+\dot{\omega}\times\mathbf{r}]\rho\,d\mathcal{V}.$$

The terms are expressed relative to a coordinate system rotating with angular velocity of $\omega\,\mathbf{e}_k$. Since the frictional torque develops in reaction to the $\mathbf{T}_{shaft}$,

$$\mathbf{T}_{friction}=-\mathbf{T}_{shaft}.$$

Assumptions   Here, we use the same assumptions as before (i.e., steady and uniform flow).

Solution   For steady rotation, the unsteady term vanishes (i.e., $\frac{\partial}{\partial t}\int \rho(\mathbf{r}\times\mathbf{V})\,d\mathcal{V}=0$).

For this problem, we also have $\int \mathbf{r}\times d\mathbf{F_s}=0$ and $\int \rho(\mathbf{r}\times\mathbf{g})\,d\mathcal{V}=0$. For constant rate of rotation $\dot{\omega}=0$, the conservation of angular momentum reduces to

$$\underbrace{\int \rho(\mathbf{r}\times\mathbf{V}_r)\mathbf{V}_r\cdot\mathbf{n}\,dA}_{I_1}=\mathbf{T}_{shaft}-\underbrace{\int_{\mathcal{V}} \mathbf{r}\times[2\omega\times\mathbf{V}_r+\omega\times(\omega\times\mathbf{r})]\rho\,d\mathcal{V}}_{I_2}.$$

The remaining terms are of two kinds: I_1, valid where mass crosses the boundary of the control volume, and I_2, to be evaluated over the entire control volume.

For an observer rotating with the sprinkler, $\omega=\omega\,\mathbf{e}_k$, $\mathbf{r}=r\,\mathbf{e}_r$. In section OB of the sprinkler arm, $\mathbf{V}_r=V_r\,\mathbf{e}_r$, and in section AB, $\mathbf{V}_{r_A}=-V_r\,\mathbf{e}_\theta$. The negative sign is included because the azimuthal velocity is positive in the counterclockwise direction. The various terms in the conservation of angular momentum are evaluated as follows:

- $\omega\times\mathbf{r}=\omega\,\mathbf{e}_k\times r\,\mathbf{e}_r=\omega r\,\mathbf{e}_\theta$
- $\omega\times(\omega\times\mathbf{r})=\omega\,\mathbf{e}_k\times\omega r\,\mathbf{e}_\theta=-r\omega^2\,\mathbf{e}_r$

- $\mathbf{r} \times [\omega \times (\omega \times \mathbf{r})] = r\,\mathbf{e}_r \times (-r\omega^2\,\mathbf{e}_r) = 0$
- $2\omega \times \mathbf{V} = 2\omega\,\mathbf{e}_k \times V_r\,\mathbf{e}_r = 2\omega V_r\,\mathbf{e}_\theta$
- $\mathbf{r} \times (2\omega \times \mathbf{V}) = r\,\mathbf{e}_r \times 2\omega V_r\,\mathbf{e}_\theta = 2r\omega V_r\,\mathbf{e}_k$

Therefore, by neglecting the effects of the small section of the sprinkler arm (AB), the integral I_2 reduces to

$$I_2 = \int_{\mathcal{V}} \mathbf{r} \times (2\omega \times \mathbf{V})\rho\,d\mathcal{V} = \int_0^R 2r\omega V_r\,\mathbf{e}_k\,A\,dr = \rho\omega R^2 V_r A\,\mathbf{e}_k.$$

Using mass conservation, $\dot{Q} = 2V_r A$. Therefore,

$$I_2 = \rho\omega R^2 \frac{\dot{Q}}{2}\,\mathbf{e}_k.$$

The term associated with the angular momentum flux at the exit of the arm is evaluated by noting that $\mathbf{r} = R\mathbf{e}_r$ and $\mathbf{V}_r = -V_r\mathbf{e}_\theta$, and by calculating

$$\mathbf{r} \times \mathbf{V}_r = R\mathbf{e}_r \times (-V_r\mathbf{e}_\theta) = -RV_r\,\mathbf{e}_k.$$

Therefore,

$$I_1 = \int_A \rho(\mathbf{r} \times \mathbf{V}_r)\mathbf{V}_r \cdot \mathbf{n}\,dA = -\int_A \rho(-RV_r\,\mathbf{e}_k - RV_r\,\mathbf{e}_k)\mathbf{V}_r \cdot \mathbf{n}\,dA = -\rho RV_r\frac{\dot{Q}}{2}.$$

Finally, by considering both arms of the sprinkler,

$$T_{friction} = -T_{shaft} = -2(I_1 + I_2) = \rho\dot{Q}R(V_r - \omega^2 R).$$

As expected, this is the same result as that obtained in Example 10.4.

10.2   Turbomachines

In the following sections, the theory of the dynamics of rotating fluids is used to study the performance of "pumps" and "turbines." These two types of machine represent the majority of turbomachine applications. However, since the flow geometry in actual turbomachines is quite complicated and the flow is almost always turbulent, their performance is evaluated experimentally. Despite the complexity of the systems, a simple finite control volume analysis is quite revealing and helpful in understanding the more complex issues. In the following paragraphs, the basic geometry of the flow and the conservation laws are introduced for a general turbomachine.

10.2.1   Theoretical Framework

The operation of many turbomachines is based on rotational effects in which the rotation of the fluid is induced by a rotating impeller (e.g., pumps, fans, blowers) or the fluid induces rotation of a runner (e.g., turbines, windmills). Figure 10.7 shows a typical schematic of the operation of such devices: the fluid guided by a set of blades enters and leaves the rotating device at two different radii. In this figure, the inlet is at the smaller radius, and the device is rotating in the clockwise direction.

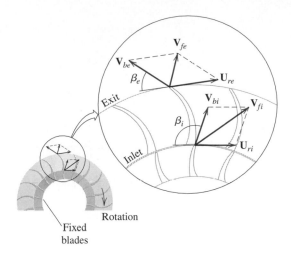

FIGURE 10.7 Typical turbomachine operation.

Flow Geometry

Under ideal conditions, the velocity *relative to the blade* is assumed to enter and leave tangent to the surface of the blade, which is known as a *shockless condition*. By considering the rotation of the blade, the *absolute fluid velocity* $\mathbf{V}_f$ is the vector sum of the relative blade velocity $\mathbf{V}_b$ and the linear velocity due to the angular rotation of the blade $\mathbf{U}_r = \omega r$, as shown in Figure 10.8:

$$\mathbf{V}_f = \mathbf{V}_b + \mathbf{U}_r.$$

It follows then that $\mathbf{V}_f$ is the actual fluid velocity as seen by a fixed observer and $\mathbf{V}_b$ is the fluid velocity seen by an observer riding on the blade (Figure 10.9). The angle β measured as shown in the figure represents the angle of the blade relative to the direction of the rotation.

The absolute fluid velocity vector can be expressed in terms of its components along the tangent $\mathbf{t}$ and normal directions $\mathbf{n}$, as shown in Figure 10.8.

$$\mathbf{V}_f = V_t\,\mathbf{t} + V_n\,\mathbf{n}.$$

The angle α, shown in the figure is defined as the angle between the velocity vector $\mathbf{V}_f$ and the normal direction.

Finally, at both the inlet and the exit boundaries, the various velocity components are expressed using the *velocity polygons* shown in Figure 10.8. According to the geometry,

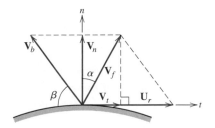

FIGURE 10.8 Velocity diagrams for turbomachinery.

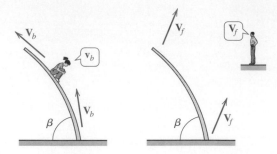

FIGURE 10.9   Relative velocity vectors.

the following relationships are valid:

$$V_{b_n} = V_b \sin\beta = V_f \cos\alpha = V_{f_n},$$

(i.e., the normal component of the absolute fluid velocity is the same as the component of the velocity relative to the blade along the normal direction). In addition, we have

$$V_{f_t} = U_r - V_b \cos\beta \quad \text{and} \quad V_{f_t} = U_r - V_{f_n}\cot\beta.$$

Following again the geometry of the velocity polygon, we have

$$U_r = V_{f_n}(\tan\alpha + \cot\beta),$$

and since $V_{f_t} = V_{f_n}\tan\alpha$,

$$U_r = V_{f_t}\left(1 + \frac{\cot\beta}{\tan\alpha}\right).$$

Simplified Theoretical Analysis

The simple analysis described below is performed using the conservation laws expressed relative to a fixed coordinate system. By selecting the entire rotating assembly as the control volume and by assuming steady flow with uniform conditions across the inlet and exit and constant density, we have the following.

Conservation of Mass: For the chosen control volume the conservation of mass,

$$\frac{\partial}{\partial t}\int \rho\, d\mathcal{V} + \int \rho \mathbf{V}\cdot\mathbf{n}\, dA = 0,$$

reduces to

$$\dot{Q} = V_{f_{ni}} A_i = V_{f_{ne}} A_e = V_{f_n} A,$$

where subscripts (i) and (e) refer to quantities at the inlet and exit, respectively, and $\dot{Q}$ to the volumetric flow rate. Therefore, for constant volumetric flow rate through the device, $\dot{Q}$

$$V_{f_t} = U_r - V_{f_n}\cot\beta = U_r - \frac{\dot{Q}}{A}\cot\beta.$$

Conservation of the Angular Momentum: The conservation of angular momentum for a fixed coordinate system is

$$\frac{\partial}{\partial t}\int \rho(\mathbf{r}\times\mathbf{V})\, d\mathcal{V} + \int \rho(\mathbf{r}\times\mathbf{V})\mathbf{V}\cdot\mathbf{n}\, dA \equiv \int \mathbf{r}\times d\mathbf{F_s} + \int \mathbf{r}\times\mathbf{g}\rho\, d\mathcal{V} + \mathbf{T}_{shaft}.$$

Using the same assumptions as previously and by neglecting the torque produced by the gravity and surface forces, it simplifies to

$$\int \rho(\mathbf{r} \times \mathbf{V})\mathbf{V} \cdot \mathbf{n} \, dA = \mathbf{T}_{shaft}.$$

For uniform flow conditions where flow crosses the boundary,

$$\mathbf{T}_{shaft} = \dot{m}[(\mathbf{r} \times \mathbf{V})_e - (\mathbf{r} \times \mathbf{V})_i].$$

Relative to the velocity polygon,

$$\mathbf{V_f} = V_{f_t}\mathbf{t} + V_{f_n}\mathbf{n},$$

and $\mathbf{r} = r\,\mathbf{n}$. Therefore, $\mathbf{r} \times \mathbf{V} = rV_{f_t}$. The shaft torque simplifies to

$$\mathbf{T}_{shaft} = \dot{m}[(rV_{f_t})_e - (rV_{f_t})_i]. \tag{10.7}$$

This is known as the *Euler turbomachine equation*. The power input/output $\dot{\mathcal{W}}_{shaft}$, is given by

$$\dot{\mathcal{W}}_{shaft} = \omega T_{shaft} = \dot{m}[(r\omega V_{f_t})_e - (r\omega V_{f_t})_i],$$

and since $r\omega = U_r$,

$$\dot{\mathcal{W}}_{shaft} = \dot{m}[(U_r V_{f_t})_e - (U_r V_{f_t})_i]. \tag{10.8}$$

By selecting the tangential velocity to be *positive in the direction of the rotation*, $\mathbf{T}_{shaft} > 0$ for a pump, and $\mathbf{T}_{shaft} < 0$ for a turbine. In pumps, the impeller imparts energy to the fluid, thus increasing its energy, whereas in a turbine, the opposite is true. The difference between producing and absorbing energy can be also deduced from the velocity diagram: when the absolute velocity vector rotates from the inlet to the exit in the direction of the rotation the device operates as a "pump", in the sense it absorbs energy. When the absolute velocity rotates in the opposite direction of the rotation the device operates as a "turbine", i.e., it produces energy.

Conservation of Energy: Finally, by applying conservation of energy to the same control volume

$$\frac{d}{dt}\int_{\mathcal{V}_{os}} \left(\frac{V^2}{2} + gz + u\right)\rho \, d\mathcal{V} + \int_s \rho\left(\frac{P}{\rho} + u + \frac{V^2}{2} + gz\right)\mathbf{V} \cdot \mathbf{n} \, ds \equiv \dot{\mathcal{Q}} - \dot{\mathcal{W}}_m,$$

and by assuming isothermal flow with no heat transfer and negligible changes in the kinetic and potential energy, we have

$$\pm\dot{\mathcal{W}}_{gained/lost} = \dot{m}\frac{\Delta P}{\rho},$$

where ΔP is the pressure change. In terms of an equivalent *pressure head*

$$\dot{\mathcal{W}}_{gained/lost} = \dot{m}gH = \rho\dot{\mathcal{Q}}gH.$$

Under ideal conditions, in the absence of losses

$$\dot{\mathcal{W}}_{gained/lost} = \omega T_{shaft} - \rho g\dot{\mathcal{Q}}H.$$

H is given by Equation (10.8),

$$H_{gained/lost} = \frac{1}{g}[(U_r V_{f_t})_e - (U_r V_{f_t})_i].$$

In real turbomachines, by accounting for losses the efficiency of "pumps" η is defined as

$$\eta = \frac{\rho g \dot{Q} H}{\omega T_{shaft}}.$$

The efficiency of turbines is defined as

$$\eta = \frac{\omega T_{shaft}}{\rho g \dot{Q} H}.$$

EXAMPLE 10.6

Problem Statement   *General turbomachine operation.* Consider the cylindrical rotating device shown in Figure 10.10, where the rotor rotates constantly at a rate of 800 rev/min. Air at a rate of 11.55 m^3/s enters the device at a radial distance of 0.8 m and exits at a radius of 1.2 m. The blade angle at the inlet is $25°$ and the angle at the exit is $55°$. If the blade height is 0.35 m, find (a) the torque required to run the device; (b) the corresponding power. Is this device operating as a pump or as a turbine?

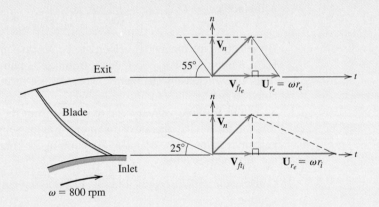

FIGURE 10.10   Schematic of Example 10.6.

Governing Equations   The shaft torque T_{shaft} and the power input/output $\dot{W}_{shaft}$ are given by

$$T_{shaft} = \dot{m}[(rV_{f_t})_e - (rV_{f_t})_i],$$

and

$$\dot{W}_{shaft} = \omega T_{shaft}.$$

If the tangential velocity is positive in the direction of the rotation, for a pump $T_{shaft} > 0$, and for a turbine $T_{shaft} < 0$.

Mass conservation gives

$$\dot{Q} = V_{n_i} A_i = V_{n_e} A_e,$$

where subscripts i and e refer to inlet and exit conditions.

Assumptions The foregoing equations were developed by assuming steady incompressible flow with uniform conditions at the inlet and exit. We also assume shockless conditions at the inlet and exit.

SOLUTION The torque is

$$\mathbf{T}_{shaft} = \dot{m}[(rV_{f_t})_e - (rV_{f_t})_i].$$

Using the geometry of the velocity polygon and mass conservation,

$$V_{f_t} = U_r - V_{f_n}\cot\beta = U_r - \frac{\dot{Q}}{A}\cot\beta.$$

At the inlet, the linear velocity of the rotor U_{r_i} is found as

$$U_{r_i} = \omega r_i = \frac{800\ rev/min \times 2\pi\ rad/rev}{60\ s/min} \times 0.8\ m = 67.02\ m/s.$$

The inlet area is $A_i = 2\pi rh = 2\pi \times 0.8\ m \times 0.35\ m = 1.76\ m^2$. Therefore,

$$V_{f_t} = U_r - \frac{\dot{Q}}{A}\cot\beta = 67.02\ m/s - \frac{11.55\ m^3/s}{1.76\ m^2}\cot 25 = 52.95\ m/s.$$

At the exit,

$$U_{r_e} = \omega r_e = \frac{800\ rev/min \times 2\pi\ rad/rev}{60\ s/min} \times 1.2\ m = 100.53\ m/s.$$

The exit area is $A_e = 2\pi rh = 2\pi \times 1.2\ m \times 0.35\ m = 2.64\ m^2$. Therefore,

$$V_{f_t} = U_r - \frac{\dot{Q}}{A}\cot\beta = 100.53\ m/s - \frac{11.55\ m^3/s}{2.64\ m^2}\cot 55 = 97.47\ m/s.$$

The torque is

$$\mathbf{T}_{shaft} = \rho\dot{Q}[(rV_{f_t})_e - (rV_{f_t})_i]$$

$$= 1.225\ kg/m^3 \times 11.55\ m^3/s \times [(1.2\ m \times 97.47\ m/s)_e - (0.8\ m \times 52.95\ m/s)_i]$$

$$= 1055.6\ Nm.$$

The power is

$$\dot{W}_{shaft} = \omega T_{shaft} = \frac{800\ rev/min \times 2\pi\ rad/rev}{60\ s/min} \times 1055.6\ Nm = 88.43\ kW.$$

Since $\mathbf{T}_{shaft} > 0$, the device is operating as a pump.

EXAMPLE 10.7

Problem Statement *General Turbomachine Operation.* Reconsider the pump described in Example 10.6, operating under the same conditions. If the blade angle is adjustable, what is the range of exit blade angles that can make the device operate as a turbine?

Governing Equations   For pump operation, $\mathbf{T}_{shaft} > 0$, and for turbine operation, $\mathbf{T}_{shaft} < 0$, where

$$\mathbf{T}_{shaft} = \dot{m}[(rV_{f_t})_e - (rV_{f_t})_i].$$

Assumptions   We use the same assumptions as in Example 10.6 (i.e., steady incompressible flow with uniform and shockless conditions at the inlet and exit).

SOLUTION   For a turbine operation,

$$\mathbf{T}_{shaft} = \dot{m}[(rV_{f_t})_e - (rV_{f_t})_i] < 0,$$

or

$$(rV_{f_t})_e < (rV_{f_t})_i,$$

by substitution

$$\left(U_r - \frac{\dot{Q}}{A}\cot\beta\right)_e < \frac{r_i}{r_e}V_{f_{t_i}},$$

and by rearranging

$$\frac{A_e}{\dot{Q}}\left[U_{r_e} - \frac{r_i}{r_e}V_{f_{t_i}}\right] < \cot\beta_e.$$

Finally, by using the values from Example 10.6, (i.e., $A_e = 2.64\ m^2$, $\dot{Q} = 11.55\ m^3/s$, $U_{r_e} = 100.53\ m/s$, $r_e = 1.2\ m$, $r_i = 0.8\ m$, and $V_{f_{t_i}} = 52.95\ m/s$), we have

$$\cot\beta_e > \frac{2.64\ m^2}{11.55\ m^3/s} \times \left[100.53\ m/s - \frac{0.8\ m}{1.2\ m} \times 52.95\ m/s\right].$$

Therefore, the device will operate as a turbine when

$$\beta_e < 3.84°.$$

10.3   Energy-Absorbing Turbomachines

A class of turbomachines are net energy-absorbing devices. Such devices are usually used to induce or sustain fluid flow to provide propulsion. Because of the energy exchange, the net effect is to increase the fluid's pressure or, equivalently, the pressure head. By convention, devices with incompressible working fluids are known as *pumps*. Fans, blowers, and compressors operate with compressible fluids or gases as working fluids. *Fans* are usually associated with low-energy input, while *compressors* are associated with relatively high energy levels. Devices that operate at moderate energy levels are known as *blowers*. Hereafter, the term "pump" will be used to signify a general energy-absorbing device operating like a pump. Therefore, the discussion is valid not only for pumps but also for fans, blowers, and compressors.

In pumps, energy is transmitted by a rotating shaft known as the *impeller*. Depending on the design of the fluid path through the pump, pumps may be classified as either *axial*, *radial*, or *mixed*. Axial pumps are those machines in which the fluid flow is mostly along the axis of the rotating shaft. Similarly, in radial pumps the fluid flows mostly in the radial direction, and in the mixed types the fluid is designed to flow in both directions.

The momentum transfer between the impeller and the fluid is controlled by a series of fixed or adjustable *vanes*. Adjustable vanes are used to accommodate changes in the operating conditions. Additionally, the operating fluid may be confined within a *casing* or a *volute*, or it may be exposed, as in the case of propellers used for propulsion.

Due to the inherent complexity of both the geometry and the flow path, pumps are almost always studied experimentally. In practice, operation information is provided by the manufacturer. However, irrespective of the apparent complexity, the simple theory of rotating fluids introduced earlier can provide insight into the operation of pumps.

EXAMPLE 10.8

Problem Statement   *Axial pump/fan.* Consider the idealized axial fan as shown in Figure 10.11, where air enters the pump radially, at a rate of $6\ m^3/s$. The hub diameter is $0.45\ m$, and the tip diameter is $0.7\ m$. The blade angle at the inlet is $32°$, and at the exit it is $68°$. For a rotor rotating at $800\ rev/min$, first construct the velocity polygons at the inlet and exit and then find (a) the shaft torque; (b) the power required.

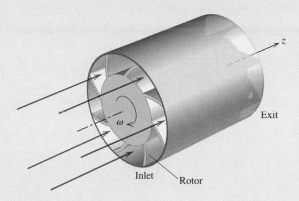

FIGURE 10.11   Schematic of Example 10.8.

Governing Equations   We select the entire fan as the control volume. Since mass crosses the boundary of the control volume, we can use the Euler turbomachine equation. In axial machines, though, the rotational velocity is a function of the radial distance at both the inlet and exit. Therefore, we will use the integral form of the equations.

$$T_{shaft} = \rho \int_e r V_{ft_e} (\mathbf{V} \cdot \mathbf{n})\, dA - \rho \int_i r V_{ft_i} (\mathbf{V} \cdot \mathbf{n})\, dA,$$

where,

$$V_{ft} = U_r - \frac{\dot{Q}}{A} \cot \beta.$$

Mass conservation yields,

$$\dot{Q} = \int_i \mathbf{V} \cdot \mathbf{n}\, dA = \int_e \mathbf{V} \cdot \mathbf{n}\, dA.$$

Assumptions   The foregoing equation were developed by assuming steady incompressible flow. We will also assume shockless entry and exit flow.

SOLUTION   For the given problem, and since conditions are uniform, mass conservation gives

$$\dot{Q} = V_{fn_i} A_i = V_{fn_e} A_e.$$

However, due to the axial operation, the flow area remains constant across the fan — that is,

$$A_i = A_e = \frac{\pi}{4}(D_e^2 - D_i^2) = \frac{\pi}{4}(0.7^2 \ m^2 - 0.45^2 \ m^2) = 0.2258 \ m^2.$$

The normal velocity is then

$$V_{fn_i} = V_{fn_e} = \frac{6 \ m^3/s}{0.2258 \ m^2} = 26.57 \ m/s.$$

The rotor is rotating at

$$\omega = \frac{800 \ rev/min \times 2\pi \ rad/rev}{60 \ s/min} = 83.78 \ rad/s.$$

Velocity Polygons

Inlet: Using the geometry of the inlet in Figure 10.12,

$$V_{ft_i} = U_r - V_{fn_i} \cot \beta_i = 83.78 \ r \ m/s - 26.57 \ m/s \times \cot 32 = (83.78 \ r - 42.52) \ m/s.$$

$$V_f = \sqrt{V_{ft_i}^2 + V_{fn_i}^2} = 83.78\sqrt{(r - 0.508)^2 + 0.101} \ m/s.$$

Exit: Using the geometry of the exit in Figure 10.12,

$$V_{ft_i} = U_r - V_{fn_i} \cot \beta_i = 83.78 \ r \ m/s - 26.57 \ m/s \times \cot 32 = (83.78 \ r - 42.52) \ m/s.$$

$$V_f = \sqrt{V_{ft_i}^2 + V_{fn_i}^2} = 83.78\sqrt{(r - 0.508)^2 + 0.101} \ m/s.$$

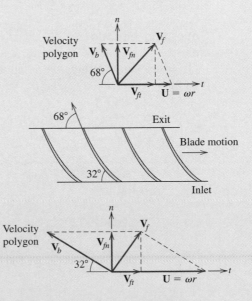

FIGURE 10.12   The geometry of the inlet and exit of the device.

Note that the absolute fluid velocity changes with the radial direction. This implies that guide vanes may be required to bring the flow close to shockless conditions. The shaft torque is found by integrating,

$$T_{shaft} = \rho \int_e r V_{ft_e} (\mathbf{V} \cdot \mathbf{n}) \, dA - \rho \int_i r V_{ft_i} (\mathbf{V} \cdot \mathbf{n}) \, dA,$$

which, by substitution gives

$$T_{shaft} = \rho \int_{r_h}^{r_t} r \left[\left(\omega r - \frac{\dot{Q}}{A} \cot \beta \right)_e - \left(\omega r - \frac{\dot{Q}}{A} \cot \beta \right)_i \right] V_{fn} \, 2\pi r \, dr,$$

$$T_{shaft} = 2\pi \rho V_{fn} \frac{\dot{Q}}{A} (\cot \beta_i - \cot \beta_e) \int_{r_h}^{r_t} r^2 \, dr,$$

$$T_{shaft} = \frac{2}{3} \pi \rho V_{fn}^2 (\cot \beta_i - \cot \beta_e)(R_t^3 - R_h^3).$$

In terms of numerical values,

$$T_{shaft} = \frac{2}{3}\pi \times 1.225 \ kg/m^3 \times 26.57^2 \ m^2/s^2$$

$$\times (\cot 32 - \cot 68)(0.35^3 \ m^3 - 0.225^3 \ m^3),$$

$$\mathbf{T}_{shaft} = 68.22 \ N \cdot m.$$

The power is

$$W = \omega T_{shaft} = 83.78 \ rad/s \times 68.22 \ N \cdot m = 5.72 \ kW.$$

EXAMPLE 10.9

Problem Statement   *Radial pump.* Water at a rate of $\dot{Q}$ enters a centrifugal pump in the radial direction as shown in Figure 10.13, and leaves at the impeller outlet with diameter $D = 20$ and blade height h. If the impeller rotates at a constant rate of ω, find the pressure head as a function of the vane angle at the exit.

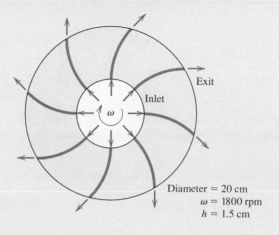

Exit

Inlet

ω

Diameter = 20 cm
ω = 1800 rpm
h = 1.5 cm

FIGURE 10.13   Schematic of Example 10.9.

Governing Equations   For the selected control volume shown in Figure 10.13, the conservation of energy for the pressure head H is

$$H_{gained/lost} = \frac{1}{g}[(U_r V_{f_t})_e - (U_r V_{f_t})_i].$$

Assumptions   The foregoing equations were developed by assuming steady incompressible flow with uniform and shockless conditions at both the inlet and exit.

SOLUTION   For purely radial entry at the inlet, $V_{ft_i} = 0$. Therefore,

$$H = \frac{1}{g} U_r V_{f_{t_e}} = \frac{1}{g} U_r \underbrace{\left(U_r - \frac{\dot{Q}}{A} \cot \beta_e \right)}_{mass\ conservation},$$

where A is the cross-sectional area $A = \pi D h$, and $U_r = 0.5\omega D$. Then,

$$H = \frac{U_r^2}{g} - \frac{U_r \dot{Q}}{g A} \cot \beta_e,$$

or

$$H = C_0 - C_1 \dot{Q},$$

where

$$C_0 \equiv \frac{U_r^2}{g}, \quad \text{and} \quad C_1 \equiv \frac{U_r}{g A} \cot \beta_e.$$

The ideal analysis predicts a linear variation of H as a function of the volumetric flow rate (Figure 10.14). The slope C_1 of the $H - \dot{Q}$ curve, however, depends on the vane angle β. When the vanes are *curved backwards* (i.e., $\beta < 90°$ the constant $C_1 > 0$), therefore, the head *decreases with* $\dot{Q}$ (Figure 10.14(a)). When the vanes are *forward curved* (i.e., $\beta > 90°$, $C_1 < 0$), the head *increases with* $\dot{Q}$ (Figure 10.14(b)). In the limit when $\beta = 90$, the head remains constant. This constant (i.e., C_0), is defined as the *shutoff head*. Pumps are usually designed with backward-curved blades. Forward-curved blades tend to develop unstable flow conditions.

In practice, because of frictional and other losses, the $H - \dot{Q}$ curve is not linear but is as shown in Figure 10.14c.

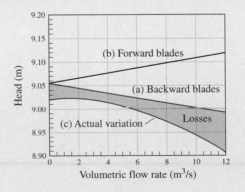

FIGURE 10.14   Head versus volumetric flow rate for a radial pump: (a) backward blades; (b) forward blades; (c) actual variation.

10.3.1 Dimensional Analysis

From the simple examples presented in the previous section, the important operational parameters are the head rise H, the shaft work $\dot{W}$, and the efficiency η. These variables depend on the effective diameter D, the roughness e, and other general geometric parameters denoted as L_i. Furthermore, pump performance depends on operational quantities (e.g., the volumetric flow rate $\dot{Q}$, the rotational speed ω) and on fluid properties (e.g., the viscosity μ and the density ρ).

Using dimensional analysis, we can reduce the variables into a family of nondimensional groups as

$$C_H = \frac{gH}{\omega^2 D^2} = f\left(\frac{L_i}{D}, \frac{e}{D}, \frac{\dot{Q}}{\omega D^3}, Re\right),$$

and

$$C_P = \frac{\dot{W}}{\rho \omega^3 D^5} = f\left(\frac{L_i}{D}, \frac{e}{D}, \frac{\dot{Q}}{\omega D^3}, Re\right),$$

where C_H is the *head rise coefficient*, C_P is the *power coefficient*, and $C_{\dot{Q}} = \dfrac{\dot{Q}}{\omega D^3}$ is the *flow coefficient*. By dividing the two coefficients and recognizing that $\omega D^3 \approx \dot{Q}$, we get the efficiency expression obtained previously. In functional form, the efficiency of pumps is given as

$$\eta = \frac{\rho g H \dot{Q}}{\dot{W}} = f\left(\frac{L_i}{D}, \frac{e}{D}, \frac{\dot{Q}}{\omega D^3}, Re\right).$$

However, since pumps operate at speeds high enough to make the flow turbulent and since the flow geometry is quite complicated, the dominant nondimensional parameter is the flow coefficient $C_{\dot{Q}}$. In practice, the performance of pumps as expressed by the coefficients C_H, C_P, and η is almost exclusively established experimentally. This information is also provided by the manufacturer. For example, Figure 10.15 shows the head rise and the efficiency of a centrifugal pump operating at 1800 *rpm* and with impeller diameter of 6 *cm*. Figure 10.16 shows the same information in terms of nondimensional parameters.

10.3.2 Scaling Properties, Specific Speed

Dimensional analysis in pump performance is quite useful and provides the means to qualitatively establish the relative importance of various flow parameters. For instance, using the similarity arguments implied by dimensional analysis, for two geometrically

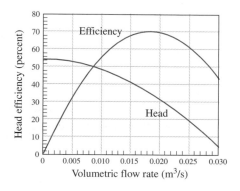

FIGURE 10.15 Pump performance in dimensional form.

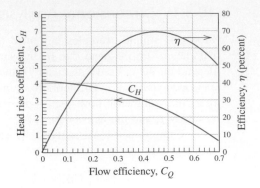

FIGURE 10.16   Pump performance in nondimensional form.

similar pumps with the same flow coefficient $\dfrac{\dot{Q}}{\omega D^3}$, we can write

$$\left(\frac{\dot{Q}}{\omega D^3}\right)_a = \left(\frac{\dot{Q}}{\omega D^3}\right)_b.$$

Therefore, according to the general functional relationship established by dimensional analysis, we also have

$$\left(\frac{gH}{\omega^2 D^2}\right)_a = \left(\frac{gH}{\omega^2 D^2}\right)_b, \quad \left(\frac{\dot{W}_{shaft}}{\rho\omega^3 D^5}\right)_a = \left(\frac{\dot{W}_{shaft}}{\rho\omega^3 D^5}\right)_b, \quad \text{and} \quad \eta_a = \eta_b.$$

These relationships are useful in predicting the performance of the same pump at different operating conditions. This is particularly important because pump information is usually given for a set of conditions that may not necessarily be those under normal operating conditions. Therefore, the same pump (i.e., for the same $C_{\dot{Q}}$ and D), operating at two different rotational speeds, the volumetric flow rate at the two conditions are related through

$$\frac{\dot{Q}_1}{\dot{Q}_2} = \frac{\omega_1}{\omega_2}.$$

The volumetric flow rate then is proportional to the rotational speed. For the same pump, dimensional analysis predicts that

$$\frac{H_1}{H_2} = \frac{\omega_1^2}{\omega_2^2}.$$

The pressure head rise is proportional to the square of the rotational speed. Finally, for the shaft work we predict that

$$\frac{\dot{W}_1}{\dot{W}_2} = \frac{\omega_1^3}{\omega_2^3},$$

indicating that the shaft work is proportional to the cube of the rotational speed.

Another interesting case is the comparison between two otherwise similar pumps, operating at the same impeller rotational speed but with different impeller diameters. In such a case, we can write

$$\frac{\dot{Q}_a}{\dot{Q}_b} = \frac{D_a^3}{D_b^3}, \quad \frac{H_a}{H_b} = \frac{D_a^2}{D_b^2} \quad \text{and} \quad \frac{\dot{W}_1}{\dot{W}_1} = \frac{D_a^5}{D_b^5}.$$

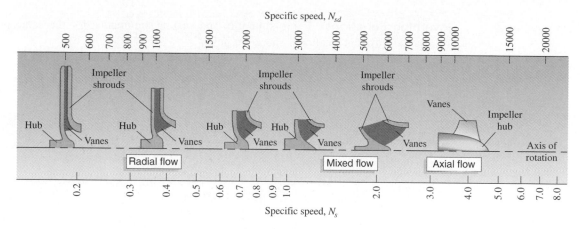

FIGURE 10.17 Classification of pumps according to N_{sd}, taken from ANSI/HI 1.1-1.2-2000, "American National Standard for Centrifugal Pumps for Nomenclature and Definitions." (Courtesy of Hydraulic Institute, 9 Sylvan Way, Parsippany, NJ 07054).

In practice, when the impeller diameter is changed, due to geometric constraints, it is usually not possible to achieve perfect geometric similarity. Therefore, the two pumps may not have the same efficiency. The empirical relation suggested by Moody and Zowski (1969) predicts with reasonable accuracy the variation in the efficiency as the impeller diameter is changed:

$$\frac{1 - \eta_a}{1 - \eta_b} = \left(\frac{D_a}{D_b} \right)^{\frac{1}{5}}.$$

By appropriately combining the flow and head coefficients, we can obtain another useful nondimensional parameter known as the *specific speed* N_s as

$$N_s = \frac{\omega \sqrt{\dot{Q}}}{(gH)^{\frac{3}{4}}}.$$

Unfortunately and despite the fact that the specific speed is nondimensional, it is usually expressed in units that are not consistent. Typically, $\dot{Q}$ is given in gallons per minute, ω in *rpm*, and the head in ft. When these units are used, the specific speed is designated by N_{sd} and it is known to be in U.S. customary units. The appropriate conversion between N_s and N_{sd} is

$$N_{sd} = 2733 N_s.$$

For centrifugal pumps, values for N_{sd} are usually in the range of $500 < N_{sd} < 4000$. Axial flow pumps are typically in the range $N_{sd} > 9000$. Since this parameter combines the effects of the two most critical variables $\dot{Q}$ and h, it is a useful design parameter. For this reason, the numerical value of N_{sd} provides the means to select among the various pump designs. Figure 10.17 shows the classification of pumps according to N_{sd}.

EXAMPLE 10.10

Problem Statement *Similarity laws.* A radial pump with average diameter of 6 *cm* is operating using air at the point of maximum efficiency with speed of 1800 *rpm*. The pump's performance is given in Figures 10.15 and 10.16. If the speed of the pump is changed to 1600 *rpm*, find the new volumetric flow rate and operating head. Find the actual power consumed by the pump.

Governing Equations Using similarity and scaling arguments, for the same pump operating at different rotational speed we have,

$$\frac{\dot{Q}_1}{\dot{Q}_2} = \frac{\omega_1}{\omega_2}, \quad \frac{H_1}{H_2} = \frac{\omega_1^2}{\omega_2^2}, \quad \eta_1 = \eta_2.$$

SOLUTION At the point of maximum efficiency, $\eta_1 \approx 70\%$, $\dot{Q} \approx 0.018\ m^3/s$, and $H_1 \approx 35\ m$. According to the similarity relations,

$$\frac{\dot{Q}_1}{\dot{Q}_2} = \frac{\omega_1}{\omega_2} \rightarrow \dot{Q}_2 = \dot{Q}_1 \frac{\omega_2}{\omega_1} = 0.018\ m^3/s \times \frac{1600\ rpm}{1800\ rpm} = 0.016\ m^3/s.$$

$$\frac{H_1}{H_2} = \frac{\omega_1^2}{\omega_2^2} \rightarrow H_2 = H_1 \frac{\omega_2^2}{\omega_1^2} = 35\ m \times \frac{1600^2\ rpm}{1800^2\ rpm} = 27.7\ m.$$

The power P_f corresponding to the fluid's head rise is

$$P_f = \rho g \dot{Q} H = 1.225\ kg/m^3 \times 9.81\ m/s^2 \times 0.016\ m^3/s \times 27.7\ m = 5.33\ W.$$

The actual power P_a used by the pump is established using the efficiency relation. For $\eta_1 = \eta_2 = 70\%$,

$$\eta_2 = \frac{P_f}{P_a} \rightarrow P_a = \frac{P_f}{\eta_2} = \frac{5.33\ W}{0.7} = 7.61\ W.$$

EXAMPLE 10.11

Problem Statement *Similarity laws.* Consider a geometrically similar pump with the one shown in Example 10.10, operating at the same rotational speed but with a characteristic diameter of 5 *cm*. Find the operating conditions and the actual power used by this pump at the point of maximum efficiency.

Governing Equations For two geometrically similar pumps, we have

$$\frac{\dot{Q}_a}{\dot{Q}_b} = \frac{D_a^3}{D_b^3}, \quad \frac{H_a}{H_b} = \frac{D_a^2}{D_b^2}.$$

However, because of geometric constraints, the efficiency is different, given by

$$\frac{1 - \eta_a}{1 - \eta_b} = \left(\frac{D_a}{D_b}\right)^{\frac{1}{5}}.$$

SOLUTION At the point of maximum efficiency, $\eta_1 \approx 70\%$, $\dot{Q} \approx 0.018\ m^3/s$, and $H_1 \approx 35\ m$. Using the similarity relations,

$$\frac{\dot{Q}_a}{\dot{Q}_b} = \frac{D_a^3}{D_b^3} \rightarrow \dot{Q}_b = \dot{Q}_1 \frac{D_b^3}{D_a^3} = 0.018\ m^3/s \times \frac{5^3}{6^3} = 0.0104\ m^3/s,$$

$$\frac{H_a}{H_b} = \frac{D_a^2}{D_b^2} \rightarrow H_b = H_a \frac{D_b^2}{D_a^2} = 35\ m \times \frac{5^2}{6^2} = 24.30\ m.$$

The power P_f, corresponding to the fluid's head rise, is

$$P_f = \rho g \dot{Q} H = 1.225 \ kg/m^3 \times 9.81 \ m/s^2 \times 0.0104 \ m^3/s \times 24.30 \ m = 3.04 \ W.$$

The actual power P_a used by the pump is established using the efficiency relation. For $\eta_a = 70\%$, the efficiency of the second pump is

$$\frac{1 - \eta_a}{1 - \eta_b} = \left(\frac{D_a}{D_b}\right)^{\frac{1}{5}} \rightarrow \eta_b = 1 - (1 - \eta_a)\left(\frac{D_b}{D_a}\right)^{\frac{1}{5}}$$

$$= 1 - (1 - 0.7)\left(\frac{5}{6}\right)^{\frac{1}{5}} = 0.71.$$

Therefore,

$$\eta_2 = \frac{P_f}{P_a} \rightarrow P_a = \frac{3.04 \ W}{\eta_2} = \frac{3.04 \ W}{0.71} = 4.28 \ W.$$

10.3.3 Pump Selection and Performance Considerations

This subsection focuses on the various factors that affect pump selection and performance in simple piping systems, using a typical problem.

EXAMPLE 10.12

Problem Statement *Simple piping system.* Find the operating conditions and the volumetric flow rate $\dot{Q}$ through the system shown in Figure 10.18 using the pump with the performance curves (Head vs. $\dot{Q}$ and Efficiency vs. $\dot{Q}$) shown in Figure 10.19.

Governing Equations The energy equation applied between the free surface and the exit of the pipe is

$$\frac{\partial E}{\partial t} + \dot{m}\left(\frac{V^2}{2} + gz + \frac{P}{\rho}\right)_1 - \dot{m}\left(\frac{V^2}{2} + gz + \frac{P}{\rho}\right)_2 + \dot{W}_{in} = \dot{m}\left(h_{major} + \sum h_{minor}\right),$$

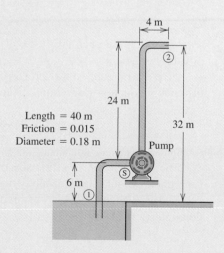

FIGURE 10.18 Schematic of the flow system in Example 10.12.

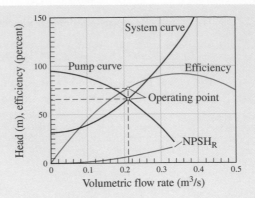

FIGURE 10.19   Performance curves for the pump use in Example 10.12.

where the major losses are calculated as

$$h_{major} = f \frac{L}{D} \frac{V^2}{2}.$$

The $\sum h_{minor}$ represents the sum of all minor losses due to the geometry of the system.

Assumptions   Assume steady flow of water at $20°C$, with constant properties.

SOLUTION   Using the conditions of the problem, with the reference elevation at point 1: $V_1 \approx 0$, $P_1 = P_2 = P_a$, $z_2 = 32 \ m$. For steady flow $\frac{\partial E}{\partial t} = 0$, the energy equation reduces to

$$\dot{W} = \dot{m} \left(\frac{V^2}{2} + gz_2 + h_{major} + \sum h_{minor} \right).$$

In this problem we assume that minor losses occur at the inlet and at the two 90° bends. For the inlet, the minor losses are expressed as

$$h_{minor} = K \frac{V^2}{2},$$

with $K = 0.5$. For each 90° bend, the losses are expressed using an equivalent length of pipe of $L_e/D = 30$.

By rearranging, and expressing the velocity in terms of the volumetric flow rate, the energy equation now simplifies to

$$\dot{W} = \rho \dot{Q} \left[gz_2 + \frac{\dot{Q}^2}{2A^2} \left(1 + f \frac{L}{D} + 2f \frac{L_e}{D} + K \right) \right].$$

The friction factor f is a function of the Re and the roughness ratio provided by the Colebrook formula (or the Moody diagram). Here, f is assumed to be constant, equal to $f = 0.015$. The head H_{pump} then required by the pump as a function of the volumetric flow rate $\dot{Q}$ is

$$H_{pump} = \frac{\dot{W}}{\rho g \dot{Q}} = z_2 + \frac{\dot{Q}^2}{2gA^2} \left(1 + f \frac{L}{D} + 2f \frac{L_e}{D} + K \right).$$

This represents the system requirement or the *system curve*. In terms of numerical values, the curve is

$$H_{pump} = 32 \ m + \frac{16\dot{Q}^2}{2 \times 9.81 \ m/s^2 \times \pi^2 \times 0.18^4 \ m^2}$$

$$\times \left(1 + 0.015 \frac{40 \ m}{0.18 \ m} + 2 \times 0.015 \times \frac{30 \ m}{0.18 \ m} + 0.5 \right).$$

$$H_{pump} = 32 \ m + 773.984\dot{Q}^2 \ m.$$

The operating point is the intersection of the system curve with the power and efficiency curves of the pump (Figure 10.19). From the figure, then, $\dot{Q} \approx 0.21 \ m^3$, $H_{pump} \approx 66 \ m$, and the efficiency of the pump is about $\approx 77\%$. The velocity in the pipe is $V = \dot{Q}/A = 0.21 \ m^3 \times 4/(\pi \times 0.18^2 \ m^2) = 8.25 \ m/s$. The power provided by the pump is

$$\dot{W} = \rho g \dot{Q} H_{pump} = \frac{998.2 \ kg/m^3 \times 9.81 \ m/s^2}{1000 \ W/kW} \times 0.21 \ m^3/s \times 66 \ m = 135.72 \ kW.$$

Cavitation

A critical issue in pump operation is that the pressure in the system may become low enough to cause the formation of bubbles, (i.e., *cavitation*). Using the energy equation, the local head is expressed as

$$\frac{P}{\rho g} + \frac{V^2}{2g} + z = H.$$

Therefore, when V and z are relatively large, the local pressure P may indeed drop below the vapor pressure P_v of the liquid. When this happens, vapor bubbles form that can damage the system and the pump. For instance, bubbles formed at the suction side of the pump can be drawn into the pump, where they eventually collapse. Upon collapse, local pressure spikes develop that damage the surface of the blades. The pressure spikes may also develop large pressure transients in the system that can destroy the piping system. Therefore, the system must be designed so that cavitation is avoided.

The potential for cavitation depends on the system parameters. However, since pumps receive liquids at relatively low pressure, as the fluid accelerates at the impeller, pumps are prone to cavitation. To avoid cavitation, we must ensure that the head at the suction is large enough to account for this pressure drop. The difference between the total head at the suction and the vapor pressure P_v must be a positive number—that is,

$$\underbrace{\underbrace{\frac{P_s}{\rho g} + \frac{V_s^2}{2g}}_{\text{Suction total Head}} - \frac{P_v}{\rho g}}_{NPSH} = H > 0.$$

H is known as the *net positive suction head*, $NPSH$. Because of the complexity of the flow, the critical $NPSH$ is different from the ideal $NPSH$ and is determined experimentally. This information is also provided by the manufacturer (Figure 10.19). This value of $NPSH$ is also known as the required $NPSH_R$ because it must be exceeded in order to avoid cavitation.

Returning to the flow in Example 10.11, the energy equation between the inlet (point 1) and an arbitrary location (point 2) in the flow gives

$$\frac{V^2}{\rho g} + \frac{P_2}{\rho g} + z_2 = \frac{P_a}{\rho g} + H_{pump} - H_{Losses_{1 \to 2}},$$

where $H_{Losses_{1 \to 2}}$ are the total losses occurring between points 1 and 2. For this problem, due to the applied boundary conditions ($P_{exit} = P_a$), the local pressure never drops below P_a. However, if the height of the delivery section and/or the losses in the system are increased, cavitation will occur.

In this problem, cavitation is initiated at the exit of the pipe when $P_2 = P_v$. For water at $20°C$, the vapor pressure is $P_v = 2.339 \ kPa$. The losses between the inlet and exit are

$$H_{Losses_{1 \to 2}} = \frac{V^2}{2g} \left(f\frac{L}{D} + 2f\frac{L_e}{D} + K \right)$$

$$= \frac{8.25^2 \ m^2/s^2}{2 \times 9.81 \ m/s^2} \times \left[0.015\frac{40 \ m}{0.18 \ m} + 2 \times 0.015\frac{30 \ m}{0.18 \ m} + 0.5 \right] = 30.64 \ m.$$

The critical elevation z_{cr}, then, is

$$z_{cr} = \frac{P_a - P_v}{\rho g} + H_{pump} - H_{Losses_{1 \to 2}} - \frac{V^2}{2g}.$$

In terms of numerical values,

$$z_{cr} = \frac{101,300 \ Pa - 2339 \ Pa}{998.2 \ kg/m^3 \times 9.81 \ m/s^2} + 66 \ m - 30.64 \ m - \frac{8.25^2 \ m^2/s^2}{2 \times 9.81 \ m/s^2} = 42 \ m.$$

Therefore, if the elevation of the exit section is raised to 42 m, bubbles will start forming at the exit of the pipe.

Now the system must be checked for the possibility of cavitation inside the pump. To avoid this, the available $NPSH_A$ must be greater than the required $NPSH_R$ for the particular pump (i.e., $NPSH_A > NPSH_R$). The $NPSH$ at the suction of the pump is determined by using the energy equation between the inlet and the suction side of the pump,

$$\underbrace{\frac{V_s^2}{\rho g} + \frac{P_s - P_v}{\rho g}}_{NPSH_A} = \frac{P_a - P_v}{\rho g} - z_s + H_{pump} - H_{Losses_{1 \to s}}.$$

Here $H_{Losses_{1 \to s}}$ represents the losses from point 1 to point s at the suction of the pump.

$$H_{Losses_{1 \to s}} = \frac{V^2}{2g} \left(f\frac{L_s}{D} + f\frac{L_e}{D} + K \right),$$

where L_s is the length of the pipe in this section. In terms of numerical values,

$$H_{Losses_{1 \to s}} = \frac{8.25^2 \ m^2/s^2}{2 \times 9.81 \ m/s^2} \times \left[0.015\frac{12 \ m}{0.18 \ m} + 0.015\frac{30 \ m}{0.18 \ m} + 0.5 \right] = 13.88 \ m.$$

Therefore,

$$NPSH_A = \frac{101300 \ Pa - 2339 \ Pa}{998.2 \ kg/m^3 \times 9.81 \ m/s^2} - 6 \ m + 66 \ m - 13.88 \ m = 56.22 \ m.$$

The available head is well above the required head to avoid cavitation in the pump (Figure 10.19). Hence, cavitation in the pump will not develop.

10.4   Energy-Producing Turbomachines

Turbines are turbomachines that produce energy by utilizing an available source of energy such as the total head of a river or the wind. Their operation, then, is essentially the reverse of that of pumps. One of the oldest known "power-producing" machines is the sail in sailing boats, by which energy extracted from the wind is changed into linear motion.

When the working fluid is an incompressible liquid (most often water), the turbines are known as *hydraulic turbines*. When the working fluid is compressible, such as steam or gas, the turbines are known, respectively, as *steam turbines* and *gas turbines*. Power is obtained either by using the impulse of a single jet impinging upon a series of *vanes* (i.e., the turbine is of the *impulse type*), or it is obtained by momentum exchanged by redirecting the flow through a series of fixed and rotating blades. In the later case, the turbines are known as *reaction turbines*. In turbines, the shaft is known as a *runner*.

Figure 10.20 shows a schematic of three basic types of turbines: (a) an impulse turbine of the *Pelton wheel* type, (b) a reaction turbine of the *Francis type*, and (c) a reaction turbine of the *Kaplan type*. In the Francis type, the flow is mostly radial, and in the Kaplan type, the flow is turned through a series of vanes in the axial direction before it enters the rotor blades, very much like the simple turbomachine discussed in Section 10.2. Impulse turbines are usually used in applications with high available head but relatively low discharge. Reaction turbines are used with large fluid volume but relatively low head. Hydraulic turbines are very efficient machines with reported efficiencies greater than 90%.

10.4.1   Impulse Turbines

The operation of impulse type turbines is quite simple: as seen in the schematic of Figure 10.21, the turbine is set in motion resulting from momentum exchange (hence the name impulse) between a high-speed liquid jet and a number of buckets mounted on the turbine shaft. Despite the complexity of the flow, a simple analysis based on the

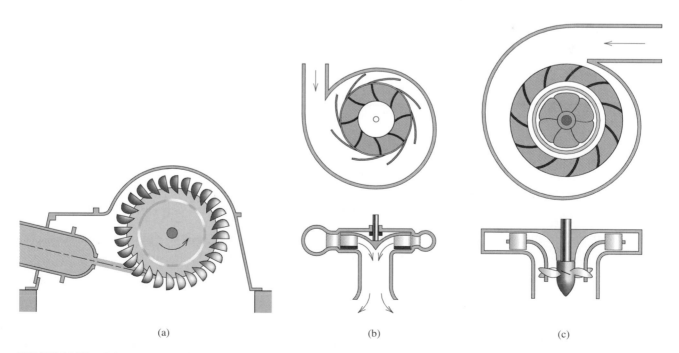

(a)   (b)   (c)

FIGURE 10.20   Schematic of (a) Pelton; (b) Francis; (c) Kaplan turbines.

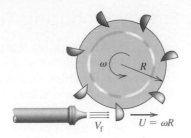

FIGURE 10.21   Schematic of an impulse type turbine (Pelton wheel).

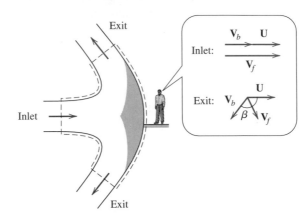

FIGURE 10.22   Velocity polygons relative to an observer rotating with the blade.

theory developed in Section 10.2.1 is quite revealing. Since the linear velocity of the blade changes in the radial direction, the analysis is based on the mean radius R_m of the blade. Selecting the bucket as a control volume, the velocity components relative to the bucket are shown in Figure 10.22.

Following the geometry shown in Figure 10.21, we have the following:

<u>Inlet</u>

$$\mathbf{U}_i = U\,\mathbf{i} = \omega R_m\,\mathbf{i}$$

$$\mathbf{V}_{f_i} = V_{f_i}\,\mathbf{i}$$

$$\mathbf{V}_{b_i} = \mathbf{V}_{f_i} - \mathbf{U}_i = (V_{f_i} - \omega R_m)\mathbf{i}$$

<u>Exit</u>

$$\mathbf{U}_e = U\,\mathbf{i} = \omega R_m\,\mathbf{i}$$

$$\mathbf{V}_{b_e} = V_{b_e}(\cos\beta\,\mathbf{i} + \sin\beta\,\mathbf{j})$$

$$\mathbf{V}_{f_e} = \mathbf{V}_{b_e} + \mathbf{U}_e = (U + V_{b_e}\cos\beta)\mathbf{i} + V_{b_e}\sin\beta\,\mathbf{j}$$

Applying conservation of mass to the selected control volume we have,

$$V_{b_e} = V_{b_i} = V_b = (V_{f_i} - U)\mathbf{i}.$$

Conservation of angular momentum applied to the same control volume yields the Euler's turbomachine equation (Equation (10.6)), which expresses the generated torque

$\mathbf{T}_{shaft}$ as a function of the inlet and exit tangential velocities (i.e., direction of rotation),

$$\mathbf{T}_{shaft} = \dot{m}[(rV_{f_t})_e - (rV_{f_t})_i].$$

Here, we have the following:

$$V_{f_{t_i}} = V_b + U,$$

and

$$V_{f_{t_e}} = U + V_b \cos \beta.$$

The shaft torque is, then,

$$\mathbf{T}_{shaft} = \dot{m} R_m [U + V_b \cos \beta - V_b - U],$$

or,

$$\mathbf{T}_{shaft} = \dot{m} R_m V_b (\cos \beta - 1).$$

Substituting $V_b = V_{f_i} - U$ and simplifying, we get

$$\mathbf{T}_{shaft} = \dot{m} R_m (1 - \cos \beta)(U - V_{f_i}).$$

The shaft torque is then a function of the deflection angle β and the difference between U and V_{f_i}. To generate torque (hence power) $V_{f_i} > U$. In this case, $\mathbf{T}_{shaft} < 0$, indicating that energy is extracted from the fluid. For maximum torque, $\beta = 180°$. However, from a practical point of view this is not possible. In practice $\beta \approx 160°$, which results in a small fractional torque reduction from the theoretical maximum.

The power generated is obtained by

$$\dot{W} = T_{shaft}\omega = \dot{m}U(1 - \cos \beta)(U - V_{f_i}). \qquad (10.9)$$

Figure 10.23 is a plot of the normalized theoretical torque and power generated as a function of the linear velocity of the buckets, at a fixed angle $\beta = 160°$. The figure shows that while the torque decreases linearly with the velocity, the generated power

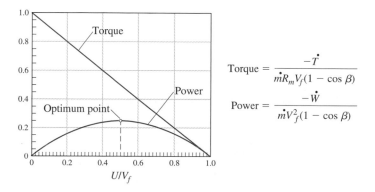

$$\text{Torque} = \frac{-\dot{T}}{\dot{m}R_m V_f(1 - \cos \beta)}$$

$$\text{Power} = \frac{-\dot{W}}{\dot{m}V_f^2(1 - \cos \beta)}$$

FIGURE 10.23   The theoretical prediction of the torque and power generated by a Pelton wheel $(\beta = 160°)$.

varies parabolically, with a definite maximum point. This maximum point occurs when $d\dot{W}/dU = 0$. Using Equation (10.9),

$$\frac{d\dot{W}}{dU} = 2U - V_{f_i} = 0.$$

Therefore, the wheel generates the maximum power when the rotational speed ω is such that

$$U = \frac{V_{f_i}}{2} \quad \text{or,} \quad \omega = \frac{V_{f_i}}{2R_m}.$$

EXAMPLE 10.13

Problem Statement *Impulse turbine.* Consider the schematic in Figure 10.24, where water from a reservoir is used to drive a Pelton wheel turbine. For the conditions shown in the figure, calculate the torque and power produced by the wheel at optimum conditions.

Governing Equations The torque generated by the wheel is given as

$$\mathbf{T}_{shaft} = \dot{m}R_m(1 - \cos\beta)(U - V_{f_i}),$$

and the power as $\dot{W} = T_{shaft}\omega$.

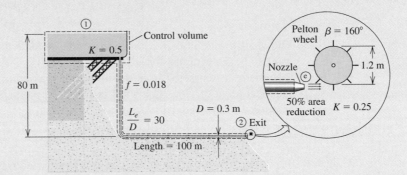

FIGURE 10.24 Schematic of Example 10.13.

The flow is delivered from the reservoir using the arrangement shown in the figure. The energy equation applied between the free surface and the exit of the pipe is

$$\frac{\partial E}{\partial t} + \dot{m}\left(\frac{V^2}{2} + gz + \frac{P}{\rho}\right)_1 - \dot{m}\left(\frac{V^2}{2} + gz + \frac{P}{\rho}\right)_2 + \dot{W}_{in} = \dot{m}\left(h_{major} + \sum h_{minor}\right),$$

where the major losses are calculated as

$$h_{major} = f\frac{L}{D}\frac{V^2}{2}.$$

The $\sum h_{minor}$ represents the sum of all minor losses due to the geometry of the system.

Assumptions Assume steady flow of water at $20°C$, with constant properties.

SOLUTION Before proceeding with the analysis of the Pelton wheel, we must determine the velocity of the jet impinging upon the buckets of the wheel. For this, we must use the energy equation. For this problem, $\dot{W} = 0$, $V_1 \approx 0$, $P_1 = P_2 = P_a$, $z_1 = 80\ m$, and

$z_2 = 0$. For steady flow $\dfrac{\partial E}{\partial t} = 0$, the energy equation reduces to

$$ gz_1 - \frac{V_2^2}{2} = \underbrace{h_{major} + \sum h_{minor}}_{Total\ Losses}, $$

where V_2 is the velocity in the pipe preceding the nozzle. The losses now include both the major and minor losses. The major losses are defined again as

$$ h_{major} = f\frac{L}{D}\frac{V_2^2}{2}. $$

In this problem, minor losses occur at the inlet to the straight section, the $90°$ turn, and the nozzle. For both the inlet and the nozzle, the minor losses are expressed as

$$ h_{minor} = K\frac{V_2^2}{2}, $$

with $K_{inlet} = 0.25$ and $K_{nozzle} = 0.5$. The loss at the bend is expressed by an equivalent length of the pipe as $L_e/D = 30$.

By rearranging the terms, the energy equation now simplifies to

$$ z_1 = \frac{V_2^2}{2g}\left[1 + f\frac{L}{D} + f\frac{L_e}{D} + K_{inlet} + K_{nozzle}\right]. $$

For the given conditions,

$$ 80\ m = \frac{V_2^2}{2 \times 9.81\ m/s^2}\left[1 + 0.018\frac{100\ m}{0.3\ m} + 0.018 \times 30 + 0.25 + 0.5\right], $$

which gives $V_2 = 13.76\ m/s$. Because of the area reduction, the flow increases in the nozzle to V_e. Using conservation of mass for the nozzle,

$$ V_e = V_2 \times \frac{A_2}{A_e} = 2V_2 = 27.52\ m/s. $$

At optimum conditions, the linear velocity of the wheel is 50% of that of the jet. Therefore,

$$ U = 0.5\ V_e = 0.5 \times 27.52\ m/s = 13.76\ m/s. $$

For this linear velocity, the rotational speed ω is

$$ \omega = \frac{U}{R} = \frac{13.76\ m/s}{0.6\ m} = 22.93\ rad/s. $$

The mass flow rate is $\dot{m} = \rho V_2 A_{pipe} = 998.2\ kg/m^3 \times 13.76\ m/s \times \pi \times 0.6^2\ m^2 = 15534\ kg/s$. At these conditions, the torque is calculated using

$$ \mathbf{T}_{shaft} = \dot{m}R_m(1 - \cos\beta)(U - V_e) = 15534\ kg/s $$

$$ \times 0.6\ m(1 - \cos 160)(13.76 - 27.52) = -248.8\ KN \cdot m. $$

The corresponding power, then, is

$$ \dot{W} = T\omega = -22.93\ rad/s \times 248.8\ KN \cdot m = -5.7\ MW. $$

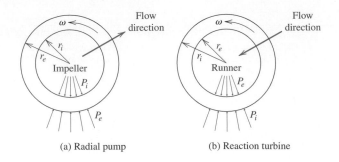

FIGURE 10.25   Comparison between pump and reaction turbine operation.

10.4.2   Reaction Turbines

The operation of reaction turbines is essentially the reverse of pumps. For instance, as shown in Figure 10.25, the operation of a Francis-type turbine is essentially the opposite of that of a centrifugal pump. Although Francis-type turbines are primarily radial flow, various existing designs are of the mixed-flow type. Kaplan turbines are mostly axial-flow turbines. Since reaction turbines are totally enclosed, unlike impulse turbines, reaction turbines are associated with both pressure and kinetic energy drop.

EXAMPLE 10.14

Problem Statement   *Reaction turbine*. A reaction turbine has inner and outer radii of 1.0 *m* and 1.5 *m*, respectively. The runner rotates at a rate of 66 *rpm*, and the inlet and exit angles are 56° and 148° respectively. If the flow rate is 10 m^3 and the blade height is 0.35 *m*, find (a) the generated torque; (b) the corresponding power; (c) the pressure drop in the runner.

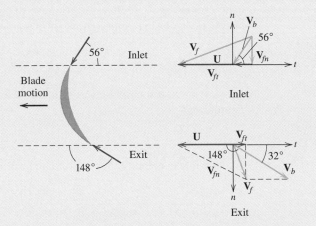

FIGURE 10.26   Schematic of Example 10.14.

Governing Equations   The power input/output $\dot{W}_{turb}$ and the shaft torque $\mathbf{T}_{turb}$, are given by

$$\dot{W}_{turb} = \omega T_{turb},$$

and

$$\mathbf{T}_{turb} = \dot{m}[(rV_{f_t})_e - (rV_{f_t})_i].$$

Mass conservation gives

$$\dot{Q} = V_{n_i} A_i = V_{n_e} A_e,$$

where subscripts i and e refer to inlet and exit conditions. The pressure drop is obtained by applying the energy equation between the inlet and exit,

$$\left(\frac{P}{\rho} + \frac{V^2}{2} + gz\right)_2 = \left(\frac{P}{\rho} + \frac{V^2}{2} + gz\right)_1 - \left|\frac{\dot{W}_{turb}}{\dot{m}}\right| - gH_{losses}.$$

The absolute value is to ensure the proper sign of the energy extracted by the turbine.

Assumptions The flow is assumed to be steady and incompressible with uniform conditions at the inlet and exit. We also assume shockless conditions at the inlet and exit.

SOLUTION The runner rotates at an angular velocity of

$$\omega = 2\pi \ rad/rev \times \frac{66 \ rpm}{60 \ s/min} = 6.91 \ rad/s.$$

Using the geometry of the velocity polygon,

Inlet

$$U_i = \omega r_i = 6.91 \ rad/s \times 1.5 \ m = 10.37 \ m/s.$$

$$V_{f_{n_i}} = \frac{\dot{Q}}{A_i} = \frac{\dot{Q}}{2\pi r_i h} = \frac{10 \ m^3}{2 \times \pi \times 1.5 \ m \times 0.35 \ m} = 3.03 \ m/s.$$

$$V_{b_i} \sin\beta_i = V_{f_n} \Rightarrow V_{b_i} = \frac{V_{f_n}}{\sin\beta_i} = \frac{3.03 \ m/s}{\sin 56} = 3.65 \ m/s,$$

$$V_{f_{t_i}} = U_i + V_{b_i} \cos\beta_i = 10.37 \ m/s + 3.65 \ m/s \times \cos 56 = 12.41 \ m/s.$$

Exit

$$U_e = \omega r_e = 6.91 \ rad/s \times 1.0 \ m = 6.91 \ m/s.$$

$$V_{f_{n_e}} = \frac{\dot{Q}}{A} = \frac{\dot{Q}}{2\pi r_e h} = \frac{10 \ m^3}{2 \times \pi \times 1.0 \ m \times 0.35 \ m} = 4.55 \ m/s.$$

$$V_{b_e} \sin\beta_e = V_{f_n} \Rightarrow V_{b_e} = \frac{V_{f_{n_e}}}{\sin\beta_e} = \frac{4.55 \ m/s}{\sin(180 - 148)} = 8.59 \ m/s,$$

$$V_{f_{t_e}} = U_e + V_{b_e} \cos\beta_e = 6.91 \ m/s + 8.59 \ m/s \times \cos 148 = -0.37 \ m/s.$$

The generated torque then is

$$
\begin{aligned}
\mathbf{T}_{turb} &= \rho \dot{Q}[(r V_{f_t})_e - (r V_{f_t})_i] \\
&= 998.2 \ kg/m^3 \times 10 \ m^3 [(-1.0 \ m \times 0.37 \ m/s)_e - (1.5 \ m \times 12.41 \ m/s)_i] \\
&= -189.5 \ kN \cdot m.
\end{aligned}
$$

The generated power is

$$\dot{W}_{turb} = \omega |T_{turb}| = 6.91 \ rad/s \times 189.5 \ kN \cdot m = 1.31 \ MW.$$

If we neglect head losses in the runner ($H_{losses} = 0$), and changes in the elevation $z_1 \approx z_2$, the energy equation gives

$$\Delta P = P_i - P_e = \rho \left[\frac{V_{f_e}^2 - V_{f_i}^2}{2} + \left| \frac{\dot{W}_{turb}}{\dot{m}} \right| \right].$$

According to the velocity polygons,

$$V_i = \sqrt{V_{f_{t_i}}^2 + V_{f_{n_i}}^2} = \sqrt{(3.03 \ m/s)^2 + (12.41 \ m/s)^2} = 12.77 \ m/s,$$

and

$$V_e = \sqrt{V_{f_{t_e}}^2 + V_{f_{n_e}}^2} = \sqrt{(0.37 \ m/s)^2 + (4.55 \ m/s)^2} = 4.57 \ m/s.$$

The pressure drop is, then,

$$\Delta P = P_i - P_e = 998.2 \ kg/m^3$$

$$\times \left[\frac{(4.57 \ m/s)^2 - (12.77 \ m/s)^2}{2} + \frac{1.31 \times 10^6 \ W}{998.2 \ kg/m^3 \times 10 \ m^3/s} \right],$$

$$\Delta P = 60 \ kPa.$$

10.5   Simple Propeller and Windmill Theory

Propellers and *windmills* are typical applications of using rotating blades attached to a shaft in order to impart energy to or extract energy from fluids. Propellers are energy-absorbing devices in which energy is used to drive the blades in order to transfer kinetic energy (and hence motion) to the surrounding fluid. In fans, this motion is used for air circulation and, hence, cooling or heating. In airplanes and ships, the fluid motion results in a reaction force responsible for propulsion. Windmills are energy-producing devices that have been used for centuries to extract energy from the wind.

Figure 10.27a shows a simple schematic of the operation of a propeller: power $\dot{W}_p$ is provided to sustain constant rotation of the blades. Because of the action of the rotating

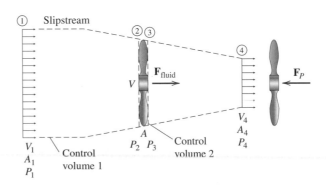

FIGURE 10.27   Schematic of propeller operation: (a) surrounding fluid motion; (b) reaction force on the propeller.

propeller, the surrounding fluid accelerates relative to the propeller and gets drawn into the blades. The fluid continues to accelerate downstream the propeller until the pressure of the stream is in equilibrium with the surrounding pressure. The fluid wake caused by the propeller (both upstream and downstream) is known as the "race" or the "slipstream." The net effect of the rotation then is a force F in the axial direction that accelerates the surrounding fluid. By reaction, then, an equal and opposite force on the propeller provides the thrust for motion (Figure 10.27b).

In the following analysis, we assume that the propeller is attached to a plane moving with velocity V_1 into stagnant air. The flow, then, relative to the plane corresponds to the schematic shown in Figure 10.27. From the description of the problem, $P_1 = P_4 = P_a$. The area across the propeller is assumed to be constant $A_2 = A_3 = A$. By assuming steady flow with uniform conditions across the boundary, the conservation laws for this problem are given as follows.

Mass Conservation: For an open system with steady and uniform flow conditions and constant density, we have

$$V_1 A_1 = V_2 A_2 = V_3 A_3 = V_4 A_4 = V A,$$

where V and A are the velocity and flow area at the propeller. Since $A_1 = A_2 = A$, then $V_2 = V_3 = V$.

Conservation of Linear Momentum: For uniform conditions at the inlet (i) and exit (e) and the same assumptions as before, conservation of linear momentum gives

$$\dot{m}(V_e - V_i) = \sum F = \sum F_{pressure} + F_{reaction}.$$

Conservation of Energy: Using the same assumptions as before and by neglecting losses, conservation of energy between point a and any point b in the downstream direction is given as

$$\dot{m}\left[\left(\frac{P}{\rho} + \frac{V^2}{2} + gz\right)_b - \left(\frac{P}{\rho} + \frac{V^2}{2} + gz\right)_a\right] = \dot{W}_p.$$

By selecting the entire slipstream area as the control volume (C.V. 1), and using $P_1 = P_4 = P_a$, the conservation of linear momentum is

$$\dot{m}(V_4 - V_1) = F_{fluid}.$$

Now, by selecting a control volume that includes only the propeller (C.V. 2) and since $V_2 = V_3 = V$,

$$-F_{prop} + P_2 A - P_3 A = 0.$$

Using $F_{fluid} = -F_{prop}$, and equating the expressions for the force

$$\dot{m}(V_4 - V_1) = A(P_3 - P_2).$$

But, since $\dot{m} = \rho V A$,

$$V(V_4 - V_1) = \frac{P_3 - P_2}{\rho}.$$

Applying the energy equation between point 1 and 2 for $z_1 = z_2$,

$$\frac{P_2}{\rho} = \frac{P_1}{\rho} + \frac{V_1^2 - V^2}{2}.$$

The energy equation for points 3 and 4 gives,

$$\frac{P_3}{\rho} = \frac{P_4}{\rho} + \frac{V_4^2 - V^2}{2}.$$

Substituting the preceding two results into the momentum equation,

$$V(V_4 - V_1) = \frac{V_4^2 - V_1^2}{2}$$

and by simplifying, we get

$$V = \frac{V_4 + V_1}{2}.$$

Therefore, the fluid velocity at the propeller is the arithmetic average of the velocities in the slipstream away from the propeller. The useful power $\dot{W}_u$, imparted to the fluid (or used for propulsion) is

$$\dot{W}_u = FV_1 = \dot{m}V_1(V_4 - V_1).$$

The net energy provided to sustain motion $\dot{W}_p$ is calculated by applying the energy equation between points 1 and 4. Using $z_1 = z_4$, and $P_1 = P_4$, we get

$$\dot{W}_p = \dot{m}\left(\frac{V_4^2 - V_1^2}{2}\right).$$

The energy loss then is the difference between $\dot{W}_p$ and $\dot{W}_u$ — that is,

$$Loss = \dot{W}_p - \dot{W}_u = \dot{m}\left[\frac{V_4^2 - V_1^2}{2} - V_1(V_4 - V_1)\right] = \dot{m}\frac{(V_4 - V_1)^2}{2},$$

which is the remaining kinetic energy in the slipstream. This "ideal loss" represents the inability of the propeller, because of operational reasons, to convert under ideal conditions all the available power into useful power. Of course, the ideal efficiency is never reached because of losses at the tip and the hub and frictional losses along the blades. Additionally, the flow is never steady but pulsating, and the effective flow area is reduced by the hub. A fraction of the available energy in swirl flow also remains unutilized in the slipstream.

The theoretical efficiency of the propeller is defined as

$$\eta_t = \frac{\dot{W}_u}{\dot{W}_P} = \frac{2V_1(V_4 - V_1)}{V_4^2 - V_1^2} = \frac{2V_1}{V_4 + V_1} = \frac{V_1}{V}.$$

Defining $\Delta V \equiv V_4 - V_1$, the theoretical efficiency can be rewritten as

$$\eta_t = \frac{1}{1 + \Delta V/2V_1}.$$

Therefore, the efficiency is maximum when ΔV is small. Plane propellers operate at about 85% efficiency. However, because of geometry concerns, ship propellers are about 60% efficient.

In view of the increased awareness of the impact of technology on the environment, alternative, environmentally friendly sources of power such as solar power and wind power are being developed. Therefore, although the concept of a windmill has been known for centuries, recently there is increased awareness of the need to improve their performance. As shown in Figure 10.28, the operation of windmills is the reverse of that of propellers. Because of extraction of energy, the fluid velocity decreases in the downstream direction and the effective slipstream area increases.

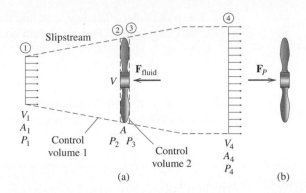

FIGURE 10.28 Schematic of windmill operation (a) surrounding fluid motion; (b) reaction force on the propeller.

The application of mass, momentum, and energy conservation equations to the entire slipstream area yields the same results obtained for propellers,

$$V = 0.5(V_4 + V_1) \quad \text{and} \quad F = \dot{m}(V_1 - V_4).$$

The useful power extracted from the wind $\dot{W}_u$ is determined by applying the energy equation between 1 and 4 with $z_1 = z_2$ and $P_1 = P_4$, as

$$\dot{W}_u = \rho V A \left(\frac{V_1^2 - V_4^2}{2} \right).$$

The efficiency of a windmill is typically defined as the ratio of the extracted power to the power available at the propeller area A at wind velocity of V_1 (i.e., $\rho A V_1 (V_1^2/2)$). Therefore,

$$\eta = \frac{V(V_1^2 - V_4^2)}{V_1^3} = \frac{(V_1 + V_4)(V_1^2 - V_4^2)}{2V_1^3}.$$

By defining the velocity ratio $V_r \equiv V_4/V_1$, the expression simplifies to

$$\eta = \frac{1}{2}[(1 + V_r)(1 - V_r^2)].$$

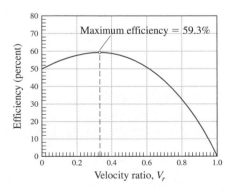

FIGURE 10.29 Efficiency of a windmill as function of the velocity ratio, V_r.

The plot of the efficiency as a function of the velocity ratio in Figure 10.29 shows a clear maximum point. This maximum occurs when

$$\frac{d\eta}{dV_r} = 1 - 2V_r - 3V_r^2 = 0,$$

or when $V_r = 1/3$. At this ratio, the efficiency of the windmill is $\eta = 59.3\%$. As a result of losses, the efficiency of modern windmills is about 50%. Dutch windmills have reported efficiencies in the range of 5% to 20%.

EXAMPLE 10.15

Problem Statement   An airplane moves through still air at 380 km/hr by discharging a total of 820 m^3/s of air, using two, 1.9 m-diameter propellers. Find (a) the thrust force; (b) the useful power; (c) the power input to the propeller; (d) the efficiency of the propeller; (e) the pressure difference across the propeller.

Governing Equations   Mass conservation for constant air density simplifies to

$$VA = \dot{Q} = constant.$$

The thrust force F and the useful power $\dot{W}_u$ are given by

$$F = \dot{m}(V_4 - V_1), \quad \text{and} \quad \dot{W}_u = FV_1.$$

The power input to the propeller $\dot{W}_p$, and pressure difference ΔP across the blades are given by

$$\dot{W}_p = \dot{m}\left(\frac{V_4^2 - V_1^2}{2}\right), \quad \text{and} \quad \Delta P = \rho\left(\frac{V_4^2 - V_1^2}{2}\right).$$

Assumptions   The preceding equations were developed by assuming steady and uniform conditions across the slipstream boundary. For the calculations here, we assume that $\rho = 1.225 \ kg/m^3$.

SOLUTION   Using mass conservation, the velocity of air relative to the airplane at the propeller is

$$V = \frac{\dot{Q}}{A} = \frac{4\dot{Q}}{\pi D^2} = \frac{4 \times 410 \ m^3/s}{\pi \times 1.9^2 \ m^2/s^2} = 144.6 \ m/s.$$

Note that since we have two propellers, the discharge was divided by two. The speed of air relative to the airplane is

$$V_1 = \frac{380 \ km/s \times 1000 \ m/km}{3600 \ s/hr} = 105.6 \ m/s.$$

In propeller operation,

$$V = \frac{V_1 + V_4}{2} \rightarrow V_4 = 2V - V_1 = 2 \times 144.60 \ m/s - 105.6 \ m/s = 183.6 \ m/s.$$

The thrust force for both propellers is then

$$F = 2\rho\dot{Q}(V_4 - V_1) = 2 \times 1.225 \ kg/m^3 \times 410 \ m^3/s \times (183.6 \ m/s - 105.6 \ m/s)$$

$$= 78.35 \ kN.$$

The total useful power imparted to the plane is

$$\dot{W}_u = FV_1 = 78.35 \ kN \times 105.6 \ m/s = 8.27 \ MW.$$

The total power input is

$$\dot{W}_p = \dot{m}\left(\frac{V_4^2 - V_1^2}{2}\right) = 1.225 \ Kg/m^3 \times 820 \ m^3/s \times \frac{(183.6 \ m/s)^2 - (105.6 \ m/s)^2}{2},$$

$$\dot{W}_p = 11.33 \ MW.$$

The theoretical efficiency then is

$$\eta_t = \frac{\dot{W}_u}{\dot{W}_p} = \frac{8.27 \ MW}{11.33 \ MW} = 0.73.$$

At these conditions the efficiency of the propeller is 73%. The pressure difference across the blades of the propeller is calculated as

$$\Delta P = \rho\left(\frac{V_4^2 - V_1^2}{2}\right) = 1.225 \ Kg/m^3 \times \frac{(183.6 \ m/s)^2 - (105.6 \ m/s)^2}{2} = 13.82 \ kPa.$$

EXAMPLE 10.16

Problem Statement   Find the power generated from a 12 m-diameter windmill operating in a 20 mph wind, having a theoretical efficiency of 40%. What is the pressure drop across the windmill?

Governing Equations   The theoretical efficiency of a windmill is given by

$$\eta = \tfrac{1}{2}[(1 + V_r)(1 - V_r^2)],$$

where $V_r = V_4/V_1$. The power input to the propeller $\dot{W}_p$ and pressure difference ΔP across the blades are given by

$$\dot{W}_p = \dot{m}\frac{V_1^2 - V_4^2}{2}, \quad \Delta P = \rho\frac{V_1^2 - V_4^2}{2}.$$

Assumptions   The preceding equations were developed by assuming steady and uniform conditions across the slipstream boundary. For the calculations here, we assume $\rho = 1.225 \ kg/m^3$.

SOLUTION   Using the data, the velocity of the wind is

$$V_1 = \frac{20 \ mph \times 1609 \ m/mile}{3600 \ s/hr} = 8.94 \ m/s.$$

Since the efficiency is 40%,

$$\eta = 0.4 = \frac{1}{2}[(1 + V_r)(1 - V_r^2)].$$

By trial and error, the efficiency equation gives

$$V_r = \frac{V_4}{V_1} = 0.734 \rightarrow V_4 = 6.56 \ m/s.$$

The velocity of the wind at the propeller is, then,

$$V = \frac{1}{2}(V_1 + V_4) = \frac{1}{2}(8.94 \ m/s + 6.56 \ m/s) = 7.75 \ m/s.$$

The total useful power generated by the windmill is

$$\dot{W}_u = \dot{m} \frac{V_1^2 - V_4^2}{2} = \rho V A \left(\frac{V_1^2 - V_4^2}{2} \right)$$

$$\dot{W}_u = 1.225 \ Kg/m^3 \times 7.75 \ m/s \times \pi \times \frac{(12 \ m)^2}{4} \times \left(\frac{(8.94 \ m)^2 - (6.56 \ m)^2}{2} \right)$$

$$= 19.80 \ KW.$$

The pressure difference across the blades of the propeller is

$$\Delta P = \rho V(V_1 - V_4) = 1.225 \ Kg/m^3 \times 7.75 \ m/s(8.94 \ m/s - 6.56 \ m/s) = 22.6 \ Pa.$$

REFERENCES

D.G. WILSON *The Design of High-Efficiency Turbomachinery and Gas Turbines*. Cambridge, MA, MIT Press, 1984.

H. ROUSE *Elementary Mechanics of Fluids*. New York, Wiley, 1946.

R.A. WALLIS *Axial Flow Fans and Ducts*. New York, Wiley, 1983.

L.F. MOODY and T. ZOWSKI "Hydraulic Machinery," in C.V. DAVIS and K.E. SORENSEN, *Handbook of Applied Hydraulics*, 3rd ed., New York, McGraw-Hill, 1969.

F.A. KRISTAL and F.A. ANNETT *Pumps: Types, Selection, Installation, Operation, and Maintenance*. New York, McGraw-Hill, 1953.

H.J. STEPANOFF *Centrifugal and Axial Flow Pumps*, 2nd ed., New York, Wiley, 1957.

Hydraulic Institute ANSI/HI 1.1-1.2-2000, "American National Standard for Centrifugal Pumps for Nomenclature and Definitions, Parsippany NJ, (2000).

H. COHEN, G.F.C. ROGERS and H.I.H. SARAVANAMUTTOO, *Gas Turbine Theory*, 3rd ed., Longman Scientific & Technical, Essex UK, and Wiley, New York, 1987.

M.P. BOYCE *Gas Turbine Engineering Handbook*, Houston, TX, Gulf Publishing, 1982.

J.C. HUNSAKER and B.G. RIGHTMIRE, *Engineering Applications of Fluid Mechanics*, New York, McGraw-Hill, 1947.

J.K. VENNARD *Elementary Fluid Mechanics*, 4th ed. New York, Wiley, 1961.

W.W. BATHIE *Fundamentals of Gas Turbines*. New York, Wiley, 1984.

PROBLEMS

1. Consider the series of blades as shown in Figure 10.30, with the absolute velocity of the air entering and leaving the blades as shown in the figure. Find: (a) the inlet and exit blade angles, (b) the velocity of the blades, and (c) the force on the blades. (Hint: because of to mass conservation, the velocity of the fluid relative to the blades remains constant.)

2. Consider the series of blades as shown in Figure 10.31, with the absolute velocity of the air entering and leaving the blades as shown in the figure. Find: (a) the inlet and exit angle of the absolute velocities, (b) the blade angles, (c) the velocity of the blades, and (d) the force on the blades. (Hint: because of to mass conservation, the velocity of the fluid relative to the blades remains constant.)

3. Find the torque required to sustain constant rotation of 4200 rev/hr of the disk shown in Figure 10.2. The gap between the rotating, and the stationary disk $H = 0.025 \ m$, is filled with oil with viscosity $\mu = 0.04 \ Pa \cdot s$. The diameter of the rotating disk is 0.5 m.

4. Consider the rotating disk in Problem 3. Find the steady angular velocity, and the time it takes for the disk to reach steady state conditions, when a constant torque of 0.3 $N \cdot m$ is applied in the vertical direction. The disk is made from aluminum ($\rho = 2700 \ kg/m^3$), and has a thickness of 3.5 cm.

5. Rework Problem 4, by selecting a coordinate system that it is attached to, and rotates with the disk.

6. Consider a sprinkler system with two arms rotating at a constant angular velocity of 46 rev/min, delivering water at a rate of 14 $liters/min$ as shown in Figure 10.5. The radius of rotation extents 22 cm, with arm diameter 0.5 cm. Find the frictional torque that develops during the operation of the sprinkler.

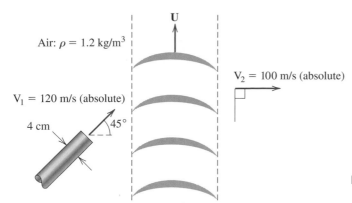

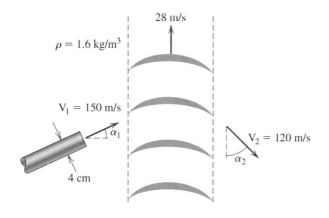

FIGURE 10.30 Schematic for Problem 1.

FIGURE 10.31 Schematic for Problem 2.

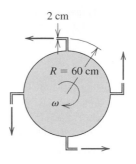

FIGURE 10.32 Schematic for Problem 8.

7. Reconsider the sprinkler in Problem 6, by selecting instead a control volume that it is attached to, and rotates with the sprinkler.

8. Consider the turbine shown in Figure 10.32. Each arm has a diameter of 2 cm and discharges 0.007 m^3/s with a radius of rotation of of 60 cm. If the shaft rotates at 120 rpm, find the generated power.

9. Using the energy equation and the velocity diagram, show that the energy equation between the inlet and exit of a general turbomachine can be expressed as

$$\frac{P_e - P_i}{\rho} + \frac{V_{b_e}^2 - V_{b_i}^2}{2} - \frac{\omega^2}{2}(r_e^2 - r_i^2)$$
$$+ g(z_e - z_i) + g H_{losses} = 0.$$

(Hint: $V_{f_t} = \omega r - V_{b_t}$, where V_{b_t} is the projection of the velocity relative to the blade V_b in the direction of rotation.)

10. Consider the cylindrical rotating device shown in Figure 10.10, where the rotor rotates constantly at a rate of

1000 rev/min. Air at a rate of 12 m^3/s enters the device at a radial distance of 0.6 m and exits at a radius of 1.0 m. The blade angle at the inlet is $28°$ and the angle at the exit is $60°$. If the blade height is 0.25 m, find (a) the torque required to run the device, (b) the corresponding power. Is this device operating as a pump or as a turbine?

11. Reconsider the pump described in Problem 10, operating under the same conditions. If the blade angle is adjustable, what is the range of exit blade angles that can make change the mode of operation of the device?

12. A centrifugal pump impeller is rotating at 480 rpm. The inlet and exit radii are 8 cm and 26 cm, respectively. The inlet and exit blade angles are $40°$ and $60°$, respectively. If the absolute velocity enters in a purely radial direction and the blade height is 5 cm, find the flowrate and the required torque and power to sustain the rotation.

13. A centrifugal pump impeller with inlet and exit diameters of 7.5 cm and 30 cm rotates at 550 rpm. If the inlet blade angle is $45°$ and the exit angle is $60°$ with blade width of 5 cm, by assuming purely radial inlet velocity, find the flow rate, the pressure difference across the impeller, the torque and the required power.

14. Consider the idealized axial fan as shown in Figure 10.11, where air enters the pump radially, at a rate of 7 m^3/s. The hub diameter is 0.40 m, and the tip diameter is 0.8 m. The blade angle at the inlet is $32°$ degrees and at the exit is $65°$ degrees. For a rotor rotating at 880 rev/min, first construct the velocity polygons at the inlet and exit and find: (a) the shaft torque and (b) the power required.

15. Water at a rate of $\dot{Q}$ m^3/s enters a centrifugal pump in the radial direction as shown in Figure 10.13 and leaves at the impeller outlet with diameter $D = 25$ cm. If the impeller rotates at a constant rate of 1850 rev/min, find the pressure head as a function of the vane angle at the exit. Use a blade height of $h = 1.5$ cm.

16. Consider a centrifugal pump that receives water at 0.06 m^3/s and rotates at 100 rad/s. The absolute velocity at the inlet (radius 5 cm), enters in the radial direction and exits at radius 15 cm at an angle of $120°$ degrees

with respect to the tangential direction. If the pump is 1.5 cm wide, find the torque and power required to run the pump.

17. Consider a cenrifugal pump that receives water at 0.1 m^3/s and rotates at 120 rad/s. The absolute velocity at the inlet (radius 6 cm), enters in the radial direction and exits at radius 18 cm at an angle of 120° degrees with respect to the tangential direction. If the pump is 2 cm wide, find the torque and power required to run the pump.

18. A centrifugal water pump operates at 180 $radians/seconds$ and requires 220 kW to run. If the outside radius is 20 cm, thickness of 2 cm and the water exits at 135° degrees with respect to the direction of the rotation, find the exit velocity and mass flow rate.

19. A water pump is tested at 2000 rpm discharges 0.2 m^3/s at a head of 100 m. At these conditions the efficiency is found to be 90%. Find the volumetric flow rate, head and power for a geometrically similar pump that it is twice as large and operates at 1600 rpm.

20. Water at velocity V_1 strikes a curved vane (angle β) that moves with velocity U as shown in Figure 10.33. For $V_1 = 12$ m/s, jet area $A = 0.1$ m^3, and $\beta = 45°$, find the reaction forces when the vane is (a) moving to the left with velocity $U = 8$ m/s, (b) moving to the right with velocity $U = 8$ m/s.

21. Repeat the same calculations for the problem described above, but use $\beta = 120°$.

22. Figure 10.34 shows an air jet 2 cm in diameter that strikes a set of blades of a turbine rotor. For the conditions shown in the figure, find the force on the rotor. What is the power transferred to the rotor? What is the velocity U of the rotor. The velocities shown in the figure are absolute and the air density is 1.2 kg/m^3.

23. Figure 10.35 shows an 90 m/s water jet 2 cm in diameter impinging on a set of blades moving with velocity U. For the conditions shown in the figure, find the force on the rotor. If the output power is 140 kW, and the approach and exit velocity angle is 40° degrees, find the velocity U of the blades.

24. The centrifugal water pump with the impeller shown in Figure 10.36 rotates at 600 rpm. By assuming purely radial

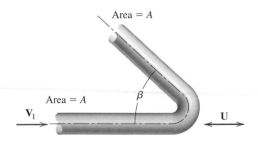

FIGURE 10.33   Schematic for Problem 20.

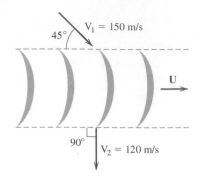

FIGURE 10.34   Schematic for Problem 22.

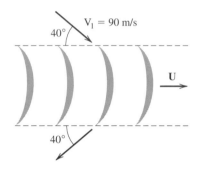

FIGURE 10.35   Schematic for Problem 23.

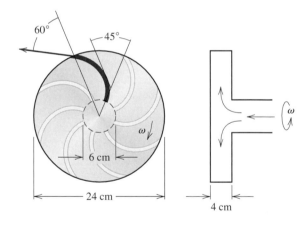

FIGURE 10.36   Schematic for Problem 24.

entry, find the flow rate, the power and the torque required to meet the conditions in the figure.

25. A centrifugal water pump impeller has an inlet radius of 8 cm and exit radius of 20 cm. If the inlet and exit blade angles are respectively 120° and 135°, for a volumetric flow rate of 0.12 m^3/s, find the rotational speed, and the required torque and power.

26. A radial pump with average diameter of 8 cm is operating using air at the point of maximum efficiency with

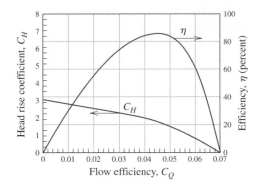

FIGURE 10.37 Schematic for Problem 28.

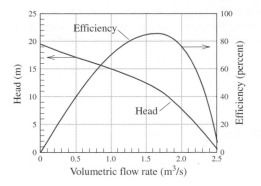

FIGURE 10.39 Performance curves for Problem 30.

speed of 1700 rpm. The pump's performance is given by Figures 10.15 and 10.16. If the speed of the pump is changed to 1500 rpm, find the new volumetric flow rate, and operating head. Find the actual power consumed by the pump.

27. Consider a geometrically similar pump with the one shown in Example 10.10, operating at the same rotational speed but with characteristic diameter of 6 cm. Find the operating conditions, and the actual power used by this pump at the point of maximum efficiency.

28. An axial air pump with average diameter of 0.6 m operating at the point of maximum efficiency with speed of 1600 rpm. The pump's performance is given by Figure 10.37. If the speed of the pump is changed to 1200 rpm, find the new volumetric flow rate, and operating head. Find the actual power consumed by the pump.

29. Consider a geometrically similar pump with the one shown in the previous problem , operating at the same rotational speed but with characteristic diameter of 0.4 m. Find the operating conditions, and the actual power used by this pump at the point of maximum efficiency.

30. Find the operating conditions, and the volumetric flow rate $\dot{Q}$ through the system shown in Figure 10.38. Using the pump with the performance curves (Head vs. $\dot{Q}$ and Efficiency vs. $\dot{Q}$) shown Figure 10.39.

31. A reaction turbine has inner and outer radii of 1.0 m and 1.4 m respectively. The runner rotates at a rate of 64 rpm, and the inlet and exit angles are 54° and 150°, respectively.

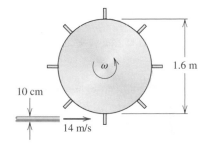

FIGURE 10.40 Schematic for Problem 33.

If the flow rate is 12 m^3, and the blade height is 0.4 m, find (a) the generated torque (b) the corresponding power, and (c) the pressure drop in the runner.

32. A reaction turbine has inlet and exit radii of 1 m and 3 m respectively. The inlet blade angle is 60° and the exit blade angle is 160°. If the velocity at the inlet is guided by a set of fixed vanes to enter at 15° degrees relative to the direction of rotation, and the flow rate is 10 m^3/s, find the rotational speed for smooth entry. Also, calculate the power, torque and the pressure drop from the inlet to the exit. Consider a thickness of 20 cm.

33. Consider a paddle wheel with flat blades as shown in Figure 10.40. If a water jet emerging from a 10 cm-diameter pipe at a speed of 14 m/s hits the blades tangentially to the rim of the wheel. If the diameter of the

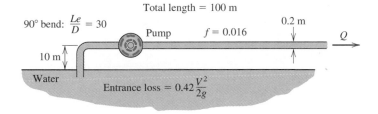

FIGURE 10.38 Schematic for Problem 30.

wheel is 1.6 m and rotates at a rate of 80 rpm, find (a) the generated torque and power. What is the rotational speed for maximum power? What is the maximum power?

34. A jet of water with velocity of 15 m/s hits a curved vane that it is moving with velocity of 10 m/s. Due to the curved vane, the jet turns 160° with respect to the original jet flow direction. If the cross-sectional area of the jet is 0.001 m^2, find the reaction forces on the vane.

35. A jet of water with velocity of 8 m/s hits a curved vane that it is moving with velocity of 6 m/s. Due to the curved vane, the jet turns 120° degrees with respect to the original jet flow direction. If the cross-sectional area of the jet is 0.001 m^2, find the reaction forces on the vane.

36. A jet of water with velocity of 20 m/s hits a curved vane that it is moving with velocity of 15 m/s. Due to the curved vane, the jet turns 130° degrees with respect to the original jet flow direction. If the cross-sectional area of the jet is 0.001 m^2, find the power produced by the vane.

37. A Pelton wheel turbine operates at the point of maximum power. The turbine has diameter of 2 m and rotates at 360 rpm. The jet diameter is 20 cm and the bucket angle is 165°. Calculated the flowrate, the generated torque and power.

38. A 3 m-diameter Pelton wheel operates at 550 rpm and develops 18 MW of power with a total available head of 1000 m, ahead of the single nozzle. Estimate the size of the nozzle.

39. A 3 m-diameter Pelton wheel develops 750 kW of power while it rotates at 1600 rpm. If the turbine operates at maximum efficiency, find the speed of the water out of the nozzle and the volumetric flowrate.

40. In a hydraulic turbine with a runner rotating at 220 rpm the inlet radius is 1 m and the velocity of the water relative to the the blade is 18 m/s. The water exit at a radius of 0.8 m with the absolute liquid velocity vector at 90° degrees with respect to the direction of rotation. If the blade angle at the exit is 160° and the pressure drop 160 kPa, find the head loss in the turbine.

41. A Francis turbine operates with inlet and exit blade velocities of 1.5 m/s and 6.2 m/s, respectively. The inlet and exit radii are 1.5 m and 1.2 m. If the runner rotates at 48 rpm, the height of the blades is 0.3 m and the absolute velocity at the exit is at 90° degrees with respect to the direction

of rotation, find the flow rate through the turbine and the generated torque.

42. A Francis turbine operates with inlet and exit radii of 1.5 m and 1.2 m with a runner width of 30 cm. If the inlet and exit velocities relative to the blade are respectively 1.52 m/s and 6.2 m/s and the absolute velocity exits at 90° with respect to the direction of rotation, find: (a) the flowrate, (b) the generated torque and power.

43. An airplane moves through still air at 280 km/hr by discharging a total of 660 m^3/s of air, using two, 1.6 m-diameter propellers. Find: (a) the thrust force, (b) the useful power, (c) the power input to the propeller, (d) the efficiency of the propeller and (e) the pressure difference across the propeller.

44. A plane is flying at 140 mph in still air ($\rho = 1.225\ kg/m^3$). The propeller has a diameter of 1.8 m. The air velocity away from the propeller is 160 mph. Calculate: (a) the air velocity at the propeller, (b) power input, (c) useful power and (d) the pressure difference across the propeller.

45. A propeller must produce a thrust force of 9800 N in order to sustain flight of an airplane traveling at 160 mph. What size propeller is required so that the theoretical efficiency is 90%?

46. A 3 m-diameter propeller driven plane is traveling at 200 mph in still air and delivers 900 kW of power. By considering an ideal propeller, calculate the slipstream velocity, the diameter of the slipstream ahead and behind the propeller. What is the corresponding thrust and efficiency.

47. Find the power generated from a 8 m-diameter windmill operating in a 18 mph wind, having theoretical efficiency of 45%. What is the pressure drop across the windmill?

48. What is the maximum power extracted from an ideal windmill with 10 m-diameter blades in a 25 mph wind? Assume standard air conditions.

49. What is the maximum power output from a 20 m diameter windmill operating at best efficiency in a 40 mph wind. What is the corresponding thrust? Assume standard air conditions.

50. An ideal windmill is operating at best efficiency in a 28 mph wind. the diameter of the blade is 15 m. What is the velocity behind the windmill? Find the thrust of the wheel, the power extracted, and the pressure difference across the blades.

11 Compressible Flow

So far we have studied the dynamics of incompressible fluids. However, many fluids are compressible — that is, under the influence of an external force, a compressible fluid occupying an initial volume $\mathcal{V}_1$ can be compressed to a smaller volume $\mathcal{V}_2 < \mathcal{V}_1$. Customarily, the term *gas* is used to denote such a compressible fluid. Gases can not only compress but can also *expand* — that is, by removing the compressive force, the volume initially occupied by the gas increases (i.e., $\mathcal{V}_2 > \mathcal{V}_1$). Compression or expansion processes produce local property changes which, as demonstrated in this chapter, are associated with phenomena quite different from those observed in incompressible fluids.

Because of the introductory level of the material, for simplicity we will assume that the gas is ideal — that is, the gas behaves according to the ideal gas constitutive relation

$$P = \rho RT,$$

where P is the pressure, T is the absolute temperature, ρ is the density, and R is the gas constant.

11.1 Sonic Speed

The ability of gases to compress or expand is directly related to the weak intermolecular forces between gas molecules. Consequently, gas molecules can be brought easily close to each other, which leads to local density changes. Because of these weak intermolecular forces, the mean free path of the molecules (i.e., the average distance between molecules) is larger than in incompressible fluids, a fact that explains the lower density in gases.

One of the most fundamental questions in compressible fluids is how a small pressure disturbance propagates in a compressible medium. Consider, for instance, a column of water subjected to an axial force F, as shown in Figure 11.1(a). Since water is essentially incompressible, the molecules cannot be pushed closer to each other. Consequently, the force is transmitted instantaneously to the other end. In this respect, the ability of incompressible fluids to transmit compressive forces is indistinguishable from that of solids.

Instead, if the same force is applied to a column of compressible gas, the behavior is different: the molecules close to the point of application compress as shown Figure 11.1(b). This compression causes a local increase in density, which attains its maximum value when the pressure within the gas at the vicinity of the disturbance reaches equilibrium with the external pressure. However, in response to this local pressure increase, adjacent molecules also compress. Similarly, then, in this progressive manner the effects of the pressure are transmitted downstream and arrive at the other end after a finite amount of time.

---

[1]G.A. Tokaty, *Philosophy of Fluid Mechanics*.

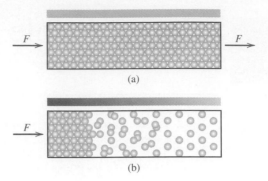

FIGURE 11.1   Schematic differences between (a) incompressible media; (b) compressible media.

To an observer, the entire phenomenon appears as a density discontinuity traveling at a finite speed. A new concept here is the fact that *disturbances in compressible gases travel at a finite speed known as the sonic speed (speed of sound)* α. This very simple concept has profound implications, and forms the basis for most compressible phenomena. *Sound is such a pressure disturbance which, once generated, travels in the surrounding air at a finite speed*. The pressure disturbance created by the sound source is translated into an audible signal in the ear.

The sonic speed is a function of local thermodynamic properties of the gas. This can be shown by using the simple flow arrangement of a channel fitted with a piston. The channel is filled with gas at an initial temperature T and pressure P as shown in Figure 11.2(a). We assume that the piston is suddenly moved to the right with a small velocity dV. The motion of the piston induces a small pressure disturbance which, according to the previous arguments, travels in the initially stagnant gas with velocity α. The compression leaves behind a slightly disturbed gas with different thermodynamic properties than the undisturbed gas.

By selecting a control volume attached to and moving with the disturbance, the fluid velocities are expressed relative to a stationary control volume (Figure 11.2(b)). For this one-dimensional open system, mass conservation is expressed as

$$\rho \alpha A = (\rho + d\rho)(\alpha - dV)A,$$

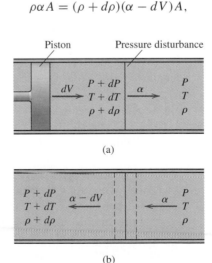

FIGURE 11.2   Propagation of a small pressure disturbance in compressible media.

where ρ is the density, A the cross-sectional area, and V the velocity of the gas. By assuming that the disturbance is small, the continuity equation is simplified, by neglecting small terms such as $d\rho dV$, to

$$\rho \alpha A = \rho \alpha A - \rho A \, dV + \alpha A \, d\rho,$$

which finally gives

$$\alpha = \frac{\rho}{d\rho} \, dV. \tag{11.1}$$

However, since dV is an arbitrary quantity, it must be eliminated so that α is expressed only in terms of local thermodynamic properties. Introducing the momentum equation

$$\dot{m} \, (\alpha - dV - \alpha) = PA - (P + dP) \, A,$$

and by simplifying,

$$\rho \alpha A \, dV = A \, dP \Rightarrow dV = \frac{dP}{\rho \alpha}. \tag{11.2}$$

Finally, by combining Equations (11.1) and (11.2) we get

$$\alpha = \frac{\rho}{d\rho} \frac{dP}{\rho \alpha} \Rightarrow \alpha^2 = \frac{dP}{d\rho}.$$

Therefore, the sonic speed is given as

$$\alpha = \sqrt{\frac{dP}{d\rho}}.$$

The speed of the disturbance, then, is a function of the rate at which pressure changes with respect to changes in the density. The more compressible the gas is (i.e., when for a given increase in pressure the corresponding density change is large), the slower the disturbance will travel. This is expected, because in this case, it will take longer for adjacent molecules to experience the pressure effects. For incompressible fluids $\rho = $ constant, we have $\alpha = \infty$. This verifies the fact that in incompressible fluids, disturbances are transmitted throughout the domain instantaneously.

Now, if we assume that the gas is ideal $P = \rho RT$ and that the compression is sufficiently small to be considered isentropic,

$$Pv^\gamma = \text{constant} = C \Rightarrow P = C\rho^\gamma;$$

therefore,

$$\frac{dP}{d\rho} = C\gamma\rho^{\gamma-1} = \frac{P}{\rho^\gamma}\gamma\rho^{\gamma-1} = \gamma\frac{P}{\rho} = \gamma RT.$$

For ideal gases, then,

$$\alpha = \sqrt{\gamma RT}, \tag{11.3}$$

showing that *the sonic speed is only a function of the local absolute temperature* measured in degrees K. For example, in ambient air with temperature of $25°C$ sound travels with velocity

$$\alpha = \sqrt{1.4 \times 287 \, J/kgK \times 298 \ K} = 346 \ m/s.$$

If the local fluid velocity is V, the nondimensional ratio

$$M = \frac{V}{\alpha}$$

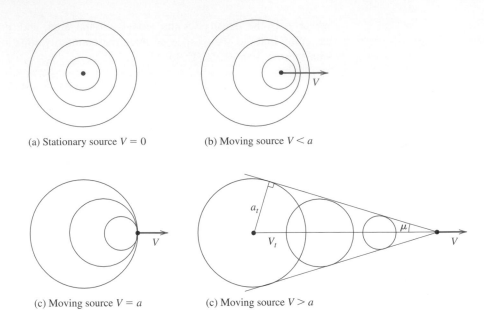

(a) Stationary source $V = 0$ (b) Moving source $V < a$

(c) Moving source $V = a$ (c) Moving source $V > a$

FIGURE 11.3   Geometry of pressure disturbances for (a) $M = 0$; (b) $M < 1$; (c) $M = 1$; (d) $M > 1$.

is known as the *Mach* number. The flow can be either *subsonic* when $V < \alpha$ or *supersonic* when $V > \alpha$. At the limit when $V = \alpha$, the flow is known as *sonic flow*. The Mach number is a measure of how important the compressibility effects are. When the flow is incompressible (i.e., $\alpha = \infty$), the Mach number $M = 0$. As the Mach number increases, the compressible effects become progressively more important. For practical applications, compressible effects are usually ignored when $M < 0.3$.

To illustrate some implications of the different flow conditions, consider first a stationary sound source. The sound emitted from the source travels in all directions with the same velocity α (Figure 11.3(a)). Therefore, the sound envelope is a set of concentric circles. This is analogous to the ripples formed when a stone is thrown into a lake. When the sound source is moving with speed smaller than the speed of sound ($M < 1$), the sound envelope is a set of circles, with the origin of each circle lying along the line of motion (Figure 11.3(b)). At the limit, when the sound source travels with the speed of sound ($M = 1$), the source and its disturbance travel together, as shown in Figure 11.3(c). However, under supersonic conditions, when the source travels faster than the speed of sound ($M > 1$), the sound envelope (i.e., the pressure disturbance) is confined to a cone behind the source. The domain outside the cone is known as the *domain of silence*. This explains the fact that the sound created by a plane flying at supersonic conditions is heard after the plane flies over an observer. Using the geometry of the cone, the half angle μ that the cone makes with respect to the horizontal as shown in Figure 11.3(d) is

$$\sin \mu = \frac{\alpha t}{V t} = \frac{1}{M} \Rightarrow \mu = \sin^{-1} \frac{1}{M}.$$

11.2   Isentropic Flow

For completeness, part of the discussion on isentropic flow from Chapter 2 is repeated here. The first law of thermodynamics for a simple compressible substance is expressed as

$$dU = \delta Q - P \, d\mathcal{V},$$

where Q is the heat transfer and $\mathcal{V}$ is the volume of the gas. This expression can be simplified further by using the definition of entropy $\delta Q = T \, d\mathcal{S}$,

$$dU = T \, d\mathcal{S} - P \, d\mathcal{V}.$$

However, $P \, d\mathcal{V} = d(P\mathcal{V}) - \mathcal{V} \, dP$. Therefore,

$$dU = T \, d\mathcal{S} - d(P\mathcal{V}) + \mathcal{V} \, dP,$$

which yields $d(U + P\mathcal{V}) = T \, d\mathcal{S} + \mathcal{V} \, dP$. Using the definition of the enthalpy H, and by dividing by the mass m,

$$T \, ds = dh - \frac{dP}{\rho}.$$

The entropy change for an ideal gas then can be expressed as

$$ds = \frac{dh}{T} - \frac{dP}{T\rho} = \frac{dh}{T} - R\frac{dP}{P} = C_p \frac{dT}{T} - R\frac{dP}{P}.$$

By integration,

$$s_2 - s_1 = \int_1^2 C_p(T) \frac{dT}{T} - R \ln \frac{P_2}{P_1}.$$

Additionally, if we assume that C_p is constant, we have

$$s_2 - s_1 = C_p \ln \frac{T_2}{T_1} - R \ln \frac{P_2}{P_1}.$$

When the entropy is constant (i.e., the flow is isentropic), $s_2 - s_1 = 0$. In such a case,

$$C_p \ln \frac{T_2}{T_1} = R \ln \frac{P_2}{P_1} \Rightarrow \ln \frac{T_2}{T_1} = \frac{R}{C_p} \ln \frac{P_2}{P_1}.$$

In terms of the specific heat ratio

$$\frac{T_2}{T_1} = \frac{\gamma - 1}{\gamma} \ln \frac{P_2}{P_1} \Rightarrow \frac{T_2}{T_1} = \left(\frac{P_2}{P_1} \right)^{\frac{\gamma-1}{\gamma}}.$$

Using the ideal gas relation, the preceding expression can be manipulated further to give the well-known expression

$$P_1 v_1^{\gamma} = P_2 v_2^{\gamma}.$$

Stagnation Conditions

A useful quantity in compressible flow is the *stagnation temperature*, defined as the temperature obtained when the flow comes to rest ($V = 0$) in an *adiabatic* process $\dot{Q} = 0$. Therefore, the stagnation temperature represents the resulting temperature when the local kinetic and thermal energies are converted into an equivalent thermal energy. Applying the energy equation between an arbitrary thermodynamic state and its corresponding stagnation state (in the absence of work, $\dot{W} = 0$), it reduces to,

$$h + \frac{V^2}{2} = h_t \Rightarrow (h_t - h) = \frac{V^2}{2},$$

where h_t is the enthalpy at stagnation conditions. By assuming constant specific heat coefficient

$$C_p(T_t - T) = \frac{V^2}{2},$$

the stagnation temperature is given as

$$T_t = T + \frac{V^2}{2C_p} = T\left(1 + \frac{V^2}{2C_p T}\right).$$

Recalling that $C_p = \gamma R/(\gamma - 1)$, then

$$T_t = T + \frac{V^2}{2C_p} = T\left(1 + \frac{V^2(\gamma - 1)}{2\gamma RT}\right).$$

Using the definition $M = V/\sqrt{\gamma RT}$, the stagnation temperature T_t is finally given in terms of M as

$$T_t = T\left(1 + \frac{(\gamma - 1)}{2}M^2\right). \tag{11.4}$$

For a given gas the ratio

$$\frac{T}{T_t} = 1 + \frac{(\gamma - 1)}{2}M^2$$

is only a function of the local Mach number M. For convenience, this ratio is presented in tabular form in Appendix B in Table B.2.

EXAMPLE 11.1

Problem Statement   Find the stagnation temperature for air at $T = 100°C$ and $M = 0.8$.

SOLUTION

$$T_t = T\left(1 + \frac{(\gamma - 1)}{2}M^2\right) = 373.15\ K \times \left(1 + \frac{(1.4 - 1)}{2}0.8^2\right),$$

$$\Rightarrow T_t = 420.9\ K.$$

Using the isentropic tables, for $M = 0.8$ the ratio $\dfrac{T}{T_t} = 0.8865$, which yields the same result as above $T_t = 420.9\ K$.

The stagnation pressure P_t is defined similarly as the pressure obtained when the flow is brought to rest *isentropically*. Using the isentropic relation

$$\frac{P_2}{P_1} = \left(\frac{T_2}{T_1}\right)^{\frac{\gamma}{\gamma - 1}},$$

and by assuming that the second thermodynamic state corresponds to the stagnation conditions — that is, $P_2 = P_t$ and $T_2 = T_t$,

$$P_t = P\left(\frac{T_t}{T}\right)^{\frac{\gamma}{\gamma - 1}}.$$

Substituting for the temperature ratio, the stagnation pressure is finally expressed as

$$P_t = P\left(1 + \frac{(\gamma - 1)}{2}M^2\right)^{\frac{\gamma}{\gamma-1}}. \tag{11.5}$$

Again, for a given gas the ratio

$$\frac{P}{P_t} = \left(1 + \frac{(\gamma - 1)}{2}M^2\right)^{-\frac{\gamma}{\gamma-1}},$$

is only a function of the local Mach number M. In the preceding expressions, T and P are respectively the *static temperature* and the *static pressure*. For convenience, this ratio is also presented in tabular form in Appendix B in Table B.2.

EXAMPLE 11.2

Problem Statement   Find the stagnation pressure for air at $P = 0.5\ MPa$, $T = 100°C$ and $M = 0.8$.

SOLUTION

$$P_t = P\left(1 + \frac{(\gamma - 1)}{2}M^2\right)^{\frac{\gamma}{\gamma-1}} = 0.5\ MPa\left(1 + \frac{(1.4 - 1)}{2}0.8^2\right)^{\frac{1.4}{1.4-1}},$$

$$\Rightarrow P_t = 0.762\ MPa.$$

The same results could be obtained by using the tables for $M = 0.8$, where the ratio

$$\frac{P}{P_t} = 0.6560 \Rightarrow P_t\frac{P}{0.6560} = 0.762\ MPa.$$

It is important to note that if the flow is adiabatic, all points in the flow have the same stagnation temperature. Similarly, if the flow is isentropic, then both the stagnation pressure and stagnation temperature are the same for all points in the flow.

11.2.1   Flow in a Channel with Variable Area

Compressible flow through a channel with variable cross-sectional area is a typical problem in which the compressible effects can be clearly demonstrated. Consider, then, flow through the diverging nozzle shown in Figure 11.4. The objective here is to determine the flow properties as a function of the local area. Selecting the fluid itself as the control volume, the continuity equation for this open system is

$$\int_1 \rho\mathbf{V}\cdot\mathbf{n}\,dS + \int_2 \rho\mathbf{V}\cdot\mathbf{n}\,dS = 0.$$

By assuming uniform conditions at the inlet and exit,

$$-\rho V A + (\rho + d\rho)(V + dV)(A + dA) = 0,$$

and by neglecting small terms such as $d\rho\,dV$, $dV\,dA$, etc., it simplifies to

$$\frac{d\rho}{\rho} + \frac{dV}{V} + \frac{dA}{A} = 0. \tag{11.6}$$

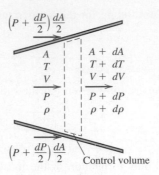

FIGURE 11.4   Compressible flow in a variable area channel.

The momentum equation

$$\int_1 \rho \mathbf{V}(\mathbf{V} \cdot \mathbf{n}) \, dS + \int_2 \rho \mathbf{V}(\mathbf{V} \cdot \mathbf{n}) \, dS = \sum \mathbf{F},$$

for one-dimensional flow with uniform conditions at the inlet and exit is given as

$$-\rho V A V + (\rho + d\rho)(V + dV)(A + dA)(V + dV)$$

$$= P A + \left(P + \frac{dP}{2} \right) dA - (P + dP)(A + dA).$$

Using the continuity equation,

$$\rho V A = (\rho + d\rho)(V + dV)(A + dA),$$

the momentum equation simplifies further to

$$dP + \rho V \, dV = 0. \tag{11.7}$$

However, for the derivation here we will rearrange the above expression as

$$\frac{dV}{V} = -\frac{dP}{\rho V^2}.$$

Using the definition of the sonic speed we get

$$\alpha^2 = \frac{dP}{d\rho},$$

$$\frac{dP}{\rho} = \alpha^2 \frac{d\rho}{\rho} \Rightarrow \frac{d\rho}{\rho} = \frac{dP}{\rho \alpha^2}.$$

By substituting the last two results into the continuity equation

$$\frac{dA}{A} - \frac{dP}{\rho V^2} + \frac{dP}{\rho \alpha^2} = 0,$$

we get

$$\frac{dA}{A} = dP \left(\frac{1}{\rho V^2} - \frac{1}{\rho \alpha^2} \right),$$

which finally yields

$$\rho V^2 \frac{dA}{A} = (1 - M^2) \, dP. \tag{11.8}$$

The preceding relation is quite revealing: for subsonic $M < 1$—that is, $(1 - M^2) > 0$, in a diverging channel $dA > 0$, Equation (11.8) implies that $dP > 0$. According then to the momentum equation (Equation (11.7))

$$dP = -\rho V \, dV,$$

the velocity decreases $dV < 0$. Therefore, subsonic flow in a diverging channel decelerates. Using similar reasoning, we can show that supersonic flow accelerates in a diverging channel.

At the limit when the $M = 1$, we observe that $dA = 0$. Therefore, the Mach number at a point of inflection (i.e., minimum area) is at sonic conditions. This point is called the *throat*. Table 11.1 shows the various combinations of flow conditions and channel geometry.

TABLE 11.1 Flow Variables as a Function of M and Changes in the Cross-Sectional Area.

Mach Number	Area	Pressure	Velocity	Comments
$M < 1$	+	+	−	Flow decelerates
$M < 1$	−	−	+	Flow accelerates
$M > 1$	+	−	+	Flow accelerates
$M > 1$	−	+	−	Flow decelerates

11.2.2 Mass Flow Rate Through a Channel

The mass flow rate through the channel with varying cross-sectional area is given as

$$\dot{m} = \rho V A.$$

Using the ideal gas relation and, whenever possible, replacing the velocity with the Mach number, we get

$$\dot{m} = \frac{P}{RT} V A = \frac{P}{RT} M \sqrt{\gamma R T} \, A = \sqrt{\frac{\gamma}{R}} \frac{P M}{\sqrt{T}} A.$$

Replacing the static with the stagnation conditions,

$$T = T_t \left[1 + \frac{(\gamma - 1)}{2} M^2 \right]^{-1},$$

and

$$P = P_t \left[1 + \frac{(\gamma - 1)}{2} M^2 \right]^{-\frac{\gamma}{\gamma - 1}},$$

we get

$$\dot{m} = \sqrt{\frac{\gamma}{R}} \frac{P_t}{\sqrt{T_t}} \frac{\left[1 + \dfrac{(\gamma - 1)}{2} M^2 \right]^{-\frac{\gamma}{\gamma - 1}}}{\left[1 + \dfrac{(\gamma - 1)}{2} M^2 \right]^{-\frac{1}{2}}} M A.$$

This expression can be simplified further to

$$\dot{m} = \sqrt{\frac{\gamma}{R}} \frac{P_t}{\sqrt{T_t}} \left[1 + \frac{(\gamma - 1)}{2} M^2 \right]^{\frac{\gamma}{1 - \gamma} + \frac{1}{2}} M A,$$

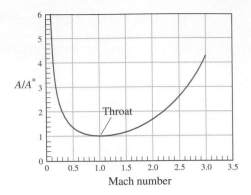

FIGURE 11.5   Mass flow rate as a function of the Mach number M.

and finally to

$$\dot{m} = \sqrt{\frac{\gamma}{R}} \frac{P_t}{\sqrt{T_t}} \left[1 + \frac{(\gamma - 1)}{2} M^2\right]^{\frac{\gamma+1}{2-2\gamma}} M A. \qquad (11.9)$$

For convenience, we introduce a reference area A^*, defined as the area where the flow is sonic $M = 1$ — that is,

$$\dot{m}^* = \sqrt{\frac{\gamma}{R}} \frac{P_t}{\sqrt{T_t}} \left[1 + \frac{(\gamma - 1)}{2}\right]^{\frac{\gamma+1}{2-2\gamma}} A^*.$$

The location of A^* may be real or fictitious depending on whether the flow actually attains sonic conditions, $M = 1$. Requiring that at A^*, the mass flow rate is the same as the flow through the channel (i.e., $\dot{m}^* = \dot{m}$), we get

$$\sqrt{\frac{\gamma}{R}} \frac{P_t}{\sqrt{T_t}} \left[1 + \frac{(\gamma - 1)}{2}\right]^{\frac{\gamma+1}{2-2\gamma}} A^* = \sqrt{\frac{\gamma}{R}} \frac{P_t}{\sqrt{T_t}} \left[1 + \frac{(\gamma - 1)}{2} M^2\right]^{\frac{\gamma+1}{2-2\gamma}} M A,$$

which yields

$$\frac{A}{A^*} = \frac{\left[1 + \dfrac{(\gamma - 1)}{2}\right]^{\frac{\gamma+1}{2-2\gamma}}}{M \left[1 + \dfrac{(\gamma - 1)}{2} M^2\right]^{\frac{\gamma+1}{2-2\gamma}}}.$$

The ratio of A/A^* is only a function of M. For convenience, the ratio A/A^* is included in tabular form in Appendix B in Table B.2. The same ratio is also plotted in Figure 11.5. The plot shows that for a given ratio, there are two possible solutions for the Mach number: one corresponding to subsonic conditions and another to supersonic conditions. The minimum area ratio occurs at $M = 1$.

EXAMPLE 11.3

Problem Statement   Air flows isentropically in a converging nozzle where the inlet and exit cross-sectional areas are 60 cm^2 and 40 cm^2, respectively. If the inlet temperature is $T_i = 320°K$, pressure $P_i = 120\ kPa$, and velocity $V_i = 100\ m/s$, find the exit Mach number M_e, temperature T_e, and pressure P_e. Assume that air behaves as an ideal gas ($\gamma = 1.4$).

SOLUTION The inlet Mach number at these conditions is

$$M_i = \frac{V_i}{\alpha_i} = \frac{100 \ m/s}{\sqrt{1.4 \times 287 \ J/kgK \times 320 \ K}} = 0.28.$$

Using Table B.2,

$$\frac{A_i}{A^*} = 2.1656 \Rightarrow A^* = \frac{60 \ cm^2}{2.1656} = 27.7 \ cm^2.$$

Therefore, for the exit

$$\frac{A_e}{A^*} = \frac{40 \ cm^2}{27.7 \ cm^2} = 1.4440.$$

By returning to the table in Appendix B, as noted earlier, we observe that this area ratio corresponds to two Mach numbers, $M \approx 0.45$ and $M \approx 1.80$. However, since this is a converging nozzle with subsonic inlet conditions, the flow at the exit cannot be supersonic. Therefore, the correct choice is the subsonic solution $M = 0.45$. For this ratio, the table in Appendix B gives

$$\frac{P_e}{P_t} = 0.8703 \quad \text{and} \quad \frac{T_e}{T_t} = 0.9611.$$

Then

$$P_e = 0.8703 \times 120 \ kPa = 104.4 \ kPa,$$

and

$$T_e = 0.9611 \times 320 \ K = 307.6 \ K.$$

EXAMPLE 11.4

Problem Statement Air is supplied from a reservoir to a converging nozzle, as shown in Figure 11.6. If the reservoir conditions are $T_r = 400 \ K$ and $P_r = 600 \ kPa$ and the nozzle exit area 40 cm^2, find the mass flow rate at different exit M. Assume isentropic flow with air behaving as an ideal gas $\gamma = 1.4$.

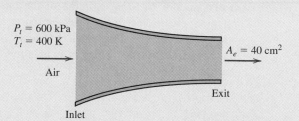

$P_t = 600$ kPa
$T_i = 400$ K

Air

Inlet

$A_e = 40$ cm^2

Exit

FIGURE 11.6 Schematic of Example 11.4.

SOLUTION According to Equation (11.9),

$$\dot{m} = \sqrt{\frac{\gamma}{R}} \frac{P_t}{\sqrt{T_t}} \left[1 + \frac{(\gamma - 1)}{2} M^2 \right]^{\frac{\gamma+1}{2-2\gamma}} M A,$$

then,

$$\dot{m} = \sqrt{\frac{1.4}{287\,J/kg\,K}} \frac{600\,kPa}{\sqrt{400K}} \left[1 + \frac{(1.4-1)}{2}M^2 \right]^{\frac{1.4+1}{2-2\times1.4}} M \frac{40cm^2}{10000cm^2/m^2}.$$

The mass flow rate, then, can be calculated for various values of M.

Since the gas originates from a reservoir, the inlet velocity is negligible (i.e., $M \approx 0$). Subsonic flow in a converging nozzle accelerates. Therefore, according to the discussion from the previous section, the flow can accelerate to a maximum $M = 1$. Indeed, by plotting the results, we observe that the mass flow rate assumes a maximum flow rate when $M = 1$ (Figure 11.7). At these conditions, the nozzle is choked (i.e., no additional mass can flow through the nozzle).

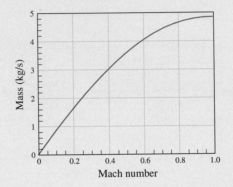

FIGURE 11.7   Mass flow rate as a function of the Mach number for Example 11.4.

According to Equation (11.5)

$$P_r = P_b \left(1 + \frac{(\gamma-1)}{2}M_e^2 \right)^{\frac{\gamma}{\gamma-1}},$$

the exit Mach is also related to the exit pressure, which is commonly referred to as the back pressure P_b. The variation of the mass flow rate through the nozzle then can be alternatively related to the back pressure: as the back pressure decreases, M increases, resulting in a higher flow rate. However, as soon as the flow at the exit becomes sonic, the nozzle chokes, for which $\dot{m}$ is maximum. This happens at the limiting back pressure $P_{b\,limit}$ corresponding to $M = 1$,

$$P_{b\,limit} = P_r \left(1 + \frac{(\gamma-1)}{2} \right)^{-\frac{\gamma}{\gamma-1}}.$$

In this problem, according to the table in Appendix B the limiting pressure is

$$P_{b\,limit} = 0.5283 \times 600\,kPa = 317\,kPa.$$

Below this back pressure, the flow within the nozzle and, of course, the mass flow rate, remain unchanged. This implies that the exit pressure is uniquely determined by $M = 1$ or $P_{exit} = P_{b\,limit}$. Therefore, when the ambient back pressure is lower than $P_{b\,limit}$, the exit pressure is different from the ambient pressure. The flow adjusts to the lower ambient pressure through a series of expansion waves, to be discussed in later sections.

11.3   Supersonic Flow–Related Phenomena

At supersonic conditions, the flow travels faster than local disturbances in the flow. Consequently, the flow is not capable of sensing downstream "obstacles" or changes in either the flow conditions or in the geometry. When the flow encounters such changes, it adjusts to them abruptly. For instance, when supersonic flow encounters a high pressure region, gas particles "crash" without prior "warning" into what appears to them as a "pressure wall" (Figure 11.8).

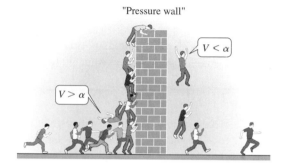

FIGURE 11.8   Normal shock dynamics represented as "molecule" individuals.

When supersonic flow is deflected because of local geometry changes, the flow properties undergo similar adjustment. This situation is in contrast with subsonic flow, where the flow adjusts smoothly to similar changes. For instance, in incompressible flow ($M = 0$) past a cylinder, upstream fluid particles can "sense" the presence of the cylinder at the downstream and, therefore, local streamlines curve appropriately to accommodate the cylinder (Figure 11.9).

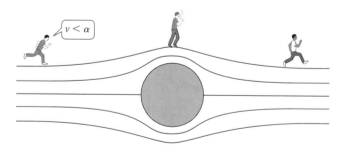

FIGURE 11.9   Flow behavior in subsonic flow.

Since changes under supersonic conditions are abrupt, they are accompanied by loss of energy. These changes also take place over a small distance, the size of which is of the same order of magnitude as the molecular free mean path. The dynamic behavior of the flow within this region is quite complex and cannot be defined exactly using classical arguments from thermodynamic equilibrium. Instead, these changes are approximated mathematically as lines of discontinuities at which the flow properties are assumed to change over an infinitesimal distance.

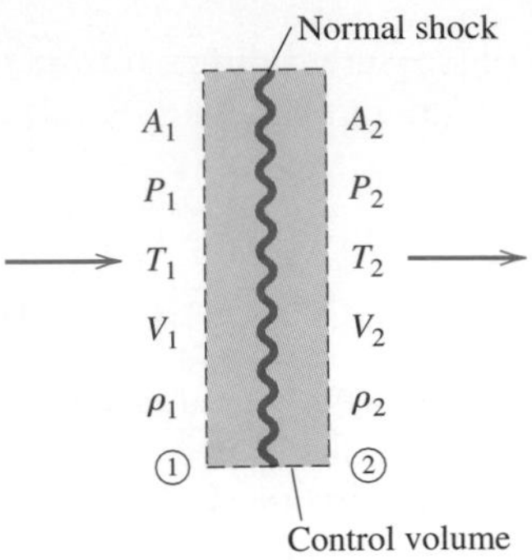

FIGURE 11.10   Schematic of a normal shock.

To summarize: because of energy loss, these process are *irreversible* and consequently they are *non-isentropic* ($ds > 0$). However, since the adjustment takes place over a small region, it is reasonable to assume that they are *adiabatic* ($\dot{Q} = 0$).

11.3.1 Normal Shocks

Because of unfavorable downstream conditions for further expansion — for instance due to a region of high pressure — supersonic flow changes to subsonic flow through a normal shock, as shown in Figure 11.10. Consistent with the physics of such change, the shock is assumed to be adiabatic $\dot{Q} = 0$, and non-isentropic $ds > 0$.

Conservation of Mass

Selecting the area around the shock as a control volume, for steady state conditions with uniform conditions along the shock, mass conservation simplifies to

$$\rho_1 V_1 A_1 = \rho_2 V_2 A_2.$$

Since $A_1 = A_2$ and by substituting the ideal gas relation,

$$\rho_1 V_1 = \rho_2 V_2 \Rightarrow \frac{P_1}{RT_1} V_1 = \frac{P_2}{RT_2} V_2.$$

Substituting $M = V\sqrt{\gamma RT}$ and simplifying,

$$\frac{P_1}{\sqrt{T_1}} M_1 = \frac{P_2}{\sqrt{T_2}} M_2.$$

Conservation of Linear Momentum

Using the same assumptions, the momentum equation is given as

$$\rho_2 V_2 V_2 A_2 - \rho_1 V_1 V_1 A_1 = P_1 A_1 - P_2 A_2.$$

By rearranging and simplifying, we get

$$\rho_1 V_1^2 + P_1 = \rho_2 V_2^2 + P_2.$$

Using the ideal gas relation $P = \rho R T$ and the definition of the Mach number, it simplifies to

$$P_1 \left(1 + \gamma M_1^2 \right) = P_2 \left(1 + \gamma M_2^2 \right).$$

In terms of the pressure ratio, we get

$$\frac{P_2}{P_1} = \frac{\left(1 + \gamma M_1^2 \right)}{\left(1 + \gamma M_2^2 \right)}. \tag{11.10}$$

Conservation of Energy

The energy equation under the same conditions gives

$$\rho_1 V_1 A_1 \left(h_1 + \frac{V_1^2}{2} \right) - \rho_2 V_2 A_2 \left(h_2 + \frac{V_2^2}{2} \right) = 0.$$

Since the mass flow rate is constant, it reduces to

$$h_1 + \frac{V_1^2}{2} = h_2 + \frac{V_2^2}{2}.$$

Rearranging,

$$h_2 - h_1 = \frac{V_1^2}{2} - \frac{V_2^2}{2},$$

and by assuming constant specific heat,

$$C_p \left(T_2 - T_1 \right) = \frac{V_1^2}{2} - \frac{V_2^2}{2},$$

or

$$T_1 + \frac{V_1^2}{2C_p} = T_2 + \frac{V_2^2}{2C_p}.$$

But since $C_p = \dfrac{\gamma R}{\gamma - 1}$, then

$$T_1 + \frac{V_1^2}{2\gamma R} (\gamma - 1) = T_2 + \frac{V_2^2}{2\gamma R} (\gamma - 1).$$

In terms of M we get

$$T_1 \left(1 + \frac{\gamma - 1}{2} M_1^2 \right) = T_2 \left(1 + \frac{\gamma - 1}{2} M_2^2 \right),$$

or

$$\frac{T_2}{T_1} = \frac{\left(1 + \frac{\gamma-1}{2} M_1^2 \right)}{\left(1 + \frac{\gamma-1}{2} M_2^2 \right)}. \tag{11.11}$$

Now, returning to the continuity equation we can write

$$M_1 \frac{P_1}{P_2} \frac{\sqrt{T_2}}{\sqrt{T_1}} = M_2.$$

Substituting the expressions obtained from the momentum and energy equations,

$$M_2 = M_1 \frac{\left(1 + \gamma M_2^2\right)}{\left(1 + \gamma M_1^2\right)} \frac{\sqrt{\left(1 + \frac{\gamma-1}{2} M_1^2\right)}}{\sqrt{\left(1 + \frac{\gamma-1}{2} M_2^2\right)}},$$

by rearranging,

$$M_2 \frac{\sqrt{\left(1 + \frac{\gamma-1}{2} M_2^2\right)}}{\left(1 + \gamma M_2^2\right)} = M_1 \frac{\sqrt{\left(1 + \frac{\gamma-1}{2} M_1^2\right)}}{\left(1 + \gamma M_1^2\right)},$$

and squaring the preceding expression we finally get

$$M_2 \frac{\left(1 + \frac{\gamma-1}{2} M_2^2\right)}{\left(1 + \gamma M_2^2\right)^2} = M_1 \frac{\left(1 + \frac{\gamma-1}{2} M_1^2\right)}{\left(1 + \gamma M_1^2\right)^2}.$$

A trivial solution to the preceding equation is that on which $M_1 = M_2$, which corresponds to the limit of isentropic flow. Defining C as the right-hand side of the above, expression we get

$$M_2 \frac{\left(1 + \frac{\gamma-1}{2} M_2^2\right)}{\left(1 + \gamma M_2^2\right)^2} = C,$$

which can be expressed in quadratic form as

$$M_2^4 \left(\gamma^2 C - \frac{\gamma-1}{2}\right) - M_2^2 \left(1 - 2\gamma C\right) + C = 0.$$

Solving for M_2^2,

$$M_2^2 = \frac{2 + (\gamma - 1) M_1^2}{(1 - \gamma) + 2\gamma M_1^2}. \tag{11.12}$$

With M_2 known, the pressure and temperature on the other side of the shock can be found using the ratios established earlier.

The solution to Equation (11.12) is shown in Figure 11.11. Since shocks develop only in supersonic flow, we consider $M_1 > 1$. Notice that the flow across the *normal*

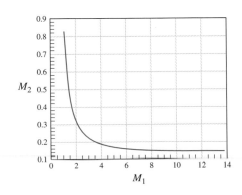

FIGURE 11.11 Mach number after a normal shock M_2 as a function of the incoming Mach number M_1.

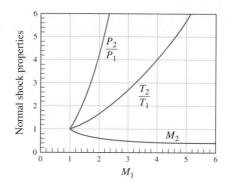

FIGURE 11.12 Mach number M_2, pressure and temperature ratio after a normal shock as a function of the Mach number before the normal shock M_1.

shock is always subsonic $M_2 < 1$. For convenience, the solution to Equation (11.12) and the property ratios are given in tabular form in Table B.3 in Appendix B.

Since the normal shock is an adiabatic process, the stagnation temperature across the shock remains constant — that is,

$$T_{t1} = T_{t2}.$$

However, since the shock is a non-isentropic process, the entropy across the shock increases. This increase in entropy can be calculated using the usual expression

$$s_2 - s_1 = C_p \ln \frac{T_2}{T_1} - R \ln \frac{P_2}{P_1}.$$

Consequently, across the shock the stagnation pressure decreases $P_{t1} > P_{t2}$ reflecting the loss of energy during the process.

Figure 11.12 shows the variation of the pressure and temperature ratio across a normal as a function of the Mach number.

EXAMPLE 11.5

Problem Statement Consider a circular diverging nozzle with inlet and exit diameters of 8 *cm* and and 18 *cm*, respectively. The nozzle receives air at $M_i = 1.2$ with static pressure and temperature of $P_i = 420\ kPa$ and $T_i = 600\ K$, respectively. If a normal shock develops at a location where the diameter is 12 *cm*, find the conditions at the exit of the nozzle.

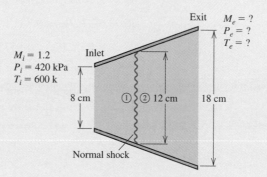

FIGURE 11.13 Schematic of Example 11.5.

SOLUTION   For the given inlet conditions, Table B.2 in Appendix B gives

$$M_i = 1.2 \frac{A_i}{A^*} = 1.030 \frac{P}{P_t} = 0.4124 \frac{P}{P_t} = 0.7764.$$

Therefore,

$$A^* = \frac{A_i}{1.030} = \frac{\pi \times 0.08^2 \; m^2}{4 \times 1.030} = 4.88 \times 10^{-3} \; m^2,$$

$$P_t^* = \frac{P_i}{0.4124} = \frac{420 \; kPa}{0.4124} = 1018.43 \; kPa,$$

$$T_t^* = \frac{T_i}{0.7764} = \frac{600 \; K}{0.7764} = 772.8 \; K.$$

First we solve the flow from the inlet of the nozzle, denoted (i), to the left-hand side of the shock denoted by (1). In this section, the flow is isentropic. The reference throat area A^* and stagnation conditions are the same as the inlet. Therefore, at this location we have

$$\frac{A_1}{A^*} = \frac{\pi D_1^2}{4A^*} = \frac{\pi \times 0.12^2 \; m^2}{4 \times 4.88 \times 10^{-3}} = 2.318.$$

Table B.2 in Appendix B gives

$$M_1 = 2.36, \quad \frac{P_1}{P_t} = 0.07281, \quad \frac{T_1}{T_t} = 0.4731.$$

Therefore,

$$P_1 = 0.07281 P_t = 0.07281 \times 1018.43 \; kPa = 74.15 \; kPa.$$

$$T_1 = 0.4731 T_t = 0.4731 \times 772.8 \; K = 365.61 \; K.$$

The conditions on the other side of the shock (2) are found in Table B.3 in Appendix B (or the normal shock equations). For $M_1 = 2.36$, Table B.3 gives $M_2 = 0.5275$. Also, we have the following ratios:

$$\frac{T_2}{T_1} = 2.002, \quad \frac{P_2}{P_1} = 6.331, \quad \frac{P_{t_2}}{P_{t_1}} = 0.5572.$$

Therefore,

$$T_2 = 2.002 T_1 = 2.002 \times 365.61 \; K = 731.95 \; K,$$

$$P_2 = 6.331 P_1 = 6.331 \times 74.15 \; kPa = 469.44 \; kPa,$$

$$P_{t_2} = 0.5572 P_{t_1} = 0.5572 \times 1018.43 \; kPa = 567.47 \; kPa.$$

Since the shock is an adiabatic process, $T_{ti} = T_{t1} = T_{t2} = 772.8 \; K$.

Finally we solve for the flow from the right-hand side of the shock (2) to the exit of the channel (e). In this section, the flow is isentropic for which Table B.2 in Appendix B is valid. At a Mach number $M_2 = 0.5275$, we have $\dfrac{A_2}{A^*} = 1.2907$. Therefore,

$$\frac{A_e}{A^*} = \frac{A_e}{A_2} \frac{A_2}{A^*} = \left(\frac{0.18}{0.12} \right)^2 \times 1.2907 = 2.9041.$$

Using Table B.2 in Appendix B, the Mach number at the exit, then, is $M_e = 0.205$. Consistent with the theory, in the diverging part of the nozzle the flow slows down. From the table we get the ratios

$$\frac{P_e}{P_t} = 0.9711, \quad \frac{T_e}{T_t} = 0.9917.$$

Finally,

$$P_e = 0.9711 P_t = 0.9711 \times 567.47 \; kPa = 551.1 \; kPa,$$

$$T_e = 0.9917$$

$$T_t = 0.9917 \times 772.8 \; K = 766.39 \; K.$$

Using the definition, the velocity at the exit is

$$V_e = M_e \sqrt{\gamma R T} = 0.205 \times \sqrt{1.4 \times 287 \; kJ/kgK \times 766.39 \; K} = 113.76 \; m/s.$$

11.3.2 Oblique Shocks

The normal shock described previously is the result of a sudden adjustment to the flow when unfavorable pressure conditions do not allow further expansion. Additionally, supersonic flow undergoes sudden change when the geometry of the flow changes abruptly. For instance, when the flow in Figure 11.14 is deflected by a small angle in the counterclockwise from the original direction, the flow adjusts to the new flow direction through an *oblique shock*. Oblique shocks are similar to normal shocks in that they are non-isentropic, adiabatic processes that take place over a short distance.

Geometry

Considering the geometry of the oblique shock, we can identify two characteristic angles both measured in the *counterclockwise direction* from the original flow direction: (a) the deflection angle β, defined as the angle of the flow after the shock; and (b) θ, the angle of the oblique shock. With respect to the same geometry, the velocity vector is decomposed into two components, one normal to the shock (subscript n) and one along the tangential direction (subscript t) — that is,

$$V_{n1} = V_1 \sin \theta \quad V_{n2} = V_2 \sin (\theta - \beta),$$
$$V_{t1} = V_1 \cos \theta \quad V_{t2} = V_2 \cos (\theta - \beta).$$

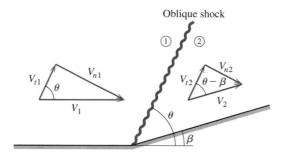

FIGURE 11.14   Schematic of an oblique shock.

Since it is more convenient to work with M, we can also write

$$M_{n1} = M_1 \sin\theta \quad M_{n2} = M_2 \sin(\theta - \beta),$$

$$M_{t1} = M_1 \cos\theta \quad M_{t2} = M_2 \cos(\theta - \beta).$$

Conservation of Mass

By considering the shock as the control volume, the continuity equation

$$\int \rho \mathbf{V} \cdot \mathbf{n}\, dA = 0$$

simplifies to

$$\rho_1 V_{n1} = \rho_2 V_{n2}. \tag{11.13}$$

Conservation of Linear Momentum

Since there is no change in pressure along the tangential direction the net force in that direction is zero. After proper simplification, the component of the momentum equation in the tangential direction then gives

$$V_{t1}\left(\rho_1 V_{n1} A_1\right) = V_{t2}\left(\rho_2 V_{n2} A_2\right),$$

which, because of mass conservation, simplifies further to

$$V_{t1} = V_{t2}.$$

This result indicates that the tangential component of the velocity remains unchanged across the shock.

The normal component of the momentum equation gives

$$P_1 + \rho_1 V_{n1}^2 = P_2 + \rho_2 V_{n2}^2. \tag{11.14}$$

Conservation of Energy

Using the same assumptions, the energy equation is given as

$$h_1 + \frac{V_1^2}{2} = h_2 + \frac{V_2^2}{2},$$

or

$$h_1 + \frac{V_{n1}^2 + V_{t1}^2}{2} = h_2 + \frac{V_{n2}^2 + V_{t2}^2}{2}.$$

Finally, since the tangential components are equal $V_{t1} = V_{t2}$,

$$h_1 + \frac{V_{n1}^2}{2} = h_2 + \frac{V_{n2}^2}{2}. \tag{11.15}$$

A closer look at the simplified forms of mass conservation, momentum, and energy equations (Equations (11.13)–(11.15)) shows that the equations written in terms of normal components are identical to those describing normal shocks. Therefore, oblique shocks can be analyzed as normal shocks, provided the equations are written in terms of the normal component of the Mach number M_{n1} and velocity $\mathbf{V}_n$.

EXAMPLE 11.6

Problem Statement Uniform supersonic air flow at $M = 2.5$ flows past a wedge, as shown in Figure 11.15. The flow adjusts to the new geometry through an oblique shock. The angle of the shock with respect to the uniform flow is $35°$. If the static temperature and pressure of the flow prior to the shock are $50°C$ and $40\ kPa$, respectively, find the conditions on the other side of the shock and the new direction of the flow.

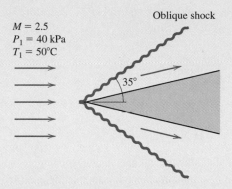

$M = 2.5$
$P_1 = 40\ kPa$
$T_1 = 50°C$

Oblique shock

$35°$

FIGURE 11.15 Schematic of Example 11.6.

SOLUTION Using the geometry

$$M_{n1} = M_1 \sin \theta = 2.5 \sin 35 = 1.43,$$

according to the *normal shock* Table B.3, the normal Mach number across the shock is $M_{n2} = 0.7274$. At the same conditions,

$$\frac{P_2}{P_1} = 2.219 \Rightarrow P_2 = 2.219 \times 40\ kPa = 88.76\ kPa,$$

and

$$\frac{T_2}{T_1} = 1.274 \Rightarrow T_2 = 1.274 \times 323\ K = 411.5\ K.$$

Since the flow is adiabatic $T_{t1} = T_{t2}$. For $M_1 = 2.5$,

$$\frac{T_1}{T_{t1}} = 0.4444 \Rightarrow T_{t1} = 323K/0.4444 = 726.8\ K = T_{t2}.$$

At the other side of the shock,

$$\frac{T_2}{T_{t2}} = \frac{411.5\ K}{726.8\ K} = 0.5662.$$

The isentropic table gives $M_2 \approx 1.96$. Using the relation

$$M_{n2} = M_2 \sin (\theta - \beta) \Rightarrow \sin (\theta - \beta) = \frac{0.7274}{1.96} = 0.3711.$$

Therefore, the new flow direction β is

$$\beta = 35° - \sin^{-1} 0.3711 = 13.22°.$$

Notice that the Mach number on the other side of the oblique shock is still supersonic. This is not a contradiction and does not violate the second law of thermodynamics since, according to the theory, only the normal component must be subsonic.

In most practical problems, however, the known variables are the incoming flow conditions and angle β. Therefore, it is more useful to express θ in terms of M_1 and β.

Since

$$V_{t1} = V_{t2} \Rightarrow M_{t1}\sqrt{\gamma R T_1} = M_{t2}\sqrt{\gamma R T_2},$$

therefore,

$$M_{t2} = M_{t1}\sqrt{\frac{T_1}{T_2}}.$$

However,

$$M_{n2} = M_{t2} \tan\theta \Rightarrow M_{n2} = M_{t1}\sqrt{\frac{T_1}{T_2}}\tan(\theta - \beta) = M_1 \cos\theta \sqrt{\frac{T_1}{T_2}}\tan(\theta - \beta).$$

By squaring the preceding expression and substituting the temperature ratio as obtained from the normal shocks

$$\frac{T_1}{T_2} = \frac{\left(1 + \frac{\gamma-1}{2}M_{n2}^2\right)}{\left(1 + \frac{\gamma-1}{2}M_{n1}^2\right)},$$

we get

$$M_{n2}^2 = M_1^2 cos^2\theta \frac{\left(1 + \frac{\gamma-1}{2}M_{n2}^2\right)}{\left(1 + \frac{\gamma-1}{2}M_{n1}^2\right)}tan^2(\theta - \beta).$$

From the normal shock relations, we have

$$M_{n2}^2 = \frac{2 + (\gamma - 1)M_{n1}^2}{(1 - \gamma) + 2\gamma M_{n1}^2}.$$

Using $M_{n1} = M_1 \sin\theta$, then, the foregoing can be combined and simplified to

$$\beta = \theta - \tan^{-1}\left[\frac{1}{\sin\theta\cos\theta}\left(\frac{\gamma-1}{\gamma+1}\sin^2\theta + \frac{2}{\gamma+1}\frac{1}{M_1^2}\right)\right]. \tag{11.16}$$

The solution to this equation is plotted in Figure 11.16. We observed that for a given incoming Mach number and deflection angle β, there are two possible solutions for the shock angle. The higher of the two angles corresponds to a *strong shock*, and the lower to a *weak shock*.

In practice, the strong shock is associated with higher pressure ratios that are not commonly observed in practical problems. Therefore, here we consider only weak shock solutions. The figure also shows that there is no solution for deflection angles beyond a certain limit. In such a case, the shock is *detached* and forms ahead of the body, as shown in Figure 11.17. Near the "nose" of the detached shock the shock is normal, thus resulting in a subsonic region just behind it. Away from the nose, however, as the shock curves and becomes more oblique, the flow behind the shock is supersonic.

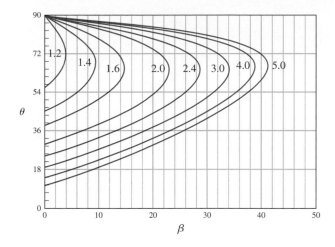

FIGURE 11.16 Deflection angle in oblique shock as a function of the freestream Mach number ($\gamma = 1.4$).

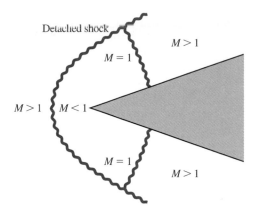

FIGURE 11.17 Schematic of the flow structure around a "blunt" nose.

11.3.3 Prandtl-Meyer Expansion

When the deflection due to the geometry changes the flow direction to a clockwise direction, as shown in Figure 11.18, the flow cannot turn the corner smoothly. Instead, it adjusts through a *Prandtl-Meyer* expansion wave. Since the effective flow area after the adjustment increases, the flow continues to expand in an essentially *isentropic* process.

Consider the schematic in Figure 11.18, where the flow due to the geometry deflects by a small angle dv in the clockwise direction. The objective here is to determine the change in the flow properties as a function of dv. Since the flow is supersonic, the disturbance caused by the corner travels in a direction defined by the Mach angle μ (Figure 11.18). According to the geometry, then,

$$V \cos \mu = (V + dV) \cos (\mu + dv).$$

Using trigonometry,

$$V \cos \mu = (V + dV) (\cos \mu \cos dv - \sin \mu \sin dv).$$

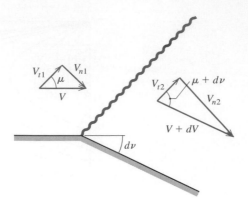

FIGURE 11.18   Schematic of a Prandtl-Meyer expansion.

For small deflection angles $\cos dv \approx 1$, and $\sin dv \approx dv$. By substituting and simplifying, we get

$$\frac{dV}{V} = \tan \mu \, dv.$$

However, since $\sin \mu = \dfrac{1}{M}$, then

$$\tan \mu = \frac{\sin \mu}{\cos \mu} = \frac{\sin \mu}{1 - \sin^2 \mu} = \frac{1}{\sqrt{M^2 - 1}}.$$

Therefore,

$$\frac{dV}{V} = \frac{1}{\sqrt{M^2 - 1}} \, dv.$$

Using the definition of the Mach number $V = M\sqrt{\gamma RT}$, upon differentiation, we get

$$\frac{dV}{V} = \frac{dM}{M} + \frac{1}{2}\frac{dT}{T}.$$

Differentiating the energy equation for adiabatic flow

$$T\left(1 + \frac{\gamma - 1}{2}M^2\right) = Constant,$$

with respect to M we get

$$\frac{dT}{T} = -\frac{2(\gamma - 1)M \, dM}{2 + (\gamma - 1)M^2}.$$

By substitution, the deflection angle in terms of M becomes

$$dv = \frac{2\sqrt{M^2 - 1}}{2 + (\gamma - 1)M^2}\frac{dM}{M}.$$

For finite deflection angles v, the preceding expression can be integrated between two different Mach numbers to yield

$$v_2 - v_1 = \left[\sqrt{\frac{\gamma + 1}{\gamma - 1}}\tan^{-1}\sqrt{\frac{\gamma - 1}{\gamma + 1}\left(M^2 - 1\right)} - \tan^{-1}\sqrt{M^2 - 1}\right]_{M_1}^{M_2} \qquad (11.17)$$

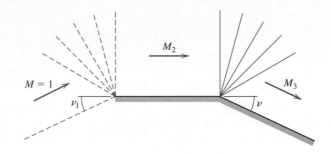

FIGURE 11.19   Reference conditions in a Prandtl-Meyer expansion.

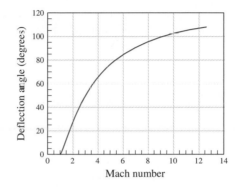

FIGURE 11.20   Deflection angle v as a function of the Mach number M ($\gamma = 1.4$).

Using the approach of earlier sections, for convenience we introduce as reference a hypothetical state where for $v_1 = 0$, the Mach number is unity ($M_1 = 1$), as shown in Figure 11.19. The simplified equation

$$v = \sqrt{\frac{\gamma + 1}{\gamma - 1}} \tan^{-1} \sqrt{\frac{\gamma - 1}{\gamma + 1} \left(M^2 - 1 \right)} - \tan^{-1} \sqrt{M^2 - 1},$$

plotted in Figure 11.20, gives the Mach number M after a finite clockwise total deflection v. For convenience, the values of v as a function of M are also tabulated in Appendix B in Table B.4.

EXAMPLE 11.7

Problem Statement   Consider uniform supersonic flow at Mach 1.5, with static pressure of 90 kPa and static temperature 300 K. Find the flow conditions after the flow has expanded through a clockwise deflection of $20°$ as shown in Figure 11.21.

SOLUTION   Using the reference conditions, we can assume that the uniform flow has resulted from a hypothetical expansion as shown in Figure 11.21. For $M = 1.5$, according to Table B.4 in Appendix B, the corresponding deflection angle is $11.905°$. Therefore, the total deflection angle with respect to the reference conditions is $20° + 11.905° = 31.905°$. Using again the tabulated results for this deflection, the $M \approx 2.2$. Therefore, as the flow deflects by $20°$ it also expands from $M = 1.5$ to $M = 2.2$.

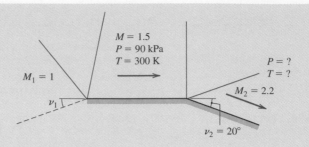

FIGURE 11.21   Schematic for Example 11.7.

Since the flow is isentropic, the other flow properties can be evaluated using the isentropic expressions developed previously:

$$\frac{T_2}{T_1} = \frac{2 + (\gamma - 1)\,M_1^2}{2 + (\gamma - 1)\,M_2^2} \tag{11.18}$$

and

$$\frac{P_2}{P_1} = \left[\frac{2 + (\gamma - 1)\,M_1^2}{2 + (\gamma - 1)\,M_2^2}\right]^{\frac{\gamma}{\gamma-1}}. \tag{11.19}$$

For the conditions given in the problem,

$$\frac{T_2}{300\ K} = \frac{2 + (1.4 - 1)\,1.5^2}{2 + (1.4 - 1)\,2.2^2} \Rightarrow T_2 = 221.04\ K,$$

and

$$\frac{P_2}{90\ kPa} = \left[\frac{2 + (1.4 - 1)\,1.5^2}{2 + (1.4 - 1)\,2.2^2}\right]^{\frac{1.4}{1.4-1}} \Rightarrow P_2 = 30.90\ kPa.$$

These are the same isentropic expressions presented in tabular form in Appendix B. Therefore, the problem could be also solved using these tables.

11.4   Flow in a Converging-Diverging Nozzle

A classical problem is compressible flow through a converging-diverging nozzle with the gas emanating from a large reservoir. This reservoir may be, for instance, the combustion chamber of a jet engine where gas is accelerated in order to generate thrust for propulsion. In such a case, the gas coming from the combustion chamber is at high temperature and pressure but relatively low velocity. As we have seen in the previous section, the only way to accelerate subsonic flow (i.e., gas at low velocity), is through a converging nozzle. Once the flow reaches sonic conditions at the throat, it can be accelerated further only by a diverging nozzle. To achieve supersonic flow then we need a combination of a converging-diverging nozzle.

Figure 11.22 shows a converging-diverging nozzle operating between a reservoir at pressure P_r and a back pressure P_b. The nozzle receives gas at relatively low velocity $M \approx 0$ and delivers it at a higher velocity at the exit. The amount of gas through the nozzle depends on the back pressure. If P_b is equal to the reservoir pressure (Figure 11.22(a)), according to the momentum equation no gas flows through the nozzle. Flow is induced

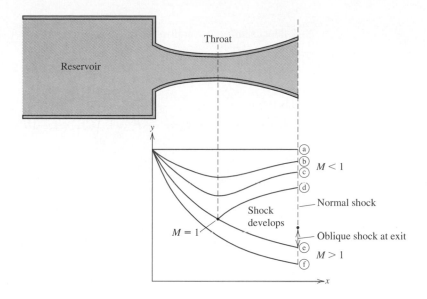

FIGURE 11.22 Property conditions in flow through a converging-diverging nozzle.

when the back pressure is lower than the reservoir pressure. Since the flow from the reservoir is subsonic, in the converging part of the nozzle the flow accelerates. However, at relatively small pressure differences $(P_r - P_b)$, the Mach number at the throat remains subsonic (Figure 11.22(b)). Under these conditions, then, the flow in the diverging part of the nozzle decelerates.

By lowering the back pressure, the velocity and the flow rate in the nozzle increases (Figure 11.22(c)). At some limiting value, the flow at the throat chokes (i.e., $M_{throat} = 1$). At this condition, if the back pressure is not low enough, the flow decelerates back to subsonic conditions (Figure 11.22(d)). At sufficiently low P_b, the flow continues to expand by accelerating to supersonic conditions (Figure 11.22(e)). As shown in the previous section, once the flow at the throat is at sonic conditions the nozzle chokes. For back pressures between points (d) and (e) in the diverging part of the nozzle, the flow encounters a pressure higher than that necessary to sustain the expansion. In such a case, as the flow is unable to "sense" this, it adjusts to this situation through a normal shock.

Very much as in the example in the previous section, by lowering the back pressure below the limit at which the flow expands isentropically to supersonic conditions, the entire flow field within the nozzle remains unchanged. Therefore, the exit pressure under choked conditions is different from the back pressure. Again, in such a case the flow adjusts to the lower back pressure through a series of expansion waves (Figure 11.22(f)).

EXAMPLE 11.8

Problem Statement   Consider a converging-diverging nozzle with a throat and exit areas of 50 cm^2 and 75 cm^2, respectively. The nozzle receives air from a reservoir at 1.2 MPa and 500 K. If the nozzle operates isentropically, find (a) the conditions at the exit of the nozzle; (b) the mass flow rate.

SOLUTION   For this nozzle,

$$\frac{A_e}{A_t^*} = \frac{75 \ cm^2}{50 \ cm^2} = 1.5.$$

Under isentropic conditions, Table B.2 gives $M_e = 1.854$. Also,

$$\frac{P_e}{P_t} = 0.1602 \quad \text{and} \quad \frac{T_e}{T_t} = 0.5926.$$

Therefore,

$$P_e = \frac{P_e}{P_t} P_t = 0.1602 \times 1.2 \; MPa = 192.24 \; kPa,$$

and

$$T_e = \frac{T_e}{T_t} T_t = 0.5926 \times 500 \; K = 296.3 \; K.$$

The velocity at the exit of the nozzle,

$$V_e = M\sqrt{\gamma RT} = 1.854\sqrt{1.4 \times 287 \; J/kgK \times 296.3 \; K} = 639.71 \; m/s.$$

The mass flow rate is calculated using the definition

$$\dot{m} = \rho V_e A_e = \frac{P_e}{RT_e} V_e A_e,$$

$$\dot{m} = \frac{192.24 \; kPa \times 1000 \; Pa/kPa}{287 \; J/kgK \times 296.3 \; K} \times 639.71 \; m/s \times \frac{75 \; cm^2}{100^2 \; cm^2/m^2} = 10.846 \; kg/s.$$

EXAMPLE 11.9

Problem Statement   A converging-diverging nozzle is designed to operate isentropically at $M_e = 1.42$. The nozzle receives air from a reservoir at a pressure of 3.5 MPa. Find (a) the exit pressure at the design conditions; (b) the maximum pressure to choke the nozzle; (c) the range of back pressures so that a normal shock develops inside the nozzle.

SOLUTION   At the design conditions $M_e = 1.42$, the table in Appendix B gives

$$\frac{A_e}{A^*} = 1.126 \quad \text{and} \quad \frac{P_e}{P_t^*} = 0.3055.$$

Therefore, the exit pressure is

$$P_e = \frac{P_e}{P_t} P_t = 0.3055 \times 3.5 \; MPa = 1.069 \; MPa.$$

The flow chokes when the back pressure is low enough to produce sonic conditions at the throat, but not low enough to produce supersonic conditions at the exit. Consequently, while $M_{throat} = 1$, the exit flow is subsonic $M_{exit} < 1$. Therefore, for $\frac{A_e}{A_*} = 1.126$, Table B.2 in Appendix B for subsonic flow gives $M_e = 0.6606$, and $\frac{P_e}{P_t} = 0.7461$. Therefore,

$$P_{e_{max}} = \frac{P_e}{P_t} P_t = 0.7461 \times 3.5 \; MPa = 2.6114 \; MPa.$$

For a normal shock to appear at the location of the throat, the back pressure must be just below the pressure required to produce sonic conditions at the throat. Therefore, for

$P_e < 2.6114\ MPa$, a normal shock will develop just after the throat. For a normal shock just inside the exit of the nozzle and since $\dfrac{A_e}{A_*} = 1.126$, $M_1 = 1.42$. The pressure at the left of the shock then is $P_1 = 1.069\ MPa$. According to the normal shock Table B.3 in Appendix B, the flow at other side of the shock is $M_2 = 0.7314$, and $\dfrac{P_2}{P_1} = 2.186$. A shock will develop at the exit of the nozzle for a back pressure of

$$P_2 = \frac{P_2}{P_1} P_1 = 2.186 \times 1.069\ MPa = 2.337\ MPa.$$

11.5 Flow with Friction

In a departure from the isentropic flow discussed earlier, in this section we consider compressible flow when friction effects are important. This flow is known as *Fanno flow*. The flow is *non-isentropic*, reflecting the fact that friction renders the process irreversible as energy is lost. To separate the issues related to heat transfer, the flow is assumed to be *adiabatic*. Subsequently, the friction effects are illustrated using flow through a channel with constant cross-sectional area.

11.5.1 Effect of Friction on Flow Properties

Consider the schematic shown in Figure 11.23. Consistent with earlier flows, the objective is to determine the effects of friction on the flow properties. Again, the Mach number M serves as the key variable that all other properties are expressed in terms of.

Applying the momentum equation to the problem shown in Figure 11.23,

$$\rho V A\,(V + dV - V) = [P - (P + dP)]\,A - \tau_f\,(Perimeter)\,dx.$$

However, since the channel may have an arbitrary cross-sectional shape, it is customary to express the cross-sectional area in terms of an *equivalent hydraulic diameter* D_h as

$$D_h = \frac{4Area}{Perimeter}.$$

The factor 4 is included so that the definition applied to a circular area reduces to the diameter of the circle. According to the definition,

$$Perimeter = \frac{4A}{D_h}.$$

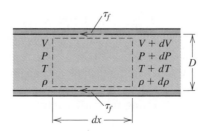

FIGURE 11.23 Compressible flow in a constant area channel with friction.

Recall also that the friction force can be expressed in terms of a *friction coefficient* f as

$$f = \frac{\tau_f}{\frac{1}{2}\rho V^2} \Rightarrow \tau_f = \frac{1}{2} f \rho V^2.$$

As discussed in earlier chapters, f is a function of the Reynolds number Re and the roughness of the channel walls.

The momentum equation then simplifies to

$$\rho V \, dV + dP + 4f \frac{dx}{D_h} \frac{1}{2} \rho V^2 = 0.$$

Dividing by ρV^2 and by substituting M whenever possible, we get

$$\frac{dV}{V} + \frac{1}{\gamma M^2} \frac{dP}{P} + 2f \frac{dx}{D_h} = 0.$$

For this problem, the continuity equation $\rho V = C$ is expressed in differential form as

$$\frac{dV}{V} + \frac{d\rho}{\rho} = 0.$$

Similarly, the ideal gas relation $P = \rho R T$ in differential form is $dP = \rho R \, dT + R T \, d\rho$, or as

$$\frac{dP}{P} = \frac{dT}{T} - \frac{dV}{V}.$$

Additionally, the definition of the Mach number $V = M\sqrt{\gamma R T}$ upon differentiation is expressed as

$$\frac{dV}{V} = \frac{dM}{M} + \frac{1}{2} \frac{dT}{T}.$$

Finally, since the flow is adiabatic, by using the energy equation

$$T \left(1 + \frac{\gamma - 1}{2} M^2 \right) = Constant,$$

by differentiation we get

$$\frac{dT}{T} = -\frac{2(\gamma - 1) M^2}{2 + (\gamma - 1) M^2} \frac{dM}{M}.$$

Using the above results, each term in the momentum equation can be expressed in terms of M as

$$f \frac{dx}{D_h} = \frac{1 - M^2}{2 + (\gamma - 1) M^2} \frac{1}{\gamma M^2} \frac{dM}{M},$$

which, upon integration gives

$$f \frac{L}{D_h} = \frac{1}{4\gamma} \left[\frac{1}{M_1^2} - \frac{1}{M_2^2} \right] + \left(\frac{1 + \gamma}{8\gamma} \right) \ln \left[\frac{M_1^2}{M_2^2} \left(\frac{2 + (\gamma - 1) M_2^2}{2 + (\gamma - 1) M_1^2} \right) \right], \qquad (11.20)$$

where L is the distance from the inlet. Equation (11.20) shows that the local Mach number M_2 is determined by the length of the channel along which friction is acting. This is expected because the friction force is proportional to the length of the channel.

Using the earlier intermediate results, the temperature changes according to

$$\frac{dT}{T} = -\frac{2(\gamma - 1) M^2}{2 + (\gamma - 1) M^2} \frac{dM}{M},$$

which upon integration gives

$$\frac{T_2}{T_1} = \frac{2 + (\gamma - 1)\, M_1^2}{2 + (\gamma - 1)\, M_2^2}.$$ (11.21)

The velocity in terms of M is

$$\frac{dV}{V} = \frac{dM}{M} + \frac{1}{2}\frac{dT}{T} = \frac{dM}{M} - \frac{1}{2}\frac{2\,(\gamma - 1)\, M\, dM}{2 + (\gamma - 1)\, M^2}.$$

This simplifies to

$$\frac{dV}{V} = \frac{2}{2 + (\gamma - 1)\, M^2}\frac{dM}{M}.$$

Upon integration, the velocity is expressed as

$$\frac{V_2}{V_1} = \frac{M_2}{M_1}\left[\frac{2 + (\gamma - 1)\, M_1^2}{2 + (\gamma - 1)\, M_2^2}\right]^{\frac{1}{2}}.$$ (11.22)

The pressure is

$$\frac{dP}{P} = \frac{dT}{T} - \frac{dV}{V},$$

which, upon substitution, simplifies to

$$\frac{dP}{P} = -\left[\frac{2 + 2\,(\gamma - 1)\, M^2}{2 + (\gamma - 1)\, M^2}\right]\frac{dM}{M}.$$

Finally, by integration we get

$$\frac{P_2}{P_1} = \frac{M_1}{M_2}\left[\frac{2 + (\gamma - 1)\, M_1^2}{2 + (\gamma - 1)\, M_2^2}\right]^{\frac{1}{2}}.$$ (11.23)

Following the approach established in previous sections, the property ratios and fL/D_h can be tabulated using reference conditions. Here, for convenience we will use as reference conditions the values of the properties at $M = 1$ (Figure 11.24). For instance,

$$f\frac{L_{max}}{D_h} = \frac{1}{4\gamma}\left[\frac{1}{M_1^2} - 1\right] + \left(\frac{1 + \gamma}{8\gamma}\right)\ln\left[M_1^2\left(\frac{\gamma + 1}{2 + (\gamma - 1)\, M_1^2}\right)\right],$$

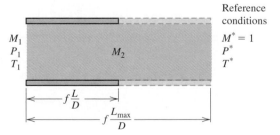

FIGURE 11.24 Reference conditions for flow with friction.

and

$$\frac{T_1}{T^*} = \frac{\gamma + 1}{2 + (\gamma - 1)\,M_1^2} \quad \text{and} \quad \frac{P_1}{P^*} = \frac{1}{M_1}\left[\frac{\gamma + 1}{2 + (\gamma - 1)\,M_1^2}\right]^{\frac{1}{2}},$$

where superscript $(*)$ indicates conditions at a location where $M_2 = 1$. L_{max} is then the length of the pipe required to choke the exit of the pipe. All properties are tabulated in Appendix B, in Table B.5.

Figure 11.25 shows the property variation associated with the Fanno flow described previously.

11.5.2 Fanno Flow Line

The expressions in the previous subsection describe the problem fully. Additional insight into the problem can be gained by evaluating the entropy variation along the channel. Therefore, before proceeding further it is useful to express the entropy change $(s - s_1)$ between the inlet state at an arbitrary point in the flow. From earlier sections we have

$$ds = C_p \ln\frac{T}{T_1} - R\ln\frac{P}{P_1},$$

in terms of the temperature. Using the ideal gas relation,

$$\frac{P}{P_1} = \frac{\rho R T}{\rho_1 R T_1} = \frac{\rho}{\rho_1}\frac{T}{T_1}.$$

But according to mass conservation $\rho V = \rho_1 V_1 = Constant$, then

$$\frac{\rho}{\rho_1} = \frac{V_1}{V}.$$

Therefore,

$$ds = C_p \ln\frac{T}{T_1} - R\ln\left(\frac{V_1}{V}\frac{T}{T_1}\right),$$

or

$$ds = C_p \ln\frac{T}{T_1} + R\ln\frac{V}{V_1} - R\ln\frac{T}{T_1} = C_v \ln\frac{T}{T_1} + R\ln\frac{V}{V_1}.$$

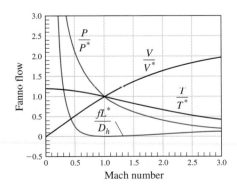

FIGURE 11.25   Property variation in Fanno flow $(\gamma = 1.4)$.

Since the flow is adiabatic, according to the energy equation the stagnation temperature T_t, (and stagnation enthalpy h_t) remains constant—that is,

$$h + \frac{V^2}{2} = h_t;$$

the velocity, then, is given as

$$V = \sqrt{2C_p \left(T_t - T\right)}.$$

Substituting the above expression for V

$$ds = C_v \ln T + R \ln \sqrt{2C_p \left(T_t - T\right)} - C_v \ln T_1 - R \ln V_1.$$

By simplifying, we get

$$ds = C_v \ln T + \frac{R}{2} \ln \left(T_t - T\right) + \frac{R}{2} ln \left(2C_p\right) - C_v ln T_1 - R \ln V_1,$$

or

$$ds = C_v \ln T + \frac{R}{2} \ln \left(T_t - T\right) + \Psi. \tag{11.24}$$

For a given gas and inlet conditions,

$$\Psi = \frac{R}{2} \ln \left(2C_p\right) - C_v \ln T_1 - R \ln V_1,$$

is a constant number. Equation (11.24) is plotted in Figure 11.26, using as typical values $T_{t1} = 400K$, $M_1 = 0.01$, and $M_2 = 5$. The line depicts the locus of all possible thermodynamic states in a channel of constant cross-sectional area and adiabatic conditions. This line is known as the *Fanno line*.

We observe that the curve $T = T(s)$ shows a clear maximum entropy point P. This point divides the curve into two branches: the upper part corresponds to subsonic flow, and the lower part corresponds to supersonic conditions. This can be seen from the energy equation for which, since the flow is adiabatic and h_t is constant, a thermodynamic state at the upper part of the branch is associated with relative low velocity and higher sonic speed (i.e., subsonic flow). For the same reasons, the lower part of the curve is associated with supersonic conditions since the relative velocity is high with correspondingly low sonic speed.

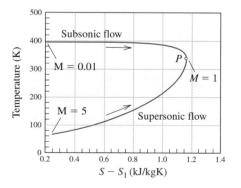

FIGURE 11.26   Fanno line.

The point of maximum entropy is determined by finding the maximum of $s = s(T)$ — that is,

$$\frac{ds}{dT} = 0 \Rightarrow \frac{C_v}{T} - \frac{R}{2\,(T_t - T)} = 0.$$

Simplifying and multiplying both sides by C_p,

$$2C_p\,(T_t - T) = \frac{C_p R}{C_v}T \Rightarrow V^2 = \gamma RT = \alpha^2.$$

The maximum entropy occurs at the point where the flow becomes sonic (i.e., $M_p = 1$).

Since the flow is adiabatic and because the process is non-isentropic, all possible thermodynamic states along the channel must be associated with increasing entropy (i.e., all points along the Fanno line must be to the right of an initial thermodynamic state). This realization is revealing: because of friction, subsonic flow accelerates (!) while supersonic flow slows down. However, from an entropy perspective, the flow cannot go beyond the point of maximum entropy (point P on the figure). When the flow reaches sonic conditions, the channel chokes, which implies that the mass flow rate $\dot{m}$ is maximum. According to Equation (11.20), this occurs when the length of the channel is L_{max}. The flow at the exit of the channel chokes for channel lengths greater than L_{max}. As the length of the channel increases, the maximum flow rate through the channel decreases. At these conditions, $M_{exit} = 1$, while the velocity at the inlet decreases to reflect the decrease in the maximum flow rate it at which can flow through the channel. Consequently, since the channel is unable to accommodate the original velocity and hence M, the reduction in the velocity results in a lower inlet Mach number.

EXAMPLE 11.10

Problem Statement   Consider a 0.4 m section of a straight pipe with diameter 4 cm and a friction factor $f = 0.015$, as shown in Figure 11.27. Find the flow conditions at the exit of the pipe when the inlet $M_i = 0.55$ and the static pressure and temperature are 160 kPa and 320 K, respectively.

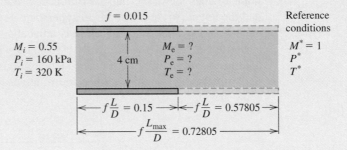

FIGURE 11.27   Schematic of Example 11.10.

SOLUTION   For the given conditions,

$$V_i = M_i\sqrt{\gamma RT_i} = 0.55 \times \sqrt{1.4 \times 287\ kJ/kgK \times 320\ K} = 197.22\ m/s,$$

and

$$f\frac{L}{D} = 0.015\frac{0.4\ m}{0.04\ m} = 0.15.$$

However, according to Table B.5 in Appendix B, for $M_i = 0.55$ the maximum length of the pipe that will choke the exit is

$$f\frac{L_{max}}{D} = 0.72805.$$

Since

$$f\frac{L_{max}}{D} > f\frac{L}{D},$$

the flow will not choke. Using Table B.5,

$$\frac{T_i}{T^*} = 1.1315, \quad \frac{P_i}{P^*} = 1.9341, \quad \frac{V_i}{V^*} = 0.58506.$$

Therefore,

$$T^* = \frac{T^*}{T_i}T_i = \frac{320\ K}{1.1315} = 282.81\ K,$$

$$P^* = \frac{P^*}{P_i}P_i = \frac{160\ kPa}{1.9341} = 82.73\ kPa.$$

and

$$V^* = \frac{V^*}{V_i}V_i = \frac{197.22\ m/s}{0.58506} = 337.09\ m/s.$$

Since the exit of the pipe has the same reference sonic conditions as the inlet, we have

$$f\frac{L_{max\,exit}}{D} = f\frac{L_{max}}{D} - f\frac{L_{max\,inlet}}{D} = 0.72805 - 0.15 = 0.57805.$$

According to Table B.5, then, $M_e = 0.58$ and

$$\frac{T_e}{T^*} = 1.1244, \quad \frac{P_e}{P^*} = 1.8282, \quad \frac{V_e}{V^*} = 0.6150.$$

Therefore,

$$T_e = \frac{T_e}{T^*}T^* = 1.1244 \times 282.81\ K = 318.K,$$

$$P_e = \frac{P_e}{P^*}P^* = 1.8282 \times 82.73\ kPa = 151.25\ kPa,$$

and

$$V_e = \frac{V_e}{V^*}V^* = 0.6150 \times 337.09\ m/s = 207.31\ m/s.$$

EXAMPLE 11.11

Problem Statement   Consider a straight 90 cm section of a pipe with 5 cm diameter. The pipe receives air at $M = 2.2$ and static pressure and temperature of 800 kPa and 580 K, respectively. The equivalent friction factor is 0.018. If a normal shock develops 0.75 m away from the inlet as shown in Figure 11.28, determine the exit flow conditions.

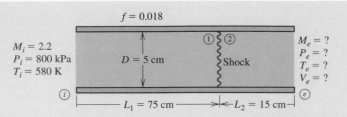

FIGURE 11.28   Schematic of Example 11.11.

SOLUTION   For the given inlet conditions,

$$V_i = M_i \sqrt{\gamma RT} = 2.2\sqrt{1.4 \times 287 \; J/kgK \times 580 \; K} = 1062.04 \; m/s.$$

First we solve for the flow from the inlet of the pipe denoted as (i), to the left hand side of the shock denoted by (1):
For the first length $L_1 = 0.75$, we have

$$f \frac{L_1}{D} = 0.018 \frac{0.75 \; m}{0.05 \; m} = 0.27.$$

According to Table B.5, for $M_i = 2.2$, $f \dfrac{L_{max}}{D} = 0.36091$. Using the table,

$$\frac{T_i}{T^*} = 0.60976, \quad \frac{P_i}{P^*} = 0.35494, \quad \frac{V_i}{V^*} = 1.7179.$$

Therefore,

$$T^* = \frac{T^*}{T_i} T_i = \frac{580 \; K}{0.60976} = 951.19 \; K,$$

$$P^* = \frac{P^*}{P_i} P_i = \frac{800 \; kPa}{0.35494} = 2253.9 \; kPa.$$

and

$$V^* = \frac{V^*}{V_i} V_i = \frac{1062.04 \; m/s}{1.7179} = 618.22 \; m/s.$$

Since state 1 has the same reference sonic conditions as the inlet, we have

$$f \frac{L}{D} = f \frac{L_{max}}{D} - f \frac{L_1}{D} = 0.3609 - 0.27 = 0.0909.$$

According to Table B.5, $M_1 = 1.375$. and

$$\frac{T_1}{T^*} = 0.87066, \quad \frac{P_1}{P^*} = 0.67376, \quad \frac{V_1}{V^*} = 1.2832.$$

Therefore,

$$T_1 = \frac{T_1}{T^*} T^* = 0.87066 \times 951.19 \; K = 828.16 \; K,$$

$$P_1 = \frac{P_1}{P^*} P^* = 0.67376 \times 2253.9 \; kPa = 1518.59 \; kPa,$$

and

$$V_1 = \frac{V_1}{V^*}V^* = 1.2832 \times 618.22 \ m/s = 793.3 \ m/s.$$

The conditions on the other side of the shock (2) are found from the normal shock equations:

For $M_1 = 1.375$, Table B.3 gives $M_2 = 0.7505$. Also, we have the following ratios,

$$\frac{T_2}{T_1} = 1.2385, \frac{P_i}{P^*} = 2.039.$$

Therefore,

$$T_2 = \frac{T_2}{T_1}T_1 = 1.2385 \times 828.16 \ K = 1025.68 \ K,$$

$$P_2 = \frac{P_2}{P_1}P_2 = 2.039 \times 1518.59 \ kPa = 3069.40 \ kPa.$$

Using the definition,

$$V_2 = M_2\sqrt{\gamma RT} = 0.7505\sqrt{1.4 \times 287 \ J/kgk \times 1025.68K} = 481.79 \ m/s.$$

Finally, we solve for the flow from the right-hand side of the shock (2) to the exit of the pipe (e):

For this length $L_2 = 0.15$ we have

$$f\frac{L_2}{D} = 018\frac{0.15 \ m}{0.05 \ m} = 0.054.$$

According to Table B.5, for $M_i = 0.7505$, $f\frac{L_{max}}{D} = 0.12728$. Using the table,

$$\frac{T_2}{T^*} = 1.07865, \quad \frac{P_2}{P^*} = 1.3848, \quad \frac{V_2}{V^*} = 0.77893.$$

Therefore,

$$T^* = \frac{T^*}{T_2}T_2 = \frac{1025.68 \ K}{1.07865} = 950.89 \ K,$$

$$P^* = \frac{P^*}{P_2}P_2 = \frac{3069.4 \ kPa}{1.3848} = 2216.5 \ kPa,$$

and

$$V^* = \frac{V^*}{V_2}V_2 = \frac{481.79 \ m/s}{0.77893} = 618.53 \ m/s.$$

Since state 2 has the same reference sonic conditions as the exit, we have

$$f\frac{L}{D} = f\frac{L_{max}}{D} - f\frac{L_2}{D} = 0.12728 - 0.054 = 0.07328.$$

According to Table B.5, $M_e = 0.8$ and

$$\frac{T_e}{T^*} = 1.06383, \quad \frac{P_e}{P^*} = 1.2892, \quad \frac{V_e}{V^*} = 0.82514.$$

Therefore,

$$T_e = \frac{T_e}{T^*} T^* = 1.06383 \times 950.89 \ K = 1011.59 \ K,$$

$$P_e = \frac{P_e}{P^*} P^* = 1.2892 \times 2216.5 \ kPa = 2857.5 \ kPa,$$

and

$$V_e = \frac{V_e}{V^*} V^* = 0.82514 \times 618.53 \ m/s = 510.37 \ m/s.$$

11.6   Flow with Heat Transfer

Another classical compressible flow problem is the *Rayleigh flow*. This is compressible flow through a *frictionless* channel heated with a uniform heat flux as shown in Figure 11.29. The objective, then, is to determine the flow properties as function of the rate of heating.

11.6.1   Effect of Heat Transfer on Flow Properties

The flow is depicted in the schematic in Figure 11.29. Again, key to the solution is the Mach number. Therefore, using the differential form of governing equations and other relevant relations, upon integration the change in the properties will be expressed in terms of M.

The energy equation applied to the control volume shown in the figure is

$$\rho V A \left[\left(h + dh + \frac{(V + dV)^2}{2} \right) - \left(h + \frac{V^2}{2} \right) \right] = \delta Q_L,$$

which simplifies to

$$dh + V dV = d \left(h + \frac{V^2}{2} \right) = \delta q.$$

However, from the definition of the stagnation conditions, $h + \frac{V^2}{2} = h_t$. Therefore, for constant specific heat,

$$C_p \, dT_t = \delta q \Rightarrow C_p \, (T_{2t} - T_{1t}) = \delta q,$$

which shows that because of heat addition the stagnation temperature increases.

The momentum equation applied to the same control volume is given as

$$\rho A V \, (V + dV - V) = (P - (P + dP)) \, A,$$

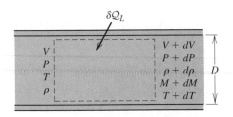

FIGURE 11.29   Compressible flow in a constant area channel with heat transfer.

or in simplified form as

$$dP + \rho V \, dV = 0.$$

Dividing by P and expressing the velocity in terms of M,

$$\frac{dP}{P} + \gamma M^2 \frac{dV}{V} = 0.$$

In deriving the property variation in terms of M, we additionally use the differential form of continuity

$$\frac{dV}{V} + \frac{d\rho}{\rho} = 0,$$

the ideal gas relation

$$\frac{dP}{P} = \frac{dT}{T} - \frac{dV}{V},$$

and the definition of the Mach number

$$\frac{dV}{V} = \frac{dM}{M} + \frac{1}{2}\frac{dT}{T}.$$

By proper substitution, the velocity can be expressed as

$$\frac{dV}{V} = \frac{2}{1 + \gamma M^2}\frac{dM}{M},$$

and upon integration as

$$\frac{V_2}{V_1} = \frac{M_2^2}{M_1^2}\frac{1 + \gamma M_1^2}{1 + \gamma M_2^2}. \tag{11.25}$$

Similarly, the pressure is expressed as

$$\frac{dP}{P} = -\frac{2\gamma M^2}{1 + \gamma M^2}\frac{dM}{M},$$

which, upon integration reduces to

$$\frac{P_2}{P_1} = \frac{1 + \gamma M_1^2}{1 + \gamma M_2^2}. \tag{11.26}$$

Finally, the temperature variation is given by

$$\frac{dT}{T} = -\frac{2\left(1 - \gamma M^2\right)}{1 + \gamma M^2}\frac{dM}{M},$$

and upon integration

$$\frac{T_2}{T_1} = \frac{M_2^2}{M_1^2}\frac{\left(1 + \gamma M_1^2\right)^2}{\left(1 + \gamma M_2^2\right)^2}. \tag{11.27}$$

Again, using the conditions at $M = 1$ as reference values, the property ratios are expressed in terms of M.

$$\frac{P_1}{P^*} = \frac{1 + \gamma}{1 + \gamma M_1^2}, \quad \frac{T_1}{T^*} = M_1^2\frac{(1 + \gamma)^2}{\left(1 + \gamma M_1^2\right)^2} \quad \text{and} \quad \frac{V_1}{V^*} = M_1^2\frac{1 + \gamma}{1 + \gamma M_1^2}. \tag{11.28}$$

For convenience, these values are given in tabular form in Appendix B, Table B.6. Figure 11.30 shows the property variation associated with the Rayleigh flow described above.

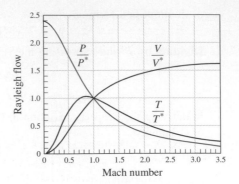

FIGURE 11.30   Property variation in Rayleigh Flow ($\gamma = 1.4$).

11.6.2   Rayleigh Flow Line

As with the Fanno flow, further insight into the flow behavior can be gained by studying the entropy variation within the flow. Because of heat addition, Rayleigh flow is obviously a non-isentropic process. However, we must distinguish between heating or cooling: with heating the entropy of the flow increases, while with cooling it decreases. The issue here, then, is to determine whether the flow has any distinguishing limiting states. Therefore, we will express $s = s(T)$. Since, however the entropy of the flow can either increase or decrease, the reference state with respect to which the change in entropy can be measured is irrelevant. Consequently, for convenience we will select as reference conditions the properties at $M = 1$. Thus

$$ds = C_p \ln \frac{T}{T^*} - R \ln \frac{P}{P^*}.$$

However, according to the relations in Equation (11.28)

$$\left(\frac{P}{P^*} \right)^2 = \frac{1}{M^2} \frac{T}{T^*} \quad \text{and} \quad M^2 = \frac{1 + \gamma}{\gamma} \frac{P^*}{P} - \frac{1}{\gamma},$$

solving for $\dfrac{P}{P^*}$ in terms of $\dfrac{T}{T^*}$,

$$\frac{P}{P^*} = \frac{\gamma + 1}{2} \pm \sqrt{\left(\frac{\gamma + 1}{2} \right)^2 - \gamma \frac{T}{T^*}},$$

substituting into the expression for the entropy

$$s - s^* = C_p \ln \frac{T}{T^*} - R \ln \left[\frac{\gamma + 1}{2} \pm \sqrt{\left(\frac{\gamma + 1}{2} \right)^2 - \gamma \frac{T}{T^*}} \right]. \tag{11.29}$$

The sketch of Equation (11.29) in Figure 11.31 reveals that the graph is divided into a supersonic branch (lower portion of the graph) and a subsonic branch (upper portion of the graph). Moreover, there are two limiting points: one corresponding to the maximum entropy in the flow (P_s) and the other to the maximum static temperature in the flow (P_t).
     Using the continuity equation,

$$\frac{d\rho}{\rho} + \frac{dV}{V} \Rightarrow V \, d\rho = -\rho \, dV.$$

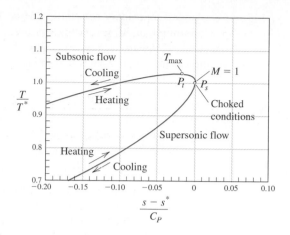

FIGURE 11.31 Rayleigh flow.

The momentum equation then gives

$$dP = V(-\rho\, dV) = V^2\, d\rho.$$

But according to the graph, the point of maximum entropy is at $ds = 0$. Recall though that for isentropic flow

$$\frac{dP}{d\rho} = \alpha^2.$$

By combining the two expressions, we find that at P_s we have $V^2 = \alpha^2$. Therefore, as with the Fanno line, the point of maximum entropy corresponds to sonic conditions (i.e., $M_{P_s} = 1$).

The Mach number at the point of maximum temperature occurs at

$$\frac{d}{dM}\left(\frac{T}{T^*}\right) = 0.$$

Differentiating the expression

$$\frac{T_1}{T^*} = M_1^2 \frac{(1+\gamma)^2}{\left(1+\gamma M_1^2\right)^2},$$

with respect to M, the derivative vanishes when

$$M_{P_t} = \sqrt{\frac{1}{\gamma}}.$$

At this Mach number the maximum temperature is

$$\frac{T_1}{T^*}_{max} = \frac{(1+\gamma)^2}{4\gamma}.$$

With the help of the diagram and the above observations, the *Rayleigh Line* indicates that with heat addition, the temperature in subsonic flow ($M < 1$) increases while the flow accelerates. The temperature reaches a maximum when $M = 1/\sqrt{\gamma}$. According to the second law of thermodynamics, the flow accelerates to a maximum velocity at the limit when $M = 1$. Further heat addition results in choking and, thus, reduction of mass flow rate. However, heat removal or cooling results in the deceleration of the flow.

Under supersonic conditions ($M > 1$), though, heat addition slows down the flow to a minimum of $M = 1$. Further, heat addition chokes the channel, thus resulting in reduction of the mass flow rate. However, contrary to subsonic flow, heat removal or cooling of the channel helps accelerate the flow.

EXAMPLE 11.12

Problem Statement Air at $M_i = 0.28$ and static temperature and pressure of $320\ K$ and $120\ kPa$, respectively, is heated in a channel with constant cross-sectional area with a total heat of $75\ kJ/kg$ (Figure 11.32). Find the flow properties at the exit of the channel.

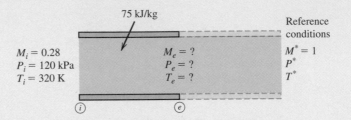

FIGURE 11.32 Schematic of Example 11.12.

SOLUTION For the given inlet Mach number $M_i = 0.28$, the isentropic Table B.2 gives

$$\frac{T_i}{T_t} = 0.9846 \Rightarrow T_t = \frac{T_i}{0.9846} = \frac{320\ K}{0.9846} = 325.0\ K.$$

The inlet velocity is obtained using the definition

$$V_i = M_i \times \sqrt{\gamma R T} = 0.28 \times \sqrt{1.4 \times 287\ J/kgK \times 320\ K} = 100.4\ m/s.$$

Using the energy equation between the inlet and exit of the channel,

$$q = C_p \left(T_{t_2} - T_{t_1} \right) \Rightarrow T_{t_2} = T_{t_1} + \frac{q}{C_P} = 325\ k + \frac{75\ kJ/kg}{1.004\ kJ/kgK} = 400\ K.$$

According to Table B.6 (Reyleigh flow), for $M_i = 0.28$,

$$\frac{T_t}{T_t^*} = 0.31035, \quad \frac{T_i}{T^*} = 0.36667, \quad \frac{P_i}{P^*} = 2.1626, \quad \frac{V_i}{V^*} = 0.16955.$$

Therefore,

$$T_t^* = \frac{T_t^*}{T_t} T_t = \frac{325\ K}{0.31035} = 1047.2\ K,$$

$$T^* = \frac{T^*}{T_i} T_i = \frac{320\ K}{0.36667} = 872.72\ K,$$

$$P^* = \frac{P^*}{P_i} P_i = \frac{120\ kPa}{2.1626} = 55.49\ kPa,$$

and

$$V^* = \frac{V^*}{V_i} V_i = \frac{100.4\ m/s}{0.16955} = 592.16\ m/s.$$

But,

$$\frac{T_{t_e}}{T_t^*} = \frac{T_{t_e}}{T_{t_i}} \times \frac{T_{t_i}}{T_t^*} = \frac{400\ K}{325\ K} \times \frac{325\ K}{1047.2\ K} = 0.38197,$$

for which Table B.6 gives $M_e = 0.32$ and

$$\frac{T_e}{T^*} = 0.45119, \qquad \frac{P_e}{P^*} = 2.0991, \qquad \frac{V_e}{V^*} = 0.21494.$$

Therefore,

$$T_e* = \frac{T_e}{T^*}T^* = 0.45119 \times 872.72\ K = 393.76\ K,$$

$$P_e = \frac{P_e}{P^*}P^* = 2.0991 \times 55.49\ kPa = 116.48\ kPa,$$

and

$$V_e = \tfrac{V_e}{V^*}V^* = 0.21494 \times 592.16\ m/s = 127.28\ m/s.$$

REFERENCES

G.A. TOKATY, *A History and Philosophy of Fluid Mechanics*, New York; Dover, 1994.

A. SHAPIRO, *The Dynamics and Thermodynamics of Compressible Fluid Flow*, **vol. 1**, New York, Ronald Press, 1953.

M.J. ZUCROW, and J.D. HOFFMAN, *Gas Dynamics*, **vol. 1**, New York, Wiley, 1976.

J.E.A. JOHN, *Gas Dynamics*, 2d ed., Boston, Allyn & Bacon, 1995.

B.K. HODGE and K. KOENING, *Compressible Fluid Dynamics*, Englewood Cliffs, NJ, Prentice Hall, 1995.

M.A. SAAD, *Compressible Fluid Flow*, Englewood Cliffs, NJ, Prentice Hall, 1985.

A.M. KUETHE, and C. CHOW, *Foundations of Aerodynamics, Bases of Aerodynamic Design*, 4th ed., Wylie, New York, 1986.

J.D. ANDERSON, *Hypersonic and High Temperature Gas Dynamics*, New York, McGraw-Hill, 1989.

J.D. ANDERSON, *Modern Compressible Flow*, New York, McGraw-Hill, 1990.

PROBLEMS

1. Find the speed of sound in air at $25°C$ and atmospheric pressure of $101\ kPa$.

2. What is the speed of sound at $20°C$ in (a) air, (b) helium, (c) hydrogen, (d) oxygen?

3. What is the speed of sound in air at (a)$30°C$, (b) $40°C$, (c)$50°C$, (d) $60°C$?

4. An aircraft flies at a Mach number of 2.2 and 8 km altitude. What is the speed of the aircraft relative to the ground?

5. Will sound travel faster in air at $25°C$ or in hydrogen at the same temperature?

6. Air is compressed in an isentropic process from an initial pressure of $150\ kPa$ and temperature of $320\ K$, to a final pressure of $450\ kPa$. Find the final temperature and density of the compressed air.

7. Find the work per unit mass required to compress helium isentropically from an initial pressure and temperature of $125\ kPa$ and $20°C$ to a final temperature of $110°C$.

8. The work per unit mass obtained from an isentropically expanding gas from an initial temperature of $600\ K$ is $485.8\ kJ/kg$. What is the gas?

9. What is the Mach number of an aircraft traveling at $495\ miles/hr$ at an altitude of $10\ km$ where the pressure is $26.5\ kPa$ and temperature $-50°C$?

10. Find the maximum velocity an aircraft can fly at an altitude of $5\ km$ before the compressible effect become important. Consider ambient conditions of $54.05\ kPa$ and $255.7\ K$.

11. What is the velocity of a projectile traveling at the speed of sound in ambient air at temperature of $26°C$ and normal atmospheric pressure conditions?

12. Find the time it takes an observer to hear the sound of an aircraft traveling at altitude of $1\ km$ and Mach number of 1.2 from the moment the aircraft passes directly above the observer. Use ambient conditions of $T = 20°C$ and $P = 100\ kPa$.

13. What are the stagnation pressure and temperature corresponding to static conditions of 410 K and 290 kPa for air flow at $M = 0.26$?

14. Find the static conditions corresponding to stagnation pressure and temperature of 1200 kPa and 700K for: (a) $M = 0.4$, $M = 1.0$, $M = 1.8$.

15. Air flows at 950 kPa, pressure 120 kPa and 200°C. What are the stagnation pressure and temperature?

16. An aircraft is flying at 450 $miles/hr$ at a height of 10 km where the pressure is 26.5 kPa and temperature −50°C. What is the pressure and temperature right at the nose of the aircraft? What are the same variables when the aircraft descends to an altitude of 5 km where the pressure is 54 kPa and the temperature −18°C?

17. What is the local Mach number when the static pressure is 50% of the stagnation pressure? What is the static temperature relative to the stagnation temperature at the same Mach number?

18. What is the stagnation pressure and temperature for air at $M = 2.6$ and static pressure and temperature of 400 kPa and 400 K?

19. What is the local Mach number when the static temperature is 50% of the stagnation temperature? What is the static pressure relative to the stagnation pressure at the same Mach number?

20. Find the stagnation density of air corresponding to a static pressure of 650 kPa and static temperature of 350 K and a local Mach number of 0.66.

21. Find the stagnation density of air corresponding to a static pressure of 250 kPa and static temperature of 300 K and a local Mach number of 1.4.

22. Air flows isentropically in a converging nozzle with an inlet temperature and pressure of 300 K and 115 kPa respectively, and inlet velocity of 75 m/s. What are the flow conditions (pressure, density, temperature, velocity) at the exit of the nozzle for an inlet area of 44 cm^2 and exit area of 35 cm^2?

23. Air undergoes an isentropic expansion in a constant diameter duct. The inlet pressure and temperature are 980 kPa and 800 K and exit pressure is 101 kPa. If the inlet velocity is 95 m/s, find the exit velocity. Assume ideal gas conditions with $R = 0.287 \ kJ/kgK$, and $\gamma = 1.4$.

24. A tank with 1 m^3 volume is filled with hydrogen at 1100 kPa and 30°C. A valve is opened, and hydrogen is allowed to escape until the pressure inside the tank is 400 kPa. If the contents of the tank underwent an isentropic process find the mass of the hydrogen left in the tank. For hydrogen $R = 4.124 \ kJ/kgK$ and $\gamma = 1.4$.

25. Air enters a compressor at 101 kPa and 25°C at a rate of 0.2 m^3/s, and leaves at 1 MPA and 50°C. If the inlet and exit diameters are equal to 5 cm, find the exit velocity. Assume that the compression process was isentropic.

26. Air flows in a converging nozzle with exit area of 40 cm^2. The stagnation pressure and temperature are 500 kPa and 400 K respectively. By assuming air behaves as an ideal gas, for isentropic flow, find the mass flow rate when the exit pressure is 300 kPa.

27. A converging nozzle with exit area of 2 cm^2 is supplied from a reservoir at 600 kPa and 800 K. If the flow is isentropic, find the mass flow rate when the exit Mach number is (a) 0.1, (b) 0.5, (c) 1.0.

28. Air is stored in a large reservoir at pressure of 800 kPa and temperature of 500 K. The air is discharged isentropically through a converging circular nozzle with a minimum diameter of 6 cm. Find the maximum mass flow rate through the nozzle. what is the pressure, temperature and velocity at the exit of the nozzle?

29. Air flows isentropically in a diverging circular nozzle with inlet diameter of 10 cm and exit diameter of 15.3 cm. If the inlet conditions are: Mach number 1.2, pressure 480 kPa and temperature 360 K, find the flow conditions (pressure, density, temperature, velocity, mass flow rate) at the exit.

30. Air at $M = 3$ enters a diverging nozzle with an exit-to-inlet ratio of 3. Determine the exit-to-inlet pressure ratio that for isentropic flow through the nozzle.

31. Air at $M = 3$ enters a diverging nozzle with an exit-to-inlet ratio of 3. Determine the exit-to-inlet pressure ratio that causes a normal shock at an area that it is twice the inlet area.

32. Air at Mach number of 0.9 flows isentropically through a circular diverging nozzle with inlet and exit diameters of 12 cm and 18.3 cm. For inlet pressure of 520 kPa and temperature of 410 K, find the Mach number, the pressure, the density, the temperature, the velocity and mass flow rate at the exit.

33. Consider isentropic air flow in a converging nozzle with inlet conditions: Mach number 2.2, pressure 140 kPa and temperature 390 K. Find the exit flow conditions for inlet and exit areas of 40 cm^2 and 22 cm^2 respectively.

34. A converging-diverging nozzle is designed to operate isentropically at a Mach number of 1.6. For reservoir conditions of 1 MPa and 300°C, find: (a) the maximum pressure so that the flow is choked at the throat, (b) the maximum flow rate, (c) the mass flow rate at an exit pressure 15% higher than the pressure found in (a). Use an exit nozzle area of 0.12 cm^2.

35. Air from a reservoir at 400 kPa and 300 K, enters a converging-diverging nozzle with exit-to-throat area ratio of 3. Under isentropic flow conditions, find the exit velocity, pressure and temperature when the mass flow rate is maximum. Is the answer unique?

36. Reconsider the conditions of the previous problem and find now the exit pressure when a normal shock develops at a section where the area ratio (with respect to the throat area) is 2.

37. Find the flow conditions (pressure, density, temperature, velocity) after air flow goes through a normal shock. The velocity prior the shock is 600 m/s, the temperature is 20°C and the pressure 340 kPa.

38. Air velocity after a normal shock is 300 m/s the temperature is 100°C and the pressure 340 kPa. What are the flow conditions prior the shock?

39. The inlet Mach number of air flowing in a diverging nozzle is 1.2 at 380 K and 500 kPa. The inlet area is 30 cm^2 and exit area 60 cm^2. If a normal shock develops at a location where the area is 45 cm^2, find the flow conditions at the exit of the nozzle.

40. The inlet Mach number of air flowing in a diverging nozzle is 1.8 at 430 K and 660 kPa. The inlet area is 20 cm^2 and exit area 50 cm^2. Find the range of back pressures that will cause a normal shock to appear right after the inlet and right before the exit.

41. Air is supplied to a converging-diverging nozzle from a large reservoir kept at 680 kPa and 200°C. If the exit to the throat area ratio is 2.8 and the back pressure is 101 kPa, find the velocity, pressure and temperature at the exit of the nozzle.

42. Air flows in a converging-diverging nozzle with exit-to-throat area ratio of 3.2. If the reservoir is at 800 kPa and 500 K, find the range of back pressures that will produce maximum flow rate through the nozzle. What are the exit conditions when the flow expands completely isentropically?

43. Air received from a reservoir at 1020 kPa and 700 K flows in a converging-diverging nozzle with exit-to-throat area ratio of 3.4. Find the range of back pressures that will generate a normal shock inside the nozzle.

44. A converging-diverging nozzle is designed to operate isentropically at $M_e = 1.6$. The nozzle receives air from a reservoir at a pressure of 3.8 MPa. Find: (a) the exit pressure at the design conditions; (b) the maximum pressure to choke the nozzle; (c) the range of back pressures so that a normal shock develops inside the nozzle.

45. The throat area of a converging-diverging nozzle is 45 cm^2 and the exit area is 100 cm^2. If air with stagnation conditions of 1200 kPa and 622 K flows through the nozzle find the exit conditions when a normal shock develops at a location where the area is 75 cm^2.

46. Find the maximum pressure that will choke the flow in a converging-diverging nozzle with exit to throat area ratio of 3.68 and a stagnation pressure of 730 kPa.

47. Uniform flow at Mach number of 2.8 is deflected 12° degrees in the counterclockwise direction by a wedge. If the static pressure and temperature are 80 kPa and 310 K find the velocity, pressure and temperature after the flow changed direction. What is the minimum angle that will cause the shock to detach from the leading edge of the wedge?

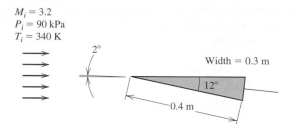

FIGURE 11.33   Schematic for Problem 48.

48. The wedge shown in Figure 11.33 is at an angle of 2° with respect to an incoming uniform air flow at a Mach number of 3.2 and static conditions of 90 kPa and 340 K. Find the drag and lift coefficients with respect to the wedge side area.

49. A wedge deflects uniform air flow at Mach number of 2.2 in a direction counterclockwise to the original flow direction. Due to the deflection an oblique shock develops at an angle of 45° degrees with respect to the incoming flow direction. If the static pressure and temperature are 25 kPa and 10°C, find the angle of the wedge and the final Mach number.

50. A uniform flow encounters a 15° wedge and gets deflected in a counterclockwise direction. If the incoming Mach number is 2.6 and the static pressure and temperature are 78 kPa and 300 K, find the flow conditions (pressure, temperature, Mach number) after the oblique shock.

51. Helium at a Mach number of 3 flows past a wedge. Experiments show that an oblique shock develops at an angle of 40° degrees. Is this possible? If yes, what is the angle of the wedge? Use $R = 2.077 \ kJ/kgK$ and $\gamma = 1.67$.

52. Uniform air flow at a Mach number of 2.4 passes over a wedge with a total angle of 18° degrees. Find the Mach number after the flow adjusts to the new flow direction imposed by the wedge.

53. Uniform air flow at Mach number of 2.4, pressure of 140 kPa and temperature of 280 K is deflected by a wedge. As a result of the deflection an oblique shock develops at a 31° angle with respect to the incoming flow. Find the angle of the wedge and the flow conditions after the oblique shock.

54. The leading edge of the wing of a supersonic aircraft can be approximated by a wedge with a total angle of 8° degrees. If the angle of attack of the aircraft is zero, determine the drag and lift force per unit span length as a function of the Mach number. Consider Mach numbers 1.8, 2.0, 2.4. Use a streamwise length of 2.0 m.

55. Uniform air flow at a Mach number of 2.2 and static pressure and temperature of 80 kPa and 260 K encounters a convex corner of 12° degrees and expands further. Determine the Mach number, pressure and temperature after the flow adjusts to the new flow conditions.

56. Supersonic air flow at Mach number of 2.4 and static pressure and temperature of 90 *kPa* and 280 *K* expands by turning through an angle of 20° degrees (clockwise direction). Find the Mach number, pressure and temperature after the expansion.

57. Consider uniform supersonic air flow at Mach 1.9, with static pressure of 95 *kPa* and static temperature 300 *K*. Find the flow conditions after the flow has expanded through a clockwise deflection of 25°.

58. Uniform supersonic air flow is deflected as shown in Figure 11.34. If after the deflection the Mach number is 1.8, with static pressure and temperature of 105 *kPa* and 285 *K*, find the flow conditions before the flow expanded.

59. A square flat plate (side 40 *cm*) is placed at an angle 8° with respect to a uniform air flow at Mach of 1.9 and static pressure and temperature of 95 *kPa* and 300 *K*. Find the drag and lift coefficients.

60. A square flat thin plate is placed at an angle 6° with respect to a uniform air flow at Mach of 2.2 and static pressure and temperature of 70 *kPa* and 260 *K*. For a 20 *cm* side plate, find the drag and lift forces.

61. A supersonic wing can be approximated by the triangular shape shown in Figure 11.35. Find the drag and lift coefficients with respect to length of the wing in the streamwise direction, for uniform air flow at a Mach number of 2.5 and 3.0, and the same static conditions of 40 *kPa* and 300 *K*.

62. Find the drag and lift coefficients for the triangular airfoil shown in Figure 11.36. The incoming uniform air flow is at a Mach number of 2.8 with static conditions of 120 *kPa* and 340 *K*.

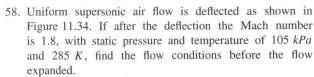

FIGURE 11.34   Schematic for Problem 58.

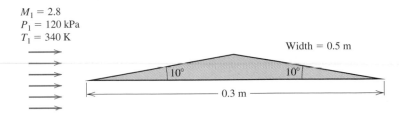

FIGURE 11.35   Schematic for Problem 61.

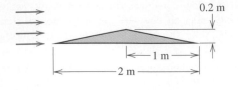

FIGURE 11.36   Schematic for Problem 62.

63. Consider a 0.45 *m* section of a straight pipe with diameter 5 *cm*, and a friction factor $f = 0.01$, as shown in Figure 11.5.2. Find the flow conditions at the exit of the pipe when the inlet $M_i = 0.64$ and the static pressure and temperature of 180 *kPa* and 330 *K* respectively.

64. Flow enters a constant area duct with Mach number of 0.6 and static pressure and temperature of 120 *kPa* and 300 *K*. For a channel length of 40 *cm*, diameter of 3 *cm* and friction factor of 0.02, find the exit Mach number and the static pressure and temperature. Assume adiabatic flow.

65. Air enters a constant area, insulated duct at a Mach number of 0.3. If the friction factor is 0.015, find the length of the duct so that the exit Mach number is (a) 0.6, (b) 1.0.

66. Supersonic air flow at Mach number of 2.2, enters a constant area pipe with a length-to-diameter ratio of 12 and friction factor of 0.01. If the stagnation pressure and temperature are respectively 500 *kPa* and 400 *K*, find the range of exit pressures for which a normal shock develops in the pipe.

67. Consider a straight 1.2 *m* section of a pipe with 5 *cm* diameter. The pipe receives air at $M = 2.1$ and static pressure and temperature of 600 *kPa* and 460 *K* respectively. The friction factor is 0.02. If a normal shock develops 0.6 *m* away from the inlet as shown in Figure 11.28, determine the exit flow conditions.

68. Air enters a constant area insulated duct with inlet Mach number of 0.58. The static pressure and temperature are 160 *kPa* and 320 *K*, respectively. The duct length is 0.40 *m* and its diameter 0.03 *m*. If the friction coefficient is 0.02, find the conditions at the exit of the channel. What is the entropy change between the inlet and exit?

69. A constant area circular duct with 5 cm diameter receives air at Mach number of 0.68, and static pressure and temperature are 160 kPa and 320 K respectively. Find the maximum length that will accommodate the imposed flow rate. What is the mass flow rate if the length of the duct is increased by 20%?

70. Find the back pressure that will produce a normal shock at the middle of a 2.4 m long insulated circular channel with 8 cm diameter. The inlet Mach number is 1.8, the static pressure and temperature are respectively 120 kPa and 315 K. Consider a friction factor of 0.01.

71. Air at $M_i = 0.22$ and static temperature and pressure of 340 K and 160 kPa respectively, is heated in a channel with constant cross-sectional area with a total heat of 60 kJ/kg, Figure 11.32. Find the flow properties at the exit of the channel.

72. Air flows into a constant area frictionless duct with inlet velocity of 110 m/s. The static pressure and temperature are 160 kPa and 410 K respectively. The diameter of the duct is 2.5 cm. Calculate the rate of heat transfer needed to choke the flow.

73. Air enters a constant area frictionless duct with inlet velocity of 90 m/s. The static pressure and temperature are 160 kPa and 380 K respectively. Calculate the maximum amount of heat that can be added without reducing the mass flow rate. Find the percentage reduction in the mass flow rate when the heat transfer is increased by 15% from this maximum heat transfer.

74. Air at supersonic conditions ($M = 1.6$) and stagnation pressure and temperature of 900 kPa and 880 K flows into a constant area frictionless duct. The diameter of the duct is 2. cm. Calculate the rate of heat transfer needed to choke the flow.

75. Air enters a circular duct at a Mach number of 0.22 and static pressure and temperature of 100 kPa and 300 K. Heat is added at a rate of 40 kJ/kg of air. By neglecting the effects of friction, find the final exit flow conditions (Mach number, pressure, temperature).

76. Air enters a channel at 110 m/s and static temperature and pressure of 380 K and 80 kPa. Find the maximum heat per unit mass of air that can be added without reducing the mass flow rate. Neglect the effects of friction.

77. Air at a Mach number of 1.6 and stagnation pressure and temperature of 800 kPa and 700 K enters a 5 cm diameter duct. Find the heat transfer to choke the exit of the duct.

Experimental Fluid Dynamics

All people by nature desire to know.[1]
—Aristotle

12.1 Basic Concepts

Experiments, and experimental observations are pivotal in the study of fluid dynamics to the extent that the two are almost synonymous terms. This is not surprising, because the natural laws, the very foundation of fluid dynamics, have been established by observation (i.e., by repeated experiments). Furthermore, as demonstrated in earlier chapters, the universal laws are made material specific by using such material constants as density and viscosity, all of which are determined experimentally. Additionally, a large number of fluid phenomena, such as many associated with turbulence, can only be studied experimentally. Ultimately, experimental observations are essential in the discovery of new fluid phenomena.

The field of experimental fluid dynamics is quite extensive, and it is not possible in a single chapter to cover the subject in great detail. Given the introductory nature of the material, we are covering the subject in sufficient detail to allow the reader to appreciate the importance, and range of issues associated with the experimental investigation of fluid dynamics. The interested reader should consult more specialized texts on this subject (Holman, 1989). Therefore, following a brief overview of some basic experimental procedures and definitions, the methods used to measure fluid properties and to characterize flow fields are discussed in greater detail. Finally, the operation of some commercially used flow-measuring devices is examined.

12.1.1 Design of Experiments

The design of proper experiments is a field that is both specialized and complicated. In addition to accurate instrumentation, experimentation requires substantial experience. To a large extent, the success of the experiment depends upon proper planning by (a) remaining focused on the scope and objective of the experiment; (b) properly identifying the relevant parameters and their expected range; (c) selecting instruments with the appropriate frequency, accuracy, and range to measure the variables; and (d) having proper control over the variables.

The dimensional analysis discussed in Chapter 6 is very helpful in the proper design of experiments. Dimensional analysis is routinely used to organize experimental variables into meaningful dimensionless groups. Moreover, similarity arguments provide the means to construct scaled experimental models that can reproduce with fidelity the behavior of full scale models or prototypes.

In experiments, one must remain open-minded to results that, at a first glance, may appear to be wrong in as much as they may be outside the range of what is expected.

---

[1] Aristotle, *Metaphysics* A.1, 980 a21.

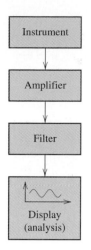

FIGURE 12.1 Schematic of a data acquisition system.

As a matter of fact, experimentalists are almost always faced with the dilemma of either accepting results as evidence of new fluid behavior or rejecting them by attributing them to experimental errors. Therefore, proper control of all aspects of the experimental setup, is critical in establishing confidence in the results. Reproducibility of the results, and proper use of statistical procedures can help identify problematic data.

12.1.2 Data Acquisition

In recent years, with the development of new electronic devices, variable sensing and data collection have been revolutionized. In general, data collection may vary from having people manually record instrument readings to recording data digitally using computers. The output from most sensing instruments is usually translated into an electric signal whose magnitude is proportional to the magnitude of the measured variable. As such, the electric signal often must be amplified by enhancing its magnitude and then conditioned by filtering out "noise" generated by the electronics. Finally, the signal is processed and saved (or displayed) as meaningful data (Figure 12.1).

Using computers, sensing and collection of data can not only be done at fast speeds, but it can also be done reliably by removing the potential for human error. Another advantage is that the system can be programmed to collect the data at precise time intervals. Moreover, with proper scanning equipment, the data acquisition system can be programmed to sequentially collect data from various measuring instruments.

12.1.3 Instrumentation Fundamentals

In fluid experiments, measuring instruments are used either to evaluate fluid parameters or to quantify fluid flow phenomena. In both cases, the validity of the results and the confidence that what is measured is correct depend on the performance and characteristics of the measuring instruments. In general, the reading from these instruments is the response or *output* of the instrument due to an *input* coming from the fluid. The *readability* of a measuring instrument refers to the size of the instrument's display for a given range of the measured variable. Obviously, larger displays can be read easier than smaller displays for the same range. The smallest subdivision of the instrument is called the *least count* or *resolution*.

In response to the input, the display pointer moves a certain linear distance determined by the *sensitivity* of the instrument, which is defined as the ratio of the linear distance to the magnitude of the measured variable. For instance, an instrument with sensitivity 1 *cm/kg* indicates that the pointer of the instrument moves 1 *cm* for every *kg* measured. The *accuracy* of the instrument measures the deviation of the reading from a known measured value. This deviation then is a measure of the instrument's *error* in the measurement. The errors may be *systematic* or *random*, depending on the source of the error. The reproducibility of the results — that is, the ability of the instrument to give the same result for the same input (irrespective of the errors) — is expressed by the *precision* of the instrument. To correct for inherent errors, instruments are always *calibrated* against a known input. Calibration helps remove the errors associated with the inherent systematic errors of the instrument.

Typically, measuring instruments produce the final "steady state" reading after a short time delay. In steady state flows, this is not a problem because if sufficient time is allowed, the instrument will indicate the proper steady value, as shown in Figure 12.2. The time delay is defined as the time it takes an instrument to reach a percentage of the final reading. For an exponential response, this *time constant* is defined as the time it takes for the instrument to reach 63.2% of its final reading.

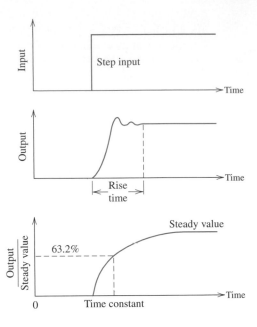

FIGURE 12.2   Time constant for transient measurements.

The response of the instrument becomes important in unsteady flow, since the measured variables vary with time. In such flows, the time delay must be much smaller than the time characteristic of changes in the phenomenon being measured. The instrument's response can also be a function of the frequency of the input. An instrument has *linear frequency response* when the ratio of the input to the output is the same irrespective of the frequency. *Linear amplitude response* refers to an instrument that maintains the same ratio independent of the magnitude of the input. Some instruments also can exhibit *hysteresis* when the instrument indicates a different reading depending on whether the reading is obtained by measuring the variable from a lower value to a higher value or by going from a higher to a lower indication.

12.1.4   Analysis of Experimental Data

The collection of experimental results is usually followed by a proper analysis of the data in order to (a) establish whether they are valid (i.e., they are not the result of experimental errors) and (b) to obtain the information they convey about the physical phenomenon being studied. Since unknown errors are always present, the question then is how confident we are that a measurement is correct, or what is the *uncertainty* associated with a particular measurement? The following subsections review some simple methods of establishing the uncertainty of experimental data and then discuss methods to analyze "raw" data.

Uncertainty Analysis

For a given instrument, the confidence in the measurements is established by repeated use of the instrument and by comparing its readings against known values. For instance, by repeatedly using the same instrument to read pressure we may develop the certainty that the real pressure is, for example, within the range of $\pm 5\ kPa$. Often, however, the final result is due to a combination of several measured quantities. Consequently, the final uncertainty is a function of the uncertainty of each variable in the problem. In such a case, we can use a simple but rather conservative approach, whereby we can assume that the maximum error in each variable occurs at the same time (which, of course, is not likely).

For example, using proper instruments we can measure the pressure P and area A to calculate the force F, according to $F = P\,A$. If the variables are measured as $P = 150 \pm 5 \ kPa$ and, $A = 5 \pm 0.05 \ m^2$, by considering the worst combination of errors, the force can be calculated as

$$F_{max} = 155 \times 5.05 \ kN = 782.75 \ kN,$$

and

$$F_{min} = 145 \times 4.95 \ kN = 717.75 \ kN,$$

where a nominal value of the force is

$$F = 150 \times 5.0 \ kN = 750 \ kN.$$

The simple analysis results in uncertainty of $+4.37\%$, -4.30%. The square-root-of-the-sum-of-squares or the RSS method proposed by Kline and McClintock (Kline and McClintock, 1953) provides a more sophisticated estimate of the total uncertainty. According to the method, the total uncertainty g_t is given by

$$g_t = \left[\left(\frac{\partial R}{\partial x_1} g_1 \right)^2 + \left(\frac{\partial R}{\partial x_2} g_2 \right)^2 + \cdots + \left(\frac{\partial R}{\partial x_n} g_n \right)^2 \right]^{\frac{1}{2}}, \qquad (12.1)$$

where the calculated variable $R = R(x_1, x_2, \ldots x_n)$ is the result that depends upon variables $x_1, x_2, \ldots x_n$, each having uncertainties $g_1, g_2, \ldots g_n$. For the preceding example, $R \equiv F$, $x_1 \equiv P$ and $x_2 \equiv A$. The criterion then becomes

$$g_t = [(A g_P)^2 + (P g_A)^2]^{\frac{1}{2}} = [(5 \times 5)^2 + (150 \times 0.05)^2]^{\frac{1}{2}} = \pm 26.10 \ kN,$$

or 3.48% of the nominal value. It is worth noting that the total uncertainty depends on the square of the uncertainties of each variable. Therefore, the larger uncertainties overshadow the smaller terms.

Statistical Reduction of Data

Often, the value of a single variable is determined from a set of individual measurements. Because of experimental errors, these readings are not necessarily the same. Therefore, a common objective in analyzing the experimental data is to use statistical methods to determine the most likely value of the variable. Additionally, statistics can also be used to eliminate data points that are likely to be erroneous. For a set of n experimental data, we can determine the *arithmetic mean* or *average* x_m by calculating

$$x_m = \frac{1}{n} \sum_{i=1}^{n} x_i. \qquad (12.2)$$

The deviation d_i of each point x_i from the mean value is defined as $d_i = x_i - x_m$. By definition, then, the sum of all deviations is zero. The *standard deviation* σ is defined as

$$\sigma = \left[\frac{1}{n} \sum_{i=1}^{n} (x_i - x_m)^2 \right]^{\frac{1}{2}}. \qquad (12.3)$$

This is called the *biased standard deviation*, reflecting the fact that this expression is strictly true for a large number of data points. In the case of limited data points σ is given instead as

$$\sigma = \left[\frac{1}{n-1} \sum_{i=1}^{n} (x_i - x_m)^2 \right]^{\frac{1}{2}}. \tag{12.4}$$

The quantity σ^2 is called the *variance*. By definition, the variance or standard deviation is a measure of the "spread" around the calculated mean value.

EXAMPLE 12.1

Problem Statement   In an experiment measuring the viscosity of glycerin at room temperature, the following data in units of $N \, s/m^2$ were obtained: $\mu = 0.995$, 1.000, 0.993, 1.009, 1.014, 0.988, 1.007, 1.003, 0.998, 0.989, 1.006, 1.010, 1.304, 1.002, 0.996, 1.008, 0.915 , 1.012, 0.993, 0.997, 1.001, 0.999. Find the statistics of these data (i.e., the average value), and the standard deviation.

SOLUTION   Using the above expressions, the average value of the viscosity is $\mu_{aver} = 1.01086 \, Ns/m^2$. The biased standard deviation for the same data is $\sigma_{biased} = 0.0679 \, Ns/m^2$. The unbiased standard deviation is $\sigma = 0.0684 \, N \, s/m^2$.

Screening of Experimental Data

At this point, we need to consider the possibility that some of the experimental measurements may likely be incorrect because of errors. If we assume that the overall random error is due to the combination of smaller errors of equal magnitude and having an equal likelihood of being either positive or negative, statistical theory shows that the probability density function of the measured data is given by the *Gaussian* or otherwise known as the *normal error distribution*:

$$P(x) = \frac{1}{\sigma \sqrt{2\pi}} e^{-\frac{(x-x_m)^2}{2\sigma^2}},$$

where σ is the standard deviation and x_m, the arithmetic mean of the experimental data. Figure 12.3 shows a schematic for the distribution function. The probability density function represents the likelihood that a measurement x lies between x and $x + dx$. By definition, then, the integration over all possible measurements (i.e., the range between

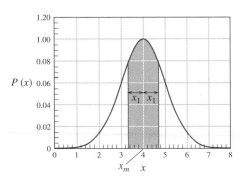

FIGURE 12.3   Probability density function.

$-\infty \leq x \leq \infty$) must be equal to unity

$$\int_{-\infty}^{\infty} P(x)\,dx = 1.0.$$

The probability that a measurement lies between $\pm x_1$ away from the mean value x_m, as shown in Figure 12.3, is

$$P = \frac{1}{\sigma\sqrt{2\pi}} \int_{x_m-x_1}^{x_m+x_1} e^{-\frac{(x-x_m)^2}{2\sigma^2}}\,dx.$$

This is, of course, the shaded area under the curve as shown in Figure 12.3.

The preceding expression can be simplified by introducing a new variable as $\eta = \frac{x - x_m}{\sigma}$ as

$$P = \frac{1}{\sqrt{2\pi}} \int_{-\eta_1}^{\eta_1} e^{-\frac{\eta^2}{2}}\,d\eta, \tag{12.5}$$

where $\eta_1 = \frac{x_1}{\sigma}$. The function given in Equation (12.5) is the error function already encountered in Chapter 7. The function is calculated and tabulated in most standard textbooks. Table 12.1 shows typical values of the distribution. In the table, η is the normalized distance to the right of the mean value $\eta = 0.0$ and the values given correspond to the area in the tail of the distribution (i.e., from η to $+\infty$).

According to the Gaussian distribution, the likelihood of having a measurement within $+1\ \sigma$ from the mean value is 68.27%, within $\pm 2\ \sigma$ from the mean value it is 95.45%; and within $\pm 3\ \sigma$ from the mean value it is 99.73%. This shows that most of the data are more likely to be $3\ \sigma$ away from x_m. The remaining issue is to decide on

TABLE 12.1 Tail Area of Unit Normal Distribution $\eta \to +\infty$

η	0.00	0.02	0.04	0.05	0.06	0.08
0.0	0.5000	0.4920	0.4840	0.4801	0.4761	0.4681
0.2	0.4207	0.4129	0.4052	0.4013	0.3974	0.3897
0.4	0.3446	0.3372	0.3300	0.3264	0.3228	0.3156
0.6	0.2743	0.2676	0.2611	0.2578	0.2546	0.2483
0.8	0.2119	0.2061	0.2005	0.1977	0.1949	0.1894
1.0	0.1587	0.1539	0.1492	0.1469	0.1446	0.1401
1.2	0.1151	0.1112	0.1075	0.1056	0.1038	0.1003
1.4	0.0808	0.0778	0.0749	0.0735	0.0721	0.0694
1.6	0.0548	0.0526	0.0505	0.0495	0.0485	0.0465
1.8	0.0359	0.0344	0.0329	0.0322	0.0314	0.0301
2.0	0.0228	0.0217	0.0207	0.0202	0.0197	0.0188
2.2	0.0139	0.0132	0.0125	0.0122	0.0119	0.0113
2.4	0.0082	0.0078	0.0073	0.0071	0.0069	0.0066
2.6	0.0047	0.0044	0.0041	0.0040	0.0039	0.0037
2.8	0.0026	0.0024	0.0023	0.0022	0.0021	0.0020
3.0	0.0013	0.0013	0.0012	0.0011	0.0011	0.0010
3.2	0.0007	0.0006	0.0006	0.0006	0.0006	0.0005
3.4	0.0003	0.0003	0.0003	0.0003	0.0003	0.0003
3.6	0.0002	0.0001	0.0001	0.0001	0.0001	0.0001
3.8	0.0001	0.0001	0.0001	0.0001	0.0001	0.0001
3.9	0.0000	0.0000	0.0000	0.0000	0.0000	0.0000

the cut-off point below which a certain measurement has low enough probability to occur and, therefore, can be eliminated form the data set. According to *Chauvenet's criterion*, any data with probability of occurrence of less than $\frac{1}{2n}$ can be assumed to be suspicious and discarded as invalid.

EXAMPLE 12.2

Problem Statement   Screen the experimental data given in Example 12.1 to decide whether any data could be eliminated.

SOLUTION   Using *Chauvenet's criterion*, for $n = 22$, any data with probability of occurrence of less than $\frac{1}{2n} = 0.02273$ can be assumed to be suspicious and discarded as invalid. In this case, the data $\mu = 1.304\ N\ s/m^2$ is more than three standard deviations away from the average value, and exceeds the criterion. Therefore, it can be eliminated with their reasonable confidence that it is possibly erroneous. Once this data point is eliminated, the new statistics of the data are $\mu_{aver} = 0.997\ N\ s/m^2$. The standard deviation for the same data is $\sigma = 0.02016\ N\ s/m^2$. Now the data point $\mu = 0.915\ N\ s/m^2$ fails the test and can be eliminated. Finally, then, $\mu_{aver} = 1.001\ N\ s/m^2$, and $\sigma = 0.00755\ N\ s/m^2$.

Regression Analysis

Suppose now we have n reliable data relating two flow parameters x_i, y_i — for instance, the drag coefficient $C_{d_i} = y_i$ and the Reynolds number $Re_i = x_i$, as shown in Figure 12.4. Consistent with the ultimate objective of the experiments to generalize the results, we seek to express the observed behavior by an empirical relation. The figure shows a scatter of the data points in a manner that suggests a linear variation. Since the trend in the figure resembles a linear function, we seek to find the linear equation $y = ax + b$, with constants a, b that follows the data points as closely as possible. This is equivalent to minimizing the total deviation of the actual points y_i from what the function $y_{ic} = ax_i + b$ predicts — that is,

$$D = \sum_{i=1}^{n}(y_i - y_{ic})^2 = \sum_{i=1}^{n}(y_i - ax_i - b)^2.$$

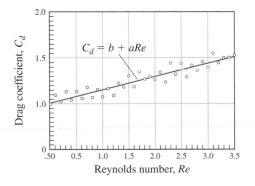

FIGURE 12.4   Plot of data given in Example 12.3.

We can find the minimum by requiring that

$$\frac{dD}{da} = 0 \Rightarrow \sum_{i=1}^{n} 2(y_i - ax_i - b)(-x_i) = 0,$$

which simplifies to

$$a \sum_{i=1}^{n} x_i^2 + b \sum_{i=1}^{n} x_i = \sum_{i=1}^{n} x_i y_i. \tag{12.6}$$

Similarly,

$$\frac{dD}{db} = 0 \Rightarrow \sum_{i=1}^{n} 2(y_i - y_{ic})(-1) = 0,$$

which reduces to

$$a \sum_{i=1}^{n} x_i + bn = \sum_{i=1}^{n} y_i. \tag{12.7}$$

The simultaneous solution of Equations (12.6) and (12.7) yields the constants a, b.

EXAMPLE 12.3

Problem Statement The drag coefficient of spherical particles settling in a liquid solution is measured experimentally as a function of the particle Reynolds number defined as $Re = D(u_p - u_f)/v$, where D is the diameter of the particles, v the viscosity of the liquid, and u_f, u_p the velocity of the liquid and the particle.

The experiment was performed for values of Re ranging from 0 to 3.5 with increments of 0.1. The following data for the drag coefficient were collected: 1.0, 1.094, 1.026, 1.132, 1.044, 1.131, 1.058, 1.177, 1.068, 1.155, 1.078, 1.165, 1.089, 1.226, 1.182, 1.305, 1.184, 1.345, 1.268, 1.293, 1.259, 1.339, 1.233, 1.440, 1.317, 1.437, 1.291, 1.413, 1.361, 1.453, 1.386, 1.547, 1.442, 1.499, 1.502, 1.527. Find a function that best describes $C_d = f(Re)$.

SOLUTION The data scatter shown in Figure 12.4 suggests that the relationship between C_d and Re is a linear function. Therefore, the procedure described previously can be used to find the best linear fit (i.e., a and b) for the data. The constants of the linear function are found by simultaneously solving:

$$a \sum_{i=1}^{n} x_i^2 + b \sum_{i=1}^{n} x_i = \sum_{i=1}^{n} x_i y_i.$$

and

$$a \sum_{i=1}^{n} x_i + bn = \sum_{i=1}^{n} y_i.$$

In this experiment, we have 36 measurements (i.e., $n = 36$). For the given data, we calculate the following $\sum_{i=1}^{n} x_i = 63$, $\sum_{i=1}^{n} x_i^2 = 149.1$, $\sum_{i=1}^{n} y_i = 45.466$, and $\sum_{i=1}^{n} x_i y_i = 85.0524$. Therefore, we have the following expressions:

$$149.1a + 63b = 85.0524,$$

and

$$63a + 36b = 45.466.$$

By solving for a and b, we get $a = 0.1412$ and $b = 1.0158$. The results then suggest the relationship,

$$C_d = 1.0158 + 0.1412 \, Re.$$

The function and the data are plotted in Figure 12.4.

The question now arises as to whether the linear fit represents the data well. The effectiveness of the resulting equation in describing the experimental data can be evaluated using the correlation coefficient r, defined as

$$r = \left[1 - \frac{\sigma_{y,x}^2}{\sigma_y^2} \right]^{\frac{1}{2}},$$

where

$$\sigma_y = \left[\frac{1}{n-1} \sum_{i=1}^{n} (y_i - y_m)^2 \right]^{\frac{1}{2}}, \quad \text{and} \quad \sigma_{y,x} = \left[\frac{1}{n-2} \sum_{i=1}^{n} (y_i - y_{ic})^2 \right]^{\frac{1}{2}},$$

where y_m is the arithmetic mean, y_i are the actual data points, and y_{ic} the calculated values for the same x_i. The closer r is to unity, the better is the linear equation in representing the data. If r is not close to unity, a different functional form must be introduced using the same procedure as before. Standard textbooks in statistical analysis give more details on other assumed functional forms (e.g., quadratic, exponential).

For the data used in Example 12.3 and the resulting linear fit $C_d = 1.0158 + 0.1412 \, Re$, we get $\sigma_y = 0.1589$ and $\sigma_{y,x} = 0.05658$. These values yield $r = 0.9344$, which is indeed close to $r \approx 1$. We conclude then that the linear fit represents the experimental data well.

12.2   Measurement of Fluid Properties

12.2.1   Volume, Mass, Density

Because of the tendency of the fluids to deform easily and assume the shape of their container, the volume of an amount of fluid can be measured very easily by using specially calibrated containers. The weight W of an incompressible fluid in a container provides an indirect way to measure the liquids mass by using $W = mg$. The measurement of density of incompressible fluids then is straightforward. However, to measure the density of compressible gases is more complicated since the density is a property that depends on the thermodynamic state of the gas. For this reason, for compressible gases density is usually measured indirectly. For example, using the ideal gas relation $P = \rho RT$ (if there is reason to believe it is valid), by measuring P and T, we can get the density of the gas. Using optical techniques, though (discussed later), the density of compressible gases can be measured directly.

12.2.2   Viscosity

As demonstrated in earlier chapters, viscosity μ is not only a critical fluid property, but is also associated with the constitutive behavior of a liquid, i.e., its *rheological behavior*.

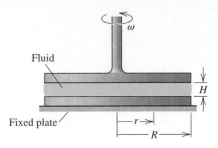

FIGURE 12.5 Schematic of a parallel-disk viscometer.

The general science dealing with this issue is called *rheology*, and the process of obtaining the constitutive fluid properties, *rheometry*.

Parallel-Disk Viscometer

The *parallel-disk viscometer* shown in Figure 12.5 is a typical instrument used to measure the viscosity of a fluid. For the measurement, fluid is placed between two disks with radius R and at a distance h apart, while the upper disk is rotating with angular velocity ω rad/s. The viscosity here (and in most other viscometers) is calculated using the definition of viscosity as the proportionality constant relating the shear stress to the corresponding fluid deformation — that is,

$$\tau_{z\theta} = \mu \Delta_{z\theta}.$$

For a small gap $R \gg h$, the variation of the azimuthal velocity v_θ along the thickness of the gap can be assumed with reasonable accuracy to vary linearly in the gap between two infinite plates — that is,

$$v_\theta = W r \frac{z}{h}.$$

The rate of deformation in cylindrical coordinates yields

$$\Delta_{z\theta} = \frac{1}{r} \frac{\partial v_z}{\partial \theta} + \frac{\partial v_\theta}{\partial z} = \omega \frac{r}{h}.$$

The shear stress is, therefore,

$$\tau_{z\theta} = \mu \omega \frac{r}{h}.$$

The total torque $\mathcal{T}$, developed by the rotation of the disk, is

$$\mathcal{T} = \int_0^R r \tau_{z\theta} 2\pi r \, dr = \int_0^R r \mu \omega \frac{r}{h} 2\pi r \, dr,$$

which upon integration gives

$$\mathcal{T} = \mu \pi \omega \frac{R^4}{2h}.$$

Finally then, by measuring the torque required to sustain the rotation, the viscosity of the fluid can be calculated by

$$\mu = \frac{2h\mathcal{T}}{\pi \omega R^4}. \tag{12.8}$$

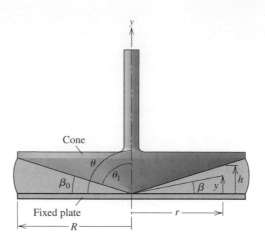

FIGURE 12.6   Schematic of a cone-plate viscometer.

Cone-Plate Viscometer

A variation of the flat-disk viscometer is the *cone-plate viscometer*, shown in Figure 12.6, where fluid is placed between a stationary plate and a rotating cone with a small angle β_0 and angular speed ω *rad/s*. Typically, the angle β_0 is in the range of 0.5° to 8°. Again, by assuming that the gap is narrow, at a radial distance r, the velocity in the direction of the rotation v_θ, varies linearly across the gap—that is,

$$v_\theta = U_c \frac{y}{h},$$

where the local linear velocity of the cone U_c is given by $U_c = \omega r$. In terms of the cone angle, the gap distance h is expressed as $h = r \sin \beta_0$, which for small angles β_0 becomes $h \approx r\theta_0$. Similarly $y \approx r\beta_0$. The velocity is then

$$v_\theta = \omega r \frac{\beta}{\beta_0},$$

and, in terms of spherical coordinates (r, θ, ϕ),

$$v_\theta = \omega r \frac{\beta}{\beta_0} = Wr \frac{\pi/2 - \theta}{\pi/2 - \beta_0}.$$

In the same coordinates, the deformation $\Delta_{\theta\phi}$ given as

$$\Delta_{\theta\phi} = \frac{\sin \theta}{r} \frac{\partial}{\partial \theta} \left(\frac{v_\phi}{\sin \theta} \right) + \frac{1}{r \sin \theta} \frac{\partial v_\theta}{\partial \phi},$$

simplifies to

$$\Delta_{\theta\phi} = -\frac{\omega}{\beta_0},$$

which shows that the rate of deformation is constant across the the thickness of the gap. Therefore, the magnitude of the shear stress is

$$\tau_{\theta\phi} = \mu \frac{\omega}{\beta_0}.$$

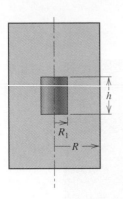

FIGURE 12.7 Schematic of a falling-cylinder viscometer.

The torque on the rotating cone is

$$T = \int \int r\tau \, dr \, ds = \int_0^R \int_0^{2\pi} \mu \frac{W}{\beta_0} r^2 \, d\theta \, dr = \frac{2\pi\mu\omega R^3}{3\beta_0}.$$

Again, by measuring the torque required to sustain the rotation, the viscosity of the fluid can be calculated by

$$\mu = \frac{3\beta_0 T}{2\pi\omega R^3}. \tag{12.9}$$

Falling-Cylinder Viscometer

As shown in Figure 12.7, a cylindrical slug is allowed to drop at a constant velocity V in a cylindrical tube filled with liquid. If the ratio of the slug diameter R_1 to the cylinder diameter R is defined as $\delta = \frac{R_1}{R}$, it can be shown (Lohrez et al., 1960) that the velocity of the fluid in the narrow gap between the slug, and the inner wall of the tube is

$$\frac{u}{V} = \frac{(1 - \eta^2) + (1 + \eta^2)\ln\eta}{(1 - \delta^2) + (1 + \delta^2)\ln\delta},$$

where $\eta = \frac{r}{R}$. For constant velocity (i.e., the acceleration is zero), the net resultant force on the slug is zero. Therefore, by balancing the various forces, we can express the viscosity of the fluid as

$$\mu = \frac{(\rho_0 - \rho)g(\delta R)^2}{2V}\left(\ln\frac{1}{\delta} - \frac{1 - \delta^2}{1 + \delta^2}\right), \tag{12.10}$$

where ρ_0 is the density of the slug and ρ the density of the fluid (Lohrez et al. 1960).

The viscometers just described are only a few of the instruments used to measure viscosity. In general, their operation is based on balancing the viscous forces to other external forces. These devices become much more complicated when they are used to measure the rheology of complex non-Newtonian fluids such as polymers. However, the rheology of such fluids is beyond the scope of this introductory textbook. Interested readers should consult more specialized books (Bird et al., 1987).

12.3   Measurement of General Flow Characteristics

12.3.1   Flow Visualization

Since fluid dynamics is a highly visual science, visualization methods are used extensively in the study of fluid phenomena. In this respect, concepts such as streamlines, streaklines, pathlines, streamfunction, and velocity potential are very useful. In general, visualization methods provide qualitative information about fluid behavior. However, more sophisticated techniques can use visual information for quantitative analysis as well.

Methods of visualizing streamlines, pathlines, and streaklines vary depending on the means used to produce them. For example, the simplest method is to inject into the flow stream either small particles or colored fluid and follow their trajectory through the flow field. Similarly, flow information can be obtained by injecting smoke into a flowing stream and by following its dynamics. For instance, Figure 12.8 is a visualization of the dynamics of a jet impinging on a solid surface using smoke that is originally injected within the boundary layer.

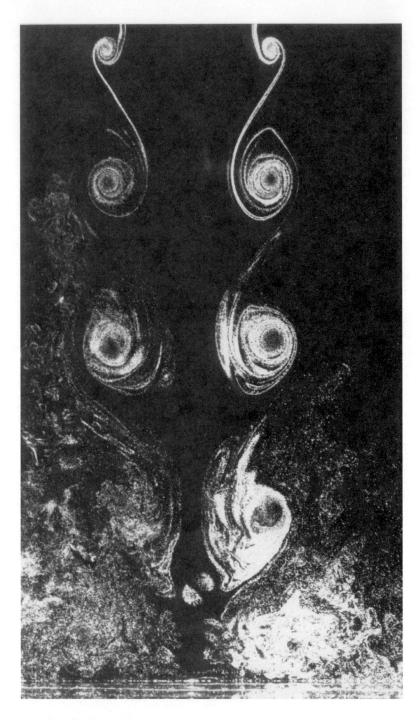

FIGURE 12.8   A 3 *cm* diameter jet forced at Strouhal number of 0.52 impinging on a solid surface 15 *cm* away. Taken from S.L. Anderson, and E.K. Logmire, (1995), *JFM*, **299**, 333–366, by permission.

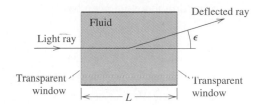

FIGURE 12.9 Light rays deflected due to density differences in a compressible medium.

"Particle" methods are used with caution so that the injected particles are small enough or their concentration is low enough that they do not interfere with the flow dynamics.

Optical Methods

Highly detailed and accurate visualizations can be also obtained by using the optical properties of light rays through compressible media. In addition to their accuracy, these methods are also attractive because they are noninvasive (i.e., they do not disturb the flow). For example, a ray of light passing through a compressible medium, as shown in Figure 12.9, deflects in the direction shown in the figure because of the density differences. The deflection ϵ is given by

$$\epsilon = \frac{\lambda}{n_1}\left(\frac{dn}{dy}\right)_{y=1},$$

where n is the refraction index, λ is the wavelength of the light source, and n_1 the refraction index outside the flow field. For small density gradients, this deflection is also related to the local density gradient $d\rho/dy$, using

$$\epsilon = \frac{\lambda}{n_1}\left(\frac{dn}{dy}\right)_{y=1} = \frac{L\beta}{\rho_s}\left(\frac{\partial\rho}{\partial y}\right)_{y=y_1}, \tag{12.11}$$

where L is the length of the flow field, ρ_s the reference density, and β a nondimensional constant which, for air, has a numerical value of 0.000292. In terms of the density the

FIGURE 12.10 Spark shadowgraph image showing transition to vapor of a nitrogen jet injected at supercritical pressure into a supersonic flow with Mach number equal to 1.84. The flow is from left to right. The supersonic stagnation pressure upstream of the jet is 4.1 *atm*. The jet injection pressure and temperature are, respectively, 58.1 *atm* and 114 *K*. The jet orifice is 0.3 *mm*, and the downstream extent of the image is 3.5 *cm*. (Hermanson *et al.*, *Journal of Propulsion and Power*, **10**(3), 1994, pp. 387–394).

refraction index is given by

$$n = \left(1 + \beta \frac{\rho}{\rho_s}\right) n_1.$$

Therefore, using the deflection of a ray of light and upon integration, the density variation in the vertical variation can by obtained. Actually, this method provides a way to measure directly the density of a gas.

***Shadowgraphs*:**   The basis of shadowgraphs follows naturally from the deflection of light rays in a compressible medium; a beam of light passing through a compressible gas with variable density deflects the light in a manner consistent with the theory. The image of the beam projected on the other side of the flow produces areas of dark and light colors indicating the density differences. This image, then, is a representation of the relative deflection of the light beam $d\epsilon/dy$ and, therefore, according to Equation (12.11) it represents the second derivative of the density $d\rho^2/dy^2$. Figure 12.10 shows a typical flow visualization using a shadowgraph technique.

***Schlieren*:**   This method is based on the *Schlieren effect*, in which a beam of light focused on a sharp edge or a knife edge produces regions of light and dark areas that are proportional to the density gradient across the viewing area. This method is used quite extensively to produce pictures of the flow as shown in Figure 12.11.

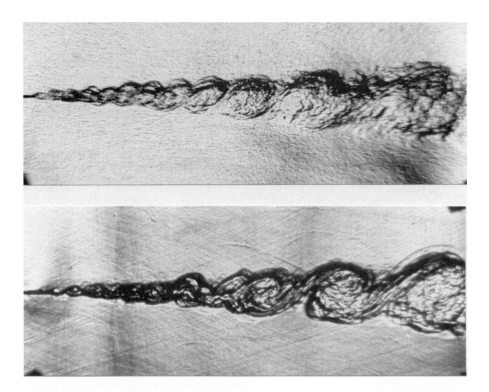

FIGURE 12.11   Spark Schlieren photograph of a turbulent, chemically reacting, mixing layer. The flow is from left to right, and the large-scale turbulent structures in the flow are clearly visible. (a) The upper stream is 1% hydrogen in nitrogen, velocity 22 m/s; the lower stream 1% fluorine in nitrogen and velocity 8.8 m/s. (b) The upper stream is 8% hydrogen in nitrogen, velocity 13 m/s; the lower stream 1% fluorine in nitrogen and velocity 5.2 m/s. For both figures the downstream extent of the image is 24 cm. (Courtesy of J.C. Hermanson, M.G. Mungal, and P.E. Dimotakis, California Institute of Technology.)

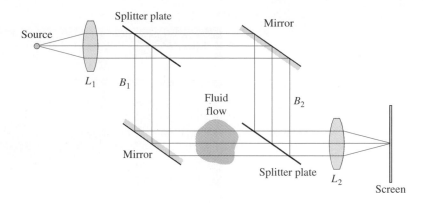

FIGURE 12.12   Basic principle of interferometry.

Interferometry: One of the most precise visualization methods is based on the change in the optical path due to changes in the refractive index as a result of density variations. For example, in the schematic of Figure 12.12, a light source is collimated using lens L_1 and then split into two beams, B_1 and B_2, before beging merged and projected onto a screen with the help of a second lens L_2. If one of the beams travels through a compressible medium, the refractive index changes according to Equation (12.11)—that is,

$$\Delta L = L(n - n_o) = \beta L \frac{\rho - \rho_o}{\rho_s}.$$

As a result of the shift in the optical path, the image on the screen will form a series of *fringes* areas of dark and light shadows. The number of fringes N then is directly related to the density variation across the flow section

$$N = \Delta \lambda = \frac{\beta L}{\lambda} \frac{\rho - \rho_o}{\rho_s},$$

measured relative to the density ρ_o at the first fringe. Figure 12.13 shows a typical flow representation using interferometry.

Fluorescent Dyes: The fluorescent dye approach is a noninvasive, remote sensing method based on the excitation of fluorescent dyes mixed in the flow using a laser beam or sheet. The flow image is constructed from the sensing of the radiative energy from the dye. Consequently, this method captures simultaneously the flow field for large domains without affecting the flow. Figure 12.14 shows a typical flow using a fluorescent dye approach.

12.3.2   Velocity

Accurate measurement of the velocity components is central to experimental fluid dynamics. The following subsections discuss various methods used to quantitatively measure the velocity.

Hot Wires: Hot wires are used to measure the velocity at a "point" in the flow field, particularly when the flow is unsteady. Hot wires are also useful for measuring the velocity fluctuations in turbulent flows. The operation of the hot wire is based on the balance of the electric power q_{elect} required to maintain a thin wire at a constant temperature T_w, while the wire is being cooled by convection at a rate of q_{conv} due to the fluid

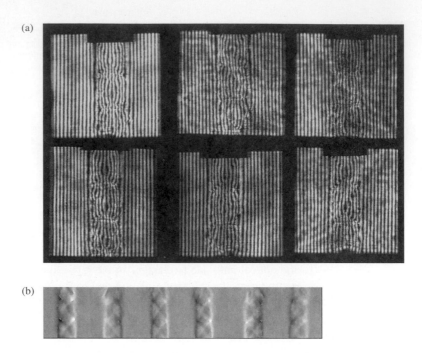

FIGURE 12.13   (a) Multiple view shearing interferograms of an underexpanded jet from 6 different directions (0, 30, 60, 90, 120, and 150 degrees) about the jet. A complex optical setup captures the six views simultaneously on a single camera. (b) Interferogram processed and data reorganized into separate views [images courtesy of David Watt and Barbara Pelliccia-Kraft of the University of New Hampshire].

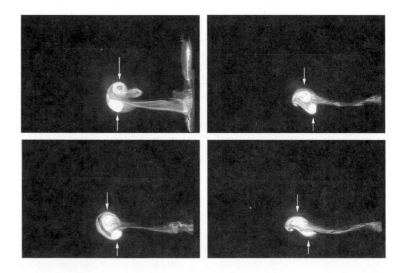

FIGURE 12.14   Temporal evolution of the decay of a vortex ring traveling horizontally in a linearly stratified environment. The vortex ring was visualized by the laser-induced fluorescence where the ring fluid contained a fluorescent dye and a vertical laser sheet was used to illuminate the flow field. Note the assymetric development of the virtex ring. (Courtesy of Hamid Johari and the American Institute of Physics.)

flow — that is,

$$q_{elect} = q_{conv}.$$

Using theory and experimental data, heat transfer by convection is given by

$$q_{conv} = (a + ba^{0.5})(T_w - T_\infty).$$

Also,

$$q_{elect} = i^2 R_0 (1 + \alpha(T_w - T_0)),$$

where i is the electric current, R_0 is the resistance of the wire, α is a constant, and T_0 is a reference temperature. Therefore, by assuming that the wire is very thin, the velocity can be evaluated by measuring the fluctuations in the current i. Hot wires can have response times in the order of 1 ms.

Laser Doppler Velocimetry: Laser Doppler methods are based on principles that are very similar to those encountered in optical methods. Here, the light from a laser beam, passing through fluid flow, scatters because of the presence of small particles. This scattering causes a *frequency shift* known as the *Doppler effect* that is proportional to the velocity of the particles in the flow. If these particles are small, the particle velocity is the same as the fluid velocity. Since the operation depends upon a small concentration of impurities in the flow, it cannot be easily used in "clean flows."

Figure 12.15 shows a typical arrangement of a laser Doppler measurement system, where the light source from a laser is split into two beams. One of the beams is passed through the flow where scattering and frequency shift take place. The other beam is passed though a neutral density filter and then recombined with the disturbed beam through a beam splitter. The fluid velocity is determined by measuring the resulting frequency shift using a photomultiplier.

Particle Tracking: A rather straightforward method to evaluate the fluid velocity is to distribute a number of solid particles in the flow and then follow their path through the flow field. Direct differentiation of the position vector of the particle with respect to time provides a measure of the fluid's velocity. Although the concept is quite simple, the practical implementation of the method is rather complicated. For instance, the particle size, concentration, and density must be selected very carefully so that the particles will not interfere with the flow and will not settle because of gravity effects. Furthermore, in a fully automated measuring arrangement, particle identification becomes quite challenging. However, there are a number of other, more sophisticated particle methods that overcome some of these difficulties.

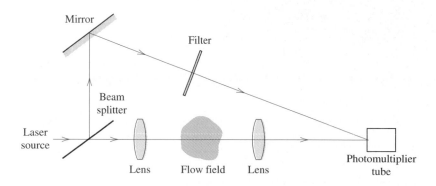

FIGURE 12.15 Schematic of a laser Doppler velocimeter.

12.3.3 Pressure

Mechanical-Based Instruments: By definition, pressure is the normal force per unit area. Therefore, a large number of pressure-measuring instruments are based on the effects of the force produced by the pressure. Consequently, many pressure-measuring instruments use mechanical principles where the force is sensed, measured, and quantified. Figures 12.16 and 12.17 show typical such devices.

Figure 12.16 shows a Bourdon-type pressure gauge in which the pressure is measured by the deflection of a properly calibrated spring. Figure 12.17 shows a diaphragm-type gauge pressure device where, because of the difference in pressure between the two sides of the meter, the diaphragm bends. The maximum deflection is then related to the actual pressure difference.

Electric-Based Instruments: Other pressure devices are based on the electric properties of piezoelectric materials. These are materials that, in response to mechanical loading (force), a voltage develops across the material. Naturally, then, this principle is used in *pressure transducers* to measure pressure. Similarly, other devices such as the *Knudsen gauge*, and the *ionization gauge* are based on other electric properties.

Hydrostatic-Based Devices: Figure 12.18 shows a typical manometer, in which the pressure is measured using the weight of the displaced liquid. The obvious concern is the density of the working liquid; consequently, the dimensions of the manometer are kept within a reasonable level.

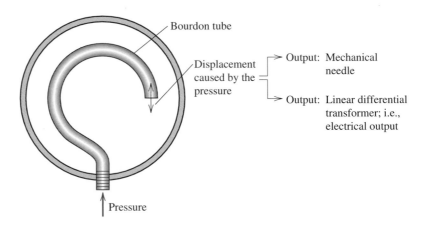

FIGURE 12.16   Schematic of a Bourdon-type pressure meter.

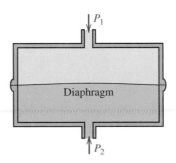

FIGURE 12.17   Schematic of a diaphragm-type pressure meter.

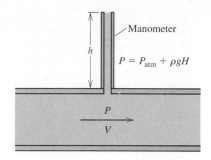

FIGURE 12.18   Schematic of a manometer device.

12.4   Standard Flow Rate Measuring Devices

12.4.1   Pitot-Type Flow Meters

The flow rate in internal flows can be measured using ideas from the operation of a static Pitot tube, shown in Figure 12.19. In this device, the velocity—and hence the volumetric flow rate—is deduced from measuring the pressure points 2 and 4. According to the schematic, if the elevation between points 3 and 4 is small, we have $P_3 \approx P_4 = P$. Using the Bernoulli equation between points 1 and 2, we have

$$P_1 + \rho \frac{V_1^2}{2} = P_2.$$

By combining the two measurements and using $\dot{Q} = VA$, where A is the cross-sectional area, we have

$$\dot{Q} = A \sqrt{2 \frac{(P_2 - P_4)}{\rho}}.$$

12.4.2   Flow-Obstruction–Based Flow Meters

Most commercial flow rate meters for internal flows are based on the dynamics of the flow when it encounters a restriction in the pipe. Three such devices are discussed, namely the *orifice plate*, the *Venturi meter* and the *flow nozzle*.

First consider the geometry of an orifice plate, shown in Figure 12.20, in which the flow area is reduced by introducing a constriction plate. Because of the constriction, the streamlines converge as shown in the figure with a corresponding increase in the velocity. Away from the constriction the flow adjusts to the original velocity distribution. The smallest dimension of the fluid jet at section 2 is called the *vena contracta*. Now assuming steady state, incompressible flow, with negligible changes in the elevation and no losses, the energy equation reduces to

$$\frac{P_1}{\rho} + \frac{V_1^2}{2} = \frac{P_2}{\rho} + \frac{V_2^2}{2},$$

or

$$P_1 - P_2 = \frac{\rho}{2}(V_2^2 - V_1^2).$$

Mass conservation requires that

$$V_1 A_1 = V_2 A_2 \Rightarrow V_1 = V_2 \frac{A_2}{A_1},$$

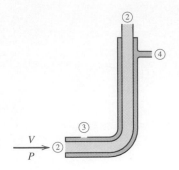

FIGURE 12.19   Schematic of a Pitot tube.

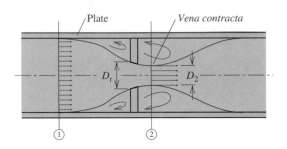

FIGURE 12.20   Schematic of flow-obstruction—based flow meters: Orifice plate.

hence

$$P_1 - P_2 = \frac{\rho V_2^2}{2}\left(1 - \frac{V_1^2}{V_2^2}\right) = \frac{\rho V_2^2}{2}\left(1 - \frac{A_2^2}{A_1^2}\right),$$

which, after rearrangement, gives the velocity V_2 as

$$V_2 = \sqrt{\frac{2\left(P_1 - P_2\right)}{\rho\left(1 - A_2^2/A_1^2\right)}}.$$

Therefore, by measuring the pressure difference $(P_1 - P_2)$ at the two sides of the contraction, we can calculate the average velocity in the pipe and hence the volumetric and mass flow rate. The foregoing expression, however, does not take into consideration several factors, including (a) the diameter of the vena contracta D_2, which is *a priori* unknown; (b) major losses; and (c) the dependence on the actual geometry of the contraction. These effects are normally accounted for by using experimentally determined empirical relations such as the *discharge coefficient C*, defined in terms of the actual and theoretical mass flow rate through the pipe as

$$C = \frac{\dot{m}_{actual}}{\dot{m}_{theoretical}}.$$

Therefore,

$$\dot{m}_{actual} = \frac{C A_t}{\sqrt{1 - (A_t/A_1)^2}}\sqrt{\rho\left(P_1 - P_2\right)},$$

where subscript t now refers to quantities at the contraction. Moreover, by using the *velocity approach coefficient β* as

$$\beta = \frac{D_t}{D_1} \Rightarrow \frac{A_t}{A_1} = \left(\frac{D_t}{D_1}\right)^2 \Rightarrow \left(\frac{A_t}{A_1}\right)^2 = \beta^4,$$

we can write

$$\dot{m}_{actual} = \frac{C A_t}{\sqrt{1 - \beta^4}} \sqrt{\rho \left(P_1 - P_2\right)},$$

and in more compact form as

$$\dot{m}_{actual} = K A_t \sqrt{\rho \left(P_1 - P_2\right)},$$

where K is an overall loss coefficient defined as

$$K = \frac{C}{\sqrt{1 - \beta^4}}.$$

This coefficient combines the effects due to losses and the geometry of the contraction. The same coefficient is usually "corrected" for the influence of the Re, using relations such as

$$K = K_o + \frac{1}{\sqrt{1 - \beta^4}} \frac{b}{Re_{D_1}^n},$$

where b and n are experimentally determined coefficients. Manufacturers usually provide installation information — that is, the appropriate location of the constriction as well as information about the range of the parameters within which the measurements are valid.

For the orifice plate meter, the coefficient C is determined experimentally to be (Miller, 1985)

$$C = 0.5959 + 0.0312\beta^{2.1} - 0.184\beta^8 + \frac{91.71\beta^{2.5}}{Re_{D_1}^{0.75}},$$

which predicts K to within $\pm 0.6\%$ for $0.2 < \beta < 0.75$ and in the range $10^4 < Re_{D_1} < 10^7$.

The same ideas can also be implemented using different geometries. For instance, the plate can be replaced by a nozzle as shown in Figure 12.21. For this *flow nozzle* geometry (Miller, 1985),

$$C = 0.9975 - \frac{6.53}{Re_{D_1}^{0.5}}\beta^{0.5},$$

which predicts K to within $\pm 2.0\%$ for $0.25 < \beta < 0.75$ and in the range $10^4 < Re_{D_1} < 10^7$.

Another popular design is the *Venturi meter*, shown in Figure 12.22. In this case, the meter is a separate piece attached to the pipe. The Venturi meter is usually cast and polished into the appropriate shape, and it is, therefore, associated with high performance. The discharged coefficient is a high as 0.980 to 0.995.

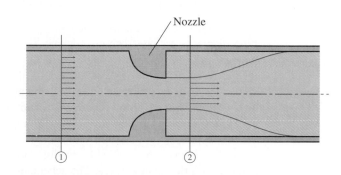

FIGURE 12.21   Schematic of a flow nozzle.

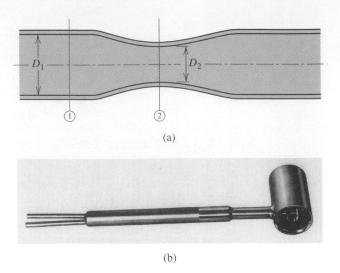

FIGURE 12.22 (a) Schematic of a Venturi meter, (b) Boost Venturi probe from United Sensor Corporation, Amherst NH 03061 (used by permission).

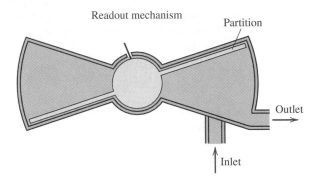

FIGURE 12.23 Schematic of a nutating-disk meter.

The selection of flow meters depends on various factors (e.g., the range of the flow parameters, accuracy, associated loss, cost). For instance, Venturi meters are associated with low head loss and excellent pressure recovery characteristics, but they are more expensive than the orifice plate and the flow nozzle.

12.4.3 Positive Displacement Flow Meters

Several flow meters are designed to operate by measuring the volume of fluid as the fluid displaces a "fixed-volume" device. For instance, Figure 12.23 shows a *nutating-disk* meter; as the fluid flows through the meter, an eccentrically mounted disk wobbles about the vertical axis. As the disk wobbles, a fixed amount of fluid flows through the meter. This amount is then recorded using a mechanical system that senses the number of wobbles. This type of meter is quite accurate in steady flows but not as accurate in unsteady flows.

Similar flow meters are shown in Figure 12.24 and Figure 12.25. The meter in Figure 12.24 is a "rotating-vane" meter whose operation is based on a rotating eccentric drum. Fluid leaks are avoided by having spring-loaded vanes so that the vane is always in contact with the inner surface. The schematic in Figure 12.25 is a *lobed-impeller meter* in which the flow rate is measured as the fluid is forced through the two rotating impellers.

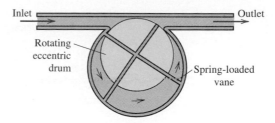

FIGURE 12.24   Schematic of a "rotating-vane" meter.

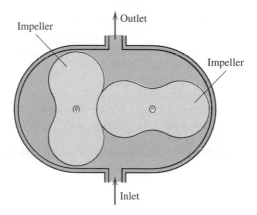

FIGURE 12.25   Schematic of a lobed-impeller meter.

12.4.4   Other Flow Meters

In commercial applications, there are a number of other designs of flow meters such as those that are based on balancing a float inside a closed pipe by using the drag of the float as the fluid flows past it (Figure 12.26). Other meters are based on the frequency of shedding from an obstacle placed in the pipe (Figure 12.27). More sophisticated meters are based on induced electromagnetc forces or the dynamics of an acoustic field using ultrasound methods.

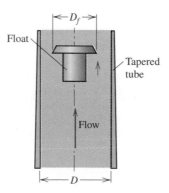

FIGURE 12.26   Schematic of a drag-based meter.

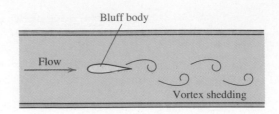

FIGURE 12.27 Schematic of a frequency of shedding–based meter.

REFERENCES

J.P. HOLMAN, *Experimental Methods for Engineers*, New York, McGraw-Hill, 1989.

A.J. WHEELER and A.R. GANJI, *Introduction to Engineering Experimentation*, Englewood Cliffs, NJ, Prentice Hall, 1996.

S.J. KLINE and F.A. McCLINTOCK, *Describing Uncertainties in Single-Sample Experiments*, Mechanical Engineering (Jan. 1953), p. 3.

R.B. BIRD, R.C. ARMSTRONG, and O. HASSAGER, *Dynamics of Polymeric Liquids*, vol. 1, *Fluid Mechanics*, 2nd ed., New York, Wiley-Interscience, 1987.

J. LOHREZ, G.W. SWIFT, and F. KURATA, AIChe J., **6**, pp. 547–550 (1960).

R.W. MILLER, *Flow Measurement Engineering Handbook*, 2nd ed., New York, McGraw-Hill, 1985.

R.W. LADENBURG (ed.), *Physical Measurements in Gas Dynamics and Combustion*, Princeton NJ, Princeton University Press, 1954.

T. ASANUMA (ed.), *Flow Visualization*, Washington D.C., Hemisphere Publishing Co., 1979.

Fluid Meters, Their Theory and Application, 6th ed., New York, ASME, 1971.

J.P. DeCARLO, *Fundamentals of Flow Measurement*, Instrument Society of America, North Carolina, 1984.

PROBLEMS

1. In an experimental run to measure the viscosity of an unknown liquid using a parallel plate viscometer, the following variables are measured with the indicated uncertainties: $h = 1.0 \pm 0.001\ mm$, $R = 25 \pm 0.001\ mm$, $\omega = 15 \pm 0.01\ rev/min$, $\mathcal{T} = 8.1 \times 10^{-4} \pm 1. \times 10^{-6}\ N \cdot m$. Find the nominal value of the measured viscosity and the uncertainty of the measurement using (a) worst-case analysis; (b) the RSS method.

2. In an experiment to measure the viscosity of an unknown liquid using a cone-plate viscometer, the following variables are measured with the indicated uncertainties: $\beta = 5 \pm 0.01^o$, $R = 40\ mm \pm 0.01\ mm$, $\omega = 9.6 \pm 0.01\ rev/min$, $\mathcal{T} = 1.018 \times 10^{-3} \pm 1 \times 10^{-6}\ N \cdot m$. Find the nominal value of the measured viscosity and the uncertainty of the measurement using (a) worst-case analysis, (b) the RSS method.

3. The volumetric water flow rate through a tube with cross-sectional area A is measured using a static Pitot tube. Find the nominal value of the flow rate and the associated uncertainty when the static (P) and stagnation (P_t) pressures are measured to be $P = 100\ kPa \pm 10\ Pa$, $P_t = 120\ kPa \pm 10\ Pa$ and the radius of the channel as $R = 0.08\ m \pm 0.001\ m$.

4. The following data were obtained from a strain gauge for the drag force (N) on a model airfoil: 12.10, 12.40, 12.20, 12.25, 12.14, 11.90, 11.95, 12.05, 12.22, 12.18, 12.21, 8.13, 12.24, 12.19, 12.23, 12.20, 12.18, 12.18, 12.16, 12.24, 9.40, 12.17, 12.19. Are there any data that are very likely to be the result of experimental errors? Using statistical analysis, find the most likely value of the force.

5. Using a hot wire, the velocity (m/s) at a point in steady flow through a sudden expansion is measured to be 1.60, 1.57, 1.58, 1.56, 1.53, 1.60, 1.56, 1.62, 1.58, 1.59, 1.58, 1.54, 1.50, 1.54, 1.69, 1.58, 1.10, 1.57, 1.58, 1.69, 1.56, 1.58, 1.57. Are there any data that are very likely to be the result of experimental errors? Using statistical analysis, find the most likely value of the force.

6. The viscosity of a viscous fluid (Pa.s) is measured to be 0.835, 0.832, 0.810, 0.815, 0.830, 0.832, 0.831, 0.834, 0.705, 0.799, 0.822, 0.826, 0.818, 0.824, 0.828, 0.826, 0.874, 0.820, 0.819, 0.842, 0.832, 0.830. Are there any data that are very likely to be the result of experimental errors? Using statistical analysis, find the most likely value of the force.

7. In calibrating an orifice plate meter, the following data were obtained for the actual mass flow rate $\dot{m}_a$ against the theoretical prediction $\dot{m}_t$.

$\dot{m}_t\ (kg/m^3)$	0.4	0.6	0.8	1.0	1.2
$\dot{m}_a\ (kg/m^3)$	0.266	0.396	0.535	0.652	0.798
$\dot{m}_t\ (kg/m^3)$	1.2	1.6	1.8	2.0	2.2
$\dot{m}_a\ (kg/m^3)$	0.916	1.064	1.187	1.332	1.452

Using the definition $\dot{m}_t = C\dot{m}_a$, find the best estimate for the constant C.

8. The following data were obtained from an experiment measuring the lift force produced by an airfoil F_L as a function of the air velocity V.

V (m/s)	10	20	30	40	50
F (N)	7.136	34.4	56.24	141.56	161.27
V (m/s)	60	70	80	90	100
F (N)	291.08	329.49	464.14	550.79	891.15

Using the definition $F_L = C_L \frac{1}{2} \rho V^2 A$, find the best estimate for the lift coefficient C_L. Use a platform area $A = 0.6\ m^2$ and density of air of $1.2\ kg/m^3$.

9. The following data were obtained from an experiment measuring the drag force produced by a flat plate F_D as a function of the air velocity V.

V (m/s)	10	20	30	40	50
F (N)	51.54	193.96	459.54	768.17	1264.93
V (m/s)	60	70	80	90	100
F (N)	1745.18	2581.99	3097.71	4188.6	4901.07

Using the definition $F_D = C_D \frac{1}{2} \rho V^2 A$, find the best estimate for the drag coefficient C_L. Use a frontal area $A = 2.4\ m^2$ and density of air of $1.2\ kg/m^3$.

13 Fundamentals of Computational Fluid Mechanics

The Navier-Stokes equations are a set of nonlinear partial differential equations the solution of which is possible only for a few simple flows in simple geometries. Most real flows, however, are both mathematically and geometrically complicated. However, realistic problems can be solved quite effectively using computational methods. This is particularly true in light of the remarkable developments in the computer industry. In recent decades, there has been an increase in both the speed and memory capacity of computing systems along with a corresponding decrease in the the cost for such systems.

For many problems, computational fluid dynamics (CFD) provides a cost-effective alternative to experimental fluid mechanics. A well developed and reliable computational code can be easily used to study flows in various geometries, boundary conditions, and flow parameters. In experimentation, these changes are usually very expensive and in many cases not possible. For instance, in computational simulations the temperature of the system can be changed almost at will, but in experiments it cannot be done easily, because of limitations in the equipment and materials. Additionally, various physical effects can be turned off, thus providing the opportunity to study the influence of in-part phenomena. The effects of gravity, for example, can be studied numerically by simply removing the gravitational term from the governing equations. In practice though, gravity effects can be removed by performing the experiment in space, where the gravity is reduced, but not eliminated! Furthermore, CFD provides much more detailed information about the flow field, because the results are obtained for a larger number of points in the flow than in a regular experiment. For instance, the numerical simulation shown in Figure 13.1 provides detailed information on the dynamics of spreading of an initially spherical droplet under the influence of gravity and surface tension forces that cannot be easily obtained experimentally (and certainly not analytically). Figure 13.2 is another demonstration of the remarkable power of numerical simulations; the simulation yields detail flow and temperature information for flow in and around an aircraft engine.

The application of CFD presumes a well-posed mathematical model of the problem, by using the conservation laws, boundary conditions, and, in the case of unsteady problems, initial conditions. Therefore, one cannot avoid the issue of the proper constitutive modeling. Since constitutive models are established by experiments, CFD cannot eliminate the need for experimental fluid mechanics. In the study of fluid mechanics, CFD is therefore both a competitor of and a natural complement to experimentation.

The mathematical model can be a set of *algebraic equations*, a set of *ordinary differential equations*, or a set of *partial differential equations*. Furthermore, the equations can be *linear* or *nonlinear*. In the case of partial differential equations, the equations

---

[1] E.R. Dodds, *The Greek and the Irrational*.

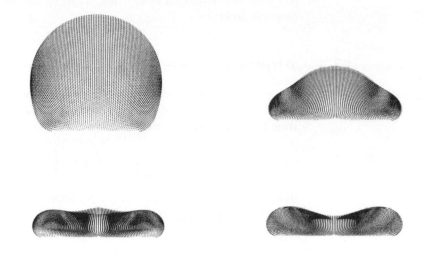

FIGURE 13.1 Spreading of an initially spherical droplet on a solid surface under the influence of gravity and surface tension forces (simulation performed by the author).

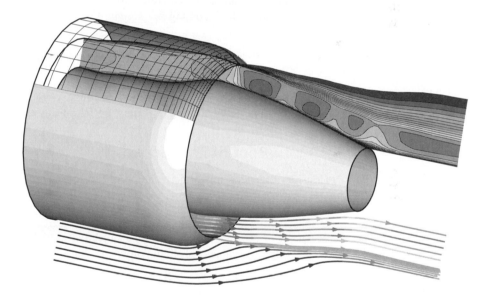

FIGURE 13.2 Simulation of the flow in an aircraft engine nozzle. Streamlines and flooded contours plotted with Tecplot data visualization software. (Courtesy of Amtec Engineering, Inc).

can be *elliptic*, *parabolic*, or *hyperbolic* in nature. These distinctions are important in the discretization of the model, where the validity of the equations is limited only at *discrete points in the solution domain*. Ultimately, the solution is obtained at these points. The discretization of the mathematical model is obtained by a number of methods (e.g., *finite difference*, *finite volume*, *finite elements*, *boundary elements*).

Since it is not possible to cover the entire field of computational fluid dynamics in a single chapter, here we introduce only the basic concepts. Nevertheless, these concepts and ideas are developed in sufficient depth so that the techniques can be readily used in practical problems.

13.1   Algebraic Equations

As encountered in several cases, mathematical models often reduce to single algebraic equations whose solution must be obtained numerically by iteration. Recall, for instance, from Chapter 8 that the friction factor could be obtained by using the Colebrook formula

$$\frac{1}{\sqrt{f}} = -0.869 \ln \left(\frac{e/D}{3.7} + \frac{2.51}{Re\sqrt{f}} \right). \tag{13.1}$$

Since f appears in both sides of the equation, the solution must be obtained numerically.

Similarly, in Prandtl-Meyer expansion, when the flow turns through a total angle v, the resulting Mach number M is a complicated function given by

$$v = \sqrt{\frac{\gamma + 1}{\gamma - 1}} \tan^{-1} \sqrt{\frac{\gamma - 1}{\gamma + 1}(M^2 - 1)} - \tan^{-1} \sqrt{M^2 - 1}. \tag{13.2}$$

The solution to this equation must also be obtained numerically.

Another example is the case of oblique shocks in which the geometry of the shock as a function of the incoming M is given by

$$\beta = \theta - \tan^{-1} \left[\frac{1}{\sin \theta \cos \theta} \left(\frac{\gamma - 1}{\gamma + 1} \sin^2 \theta + \frac{2}{\gamma + 1} \frac{1}{M_1^2} \right) \right]. \tag{13.3}$$

Given the complexity of the expression, the angle of the shock θ as a function of the Mach number M and deflection angle β must be found numerically.

13.1.1   Root of Equations

Consider a general algebraic equation of the form

$$y = f(x).$$

For most of such equations, the final objective is to find the root of the equation—that is, the value (or values) of the unknown x_i that satisfies

$$F = y - f(x_i) = 0. \tag{13.4}$$

The problem is shown schematically in Figure 13.3. As shown in the figure, depending on the order of F, the function F can have multiple solutions. An obvious choice to find the roots of Equation (13.4) is to start guessing values of x_i until we find those that satisfy $F(x_i) = 0$. However, this approach can be inefficient and lengthy. Fortunately, the search for the roots can be accelerated by using a number of numerical procedures. These methods can also be implemented easily in computer form. We review two such methods: (a) the *bisection method* and (b) the *Newton-Raphson method*.

Bisection Method

The bisection method formalizes the search for the root that lies in the range of (x_{min}, x_{max}), where x_{min}, and x_{max} are initial limits set by the user. During the iteration procedure,

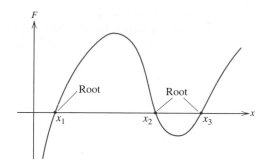

FIGURE 13.3 Multiple roots of a function.

the estimate for the root x_g is assumed to be the midpoint of the current range $x_g = \dfrac{x_{min} + x_{max}}{2}$, while the limits x_{min} and x_{max} are updated depending on the sign of $F(x_g)$ using

$$\text{If } F(x_g) > 0 \quad x_{min} = x_g,$$

and

$$\text{If } F(x_g) < 0 \quad x_{max} = x_g.$$

However, the above update is not unique and depends on the functional form of F. For instance, the proper update may be the opposite — that is,

$$\text{If } F(x_g) > 0 \quad x_{max} = x_g,$$

and

$$\text{If } F(x_g) < 0 \quad x_{min} = x_g.$$

The proper criterion for updating the range must be determined on a case-by-case basis, by keeping track of x_g: if x_g between iterations remains unchanged, the criterion must be reversed.

The procedure is terminated when $|F(x_g)| < \epsilon$, where ϵ is a predetermined small number typically of the same order as the machine accuracy. In general, the method works well and yields the root of the expression provided the range within which the solution lies is known.

EXAMPLE 13.1

Problem Statement Using the bisection method, find the friction factor by solving the Colebrook formula,

$$\frac{1}{\sqrt{f}} = -0.869 \ln \left(\frac{e/D}{3.7} + \frac{2.51}{Re\sqrt{f}} \right),$$

for flow through a pipe with $Re = 298{,}805$ and roughness ratio $e/D = 0.0004$.

SOLUTION For the solution procedure, the equation is expressed as

$$F = \frac{1}{\sqrt{f}} + 0.869 \ln \left(\frac{e/D}{3.7} + \frac{2.51}{Re\sqrt{f}} \right).$$

The root of $F(f)$ is the unknown friction factor f. Here we use $\epsilon = 10^{-6}$.

The method is implemented in the Fortran program found in Appendix D. Note the proper criterion for updating the limits f_{min} and f_{max}. For the given conditions with the limits initially at $f_{min} = 0$ and $f_{max} = 0.2$, the program gives the following intermediate estimates for the friction factor $0.1, 0.05, 0.025 \ldots$ until it converges in 21 iterations to $f = 0.0176$.

Newton-Raphson Method

The Newton-Raphson method is based on Taylor's expansion series: if x_i is an estimate for the root x_r, the function expanded around x_i is then given as

$$F(x_r) = F(x_i) + F'(x_i)\,dx + F''(x_i)(dx)^2 + \cdots,$$

where primes such as F' denote differentiation with respect to x.

By definition, if x_r is the root of $F(x)$, then $F(x_r) = 0$. Therefore, by considering only the first two terms of the foregoing series, an appropriate correction $dx = x_{i+1} - x_i$ of the current estimate x_i is given as

$$dx = x_{i+1} - x_i = -\frac{F(x_i)}{F'(x_i)}.$$

As shown in Figure 13.4, geometrically the method is equivalent to approximating the function by a linear function using the local tangent. Formally, then, using an initial estimate of x_i, the iteration proceeds by correcting the estimate according to

$$x_{i+1} = x_i - \frac{F(x_i)}{F'(x_i)},$$

until $|x_{i+1} - x_i| < \epsilon$, where ϵ is again a small tolerance number (usually 10^{-6}).

In general, the method works very well having one of the fastest convergence rates (quadratic). However, the method has two limitations: (a) unless the initial guess is sufficiently close to the root, there is no guarantee that the procedure will converge; and (b) when $F' = 0$, the method breaks down.

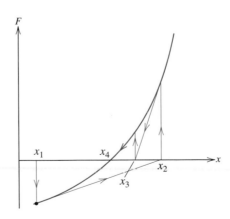

FIGURE 13.4   Geometric description of the Newton-Raphson method.

EXAMPLE 13.2

Problem Statement Using the Newton-Raphson method, find the friction factor by solving the Colebrook formula,

$$\frac{1}{\sqrt{f}} = -0.869 \ln \left(\frac{e/D}{3.7} + \frac{2.51}{Re\sqrt{f}} \right),$$

for flow through a pipe with $Re = 298,805$ and roughness ratio $e/D = 0.0004$.

SOLUTION For the solution procedure, the equation is again expressed as

$$F = \frac{1}{\sqrt{f}} + 0.869 \ln \left(\frac{e/D}{3.7} + \frac{2.51}{Re\sqrt{f}} \right).$$

Again, the root of $F(f)$ is the unknown friction factor f. The tolerance is selected again as $\epsilon = 10^{-6}$. This method is implemented in the Fortran program found in Appendix D. For the given conditions and initial estimate (as discussed in Chapter 8) given by

$$f_1 = 0.25 \left[0.434 \ln \left(\frac{e/D}{3.7} + \frac{5.74}{Re^{0.9}} \right) \right]^{-2},$$

the program after three (!) iterations converges to $f = 0.0176$. This example shows the faster convergence of the Newton-Raphson iteration procedure as compared with the bisection method.

13.1.2 Numerical Integration

In many fluid problems, the solution to the problem is the integral of a function $f(x)$ over a certain domain. For instance, in hydrostatics in order to determine the total force, we must integrate the pressure over submerged surfaces. Therefore, when the function $f(x)$ is complicated, the integral must be performed numerically. For a given function $f(x)$ the integral

$$A = \int_{x_1}^{x_2} f(x)\,dx$$

is the area under the graph, as shown in Figure 13.5.

An obvious approach is to subdivide the domain into smaller strips and add the area of each one of them. Indeed, numerical integration starts by dividing the range between x_1 and x_2 into n strips. Including the two ends, then, we have $n + 1$ points along the x-axis.

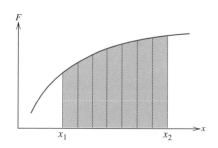

FIGURE 13.5 Schematic of numerical integration.

Therefore,

$$x_i = x_1 + (i - 1)h \quad \text{where} \quad h = \frac{x_2 - x_1}{n} \quad \text{and} \quad i = 1, 2, \dots n + 1.$$

The majority of the numerical integration methods are distinguished by the manner in which the area of each strip is evaluated. In the following subsections, we consider two such methods: (a) the *trapezoidal rule* and (b) *Simpson's rule*.

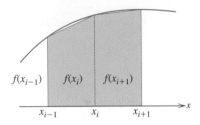

FIGURE 13.6   Discretization of the area under the curve $f(x)$.

Trapezoidal Rule

Using the trapezoidal method, each strip is assumed to be a trapezoid as shown in Figure 13.6. The area of each strip is then

$$A_{strip} = [f(x_i) + f(x_{i-1})] \left(\frac{x_i - x_{i-1}}{2} \right).$$

The integral A is obtained by summation as

$$A = \sum A_{strip} = \sum_{k=1}^{k=n} [f(x_k) + f(x_{k+1})] \left(\frac{x_{k+1} - x_k}{2} \right).$$

Obviously, then, the accuracy of the integration depends on the number of segments n. As the number increases, the approximation of each strip as trapezoids is more accurate (i.e., curved surfaces are more accurately represented as a collection of straight sections).

Simpson's Rule

Simpson's rule is an improvement over the trapezoidal rule in which strips are considered in pairs, as shown in Figure 13.7. Instead of the linear approximation of the trapezoidal rule, a parabola now is passed through the three points marking off the two strips. Obviously, this

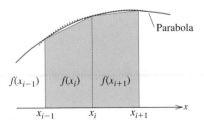

FIGURE 13.7   Geometry of Simpson's rule.

approximation performs better in representing curved geometries. However, since strips are considered as pairs, the method requires an even number of strips. By summation, the integral is given by

$$A - \int_{x_1}^{x_2} f(x)\,dx = \frac{\Delta x}{3}[f(x_1) + 4f(x_2) + 2f(x_3) + 4f(x_4) + \cdots$$

$$+\, 2f(x_{n-1}) + 4f(x_n) + f(x_{n+1})],$$

where Δx is the size of the strip.

EXAMPLE 13.3

Problem Statement   Using the trapezoidal rule and Simpson's rule, calculate numerically the total force on a $2\ m \times 2\ m$–square flat plate due to a pressure given according to

$$P = 100 + 20e^x\ kPa.$$

Compare the numerical result for different numbers of discrete subdomains to the exact analytic solution.

Governing Equations   By definition, the force is obtained by integration as

$$F = -\int P\,\mathbf{n}\,W\,dx,$$

where P is the pressure, $\mathbf{n}$ the outward pointing unit normal vector, and W the width of the plate.

SOLUTION   Using the definition, the total force is obtained analytically by integrating

$$F = -\int_{x=0}^{x=2\ m} P\,\mathbf{n}\,W\,dx = -\int_{x=0}^{x=2\ m} [100 + 20e^x]2\,dx\,\mathbf{j},$$

$$F = -[400 + 40(e^2 - 1)]kN\,\mathbf{j} = -655.56\ kN\,\mathbf{j}.$$

Using the procedures described previously, the results from the numerical integration are shown in Table 13.1. Note that for a sufficiently large number of subdivisions (here 200 for the trapezoidal rule and 10(!) for Simpson's rule), the numerical results are identical to the analytic solution. The results also show the better accuracy of Simpson's rule because fewer subdivisions are required. The Fortran programs used for this problem are found in Appendix D.

13.2   Ordinary Differential Equations

The next level of sophistication in the mathematical models is the need to solve numerically first-order ordinary differential equations (ODE) of the type

$$y' = F(x, y), \tag{13.5}$$

TABLE 13.1 Results from Numerical Integration as a Function of the Number of Subdivisions

n	Exact	Trapezoidal	Simpson's
2	$-655.56\ kN$	$-676.51\ kN$	$-656.83\ kN$
4	$-655.56\ kN$	$-660.86\ kN$	$-655.65\ kN$
10	$-655.56\ kN$	$-656.41\ kN$	$-655.56\ kN$
20	$-655.56\ kN$	$-655.77\ kN$	$-655.56\ kN$
50	$-655.56\ kN$	$-655.60\ kN$	$-655.56\ kN$
100	$-655.56\ kN$	$-655.57\ kN$	$-655.56\ kN$
200	$-655.56\ kN$	$-655.56\ kN$	$-655.56\ kN$

where y' represents the derivative with respect to x. A typical example is, for instance, the solution to Example 2.1 where the velocity of the ball is evaluated by integrating

$$\frac{du}{dt} = 0 \quad \text{and} \quad \frac{dv}{dt} = -g,$$

and the trajectory as

$$\frac{dx}{dt} = u \quad \text{and} \quad \frac{dy}{dt} = v.$$

In this case, the integration is straightforward. However, if we consider the drag forces acting on the ball in the form

$$\frac{du}{dt} = -F_x(u, v) \quad \text{and} \quad \frac{dv}{dt} = -F_y(u, v),$$

where $F_x(u, v)$ and $F_y(u, v)$ are functions of the local velocity components, the integration must be performed numerically.

Ordinary differential equations can also be of higher order of the form

$$y^n = F(x, y, y', y'', \ldots), \tag{13.6}$$

For instance, this is the form of the boundary layer equation obtained using the Blasius transformation — that is,

$$2f''' + ff'' = 0.$$

Higher-order ODEs, however, can be expressed as a system of first-order ODEs by introducing additional variables of the form $y_1 = y'$, $y_2 = y''$ etc.

By definition, first-order differential equations are *initial-value problems* in which, by starting from a lower limit x_1 and a known initial value $y(x = 0, y = 0)$, the equations are integrated to an upper limit x_2 (Figure 13.8(a)). Higher-order ODEs can be *boundary-value problems* with boundary conditions on both ends of the domain. In such a case, the integration scheme must be such that these conditions are satisfied. Often this is achieved by trial and error (or by a *shooting* method): the equations are integrated using an estimate for the value of the variable at the lower limit. The process is repeated until the boundary condition at the upper limit is satisfied (Figure 13.8(b)).

Ultimately, however, the numerical integration depends on the approximation of the derivative y'. Numerical integration methods then are distinguished by the manner in which they handle this derivative. As discussed subsequently, these methods have different *accuracy* and *stability* characteristics.

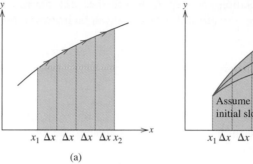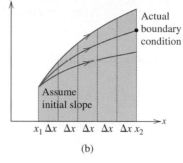

(a)   (b)

FIGURE 13.8   Integration of (a) initial value and (b) boundary value ordinary differential equations.

13.2.1   Integration Schemes for ODEs

The simplest integration scheme is the *forward Euler's* method in which the equation is approximated as

$$\frac{y^{n+1} - y^n}{\Delta x} = F(y^n, x^n).$$

Therefore, by starting from an initial value y^1, the solution at each subsequent step Δx away is obtained using the recursive formula

$$y^{n+1} = y^n + \Delta x\, F(y^n, x^n),$$

where n is the step counter. This method is *explicit*, because the equation is integrated using the value of F at the previous step. The method is easy to implement but is not as accurate as other methods. This method is also *first-order accurate*. This means that the error in the approximation is of order $O(\Delta x)$.

An alternative approximation is the *backward Euler's* method

$$y^{n+1} = y^n + \Delta x\, F(y^{n+1}, x^{n+1}).$$

This is an *implicit* method, because the right-hand side of the equation is a function of the unknown y^{n+1}. The equation is as accurate as the forward Euler's method ($O(\Delta x)$) but more complicated to solve. With an implicit method, then, the equation at each discrete step results in a set of algebraic equations to be solved simultaneously for all unknowns. Implicit methods, however, are more *stable* than explicit methods.

A natural variation of these two methods is their combination into what is known as the *weighted Euler's* method,

$$y^{n+1} = y^n + \Delta x[\theta F(y^{n+1}, x^{n+1}) + (1 - \theta)F(y^n, x^n)],$$

where θ is a constant $0 \le \theta \le 1$. At the two limits $\theta = 0$ and $\theta = 1$, the forward and backward Euler's methods are recovered, respectively. Again, this method is implicit (for $\theta \ne 0$), but being second-order accurate, it is more accurate than the forward Euler's method.

When the solution is obtained for one discrete step at a time, in a single stage or calculation, the method is known as a *single-step, single-stage* method. Explicit methods, for instance, are such methods. Implicit methods, though, by definition are *multistep* methods.

Among various integration schemes, explicit methods are the most attractive to use in as much as they are easy to implement and require a single calculation to advance the solution one step. However, they lack the accuracy of implicit methods. The accuracy of the forward Euler's method can be improved by introducing the *corrected Euler's* method

$$y^* = y^n + \Delta x F(y^n, x^n),$$

$$y^{n+1} = y^n + \frac{\Delta x}{2}[F(y^n, x^n) + F(y^*, x^{n+1})].$$

This is a *single-step, two-stage* method; while being explicit, it is second-order accurate. Because of the extra stage, though, the method is twice as "expensive" as the single-stage method.

The foregoing discussion introduces the *Runge-Kutta* methods, a class of higher-order explicit, one-step, multistage methods. The most popular of the Runge-Kutta methods is the fourth-order accurate scheme $O((\Delta x)^4)$ given as

$$y^* = y^n + \frac{\Delta x}{2} F^n$$

$$y^{**} = y^n + \frac{\Delta x}{2} F^*$$

$$y^{***} = y^n + \Delta x F^{**}$$

$$y^{n+1} = y^n + \frac{\Delta x}{6}[F^n + 2F^* + 2F^{**} + F^{***}],$$

where

$$F^* = F\left(y^*, x^n + \frac{\Delta x}{2}\right), \; F^* = F\left(y^{**}, x^n + \frac{\Delta x}{2}\right), \; F^{***} = F(y^{***}, x^n + \Delta x).$$

Because of the discrete nature of numerical solutions, accuracy is a function of the step size Δx. Therefore, in *all* numerical solutions, the step size must be selected such that the final solution is independent of Δx. However, for step sizes above a critical value, the solution may be *unstable*. In such a case, the numerical solution may diverge significantly from a smooth solution by giving erratic results. Therefore, we must select a Δx that gives both stable and accurate solutions.

EXAMPLE 13.4

Problem Statement   Consider the cylindrical reservoir with diameter D, shown in Figure 3.22, where the liquid at an initial height H_0, is draining through a small hole with diameter d, at the bottom. Using numerical integration, find the time it takes for the reservoir to empty completely. Compare the results with the exact analytic solution. Use a ratio of D/d of 10 and $H_0 = 1.0$.

Basic Assumptions   The flow is assumed to be frictionless and adiabatic. Moreover, it is assumed that since the reservoir drains slowly, the acceleration of the system is small and can be neglected.

Governing Equations   Using the basic assumptions and conservation principles as shown in Example 3.17, the rate of change of the liquid level is given by the ODE

$$\frac{dh}{dt} = -\sqrt{2gh}\left(\frac{D^4}{d^4} - 1\right)^{-\frac{1}{2}}.$$

SOLUTION   The rate of draining as shown above is a first-order differential equation of the form given in Equation (13.5). As shown in Example 3.17, the equation could be solved analytically. By rearranging the result obtained in Example 3.17, we get

$$h(t) = \left[\sqrt{H_0} - \sqrt{\frac{g}{2\left(\dfrac{D^4}{d^4} - 1\right)}} \, t \right]^2 .$$

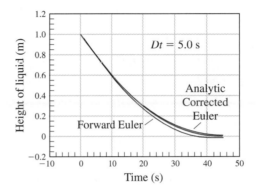

FIGURE 13.9   Draining tank: comparison of two numerical solutions to the exact solution at the same time step $\Delta t = 5.0\ sec$.

Numerically we solve the problem using the forward method and the corrected Euler's method. Figure 13.9 compares the numerical results from the two numerical methods with the exact analytic solution at the same step Δt. As shown in the figure, the corrected Euler's method is more accurate than the forward method. Figure 13.10 shows that for a sufficiently small Δt, the solutions are indistinguishable from each other. The Fortran program used for this problem is found in Appendix D.

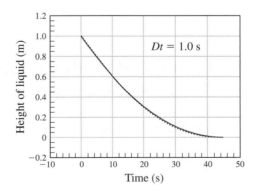

FIGURE 13.10   Draining tank: numerical solution at sufficiently small time step ($\Delta t = 1.0$).

EXAMPLE 13.5

Problem Statement Solve the similarity boundary layer equation obtained in Chapter 8 from Blasius's transformation, using a fourth-order Runge-Kutta integration.

Governing Equations The similarity equation is given in Chapter 8 as

$$2f''' + ff'' = 0,$$

with the following boundary conditions:

$$\text{At } y = 0 \Rightarrow \eta = 0 \text{ then } u = 0 = Uf' \Rightarrow f' = 0.$$

At a distance far away from the boundary layer,

$$\eta = \text{large } u = U = Uf' \Rightarrow f' = 1.$$

SOLUTION The preceding equation is a third-order ODE. It can be reduced to a set of first-order ODEs by introducing $f_1 = f'$ and $f_2 = f''$ —that is,

$$f_2' = -\frac{1}{2}ff_2$$

$$f_1' = f_2$$

$$f' = f_1$$

The foregoing equations must be integrated simultaneously subject to the boundary conditions,

$$f(0) = 0, \quad f(0)' = 0 \quad \text{and} \quad f(\eta = \text{large})' = 1.$$

Unfortunately, this is a boundary-value problem with boundary conditions at both ends of the domain. In this case, since $f''(0)$ is unknown, a typical step-by-step integration cannot be initiated. This is resolved by using a shooting method: the equations are integrated by guessing the value of $f''(0)$ until the correct guess satisfies the boundary condition $f'(\eta = \text{large}) = 1$ at sufficiently large η (about $\eta = 6$). For this problem, the proper value is $f''(0) = 0.33206$. Upon integration, the second boundary condition is satisfied at about $\eta = 5$.

The results of the integration are tabulated in Table B.1 and plotted in Figure 8.6. The Fortran program used for this problem is found in Appendix D.

13.3   Partial Differential Equations

The conservation laws in differential form are a set of nonlinear partial differential equations (PDEs), the solution of which is possible only for simple flows. A general 2-D PDE is given in the form of

$$a\frac{\partial^2 u}{\partial x^2} + 2b\frac{\partial^2 u}{\partial x \partial y} + c\frac{\partial^2 u}{\partial y^2} + d\frac{\partial u}{\partial x} + e\frac{\partial u}{\partial y} + fu + g = 0.$$

According to the sign of $(b^2 - ac)$, PDEs can be classified as:

- *Elliptic equations*, when ($b^2 - ac < 0$). A typical elliptic equation is the well-known Laplace's equation $\nabla^2 f = 0$. In elliptic problems, there are no initial conditions, but we do have boundary conditions. Boundary conditions can be of three types: (a) first type or Dirichlet boundary condition, when the value of the unknown variable is fixed at a boundary; (b) flux type, when the derivative of the unknown is fixed; and (c) mixed or Cauchy boundary condition, where the boundary condition is a combination of the preceding two conditions. For a unique solution, the value of the variable must be specified somewhere in the domain. These equations are the easiest to solve numerically. Almost universally, all numerical methods are introduced and demonstrated using prototype elliptic problems (e.g., such as inviscid flow, heat conduction).

- *Parabolic equations*, when ($b^2 - ac = 0$). Typical parabolic equations are the boundary-layer equations. For this type of equation, we must specify both initial (if unsteady) and boundary conditions. Because of the nature of parabolic equations, boundary equations are not required downstream of the flow (i.e., no boundary conditions are required when the integration is performed into an *open domain*). At *closed boundaries*, boundary conditions may be of any of the three types previously specified. This type of equation is fairly easy to solve, especially because downstream boundary conditions are not required.

- *Hyperbolic equations*, when ($b^2 - ac > 0$). In these equations, both initial and boundary conditions are required. These are the most difficult equations to solve, because "flow information" travels in distinct directions, known as characteristic directions. Smooth solutions are normally obtained along these directions. Elsewhere, the solution is prone to instabilities, therefore, requiring specialized numerical methods.

The theoretical justification for the behavior of the various types of PDEs is beyond the scope of this book. Interested readers can consult more specialized books on this topic. Depending on the physical problem, the Navier-Stokes equations can be any of the preceding three types. Additionally, since the constants a, b, c in the equations are variable, the equations change type according to the solution.

13.3.1 Discretization Methods

The solution to PDEs is obtained numerically at discrete points. These discrete points are obtained by subdividing the flow domain into small *computational cells*. The resulting collection of these cells is known as the *computational mesh*, or simply the mesh. Distinct points on these cells represent the computational points at which the solution is obtained. The size and distribution of the cells determines also the accuracy of the solution.

In most discretizations, the solution to the discrete mathematical problem reduces to a set of algebraic equations to be solved for the unknowns at the selected points. For nonlinear problems, the solution is obtained incrementally by *iteration* (i.e., after successive improvements on an assumed solution). Discretization methods are distinguished by the way they represent in discrete form terms such as

$$\frac{\partial}{\partial x}, \frac{\partial}{\partial y}, \frac{\partial^2}{\partial x^2}, \text{ etc.,}$$

that normally appear in PDEs. Typical discretization schemes include (a) *finite difference methods*, (b) *volume methods*, (c) *finite element methods*, (d) *boundary integral methods*, (e) *spectral methods*. Here, because of the introductory nature of the material, we review the salient features and theoretical basis of the finite difference method.

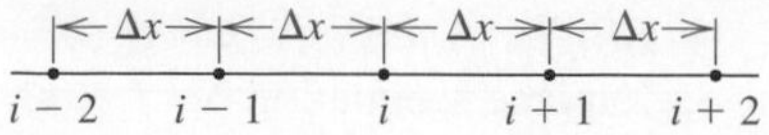

$$\stackrel{|\leftarrow \Delta x \rightarrow|\leftarrow \Delta x \rightarrow|\leftarrow \Delta x \rightarrow|\leftarrow \Delta x \rightarrow|}{\underset{i-2 \quad\quad i-1 \quad\quad i \quad\quad i+1 \quad\quad i+2}{\bullet \quad\quad \bullet \quad\quad \bullet \quad\quad \bullet \quad\quad \bullet}}$$

FIGURE 13.11   One dimensional mesh.

13.3.2   Finite Difference Method

The theoretical foundation of the finite difference method is the Taylor's expansion series. Consider a typical one-dimensional mesh with the nodes numbered as shown in Figure 13.11: $\ldots i-2, i-1, i, i+1, i+2, \ldots$. Using Taylor's series, the function f_i expanded in the positive direction gives

$$f_{i+1} = f_i + \frac{\partial f}{\partial x}\Delta x + \frac{\partial^2 f}{\partial x^2}\frac{(\Delta x)^2}{2} + \cdots.$$

By neglecting small terms such as $(\Delta x)^2$, we get the *forward difference* approximation for the first derivative,

$$\frac{\partial f}{\partial x} \approx \frac{f_{i+1} - f_i}{\Delta x},$$

which is first order (i.e., $O(\Delta x)$). Alternatively, if the function is expanded backwards,

$$f_{i-1} = f_i - \frac{\partial f}{\partial x}\Delta x + \frac{\partial^2 f}{\partial x^2}\frac{(\Delta x)^2}{2} + \cdots.$$

By neglecting again small terms we get the *backward difference* approximation

$$\frac{\partial f}{\partial x} \approx \frac{f_i - f_{i-1}}{\Delta x},$$

which is again first order (i.e., $O(\Delta x)$).

If now we subtract the preceding two expansions, we get the *central difference* approximation,

$$\frac{\partial f}{\partial x} \approx \frac{f_{i+1} - f_{i-1}}{2\Delta x}.$$

This approximation is second-order accurate (i.e., $O((\Delta x)^2)$).

Proceeding in a similar manner, by adding the two series, for the second derivative we get the *central difference* approximation

$$\frac{\partial^2 f}{\partial x^2} \approx \frac{f_{i+1} - 2f_i + f_{i-1}}{(\Delta x)^2},$$

which is also second-order accurate $O((\Delta x)^2)$. The approximation of these derivatives in other directions is handled in exactly the same manner.

13.4   Inviscid Flow

Because of the absence of the diffusive viscous effects, the Euler equations are a set hyperbolic equations, the solution of which is difficult to obtain because they are inherently unstable. Fortunately, the equivalent streamfunction or velocity-potential formulations reduce the problem to an elliptic Laplace equation. In the worst case, for rotational inviscid flows the solution reduces a Poisson equation $\nabla^2 \psi = -\omega$. The solution is completed by integrating the Bernoulli equation along streamlines constructed from the streamfunction ψ.

13.4.1   Finite Difference Solution of $\nabla^2 \psi = 0$

Using a second-order accurate central difference discretization scheme

$$\frac{\partial^2 \psi}{\partial x^2} = \frac{\psi_{i-1,j} - 2\psi_{i,j} + \psi_{i+1,j}}{(\Delta x)^2},$$

$$\frac{\partial^2 \psi}{\partial y^2} = \frac{\psi_{i,j-1} - 2\psi_{i,j} + \psi_{i,j+1}}{(\Delta y)^2},$$

by combining the two, the original equation becomes

$$\frac{\psi_{i-1,j} - 2\psi_{i,j} + \psi_{i+1,j}}{(\Delta x)^2} + \frac{\psi_{i,j-1} - 2\psi_{i,j} + \psi_{i,j+1}}{(\Delta y)^2} = 0,$$

which simplifies to

$$\psi_{i,j} = \frac{(\Delta y)^2(\psi_{i-1,j} + \psi_{i+1,j}) + (\Delta x)^2(\psi_{i,j-1} + \psi_{i,j+1})}{2((\Delta x)^2 + (\Delta y)^2)}. \tag{13.7}$$

When $\Delta x = \Delta y$, $\psi_{i,j}$ is simply the arithmetic mean of the value of ψ at the four neighboring nodes — that is,

$$\psi_{i,j} = \frac{\psi_{i-1,j} + \psi_{i+1,j} + \psi_{i,j-1} + \psi_{i,j+1}}{4},$$

the simplest way to solve the discrete system is by successive iterations where $\psi_{i,j}$ at each node is evaluated using Equation (13.7) according to

$$\psi_{i,j}^{n+1} = \frac{(\Delta y)^2(\psi_{i-1,j}^n + \psi_{i+1,j}^n) + (\Delta x)^2(\psi_{i,j-1}^n + \psi_{i,j+1}^n)}{2((\Delta x)^2 + (\Delta y)^2)},$$

where superscript n is the iteration counter. The iterations are terminated when the problem converges — that is, $max(|\psi_{i,j}^{n+1} - \psi_{i,j}^n|) < \epsilon$ where ϵ is a small tolerance number ($10^{-4} - 10^{-6}$). The rate of convergence is usually improved by using a weighted update between successive iterations as

$$\psi_{i,j}^{n+1} = \alpha \psi_{i,j}^{n+1} + (1 - \alpha)\psi_{i,j}^n,$$

where $0 < \alpha \leq 1$.

EXAMPLE 13.6

Problem Statement   *Inviscid flow in a 90° corner.* Find the streamline distribution and flow field in an inviscid flow, irrotational through a 90° corner with the geometry and boundary conditions as shown in Figure 13.12.

Governing equations   For irrotational inviscid flows as established earlier, we can write $\nabla^2 \psi = 0$, or in Cartesian form

$$\frac{\partial^2 \psi}{\partial x^2} + \frac{\partial^2 \psi}{\partial y^2} = 0.$$

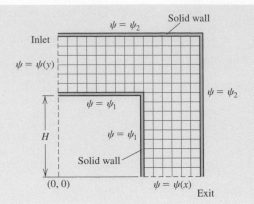

FIGURE 13.12   Schematic of Example 13.7.

Boundary conditions   At the inlet and exit areas, ψ is assumed to vary linearly. Along the upper and lower boundaries, ψ is assumed to be constant equal to $\psi = 0$ and $\psi = 1$, respectively.

SOLUTION   The flow domain is discretized as shown in Figure 13.12. The governing equation is approximated using the central difference scheme previously discussed and implemented in computer form. For the boundary conditions shown in Figure 13.12, a discrete solution is given in Figure 13.13. The Fortran program used to obtain the results is given in Appendix D.

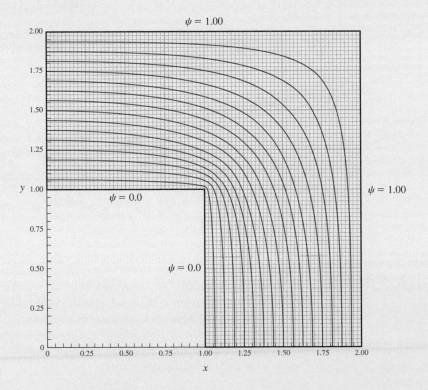

FIGURE 13.13   Finite difference solution for inviscid flow in a 90° corner, $Dx = Dy = 0.025$, $NEX1 = NEX2 = NEY1 = NEY2 = 40$.

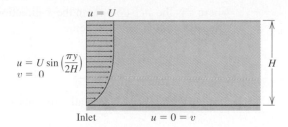

FIGURE 13.14   Boundary layer flow along a flat plate.

13.5   Viscous Flow

For viscous flows, the PDEs describing fluid flow can have different forms. Except for elliptic problems, parabolic problems for laminar incompressible flows may be the easiest to solve. For instance, the boundary layer equations in differential form represent a set of parabolic equations. As shown subsequently, these equations can be discretized and solved using any of the classical discretization methods. Unfortunately, however, these equations are nonlinear and their solution must be obtained by iteration.

13.5.1   Boundary layer

Consider the boundary layer flow of a laminar incompressible fluid with the boundary conditions shown in Figure 13.14. Because of the parabolic nature of the theory, boundary conditions are not required at the downstream. Instead, conditions are specified at the inlet, along the solid surface and the along the outer edge of the boundary layer. The governing equations are then,

Continuity

$$\frac{\partial u}{\partial x} + \frac{\partial v}{\partial y} = 0. \tag{13.8}$$

x-momentum

$$\frac{\partial u}{\partial t} + u\frac{\partial u}{\partial x} + v\frac{\partial u}{\partial y} = -\frac{1}{\rho}\frac{\partial P}{\partial x} + \frac{\mu}{\rho}\frac{\partial^2 u}{\partial y^2}. \tag{13.9}$$

The pressure in the streamwise direction x (the direction along the main flow) is determined by writing the momentum equation at the edge of the boundary layer, where the viscous stresses vanish — that is,

$$\frac{\partial U}{\partial t} + U\frac{\partial U}{\partial x} = -\frac{1}{\rho}\frac{\partial P}{\partial x}. \tag{13.10}$$

Since the solution does not depend on the downstream conditions, we can employ a convenient solution strategy in which the solution is obtained in a stepwise manner by solving simultaneously for all the nodes at a constant x-direction. We will demonstrate the solution procedure by solving steady boundary-layer flow past a flat plate.

The various terms in the differential equations are approximated in the mesh shown in Figure 13.15. Whenever possible, the terms are approximated using the highest accuracy: derivatives, for instance, in the x-direction are evaluated using the first-order difference scheme,

$$\frac{\partial u}{\partial x} = \frac{u_i - u_i^n}{\Delta x},$$

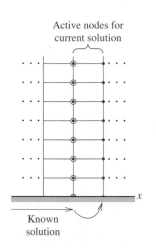

FIGURE 13.15
Schematic of a typical mesh and the solution procedure. Because of the parabolic nature of the equations, solution is obtained for a vertical column at a time.

where n is the step counter in the x-direction, but derivatives with respect to y are evaluated using the second-order scheme,

$$\frac{\partial u}{\partial y} = \frac{u_{i+1} - u_{i-1}}{2\Delta y}.$$

Similarly, the second derivative is approximated by the second-order central difference,

$$\frac{\partial^2 u}{\partial y^2} = \frac{u_{i+1} - 2u_i + u_{i-1}}{(\Delta y)^2}.$$

By discretizing and combining all the terms,

$$F_I = \left(\frac{u_i - u_i^n}{\Delta x} \right) + \left(\frac{v_{i+1} - v_{i-1}}{2\Delta y} \right).$$

$$F_{II} = u_i \left(\frac{u_i - u_i^n}{\Delta x} \right) + v_i \left(\frac{u_{i+1} - u_{i-1}}{2\Delta y} \right) + U \frac{dU}{dx} - \frac{\mu}{\rho} \left(\frac{u_{i+1} - 2u_i + u_{i-1}}{(\Delta y)^2} \right).$$

In the foregoing expression, U is the velocity away from the edge of the boundary layer. In the case of flow past a flat plate, $dU/dx = 0$. The boundary conditions are $u(x, y = 0) = v(x, y = 0) = 0$ and $u(x, y = \text{large}) = U$. At an arbitrary location away from the edge of the plate, the velocity is fixed by assuming that it varies sinusoidally.

The discrete equations written for the N nodes in the y-direction represent a set of $2N$ coupled, nonlinear algebraic equations. Therefore, the solution must be obtained by iteration. Here, we use the Newton-Raphson iteration procedure,

$$\mathbf{A} \, \mathcal{X} = -B,$$

where $\mathcal{X}$ is the vector of the correction on the unknowns during the iteration procedure $\mathcal{X} = \mathcal{X}(\Delta v_1, \Delta u_1, \Delta v_2, \Delta u_2 \ldots \Delta v_n, \Delta u_n)$, B is the right-hand side vector $B = B(F_{I_1}, F_{II_1}, F_{I_2}, F_{II_2}, \ldots F_{I_N}, F_{II_N})$. Matrix $\mathbf{A}$ represents the derivatives of the algebraic equations with respect to the unknowns. Therefore, at each iteration step, the corrections on the unknowns are obtained simultaneously by solving

$$
\begin{bmatrix}
\dfrac{\partial F_{I_1}}{\partial v_1} & \dfrac{\partial F_{I_1}}{\partial u_1} & \dfrac{\partial F_{I_1}}{\partial v_2} & \dfrac{\partial F_{I_1}}{\partial u_2} & \cdots & \\
\dfrac{\partial F_{II_1}}{\partial v_1} & \dfrac{\partial F_{II_1}}{\partial u_1} & \dfrac{\partial F_{II_1}}{\partial v_2} & \dfrac{\partial F_{II_1}}{\partial u_2} & \cdots & \\
\vdots & \vdots & \vdots & \vdots & \vdots & \\
\cdots & \dfrac{\partial F_{I_N}}{\partial v_{N-1}} & \dfrac{\partial F_{I_N}}{\partial u_{N-1}} & \dfrac{\partial F_{I_N}}{\partial v_N} & \dfrac{\partial F_{I_N}}{\partial u_N} & \\
\cdots & \dfrac{\partial F_{II_N}}{\partial v_{N-1}} & \dfrac{\partial F_{II_N}}{\partial u_{N-1}} & \dfrac{\partial F_{II_N}}{\partial v_N} & \dfrac{\partial F_{II_N}}{\partial u_N} &
\end{bmatrix}
\begin{pmatrix}
\Delta v_1 \\ \Delta u_2 \\ \Delta v_1 \\ \vdots \\ \Delta v_N \\ \Delta u_N
\end{pmatrix}
= -
\begin{pmatrix}
F_{I_1} \\ F_{II_1} \\ F_{I_2} \\ \vdots \\ F_{I_N} \\ F_{II_N}
\end{pmatrix}.
$$

For the discrete equations,

$$\frac{\partial F_I}{\partial u_i} = \frac{1.0}{\Delta x}, \quad \frac{\partial F_I}{\partial v_i} = 0.0, \quad \frac{\partial F_I}{\partial v_{i+1}} = \frac{0.5}{\Delta y}, \quad \frac{\partial F_I}{\partial v_{i-1}} = -\frac{0.5}{\Delta y},$$

$$\frac{\partial F_{II}}{\partial u_{i-1}} = -\frac{0.5 v_i}{\Delta y} - \frac{\mu}{\rho} \left(\frac{1.0}{(\Delta y)^2} \right), \quad \frac{\partial F_{II}}{\partial u_{i+1}} = \frac{0.5 v_i}{\Delta y} - \frac{\mu}{\rho} \left(\frac{1.0}{(\Delta y)^2} \right).$$

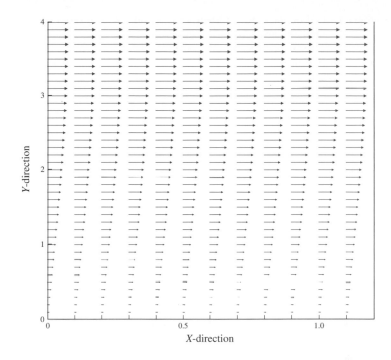

FIGURE 13.16 Velocity field obtained numerically ($\mu = 0.001$, $\rho = 1.0$, $\Delta x = \Delta y = 0.1$, $N_y = 60$).

$$\frac{\partial F_{II}}{\partial u_i} = \left(\frac{u_i - u_i^n}{\Delta x}\right) + \frac{u_i}{\Delta x} - \frac{\mu}{\rho}\left(-\frac{2.0}{(\Delta y)^2}\right).$$

$$\frac{\partial F_{II}}{\partial v_{i-1}} = 0.0, \qquad \frac{\partial F_{II}}{\partial v_i} = \frac{u_{i+1} - u_{i-1}}{2\Delta y}, \frac{\partial F_{II}}{\partial v_{i+1}} = 0.0.$$

Figure 13.16 shows typical velocity profiles as function of distance along the plate.

REFERENCES

D.A. ANDERSON, J.C. TANNEHILL, and R.H. FLETCHER, *Computational Fluid Mechanics and Heat Transfer*, New York, Hemisphere Publishing Corp., 1984.

L. LAPIDUS, and G.F. PINDER, *Numerical Solution of Partial Differential Equations in Science and Engineering*, New York, Wiley Interscience, 1982.

J.H. FERZIGER, *Numerical Methods for Engineering Application*, New York, Wiley, 1981.

M.A. CELIA, and W.G. GRAY, *Numerical Methods for Differential Equations*, Englewood Cliffs, NJ, Prentice Hall, 1992.

C.A.J. FLETCHER, *Computational Techniques for Fluid Mechanics*, vol. I, *Fundamental and General Techniques*, 2nd ed., Berlin, Germany, Springer Verlag, 1990.

J.H. MATHEWS, *Numerical Methods for Science and Engineering*, 2nd ed., Englewood Cliffs, NJ, Prentice Hall, 1992.

R.W. HAMMING, *Numerical Methods for Scientists and Engineers*, 2nd ed., New York, Dover, 1986.

W.H. PRESS, B.P. FLANNERY, S.A. TEUKOLSKY, and W.T. VETTERLING, *Numerical Recipes, The Art of Scientific Computing*, Cambridge, England, Cambridge University Press, 1986.

PROBLEMS

1. Write a program based on the bisection method to solve the Prandtl-Meyer expansion described in Equation (13.2). Verify the accuracy of the results using the tabulated values in the Appendix B.

2. Write a program based on the Newton-Raphson method to solve the Prandtl-Meyer expansion described in Equation (13.2) Verify the accuracy of the results using the tabulated values in the Appendix B.

3. Write a computer code to determine the power required to pump liquid through a pipe of a given roughness. Verify the results with the solution of example problems in Chapter 8.

4. Write a program based on the Forward Euler's method to integrate Blasius' similarity form of the boundary layer equations. Verify the accuracy of the results using the tabulated values in the Chapter 8.

5. Use the fourth-order Runge-Kutta scheme to integrate the trajectory of the ball obtained in Example 2.1, under general conditions, (i.e., angle, initial velocity, etc.). Improve the physical model by including the effects of the drag force.

6. Write a computer code using the fourth order Runge-Kutta integration scheme to predict the trajectory of a ball spinning with angular velocity ω, by assuming an initial velocity V_0, and by including the effects of drag forces.

7. Write a computer code using Forward Euler's integration method to predict the trajectory of a ball spinning with angular velocity ω, by assuming an initial velocity V_0, and by including the effects of drag forces.

8. A solid body quenched in liquid cools according to

$$mC\frac{dT}{dt} = h(T - T_L),$$

where T is the temperature, m is the mass of the body, C the solid's specific heat, h is a heat transfer coefficient, T_L is the temperature of the liquid and t is time. Develop a computer code based on the Forward Euler's method that can predict the temperature as a function of time. For practice consider a spherical aluminum ball with diameter of 10 cm being quenched from $500°C$ to the temperature of the liquid of $100°C$. For aluminum $\rho = 2707 \ kg/m^3$ and $C = 0.896 \ kJ/kg°C$. For h use a range of values to determine its effect on the cooling rate. Check your results by comparing them to the exact solution.

9. For the problem described above developed a Runge-Kutta integration code to determine the temperature as a function of time. Use the same conditions as above.

10. Develop a finite difference-based code to solve inviscid flow in a two-dimensional sudden expansion by using a point-by-point successive iteration scheme. What is the effect on the rate of convergence when relaxation is used?

11. Develop a finite difference-based code to solve inviscid flow in a two-dimensional sudden expansion by using a line-by-line successive iteration scheme.

12. Develop a finite difference based code to solve inviscid flow in a two-dimensional sudden expansion by using a fully implicit solution.

13. Modify the boundary layer computer code given in Appendix D so that the spacing of the nodes in the y-direction is non-uniform.

14. Modify the boundary layer computer code given in Appendix D and solve for the flow past a surface where the freestream velocity U is given by $U = ax^n$ where a and n are constants and x the distance along the surface.

15. The conservation of linear momentum for steady flow in a straight three dimensional channel with a rectangular cross-section is

$$0 = -\frac{dp}{dx} + \mu\left(\frac{\partial^2 u}{\partial x^2} + \frac{\partial^2 u}{\partial y^2}\right).$$

Develop a finite difference code based on a point by point iteration procedure to solve for the velocity distribution in the channel.

16. The conservation of linear momentum for steady flow in a straight three dimensional channel with a rectangular cross-section is

$$0 = -\frac{dp}{dx} + \mu\left(\frac{\partial^2 u}{\partial x^2} + \frac{\partial^2 u}{\partial y^2}\right).$$

Develop a finite difference code based on a line-by-line iteration procedure to solve for the velocity distribution in the channel.

17. The conservation of linear momentum for steady flow in a straight three dimensional channel with a rectangular cross-section is

$$0 = -\frac{dp}{dx} + \mu\left(\frac{\partial^2 u}{\partial x^2} + \frac{\partial^2 u}{\partial y^2}\right).$$

Develop a finite difference based procedure to solve directly (without using an iteration procedure)for the velocity distribution in the channel.

18. Develop a finite difference code to determine steady, fully developed laminar flow in a straight, three-dimensional channel with a cross section that is an equilateral triangle with side a. Compare the results to the exact solution

$$u_x = -\frac{1}{36\mu}\frac{dP}{dx}\frac{1}{a}\left(\sqrt{3}z + 3y + a\right)\left(\sqrt{3}z + 3y - a\right)$$
$$\left(\sqrt{3}z - 3y - a\right),$$

where μ is the viscosity and dP/dx is the pressure gradient.

19. Develop a finite difference code to solve for fully developed flow in a triangular channel as described above with a point-by-point iteration with over relaxation. Compare the results to the exact solution given above.

20. *Stagnation point flow.* Consider a two-dimensional flow around a cylindrical obstacle, where the boundary layer starts at the stagnation point ($x = 0$) and extends on either side of the cylinder x-direction and in the normal direction z. The velocity away from the boundary layer is $U = (x/l)U_1$, where l and U_1 are respectively characteristic length and velocity scales. A similarity solution for this problem is

$$f''' + ff'' - f'^2 + 1 = 0,$$

where derivatives are taken with respect to $\eta \equiv (U_1/\nu l)^{0.5}z$. The boundary conditions are $f(0) = f'(0) = 0$, and $f'(\infty) = 1$. The actual velocity components

u, w, are recovered using $u = Uf'(\eta)$, and $w = -(\nu U_1/l)^{0.5} f(\eta)$. In the above expressions ν is the kinematic viscosity. Modify the Runge-Kutta boundary layer computer code given in Appendix to determine this flow.

21. Write a program based on the Forward Euler's method to integrate the similarity equation for the stagnation point flow described above.

22. *Flow past a wedge.* Consider a two-dimensional flow past a wedge where the velocity away from the boundary layer is $U = u_1 x^m$, with u_1 and m being constants; m depends on the angle of the wedge. A similarity solution for this problem is

$$f''' + ff'' + \beta(1 - f'^2) = 0,$$

where derivatives are taken with respect to $\eta \equiv y[(m + 1)U/(2\nu x)]^{0.5}$. The constant β is given by $\beta = 2m/(m + 1)$. The boundary conditions are $f(0) = f'(0) = 0$, and $f'(\infty) = 1$. The actual velocity components u, v, are recovered using $u = Uf'(\eta)$, and $v = -(0.5(m+1)\nu u_1 x^{m-1})^{0.5}(f(\eta) + (m-1)/(m+1)\eta f(\eta)')$. In the above expressions ν is the kinematic viscosity. Modify the Runge-Kutta boundary layer computer code given in Appendix D to determine this flow.

23. Write a program based on the Forward Euler's method to integrate the similarity equation for flow past a wedge described above.

24. *Flow in a convergent channel.* Consider two dimensional flow in a convergent channel, similar to sink flow described in Chapter 9. However, in this case the flow is confined between two solid surfaces. Therefore, a boundary layer develops where the velocity changes from zero at the surface, to $U = -u_1/x$ away from it. The distance x is measured from the origin. Positive direction is considered the direction opposite to that of the flow. Therefore, flow in the channel is considered negative. A similarity solution for this problem is

$$f''' - f'^2 + 1 = 0,$$

where derivatives are taken with respect to $\eta \equiv (y/x)[u_1/\nu]^{0.5}$. The boundary conditions are $f'(0) = 0$, $f'(\infty) = 1$ and $f''(\infty) = 0$. The actual velocity components u, v, are recovered using $u = Uf'(\eta)$, and $v = -(\eta/x)(\nu u_1)^{0.5} f(\eta)'$. In the above expressions ν is the kinematic viscosity. Modify the Runge-Kutta boundary layer computer code given in Appendix D to determine this flow.

25. Write a program based on the Forward Euler's method to integrate the similarity equation for flow in a convergent channel as described above.

26. A classic problem is flow induced by a long oscillating plate with velocity $u_{plate} = U_0 \cos(\omega t)$ where the fluid extends far away from the plate. In this case the resulting flow is unsteady given by

$$\frac{\partial u}{\partial t} = \nu \frac{\partial^2 u}{\partial z^2},$$

where ν is the fluid's kinematic viscosity, and z the direction normal to the plate. Develop an unsteady finite difference code to determine the velocity as a function of time.

27. Consider the Example 7.7, due to the action of wind-induced shear the water at the surface is suddenly set to motion with velocity u_o. By reaction, due to the action of viscosity the shear is transmitted to the lower fluid layers. The flow is governed by

$$\frac{\partial u}{\partial t} = \nu \frac{\partial^2 u}{\partial y^2}$$

with a known velocity at the free surface and zero velocity far away from the free surface. Develop an unsteady computer code based on a fully explicit method to calculate the velocity ad a function of the depth and time. Compare the solution to the exact solution described in Chapter 7.

28. Develop a fully implicit code to solve the above problem. Compare the solution to the exact solution described in Chapter 7.

Fluid Properties

TABLE A.1 Properties of Various Liquids (at 1.0 Atm and 300 K)

Liquid	Density ρ (kg/m^3)	Dyn. Visc. μ $(Pa \cdot s)$	Specific Heat C_p $(kJ/Kg\ K)$	Cond. κ $(W/m\ K)$
Alcohol (ethyl)	785.1	0.001095	2.44	0.171
Castor Oil	956.1	0.650	1.97	0.180
Ethylene Glycol	1097.0	0.0162	2.36	0.258
Glycerin	1259.0	0.950	2.62	0.287
Kerosene	820.1	0.0331	1.84	—
Mercury	13560.6	0.00153	0.139	—

Source: *Handbook of Chemistry and Physics*, 69th ed., CRC Press, 1988.

TABLE A.2 Surface Tension (in contact with air)

Fluid	Surface Tension σ (N/m)	Fluid	Surface Tension σ (N/m)
Acetone	0.0231	Glycol	0.0477
Benzene	0.0282	Glycerol	0.063
Carbon Tetrachloride*	0.0263	Mercury	0.484
Ethyl Alcohol*	0.0224	Methyl Alcohol	0.0222
Chloroform	0.0271	Water	0.072

* in contact with vapor

Source: *Handbook of Chemistry and Physics*, 69th ed., CRC Press, 1988.

TABLE A.3 Physical Properties of Water

Temp. T $(^\circ C)$	Density ρ (kg/m^3)	Dyn. Visc. μ $(N \cdot s /m^2)$	Kin. Visc. ν (m^2/s)	Surface Tension β (N/m)
0	999.9	$1.787\ E-3$	$1.787\ E-6$	$7.56\ E-2$
5	1000.0	$1.519\ E-3$	$1.519\ E-6$	$7.49\ E-2$
10	999.7	$1.307\ E-3$	$1.307\ E-6$	$7.42\ E-2$
20	998.2	$1.002\ E-3$	$1.004\ E-6$	$7.28\ E-2$
30	995.7	$7.975\ E-4$	$8.009\ E-7$	$7.12\ E-2$
40	992.2	$6.529\ E-4$	$6.580\ E-7$	$6.96\ E-2$
50	988.1	$5.468\ E-4$	$5.534\ E-7$	$6.79\ E-2$
60	983.2	$4.665\ E-4$	$4.745\ E-7$	$6.62\ E-2$
70	977.8	$4.042\ E-4$	$4.134\ E-7$	$6.44\ E-2$
80	971.8	$3.547\ E-4$	$3.650\ E-7$	$6.26\ E-2$
90	965.3	$3.147\ E-4$	$3.260\ E-7$	$6.08\ E-2$
100	958.4	$2.818\ E-4$	$2.940\ E-7$	$5.89\ E-2$

Source: *Handbook of Chemistry and Physics*, 69th ed., CRC Press, 1988.

TABLE A.4   Air Properties at Standard Conditions

Temp. T ($^\circ C$)	Density ρ (kg/m^3)	Dyn. Visc. μ ($N \cdot s/m^2$)	Kin. Visc. ν (m^2/s)	Sp. Heat R. $\gamma = C_p/C_v$
−40	1.514	$1.57\,E-5$	$1.04\,E-5$	1.401
−20	1.395	$1.63\,E-5$	$1.17\,E-5$	1.401
0	1.292	$1.71\,E-5$	$1.32\,E-5$	1.401
5	1.269	$1.73\,E-5$	$1.36\,E-5$	1.401
10	1.247	$1.76\,E-5$	$1.41\,E-5$	1.401
15	1.225	$1.80\,E-5$	$1.47\,E-5$	1.401
20	1.204	$1.82\,E-5$	$1.51\,E-5$	1.401
25	1.184	$1.85\,E-5$	$1.56\,E-5$	1.401
30	1.165	$1.86\,E-5$	$1.60\,E-5$	1.400
40	1.127	$1.87\,E-5$	$1.66\,E-5$	1.400
50	1.109	$1.95\,E-5$	$1.76\,E-5$	1.400
60	1.060	$1.97\,E-5$	$1.86\,E-5$	1.399
70	1.029	$2.03\,E-5$	$1.97\,E-5$	1.399
80	0.9996	$2.07\,E-5$	$2.07\,E-5$	1.399
90	0.9721	$2.14\,E-5$	$2.20\,E-5$	1.398
100	0.9461	$2.17\,E-5$	$2.29\,E-5$	1.397
200	0.7461	$2.53\,E-5$	$3.39\,E-5$	1.390

Source: R.D. Blevins, *Applied Fluid Mechanics Handbook*, Van Nostrand Reinhold Co., *New York* 1984.

TABLE A.5   Properties of the U.S. Standard Atmosphere

Alt. (m)	Temp. T ($^\circ C$)	Accel. g (m/s^2)	Pres. P (N/m^2) Abs.	Dens. ρ (kg/m^3)	Dyn. Visc. μ ($N \cdot s/m^2$)
−1000	21.50	9.810	$1.139\,E+5$	$1.347\,E+0$	$1.821\,E-5$
0	15.00	9.807	$1.013\,E+5$	$1.225\,E+0$	$1.789\,E-5$
1000	8.50	9.804	$8.988\,E+4$	$1.112\,E+0$	$1.758\,E-5$
2000	2.00	9.801	$7.950\,E+4$	$1.007\,E+0$	$1.726\,E-5$
3000	−4.49	9.797	$7.012\,E+4$	$9.093\,E-1$	$1.694\,E-5$
4000	−10.98	9.794	$6.166\,E+4$	$8.194\,E-1$	$1.661\,E-5$
5000	−17.47	9.791	$5.405\,E+4$	$7.364\,E-1$	$1.628\,E-5$
6000	−23.96	9.788	$4.722\,E+4$	$6.601\,E-1$	$1.595\,E-5$
7000	−30.45	9.785	$4.111\,E+4$	$5.900\,E-1$	$1.561\,E-5$
8000	−36.94	9.782	$3.565\,E+4$	$5.258\,E-1$	$1.527\,E-5$
9000	−43.42	9.779	$3.080\,E+4$	$4.671\,E-1$	$1.493\,E-5$
10000	−49.90	9.776	$2.650\,E+4$	$4.135\,E-1$	$1.458\,E-5$
15000	−56.50	9.761	$1.211\,E+4$	$1.948\,E-1$	$1.422\,E-5$
20000	−56.50	9.745	$5.529\,E+3$	$8.891\,E-2$	$1.422\,E-5$
25000	−51.60	9.730	$2.549\,E+3$	$4.008\,E-2$	$1.448\,E-5$
30000	−46.64	9.715	$1.197\,E+3$	$1.841\,E-2$	$1.475\,E-5$

Source: *U.S. Standard Atmosphere*, 1976, U.S. Government Printing Office, Washington, D.C.

Compressible Flow Tables

TABLE B.1 Taken from Hilsenrath *et al.*, *Tables of Thermodynamic and Transport Properties*, Elmsford, N.Y., Pergamon Press, 1960.

Thermodynamic Properties of Air at 1 *Atm.*

$T(K)$	C_p/R	γ	$(h - h_{ref})$ kJ/kg	$(s - s_{ref})/R$
200	3.506	1.406	199.7	22.497
300	3.506	1.402	300.2	23.917
400	3.533	1.396	401.2	24.929
500	3.588	1.387	503.3	25.723
600	3.663	1.376	607.4	26.383
700	3.746	1.365	713.7	26.954
800	3.828	1.354	822.4	27.460
900	3.906	1.345	933.4	27.915
1000	3.979	1.336	1046.6	28.330
1200	4.109	1.322	1278.8	29.068
1400	4.230	1.310	1518.2	29.711
1600	4.352	1.299	1764.5	30.284
1800	4.487	1.288	2018.2	30.804
2000	4.662	1.274	2280.6	31.284

TABLE B.2 Data Generated Using Isentropic Relations

Isentropic Flow $\gamma = 1.4$

M	T/T_t	P/P_t	A/A^*	M	T/T_t	P/P_t	A/A^*
0.0000	1.0000	1.0000	∞	0.3400	0.9774	0.9231	1.8229
0.0200	0.9999	0.9997	28.9421	0.3600	0.9747	0.9143	1.7358
0.0400	0.9997	0.9989	14.4815	0.3800	0.9719	0.9052	1.6587
0.0600	0.9993	0.9975	9.6659	0.4000	0.9690	0.8956	1.5901
0.0800	0.9987	0.9955	7.2616	0.4200	0.9659	0.8857	1.5289
0.1000	0.9980	0.9930	5.8218	0.4400	0.9627	0.8755	1.4740
0.1200	0.9971	0.9900	4.8643	0.4600	0.9594	0.8650	1.4246
0.1400	0.9961	0.9864	4.1824	0.4800	0.9559	0.8541	1.3801
0.1600	0.9949	0.9823	3.6727	0.5000	0.9524	0.8430	1.3398
0.1800	0.9936	0.9776	3.2779	0.5200	0.9487	0.8317	1.3034
0.2000	0.9921	0.9725	2.9635	0.5400	0.9449	0.8201	1.2703
0.2200	0.9904	0.9668	2.7076	0.5600	0.9410	0.8082	1.2403
0.2400	0.9886	0.9607	2.4956	0.5800	0.9370	0.7962	1.2130
0.2600	0.9867	0.9541	2.3173	0.6000	0.9328	0.7840	1.1882
0.2800	0.9846	0.9470	2.1656	0.6200	0.9286	0.7716	1.1656
0.3000	0.9823	0.9395	2.0351	0.6400	0.9243	0.7591	1.1451
0.3200	0.9799	0.9315	1.9219	0.6600	0.9199	0.7465	1.1265

TABLE B.2 (*Continued*)

Isentropic Flow $\gamma = 1.4$							
M	**T/T_t**	**P/P_t**	**A/A^***	**M**	**T/T_t**	**P/P_t**	**A/A^***
0.6800	0.9153	0.7338	1.1097	1.6800	0.6392	0.2088	1.3190
0.7000	0.9107	0.7209	1.0944	1.7000	0.6337	0.2026	1.3376
0.7200	0.9061	0.7080	1.0806	1.7200	0.6283	0.1966	1.3567
0.7200	0.9061	0.7080	1.0806	1.7400	0.6229	0.1907	1.3764
0.7400	0.9013	0.6951	1.0681	1.7600	0.6175	0.1850	1.3967
0.7600	0.8964	0.6821	1.0570	1.7800	0.6121	0.1794	1.4175
0.7800	0.8915	0.6690	1.0471	1.8000	0.6068	0.1740	1.4390
0.8000	0.8865	0.6560	1.0382	1.8200	0.6015	0.1688	1.4610
0.8200	0.8815	0.6430	1.0305	1.8400	0.5963	0.1637	1.4836
0.8400	0.8763	0.6300	1.0237	1.8600	0.5910	0.1587	1.5069
0.8600	0.8711	0.6170	1.0179	1.8800	0.5859	0.1539	1.5308
0.8800	0.8659	0.6041	1.0129	1.9000	0.5807	0.1492	1.5553
0.9000	0.8606	0.5913	1.0089	1.9200	0.5756	0.1447	1.5804
0.9200	0.8552	0.5785	1.0056	1.9400	0.5705	0.1403	1.6062
0.9400	0.8498	0.5658	1.0031	1.9600	0.5655	0.1360	1.6326
0.9600	0.8444	0.5532	1.0014	1.9800	0.5605	0.1318	1.6597
0.9800	0.8389	0.5407	1.0003	2.0000	0.5556	0.1278	1.6875
1.0000	0.8333	0.5283	1.0000	2.0200	0.5506	0.1239	1.7160
1.0200	0.8278	0.5160	1.0003	2.0400	0.5458	0.1201	1.7451
1.0400	0.8222	0.5039	1.0013	2.0600	0.5409	0.1164	1.7750
1.0600	0.8165	0.4919	1.0029	2.0800	0.5361	0.1128	1.8056
1.0800	0.8108	0.4800	1.0051	2.1000	0.5313	0.1094	1.8369
1.1000	0.8052	0.4684	1.0079	2.1200	0.5266	0.1060	1.8690
1.1200	0.7994	0.4568	1.0113	2.1400	0.5219	0.1027	1.9018
1.1400	0.7937	0.4455	1.0153	2.1600	0.5173	0.0996	1.9354
1.1600	0.7879	0.4343	1.0198	2.1800	0.5127	0.0965	1.9698
1.1800	0.7822	0.4232	1.0248	2.2000	0.5081	0.0935	2.0050
1.2000	0.7764	0.4124	1.0304	2.2200	0.5036	0.0906	2.0409
1.2200	0.7706	0.4017	1.0366	2.2400	0.4991	0.0878	2.0777
1.2400	0.7648	0.3912	1.0432	2.2600	0.4947	0.0851	2.1153
1.2600	0.7590	0.3809	1.0504	2.2800	0.4903	0.0825	2.1538
1.2800	0.7532	0.3708	1.0581	2.3000	0.4859	0.0800	2.1931
1.3000	0.7474	0.3609	1.0663	2.3200	0.4816	0.0775	2.2333
1.3200	0.7416	0.3512	1.0750	2.3400	0.4773	0.0751	2.2744
1.3400	0.7358	0.3417	1.0842	2.3600	0.4731	0.0728	2.3164
1.3600	0.7300	0.3323	1.0940	2.3800	0.4688	0.0706	2.3593
1.3800	0.7242	0.3232	1.1042	2.4000	0.4647	0.0684	2.4031
1.4000	0.7184	0.3142	1.1149	2.4200	0.4606	0.0663	2.4479
1.4200	0.7126	0.3055	1.1262	2.4400	0.4565	0.0643	2.4936
1.4400	0.7069	0.2969	1.1379	2.4600	0.4524	0.0623	2.5403
1.4600	0.7011	0.2886	1.1501	2.4800	0.4484	0.0604	2.5880
1.4800	0.6954	0.2804	1.1629	2.5000	0.4444	0.0585	2.6367
1.5000	0.6897	0.2724	1.1762	2.5200	0.4405	0.0567	2.6865
1.5200	0.6840	0.2646	1.1899	2.5400	0.4366	0.0550	2.7372
1.5400	0.6783	0.2570	1.2042	2.5600	0.4328	0.0533	2.7891
1.5600	0.6726	0.2496	1.2190	2.5800	0.4289	0.0517	2.8420
1.5800	0.6670	0.2423	1.2344	2.6000	0.4252	0.0501	2.8960
1.6000	0.6614	0.2353	1.2502	2.6200	0.4214	0.0486	2.9511
1.6200	0.6558	0.2284	1.2666	2.6400	0.4177	0.0471	3.0073
1.6400	0.6502	0.2217	1.2836	2.6600	0.4141	0.0457	3.0647
1.6600	0.6447	0.2151	1.3010	2.6800	0.4104	0.0443	3.1233

TABLE B.2   (*Continued*)

Isentropic Flow $\gamma = 1.4$

M	T/T_t	P/P_t	A/A^*	M	T/T_t	P/P_t	A/A^*
2.7000	0.4068	0.0430	3.1830	3.5000	0.2899	0.0131	6.7896
2.7200	0.4033	0.0417	3.2440	3.5200	0.2875	0.0127	6.9172
2.7400	0.3998	0.0404	3.3061	3.5400	0.2852	0.0124	7.0471
2.7600	0.3963	0.0392	3.3695	3.5600	0.2829	0.0120	7.1791
2.7800	0.3928	0.0380	3.4342	3.5800	0.2806	0.0117	7.3135
2.8000	0.3894	0.0368	3.5001	3.6000	0.2784	0.0114	7.4501
2.8200	0.3860	0.0357	3.5674	3.6200	0.2762	0.0111	7.5891
2.8400	0.3827	0.0347	3.6359	3.6400	0.2740	0.0108	7.7305
2.8600	0.3794	0.0336	3.7058	3.6600	0.2718	0.0105	7.8742
2.8800	0.3761	0.0326	3.7771	3.6800	0.2697	0.0102	8.0204
2.9000	0.3729	0.0317	3.8498	3.7000	0.2675	0.0099	8.1691
2.9200	0.3696	0.0307	3.9238	3.7200	0.2654	0.0096	8.3202
2.9400	0.3665	0.0298	3.9993	3.7400	0.2633	0.0094	8.4739
2.9600	0.3633	0.0289	4.0763	3.7600	0.2613	0.0091	8.6302
2.9800	0.3602	0.0281	4.1547	3.7800	0.2592	0.0089	8.7891
3.0000	0.3571	0.0272	4.2346	3.8000	0.2572	0.0086	8.9506
3.0200	0.3541	0.0264	4.3160	3.8200	0.2552	0.0084	9.1148
3.0400	0.3511	0.0256	4.3990	3.8400	0.2532	0.0082	9.2817
3.0600	0.3481	0.0249	4.4835	3.8600	0.2513	0.0080	9.4513
3.0800	0.3452	0.0242	4.5696	3.8800	0.2493	0.0077	9.6237
3.1000	0.3422	0.0234	4.6573	3.9000	0.2474	0.0075	9.7990
3.1200	0.3393	0.0228	4.7467	3.9200	0.2455	0.0073	9.9771
3.1400	0.3365	0.0221	4.8377	3.9400	0.2436	0.0071	10.1581
3.1600	0.3337	0.0215	4.9304	3.9600	0.2418	0.0069	10.3420
3.1800	0.3309	0.0208	5.0248	3.9800	0.2399	0.0068	10.5289
3.2000	0.3281	0.0202	5.1210	4.0000	0.2381	0.0066	10.7188
3.2200	0.3253	0.0196	5.2189	4.0200	0.2363	0.0064	10.9117
3.2400	0.3226	0.0191	5.3186	4.0400	0.2345	0.0062	11.1077
3.2600	0.3199	0.0185	5.4201	4.0600	0.2327	0.0061	11.3068
3.2800	0.3173	0.0180	5.5234	4.0800	0.2310	0.0059	11.5091
3.3000	0.3147	0.0175	5.6286	4.1000	0.2293	0.0058	11.7146
3.3200	0.3121	0.0170	5.7358	4.1200	0.2275	0.0056	11.9234
3.3400	0.3095	0.0165	5.8448	4.1400	0.2258	0.0055	12.1354
3.3600	0.3069	0.0160	5.9558	4.1600	0.2242	0.0053	12.3508
3.3800	0.3044	0.0156	6.0687	4.1800	0.2225	0.0052	12.5695
3.4000	0.3019	0.0151	6.1837	4.2000	0.2208	0.0051	12.7916
3.4200	0.2995	0.0147	6.3007	4.2200	0.2192	0.0049	13.0172
3.4400	0.2970	0.0143	6.4198	4.2400	0.2176	0.0048	13.2463
3.4600	0.2946	0.0139	6.5409	4.2600	0.2160	0.0047	13.4789
3.4800	0.2922	0.0135	6.6642	4.2800	0.2144	0.0046	13.7151

TABLE B.3   Data Generated Using Normal Shock Relations

Normal Shock $\gamma = 1.4$

M_1	M_2	T_2/T_1	P_2/P_1	M_1	M_2	T_2/T_1	P_2/P_1
1.0000	1.0000	1.0000	1.0000	2.0000	0.5774	1.6875	4.5000
1.0200	0.9805	1.0132	1.0471	2.0200	0.5740	1.7038	4.5938
1.0400	0.9620	1.0263	1.0952	2.0400	0.5707	1.7203	4.6885
1.0600	0.9444	1.0393	1.1442	2.0600	0.5675	1.7369	4.7842
1.0800	0.9277	1.0522	1.1941	2.0800	0.5643	1.7536	4.8808
1.1000	0.9118	1.0649	1.2450	2.1000	0.5613	1.7705	4.9783
1.1200	0.8966	1.0776	1.2968	2.1200	0.5583	1.7875	5.0768
1.1400	0.8820	1.0903	1.3495	2.1400	0.5554	1.8046	5.1762
1.1600	0.8682	1.1029	1.4032	2.1600	0.5525	1.8219	5.2765
1.1800	0.8549	1.1154	1.4578	2.1800	0.5498	1.8393	5.3778
1.2000	0.8422	1.1280	1.5133	2.2000	0.5471	1.8569	5.4800
1.2200	0.8300	1.1405	1.5698	2.2200	0.5444	1.8746	5.5831
1.2400	0.8183	1.1531	1.6272	2.2400	0.5418	1.8924	5.6872
1.2600	0.8071	1.1657	1.6855	2.2600	0.5393	1.9104	5.7922
1.2800	0.7963	1.1783	1.7448	2.2800	0.5368	1.9285	5.8981
1.3000	0.7860	1.1909	1.8050	2.3000	0.5344	1.9468	6.0050
1.3200	0.7760	1.2035	1.8661	2.3200	0.5321	1.9652	6.1128
1.3400	0.7664	1.2162	1.9282	2.3400	0.5297	1.9838	6.2215
1.3600	0.7572	1.2290	1.9912	2.3600	0.5275	2.0025	6.3312
1.3800	0.7483	1.2418	2.0551	2.3800	0.5253	2.0213	6.4418
1.4000	0.7397	1.2547	2.1200	2.4000	0.5231	2.0403	6.5533
1.4200	0.7314	1.2676	2.1858	2.4200	0.5210	2.0595	6.6658
1.4400	0.7235	1.2807	2.2525	2.4400	0.5189	2.0788	6.7792
1.4600	0.7157	1.2938	2.3202	2.4600	0.5169	2.0982	6.8935
1.4800	0.7083	1.3069	2.3888	2.4800	0.5149	2.1178	7.0088
1.5000	0.7011	1.3202	2.4583	2.5000	0.5130	2.1375	7.1250
1.5200	0.6941	1.3336	2.5288	2.5200	0.5111	2.1574	7.2421
1.5400	0.6874	1.3470	2.6002	2.5400	0.5092	2.1774	7.3602
1.5600	0.6809	1.3606	2.6725	2.5600	0.5074	2.1976	7.4792
1.5800	0.6746	1.3742	2.7458	2.5800	0.5056	2.2179	7.5991
1.6000	0.6684	1.3880	2.8200	2.6000	0.5039	2.2383	7.7200
1.6200	0.6625	1.4018	2.8951	2.6200	0.5022	2.2590	7.8418
1.6400	0.6568	1.4158	2.9712	2.6400	0.5005	2.2797	7.9645
1.6600	0.6512	1.4299	3.0482	2.6600	0.4988	2.3006	8.0882
1.6800	0.6458	1.4440	3.1261	2.6800	0.4972	2.3217	8.2128
1.7000	0.6405	1.4583	3.2050	2.7000	0.4956	2.3429	8.3383
1.7200	0.6355	1.4727	3.2848	2.7200	0.4941	2.3642	8.4648
1.7400	0.6305	1.4873	3.3655	2.7400	0.4926	2.3858	8.5922
1.7600	0.6257	1.5019	3.4472	2.7600	0.4911	2.4074	8.7205
1.7800	0.6210	1.5167	3.5298	2.7800	0.4896	2.4292	8.8498
1.8000	0.6165	1.5316	3.6133	2.8000	0.4882	2.4512	8.9800
1.8200	0.6121	1.5466	3.6978	2.8200	0.4868	2.4733	9.1111
1.8400	0.6078	1.5617	3.7832	2.8400	0.4854	2.4955	9.2432
1.8600	0.6036	1.5770	3.8695	2.8600	0.4840	2.5179	9.3762
1.8800	0.5996	1.5924	3.9568	2.8800	0.4827	2.5405	9.5101
1.9000	0.5956	1.6079	4.0450	2.9000	0.4814	2.5632	9.6450
1.9200	0.5918	1.6236	4.1341	2.9200	0.4801	2.5861	9.7808
1.9400	0.5880	1.6394	4.2242	2.9400	0.4788	2.6091	9.9175
1.9600	0.5844	1.6553	4.3152	2.9600	0.4776	2.6322	10.0552
1.9800	0.5808	1.6713	4.4071	2.9800	0.4764	2.6555	10.1938

TABLE B.3 (*Continued*)

Normal Shock $\gamma = 1.4$

M_1	M_2	T_2/T_1	P_2/P_1	M_1	M_2	T_2/T_1	P_2/P_1
3.0000	0.4752	2.6790	10.3333	3.4000	0.4552	3.1802	13.3200
3.0200	0.4740	2.7026	10.4738	3.4200	0.4544	3.2069	13.4791
3.0400	0.4729	2.7264	10.6152	3.4400	0.4535	3.2337	13.6392
3.0600	0.4717	2.7503	10.7575	3.4600	0.4527	3.2607	13.8002
3.0800	0.4706	2.7744	10.9008	3.4800	0.4519	3.2878	13.9621
3.1000	0.4695	2.7986	11.0450	3.5000	0.4512	3.3151	14.1250
3.1200	0.4685	2.8230	11.1901	3.5200	0.4504	3.3425	14.2888
3.1400	0.4674	2.8475	11.3362	3.5400	0.4496	3.3701	14.4535
3.1600	0.4664	2.8722	11.4832	3.5600	0.4489	3.3978	14.6192
3.1800	0.4654	2.8970	11.6311	3.5800	0.4481	3.4257	14.7858
3.2000	0.4643	2.9220	11.7800	3.6000	0.4474	3.4537	14.9533
3.2200	0.4634	2.9471	11.9298	3.6200	0.4467	3.4819	15.1218
3.2400	0.4624	2.9724	12.0805	3.6400	0.4460	3.5103	15.2912
3.2600	0.4614	2.9979	12.2322	3.6600	0.4453	3.5388	15.4615
3.2800	0.4605	3.0234	12.3848	3.6800	0.4446	3.5674	15.6328
3.3000	0.4596	3.0492	12.5383	3.7000	0.4439	3.5962	15.8050
3.3200	0.4587	3.0751	12.6928	3.7200	0.4433	3.6252	15.9781
3.3400	0.4578	3.1011	12.8482	3.7400	0.4426	3.6543	16.1522
3.3600	0.4569	3.1273	13.0045	3.7600	0.4420	3.6836	16.3272
3.3800	0.4560	3.1537	13.1618	3.7800	0.4414	3.7130	16.5031

TABLE B.4   Data Generated Using the Prandtl Meyer Expansion Relations

Prandtl Meyer Expansion, $\gamma = 1.4$ (ν in degrees.)

M	ν	M	ν	M	ν	M	ν
1.0200	0.1257	1.7200	18.3964	1.4600	10.7305	2.1600	30.6884
1.0400	0.3510	1.7400	18.9814	1.4800	11.3169	2.1800	31.2122
1.0600	0.6367	1.7600	19.5646	1.5000	11.9052	2.2000	31.7325
1.0800	0.9680	1.7800	20.1458	1.5200	12.4949	2.2200	32.2494
1.1000	1.3362	1.8000	20.7251	1.5400	13.0856	2.2400	32.7629
1.1200	1.7350	1.8200	21.3021	1.5600	13.6770	2.2600	33.2730
1.1400	2.1600	1.8400	21.8768	1.5800	14.2686	2.2800	33.7796
1.1600	2.6073	1.8600	22.4492	1.6000	14.8604	2.3000	34.2828
1.1800	3.0743	1.8800	23.0190	1.6200	15.4518	2.3200	34.7825
1.2000	3.5582	1.9000	23.5861	1.6400	16.0427	2.3400	35.2787
1.2200	4.0572	1.9200	24.1506	1.6600	16.6328	2.3600	35.7715
1.2400	4.5694	1.9400	24.7123	1.6800	17.2220	2.3800	36.2607
1.2600	5.0931	1.9600	25.2711	1.7000	17.8099	2.4000	36.7465
1.2800	5.6272	1.9800	25.8269	2.4200	37.2289	3.1200	52.0195
1.3000	6.1703	2.0000	26.3798	2.4400	37.7077	3.1400	52.3864
1.3200	6.7213	2.0200	26.9296	2.4600	38.1831	3.1600	52.7505
1.3400	7.2794	2.0400	27.4762	2.4800	38.6551	3.1800	53.1118
1.3600	7.8435	2.0600	28.0197	2.5000	39.1236	3.2000	53.4703
1.3800	8.4130	2.0800	28.5600	2.5200	39.5886	3.2200	53.8261
1.4000	8.9870	2.1000	29.0971	2.5400	40.0503	3.2400	54.1791
1.4200	9.5650	2.1200	29.6308	2.5600	40.5085	3.2600	54.5294
1.4400	10.1464	2.1400	30.1613	2.5800	40.9633	3.2800	54.8770

TABLE B.4   (*Continued*)

Prandtl Meyer Expansion, $\gamma = 1.4$ (ν in degrees.)

M	ν	M	ν	M	ν	M	ν
2.6000	41.4147	3.3000	55.2220	4.2600	69.0628	4.9600	76.5435
2.6200	41.8628	3.3200	55.5643	4.2800	69.3026	4.9800	76.7325
2.6400	42.3074	3.3400	55.9040	4.3000	69.5406	5.0000	76.9202
2.6600	42.7488	3.3600	56.2411	4.3200	69.7770	5.0200	77.1068
2.6800	43.1868	3.3800	56.5756	4.3400	70.0118	5.0400	77.2921
2.7000	43.6215	3.4000	56.9075	4.3600	70.2449	5.0600	77.4762
2.7200	44.0529	3.4200	57.2369	4.3800	70.4763	5.0800	77.6591
2.7400	44.4810	3.4400	57.5639	4.4000	70.7062	5.1000	77.8409
2.7600	44.9059	3.4600	57.8883	4.4200	70.9344	5.1200	78.0215
2.7800	45.3275	3.4800	58.2102	4.4400	71.1611	5.1400	78.2009
2.8000	45.7459	3.5000	58.5298	4.4600	71.3862	5.1600	78.3792
2.8200	46.1611	3.5200	58.8468	4.4800	71.6097	5.1800	78.5564
2.8400	46.5731	3.5400	59.1615	4.5000	71.8317	5.2000	78.7324
2.8600	46.9820	3.5600	59.4739	5.2200	78.9074	5.9200	84.3979
2.8800	47.3877	3.5800	59.7838	5.2400	79.0812	5.9400	84.5385
2.9000	47.7903	3.6000	60.0914	5.2600	79.2539	5.9600	84.6783
2.9200	48.1898	3.6200	60.3968	5.2800	79.4256	5.9800	84.8173
2.9400	48.5863	3.6400	60.6998	5.3000	79.5962	6.0000	84.9555
2.9600	48.9796	3.6600	61.0005	5.3200	79.7657	6.0200	85.0929
2.9800	49.3700	3.6800	61.2990	5.3400	79.9341	6.0400	85.2295
3.0000	49.7574	3.7000	61.5953	5.3600	80.1015	6.0600	85.3653
3.0200	50.1417	3.7200	61.8893	5.3800	80.2679	6.0800	85.5004
3.0400	50.5231	3.7400	62.1812	5.4000	80.4332	6.1000	85.6347
3.0600	50.9016	3.7600	62.4709	5.4200	80.5975	6.1200	85.7682
3.0800	51.2771	3.7800	62.7584	5.4400	80.7608	6.1400	85.9010
3.1000	51.6497	3.8000	63.0438	5.4600	80.9231	6.1600	86.0330
3.8200	63.3271	4.5200	72.0522	5.4800	81.0845	6.1800	86.1643
3.8400	63.6083	4.5400	72.2712	5.5000	81.2448	6.2000	86.2948
3.8600	63.8874	4.5600	72.4887	5.5200	81.4041	6.2200	86.4246
3.8800	64.1645	4.5800	72.7046	5.5400	81.5625	6.2400	86.5537
3.9000	64.4395	4.6000	72.9192	5.5600	81.7199	6.2600	86.6820
3.9200	64.7125	4.6200	73.1322	5.5800	81.8764	6.2800	86.8097
3.9400	64.9836	4.6400	73.3438	5.6000	82.0319	6.3000	86.9366
3.9600	65.2526	4.6600	73.5540	5.6200	82.1865	6.3200	87.0628
3.9800	65.5197	4.6800	73.7628	5.6400	82.3401	6.3400	87.1884
4.0000	65.7848	4.7000	73.9701	5.6600	82.4929	6.3600	87.3132
4.0200	66.0480	4.7200	74.1761	5.6800	82.6447	6.3800	87.4374
4.0400	66.3093	4.7400	74.3807	5.7000	82.7956	6.4000	87.5608
4.0600	66.5688	4.7600	74.5839	5.7200	82.9456	6.4200	87.6836
4.0800	66.8263	4.7800	74.7858	5.7400	83.0947	6.4400	87.8058
4.1000	67.0820	4.8000	74.9863	5.7600	83.2429	6.4600	87.9272
4.1200	67.3359	4.8200	75.1855	5.7800	83.3903	6.4800	88.0480
4.1400	67.5879	4.8400	75.3833	5.8000	83.5368	6.5000	88.1682
4.1600	67.8381	4.8600	75.5799	5.8200	83.6824	6.5200	88.2877
4.1800	68.0866	4.8800	75.7752	5.8400	83.8272	6.5400	88.4065
4.2000	68.3332	4.9000	75.9692	5.8600	83.9711	6.5600	88.5247
4.2200	68.5782	4.9200	76.1619	5.8800	84.1142	6.5800	88.6423
4.2400	68.8214	4.9400	76.3533	5.9000	84.2565	6.6000	88.7592

TABLE B.5   Data Generated Using the Fanno Flow Relations

Fanno Flow, $\gamma = 1.4$

M	T/T^*	P/P^*	P_t/P_t^*	V/V^*	fl_{max}/D
0.0000	1.2000	∞	∞	0.0000	∞
0.0200	1.1999	54.7701	28.9421	0.0219	444.6125
0.0400	1.1996	27.3817	14.4815	0.0438	110.0881
0.0600	1.1991	18.2508	9.6659	0.0657	48.2578
0.0800	1.1985	13.6843	7.2616	0.0876	26.6796
0.1000	1.1976	10.9435	5.8218	0.1094	16.7304
0.1200	1.1966	9.1156	4.8643	0.1313	11.3520
0.1400	1.1953	7.8093	4.1824	0.1531	8.1278
0.1600	1.1939	6.8291	3.6727	0.1748	6.0495
0.1800	1.1923	6.0662	3.2779	0.1965	4.6357
0.2000	1.1905	5.4554	2.9635	0.2182	3.6333
0.2200	1.1885	4.9554	2.7076	0.2398	2.8990
0.2400	1.1863	4.5383	2.4956	0.2614	2.3466
0.2600	1.1840	4.1851	2.3173	0.2829	1.9219
0.2800	1.1815	3.8820	2.1656	0.3043	1.5893
0.3000	1.1788	3.6191	2.0351	0.3257	1.3248
0.3200	1.1759	3.3887	1.9219	0.3470	1.1117
0.3400	1.1729	3.1853	1.8229	0.3682	0.9380
0.3600	1.1697	3.0042	1.7358	0.3893	0.7950
0.3800	1.1663	2.8420	1.6587	0.4104	0.6764
0.4000	1.1628	2.6958	1.5901	0.4313	0.5771
0.4200	1.1591	2.5634	1.5289	0.4522	0.4936
0.4400	1.1553	2.4428	1.4740	0.4729	0.4229
0.4600	1.1513	2.3326	1.4246	0.4936	0.3627
0.4800	1.1471	2.2313	1.3801	0.5141	0.3113
0.5000	1.1429	2.1381	1.3398	0.5345	0.2673
0.5200	1.1384	2.0519	1.3034	0.5548	0.2294
0.5400	1.1339	1.9719	1.2703	0.5750	0.1967
0.5600	1.1292	1.8975	1.2403	0.5951	0.1684
0.5800	1.1244	1.8282	1.2130	0.6150	0.1439
0.6000	1.1194	1.7634	1.1882	0.6348	0.1227
0.6200	1.1143	1.7026	1.1656	0.6545	0.1043
0.6400	1.1091	1.6456	1.1451	0.6740	0.0883
0.6600	1.1038	1.5919	1.1265	0.6934	0.0745
0.6800	1.0984	1.5413	1.1097	0.7127	0.0624
0.7000	1.0929	1.4935	1.0944	0.7318	0.0520
0.7200	1.0873	1.4482	1.0806	0.7508	0.0430
0.7400	1.0815	1.4054	1.0681	0.7696	0.0353
0.7600	1.0757	1.3647	1.0570	0.7883	0.0286
0.7800	1.0698	1.3261	1.0471	0.8068	0.0229
0.8000	1.0638	1.2893	1.0382	0.8251	0.0181
0.8200	1.0578	1.2542	1.0305	0.8433	0.0140
0.8400	1.0516	1.2208	1.0237	0.8614	0.0106
0.8600	1.0454	1.1889	1.0179	0.8793	0.0077
0.8800	1.0391	1.1583	1.0129	0.8970	0.0054
0.9000	1.0327	1.1291	1.0089	0.9146	0.0036
0.9200	1.0263	1.1011	1.0056	0.9320	0.0022
0.9400	1.0198	1.0743	1.0031	0.9493	0.0012
0.9600	1.0132	1.0485	1.0014	0.9663	0.0005
0.9800	1.0066	1.0238	1.0003	0.9832	0.0001

TABLE B.5   (*Continued*)

Fanno Flow, $\gamma = 1.4$

M	T/T^*	P/P^*	P_t/P_t^*	V/V^*	fl_{max}/D
1.0000	1.0000	1.0000	1.0000	1.0000	0.0000
1.0200	0.9933	0.9771	1.0003	1.0166	0.0001
1.0400	0.9866	0.9551	1.0013	1.0330	0.0004
1.0600	0.9798	0.9338	1.0029	1.0492	0.0010
1.0800	0.9730	0.9133	1.0051	1.0653	0.0016
1.1000	0.9662	0.8936	1.0079	1.0812	0.0025
1.1200	0.9593	0.8745	1.0113	1.0970	0.0035
1.1400	0.9524	0.8561	1.0153	1.1126	0.0045
1.1600	0.9455	0.8383	1.0198	1.1280	0.0057
1.1800	0.9386	0.8210	1.0248	1.1432	0.0070
1.2000	0.9317	0.8044	1.0304	1.1583	0.0084
1.2200	0.9247	0.7882	1.0366	1.1732	0.0099
1.2400	0.9178	0.7726	1.0432	1.1879	0.0114
1.2600	0.9108	0.7574	1.0504	1.2025	0.0129
1.2800	0.9038	0.7427	1.0581	1.2169	0.0146
1.3000	0.8969	0.7285	1.0663	1.2311	0.0162
1.3200	0.8899	0.7147	1.0750	1.2452	0.0179
1.3400	0.8829	0.7012	1.0842	1.2591	0.0196
1.3600	0.8760	0.6882	1.0940	1.2729	0.0214
1.3800	0.8690	0.6755	1.1042	1.2864	0.0231
1.4000	0.8621	0.6632	1.1149	1.2999	0.0249
1.4200	0.8551	0.6512	1.1262	1.3131	0.0267
1.4400	0.8482	0.6396	1.1379	1.3262	0.0285
1.4600	0.8413	0.6282	1.1501	1.3392	0.0304
1.4800	0.8344	0.6172	1.1629	1.3520	0.0322
1.5000	0.8276	0.6065	1.1762	1.3646	0.0340
1.5200	0.8207	0.5960	1.1899	1.3770	0.0358
1.5400	0.8139	0.5858	1.2042	1.3894	0.0377
1.5600	0.8071	0.5759	1.2190	1.4015	0.0395
1.5800	0.8004	0.5662	1.2344	1.4135	0.0413
1.6000	0.7937	0.5568	1.2502	1.4254	0.0431
1.6200	0.7869	0.5476	1.2666	1.4371	0.0449
1.6400	0.7803	0.5386	1.2836	1.4487	0.0467
1.6600	0.7736	0.5299	1.3010	1.4601	0.0484
1.6800	0.7670	0.5213	1.3190	1.4713	0.0502
1.7000	0.7605	0.5130	1.3376	1.4825	0.0520
1.7200	0.7539	0.5048	1.3567	1.4935	0.0537
1.7400	0.7474	0.4969	1.3764	1.5043	0.0554
1.7600	0.7410	0.4891	1.3967	1.5150	0.0571
1.7800	0.7345	0.4815	1.4175	1.5256	0.0588
1.8000	0.7282	0.4741	1.4390	1.5360	0.0605
1.8200	0.7218	0.4668	1.4610	1.5463	0.0621
1.8400	0.7155	0.4597	1.4836	1.5564	0.0638
1.8600	0.7093	0.4528	1.5069	1.5664	0.0654
1.8800	0.7030	0.4460	1.5308	1.5763	0.0670
1.9000	0.6969	0.4394	1.5553	1.5861	0.0686
1.9200	0.6907	0.4329	1.5804	1.5957	0.0702
1.9400	0.6847	0.4265	1.6062	1.6052	0.0717

TABLE B.5   (*Continued*)

Fanno Flow, $\gamma = 1.4$

M	T/T^*	P/P^*	P_t/P_t^*	V/V^*	fl_{max}/D
1.9600	0.6786	0.4203	1.6326	1.6146	0.0732
1.9800	0.6726	0.4142	1.6597	1.6239	0.0748
2.0000	0.6667	0.4082	1.6875	1.6330	0.0762
2.0200	0.6608	0.4024	1.7160	1.6420	0.0777
2.0400	0.6549	0.3967	1.7451	1.6509	0.0792
2.0600	0.6491	0.3911	1.7750	1.6597	0.0806
2.0800	0.6433	0.3856	1.8056	1.6683	0.0821
2.1000	0.6376	0.3802	1.8369	1.6769	0.0835
2.1200	0.6320	0.3750	1.8690	1.6853	0.0849
2.1400	0.6263	0.3698	1.9018	1.6936	0.0862
2.1600	0.6208	0.3648	1.9354	1.7018	0.0876
2.1800	0.6152	0.3598	1.9698	1.7099	0.0889
2.2000	0.6098	0.3549	2.0050	1.7179	0.0902
2.2200	0.6043	0.3502	2.0409	1.7258	0.0915
2.2400	0.5989	0.3455	2.0777	1.7336	0.0928
2.2600	0.5936	0.3409	2.1153	1.7412	0.0941
2.2800	0.5883	0.3364	2.1538	1.7488	0.0953
2.3000	0.5831	0.3320	2.1931	1.7563	0.0966
2.3200	0.5779	0.3277	2.2333	1.7637	0.0978
2.3400	0.5728	0.3234	2.2744	1.7709	0.0990
2.3600	0.5677	0.3193	2.3164	1.7781	0.1002
2.3800	0.5626	0.3152	2.3593	1.7852	0.1013
2.4000	0.5576	0.3111	2.4031	1.7922	0.1025
2.4200	0.5527	0.3072	2.4479	1.7991	0.1036
2.4400	0.5478	0.3033	2.4936	1.8059	0.1047
2.4600	0.5429	0.2995	2.5403	1.8126	0.1058
2.4800	0.5381	0.2958	2.5880	1.8192	0.1069
2.5000	0.5333	0.2921	2.6367	1.8257	0.1080
2.5200	0.5286	0.2885	2.6864	1.8322	0.1091
2.5400	0.5239	0.2850	2.7372	1.8386	0.1101
2.5600	0.5193	0.2815	2.7891	1.8448	0.1111
2.5800	0.5147	0.2781	2.8420	1.8510	0.1121
2.6000	0.5102	0.2747	2.8960	1.8571	0.1131
2.6200	0.5057	0.2714	2.9511	1.8632	0.1141
2.6400	0.5013	0.2682	3.0073	1.8691	0.1151
2.6600	0.4969	0.2650	3.0647	1.8750	0.1161
2.6800	0.4925	0.2619	3.1233	1.8808	0.1170
2.7000	0.4882	0.2588	3.1830	1.8865	0.1180
2.7200	0.4839	0.2558	3.2439	1.8922	0.1189
2.7400	0.4797	0.2528	3.3061	1.8978	0.1198
2.7600	0.4755	0.2498	3.3695	1.9033	0.1207
2.7800	0.4714	0.2470	3.4342	1.9087	0.1216
2.8000	0.4673	0.2441	3.5001	1.9140	0.1224
2.8200	0.4632	0.2414	3.5674	1.9193	0.1233
2.8400	0.4592	0.2386	3.6359	1.9245	0.1242
2.8600	0.4552	0.2359	3.7058	1.9297	0.1250

TABLE B.6   Data Generated Using the Rayleigh Flow Relations

Rayleigh Flow, $\gamma = 1.4$

M	T/T^*	T_t/T_t^*	P/P^*	P_t/P_t^*	V/V^*
0.0000	0.0000	0.0000	2.4000	1.2679	0.0000
0.0200	0.0023	0.0019	2.3987	1.2675	0.0010
0.0400	0.0092	0.0076	2.3946	1.2665	0.0038
0.0600	0.0205	0.0171	2.3880	1.2647	0.0086
0.0800	0.0362	0.0302	2.3787	1.2623	0.0152
0.1000	0.0560	0.0468	2.3669	1.2591	0.0237
0.1200	0.0797	0.0666	2.3526	1.2554	0.0339
0.1400	0.1069	0.0895	2.3359	1.2510	0.0458
0.1600	0.1374	0.1151	2.3170	1.2461	0.0593
0.1800	0.1708	0.1432	2.2959	1.2406	0.0744
0.2000	0.2066	0.1736	2.2727	1.2346	0.0909
0.2200	0.2445	0.2057	2.2477	1.2281	0.1088
0.2400	0.2841	0.2395	2.2209	1.2213	0.1279
0.2600	0.3250	0.2745	2.1925	1.2140	0.1482
0.2800	0.3667	0.3104	2.1626	1.2064	0.1696
0.3000	0.4089	0.3469	2.1314	1.1985	0.1918
0.3200	0.4512	0.3837	2.0991	1.1904	0.2149
0.3400	0.4933	0.4206	2.0657	1.1822	0.2388
0.3600	0.5348	0.4572	2.0314	1.1737	0.2633
0.3800	0.5755	0.4935	1.9964	1.1652	0.2883
0.4000	0.6151	0.5290	1.9608	1.1566	0.3137
0.4200	0.6535	0.5638	1.9247	1.1480	0.3395
0.4400	0.6903	0.5975	1.8882	1.1394	0.3656
0.4600	0.7254	0.6301	1.8515	1.1308	0.3918
0.4800	0.7587	0.6614	1.8147	1.1224	0.4181
0.5000	0.7901	0.6914	1.7778	1.1141	0.4444
0.5200	0.8196	0.7199	1.7409	1.1059	0.4708
0.5400	0.8469	0.7470	1.7043	1.0979	0.4970
0.5600	0.8723	0.7725	1.6678	1.0901	0.5230
0.5800	0.8955	0.7965	1.6316	1.0826	0.5489
0.6000	0.9167	0.8189	1.5957	1.0753	0.5745
0.6200	0.9358	0.8398	1.5603	1.0682	0.5998
0.6400	0.9530	0.8592	1.5253	1.0615	0.6248
0.6600	0.9682	0.8771	1.4908	1.0550	0.6494
0.6800	0.9814	0.8935	1.4569	1.0489	0.6737
0.7000	0.9929	0.9085	1.4235	1.0431	0.6975
0.7200	1.0026	0.9221	1.3907	1.0376	0.7209
0.7400	1.0106	0.9344	1.3585	1.0325	0.7439
0.7600	1.0171	0.9455	1.3270	1.0278	0.7665
0.7800	1.0220	0.9553	1.2961	1.0234	0.7885
0.8000	1.0255	0.9639	1.2658	1.0193	0.8101
0.8200	1.0276	0.9715	1.2362	1.0157	0.8313
0.8400	1.0285	0.9781	1.2073	1.0124	0.8519
0.8600	1.0283	0.9836	1.1791	1.0095	0.8721
0.8800	1.0269	0.9883	1.1515	1.0070	0.8918
0.9000	1.0245	0.9921	1.1246	1.0049	0.9110
0.9200	1.0212	0.9951	1.0984	1.0031	0.9297
0.9400	1.0170	0.9973	1.0728	1.0017	0.9480
0.9600	1.0121	0.9988	1.0479	1.0008	0.9658
0.9800	1.0064	0.9997	1.0236	1.0002	0.9831

TABLE B.6 (*Continued*)

Rayleigh Flow, $\gamma = 1.4$					
M	T/T^*	T_t/T_t^*	P/P^*	P_t/P_t^*	V/V^*
1.0000	1.0000	1.0000	1.0000	1.0000	1.0000
1.0200	0.9930	0.9997	0.9770	1.0002	1.0164
1.0400	0.9855	0.9989	0.9546	1.0008	1.0325
1.0600	0.9776	0.9977	0.9327	1.0017	1.0480
1.0800	0.9691	0.9960	0.9115	1.0031	1.0632
1.1000	0.9603	0.9939	0.8909	1.0049	1.0780
1.1200	0.9512	0.9915	0.8708	1.0070	1.0923
1.1400	0.9417	0.9887	0.8512	1.0095	1.1063
1.1600	0.9320	0.9856	0.8322	1.0124	1.1198
1.1800	0.9220	0.9823	0.8137	1.0157	1.1330
1.2000	0.9118	0.9787	0.7958	1.0194	1.1459
1.2200	0.9015	0.9749	0.7783	1.0235	1.1584
1.2400	0.8911	0.9709	0.7613	1.0279	1.1705
1.2600	0.8805	0.9668	0.7447	1.0328	1.1823
1.2800	0.8699	0.9624	0.7287	1.0380	1.1938
1.3000	0.8592	0.9580	0.7130	1.0437	1.2050
1.3200	0.8484	0.9534	0.6978	1.0497	1.2159
1.3400	0.8377	0.9487	0.6830	1.0561	1.2264
1.3600	0.8269	0.9440	0.6686	1.0629	1.2367
1.3800	0.8161	0.9391	0.6546	1.0701	1.2467
1.4000	0.8054	0.9343	0.6410	1.0777	1.2564
1.4200	0.7947	0.9293	0.6278	1.0856	1.2659
1.4400	0.7840	0.9243	0.6149	1.0940	1.2751
1.4600	0.7735	0.9193	0.6024	1.1028	1.2840
1.4800	0.7629	0.9143	0.5902	1.1120	1.2927
1.5000	0.7525	0.9093	0.5783	1.1215	1.3012
1.5200	0.7422	0.9042	0.5668	1.1315	1.3095
1.5400	0.7319	0.8992	0.5555	1.1419	1.3175
1.5600	0.7217	0.8942	0.5446	1.1527	1.3253
1.5800	0.7117	0.8892	0.5339	1.1640	1.3329
1.6000	0.7017	0.8842	0.5236	1.1756	1.3403
1.6200	0.6919	0.8792	0.5135	1.1877	1.3475
1.6400	0.6822	0.8743	0.5036	1.2002	1.3546
1.6600	0.6726	0.8694	0.4940	1.2131	1.3614
1.6800	0.6631	0.8645	0.4847	1.2264	1.3681
1.7000	0.6538	0.8597	0.4756	1.2402	1.3746
1.7200	0.6445	0.8549	0.4668	1.2545	1.3809
1.7400	0.6355	0.8502	0.4581	1.2692	1.3870
1.7600	0.6265	0.8455	0.4497	1.2843	1.3931
1.7800	0.6176	0.8409	0.4415	1.2999	1.3989
1.8000	0.6089	0.8363	0.4335	1.3159	1.4046
1.8200	0.6004	0.8317	0.4257	1.3324	1.4102
1.8400	0.5919	0.8273	0.4181	1.3494	1.4156
1.8600	0.5836	0.8228	0.4107	1.3669	1.4209
1.8800	0.5754	0.8185	0.4035	1.3849	1.4261
1.9000	0.5673	0.8141	0.3964	1.4033	1.4311
1.9200	0.5594	0.8099	0.3895	1.4222	1.4360
1.9400	0.5516	0.8057	0.3828	1.4417	1.4408
1.9600	0.5439	0.8015	0.3763	1.4616	1.4455
1.9800	0.5364	0.7974	0.3699	1.4821	1.4501

TABLE B.6   (*Continued*)

Rayleigh Flow, $\gamma = 1.4$

M	T/T^*	T_t/T_t^*	P/P^*	P_t/P_t^*	V/V^*
2.0000	0.5289	0.7934	0.3636	1.5031	1.4545
2.0200	0.5216	0.7894	0.3575	1.5246	1.4589
2.0400	0.5144	0.7855	0.3516	1.5467	1.4632
2.0600	0.5074	0.7816	0.3458	1.5693	1.4673
2.0800	0.5004	0.7778	0.3401	1.5924	1.4714
2.1000	0.4936	0.7741	0.3345	1.6162	1.4753
2.1200	0.4868	0.7704	0.3291	1.6404	1.4792
2.1400	0.4802	0.7667	0.3238	1.6653	1.4830
2.1600	0.4737	0.7631	0.3186	1.6908	1.4867
2.1800	0.4673	0.7596	0.3136	1.7168	1.4903
2.2000	0.4611	0.7561	0.3086	1.7434	1.4938
2.2200	0.4549	0.7527	0.3038	1.7707	1.4973
2.2400	0.4488	0.7493	0.2991	1.7986	1.5007
2.2600	0.4428	0.7460	0.2945	1.8271	1.5040
2.2800	0.4370	0.7428	0.2899	1.8562	1.5072
2.3000	0.4312	0.7395	0.2855	1.8860	1.5103
2.3200	0.4256	0.7364	0.2812	1.9165	1.5134
2.3400	0.4200	0.7333	0.2769	1.9476	1.5165
2.3600	0.4145	0.7302	0.2728	1.9794	1.5194
2.3800	0.4091	0.7272	0.2688	2.0119	1.5223
2.4000	0.4038	0.7242	0.2648	2.0451	1.5252
2.4200	0.3986	0.7213	0.2609	2.0789	1.5279
2.4400	0.3935	0.7184	0.2571	2.1136	1.5306
2.4600	0.3885	0.7156	0.2534	2.1489	1.5333
2.4800	0.3836	0.7128	0.2497	2.1850	1.5359
2.5000	0.3787	0.7101	0.2462	2.2218	1.5385
2.5200	0.3739	0.7074	0.2427	2.2594	1.5410
2.5400	0.3692	0.7047	0.2392	2.2978	1.5434
2.5600	0.3646	0.7021	0.2359	2.3370	1.5458
2.5800	0.3601	0.6995	0.2326	2.3770	1.5482
2.6000	0.3556	0.6970	0.2294	2.4177	1.5505
2.6200	0.3512	0.6945	0.2262	2.4593	1.5527
2.6400	0.3469	0.6921	0.2231	2.5018	1.5549
2.6600	0.3427	0.6896	0.2201	2.5451	1.5571
2.6800	0.3385	0.6873	0.2171	2.5892	1.5592
2.7000	0.3344	0.6849	0.2142	2.6343	1.5613
2.7200	0.3304	0.6826	0.2113	2.6802	1.5634
2.7400	0.3264	0.6804	0.2085	2.7270	1.5654
2.7600	0.3225	0.6781	0.2058	2.7748	1.5673
2.7800	0.3186	0.6760	0.2030	2.8235	1.5693
2.8000	0.3149	0.6738	0.2004	2.8731	1.5711
2.8200	0.3111	0.6717	0.1978	2.9237	1.5730
2.8400	0.3075	0.6696	0.1953	2.9752	1.5748
2.8600	0.3039	0.6675	0.1927	3.0278	1.5766
2.8800	0.3004	0.6655	0.1903	3.0813	1.5784
2.9000	0.2969	0.6635	0.1879	3.1359	1.5801
2.9200	0.2934	0.6615	0.1855	3.1914	1.5818

Differential Form of the Governing Equations

In Cartesian coordinates $\mathbf{V} = \mathbf{V}(x, y, z)$, the equations for an incompressible Newtonian fluid are given as

Continuity

$$\frac{\partial u}{\partial x} + \frac{\partial v}{\partial y} + \frac{\partial w}{\partial z} = 0$$

x-momentum

$$\rho\left(\frac{\partial u}{\partial t} + u\frac{\partial u}{\partial x} + v\frac{\partial u}{\partial y} + w\frac{\partial u}{\partial z}\right) = \rho g_x - \frac{\partial P}{\partial x} + \mu\left[\frac{\partial^2 u}{\partial x^2} + \frac{\partial^2 u}{\partial y^2} + \frac{\partial^2 u}{\partial z^2}\right]$$

y-momentum

$$\rho\left(\frac{\partial v}{\partial t} + u\frac{\partial v}{\partial x} + v\frac{\partial v}{\partial y} + w\frac{\partial v}{\partial z}\right) = \rho g_y - \frac{\partial P}{\partial y} + \mu\left[\frac{\partial^2 v}{\partial x^2} + \frac{\partial^2 v}{\partial y^2} + \frac{\partial^2 v}{\partial z^2}\right]$$

z-momentum

$$\rho\left(\frac{\partial w}{\partial t} + u\frac{\partial w}{\partial x} + v\frac{\partial w}{\partial y} + w\frac{\partial w}{\partial z}\right) = \rho g_z - \frac{\partial P}{\partial z} + \mu\left[\frac{\partial^2 w}{\partial x^2} + \frac{\partial^2 w}{\partial y^2} + \frac{\partial^2 w}{\partial z^2}\right]$$

In cylindrical coordinates $\mathbf{V} = \mathbf{V}(r, \theta, z)$,

Continuity

$$\frac{1}{r}\frac{\partial}{\partial r}(rv_r) + \frac{1}{r}\frac{\partial}{\partial \theta}(v_\theta) + \frac{\partial w}{\partial z} = 0$$

r-momentum

$$\rho\left(\frac{\partial v_r}{\partial t} + v_r\frac{\partial v_r}{\partial r} + \frac{v_\theta}{r}\frac{\partial v_r}{\partial \theta} - \frac{v_\theta^2}{r} + v_z\frac{\partial v_r}{\partial z}\right) = \rho g_r - \frac{\partial P}{\partial r}$$

$$+ \mu\left[\frac{\partial}{\partial r}\left(\frac{1}{r}\frac{\partial}{\partial r}(rv_r)\right) + \frac{1}{r^2}\frac{\partial^2 v_r}{\partial \theta^2} - \frac{2}{r^2}\frac{\partial v_\theta}{\partial \theta} + \frac{\partial^2 v_r}{\partial z^2}\right]$$

θ-momentum

$$\rho\left(\frac{\partial v_\theta}{\partial t} + v_r\frac{\partial v_\theta}{\partial r} + \frac{v_\theta}{r}\frac{\partial v_\theta}{\partial \theta} + \frac{v_r v_\theta}{r} + v_z\frac{\partial v_\theta}{\partial z}\right) = \rho g_r - \frac{1}{r}\frac{\partial P}{\partial \theta}$$

$$+ \mu\left[\frac{\partial}{\partial r}\left(\frac{1}{r}\frac{\partial}{\partial r}(rv_\theta)\right) + \frac{1}{r^2}\frac{\partial^2 v_\theta}{\partial \theta^2} + \frac{2}{r^2}\frac{\partial v_r}{\partial \theta} + \frac{\partial^2 v_\theta}{\partial z^2}\right]$$

z-momentum

$$\rho\left(\frac{\partial v_z}{\partial t} + v_r\frac{\partial v_z}{\partial r} + \frac{v_\theta}{r}\frac{\partial v_z}{\partial \theta} + v_z\frac{\partial v_z}{\partial z}\right) = \rho g_r - \frac{\partial P}{\partial z}$$

$$+ \mu\left[\frac{1}{r}\frac{\partial}{\partial r}\left(r\frac{\partial}{\partial r}(v_z)\right) + \frac{1}{r^2}\frac{\partial^2 v_z}{\partial \theta^2} + \frac{\partial^2 v_z}{\partial z^2}\right]$$

In spherical coordinates $\mathbf{V} = \mathbf{V}(r, \theta, \phi)$,

Continuity

$$\frac{1}{r^2}\frac{\partial}{\partial r}(r^2 v_r) + \frac{1}{r \sin \theta}\frac{\partial}{\partial \theta}(v_\theta \sin \theta) + \frac{1}{r \sin \theta}\frac{\partial v_\phi}{\partial \phi} = 0$$

r-momentum

$$\rho\left(\frac{\partial v_r}{\partial t} + v_r \frac{\partial v_r}{\partial r} + \frac{v_\theta}{r}\frac{\partial v_r}{\partial \theta} + \frac{v_\phi}{r \sin \theta}\frac{\partial v_r}{\partial \phi} - \frac{v_\theta^2 + v_\phi^2}{r}\right) = \rho g_r - \frac{\partial P}{\partial r}$$

$$+ \mu\left[\frac{1}{r^2}\frac{\partial^2}{\partial r^2}(r^2 v_r) + \frac{1}{r^2 \sin \theta}\frac{\partial}{\partial \theta}\left(\sin \theta \frac{\partial v_r}{\partial \theta}\right) + \frac{1}{r^2 \sin_\theta^2}\frac{\partial^2 v_r}{\partial \phi^2}\right]$$

θ-momentum

$$\rho\left(\frac{\partial v_\theta}{\partial t} + v_r \frac{\partial v_\theta}{\partial r} + \frac{v_\theta}{r}\frac{\partial v_\theta}{\partial \theta} + \frac{v_\phi}{r \sin \theta}\frac{\partial v_\theta}{\partial \phi} + \frac{v_r v_\theta}{r} - \frac{v_\phi^2}{r}\cot \theta\right) = \rho g_\theta - \frac{1}{r}\frac{\partial P}{\partial \theta}$$

$$+ \mu\left[\frac{1}{r^2}\frac{\partial}{\partial r}\left(r^2 \frac{\partial v_\theta}{\partial r}\right) + \frac{1}{r^2}\frac{\partial}{\partial \theta}\left(\frac{1}{\sin \theta}\frac{\partial}{\partial \theta}(v_\theta \sin \theta)\right)\right.$$

$$\left. + \frac{1}{r^2 \sin^2 \theta}\frac{\partial^2 v_\theta}{\partial \phi^2} + \frac{2}{r^2}\frac{\partial v_r}{\partial \theta} - \frac{2 \cos \theta}{r^2 \sin^2 \theta}\frac{\partial v_\phi}{\partial \phi}\right]$$

φ-momentum

$$\rho\left(\frac{\partial v_\phi}{\partial t} + v_r \frac{\partial v_\phi}{\partial r} + \frac{v_\theta}{r}\frac{\partial v_\phi}{\partial \theta} + \frac{v_\phi}{r \sin \theta}\frac{\partial v_\phi}{\partial \phi} + \frac{v_\phi v_r}{r} + \frac{v_\phi v_\theta}{r}\cot \theta\right) = \rho g_\phi - \frac{1}{r \sin \theta}\frac{\partial P}{\partial \phi}$$

$$+ \mu\left[\frac{1}{r^2}\frac{\partial}{\partial r}\left(r^2 \frac{\partial v_\phi}{\partial r}\right) + \frac{1}{r^2}\frac{\partial}{\partial \theta}\left(\frac{1}{\sin \theta}\frac{\partial}{\partial \theta}(v_\phi \sin \theta)\right)\right.$$

$$\left. + \frac{1}{r^2 \sin^2 \theta}\frac{\partial^2 v_\phi}{\partial \phi^2} + \frac{2}{r^2 \sin \theta}\frac{\partial v_r}{\partial \phi} + \frac{2 \cos \theta}{r^2 \sin^2 \theta}\frac{\partial v_\theta}{\partial \phi}\right]$$

Computer Programs

D.1 Numerical Solution of Algebraic Equations

D.1.1 Calculation of Friction Factor f: Bisection Method

```
C---------------------------------------------
C Program to Calculate the Friction Factor Using
C Colebrook's Formula
C Numerical Method: Bisection
C Written by A. Alexandrou
C---------------------------------------------
C Enter Data
C---------------------------------------------
      WRITE(6,*)'Enter Roughness Ratio (e/d)'
      READ(5,*)ED
      WRITE(6,*)'Enter Reynolds Number (Re)'
      READ(5,*)Re
C---------------------------------------------
C Establish Initial Range
C---------------------------------------------
      WRITE(6,*)'Enter Minimum Value'
      READ(5,*)FMIN
      WRITE(6,*)'Enter Maximum Value'
      READ(5,*)FMAX
C---------------------------------------------
C Initiate Iteration Procedure
C Note: the solution is setup for 1/F
C---------------------------------------------
10    ITER=ITER+1
      FR = 0.5*(FMIN+FMAX)
      F=1.+SQRT(FR)*0.869*ALOG(ED/3.7+2.51/(RE*SQRT(FR)))
      IF(F.GT.0.0)FMIN=FR
      IF(F.LT.0.0)FMAX=FR
      WRITE(*,*)'FR=', FR
      IF(ABS(F).GT.1.E-6) GO TO 10
C---------------------------------------------
C PRINT THE SOLUTION
C---------------------------------------------
      WRITE(6,*)'THE FRICTION FACTOR IS', FR
      STOP
      END
```

D.1.2 Calculation of Friction Factor f: Newton-Raphson Method

```
C---------------------------------------------
C Program to Calculate the Friction Factor Using
C Colebrook's Formula
```

```
C Numerical Method: Newton-Rapshon
C Written By Prof. A. Alexandrou
C----------------------------------------
C Enter Data
C----------------------------------------
      WRITE(6,*)'Enter Roughness Ratio (e/d)'
      READ(5,*)ED
      WRITE(6,*)'Enter Reynolds Number (Re)'
      READ(5,*)Re
C----------------------------------------
C Establish Initial Guess For F
C----------------------------------------
      F= 0.25/(0.434*ALOG(ED/3.7 + 5.74/RE**0.9 )**2 )
      F=1/SQRT(F)
      ITER=0
C----------------------------------------
C Initiate Iteration Procedure
C Note: the solution is set up for 1/F
C----------------------------------------
10    ITER=ITER+1
      G= ED/3.7+2.51*F/RE
      RF=F +0.869*ALOG(G)
      DFR= 1. + 0.869/G *2.51/RE
      DF = - RF/DFR
      F= F+ DF
      IF(ABS(DF).GT.1.E-6) GO TO 10
      F=1./F**2
C----------------------------------------
C Print The Solution
C----------------------------------------
      WRITE(6,*)'THE FRICTION FACTOR IS', F
      STOP
      END
```

D.2   Numerical Integration

D.2.1   Trapezoidal Integration

```
C----------------------------------------
C Program for Example 13.4
C Numerical Method: Trapezoidal Integration
C Written by A. Alexandrou
C----------------------------------------
      DIMENSION P(10000), X(10000)
C----------------------------------------
C Enter Data
C----------------------------------------
      WRITE(6,*)'Enter Range of Integration'
      WRITE(6,*)
      WRITE(*,*)'Xmin ?'
      READ(5,*)XMIN
```

```
          WRITE(6,*)
          WRITE(*,*)'Xmax ?'
          READ(5,*)XMAX
          WRITE(6,*)'Enter Number of Strips'
          READ(5,*)N
          DX=(XMAX-XMIN)/FLOAT(N)
          NPOINTS=N+1
C---------------------------------------------
C Initialize Pressure and Coordinates
C---------------------------------------------
          DO I=1,NPOINTS
          X(I)= FLOAT(I-1)*DX
          P(I)=100.+ 20.*EXP(X(I))
          END DO
C---------------------------------------------
C Integrate for the Force
C---------------------------------------------
          FORCE=0.0
          DO I=1,N
          FORCE=FORCE-2*(P(I)+P(I+1))*(X(I+1)-X(I))*0.5
          END DO
C---------------------------------------------
C Print Solution
C---------------------------------------------
          WRITE(6,*)'THE FORCE IS ', FORCE
          STOP
          END
```

D.2.2   Simpson's Integration

```
C---------------------------------------------
C Program for Example 13.4
C Numerical Method: Simpson's Integration
C Written by A. Alexandrou
C---------------------------------------------
          DIMENSION P(10000), X(10000)
C---------------------------------------------
C Enter Data
C---------------------------------------------
          WRITE(6,*)'Enter Range of Integration'
          WRITE(6,*)
          WRITE(*,*)'Xmin ?'
          READ(5,*)XMIN
          WRITE(6,*)
          WRITE(*,*)'Xmax ?'
          READ(5,*)XMAX
          WRITE(6,*)'Enter Number of Strips'
          READ(5,*)N
          DX=(XMAX-XMIN)/FLOAT(N)
          NPOINTS=N+1
C---------------------------------------------
```

```
C Initialize Pressure and Coordinates
C------------------------------------------
        DO I=1,NPOINTS
        X(I)= FLOAT(I-1)*DX
        P(I)=100.+ 20.*EXP(X(I))
        END DO
C------------------------------------------
C Integrate for the Force
C------------------------------------------
        FORCE=0.0
        DO I=1,NPOINTS
        CONS=2.
        IF( (I/2)*2.EQ.I) CONS=4.
        IF((I.EQ.1).OR.(I.EQ.NPOINTS))CONS=1.
        FORCE=FORCE- 2.0* (DX/3.)*CONS*P(I)
        END DO
C------------------------------------------
C Print Solution
C------------------------------------------
        WRITE(6,*)'THE FORCE IS ', FORCE
        STOP
        END
```

D.3   Numerical Integration of ODEs

D.3.1   Euler's Methods

```
C------------------------------------------
C Program for Example 13.5
C Numerical Method: Various Euler's Methods
C Written by A. Alexandrou
C------------------------------------------
C Enter Data
C------------------------------------------
        WRITE(6,*)'Enter initial liquid level'
        WRITE(6,*)
        WRITE(*,*)'H ?'
        READ(5,*)H0
        WRITE(6,*)'Enter diameter ratio'
        WRITE(6,*)
        WRITE(*,*)'D/d ?'
        READ(5,*)DD
        WRITE(6,*)'Enter step size (dt)'
        READ(5,*)DT
        G=9.81
        CONS0=1./(DD**4. -1.)
        CONS1=SQRT(G*CONS0/2.0)
        HF=H0
        HC=H0
C------------------------------------------
C Integration Loop
```

```
C-------------------------------------------
        WRITE(6,*)' TIME  ', ' HEIGHT'
        WRITE(6,*)TIME,H0
        TIME=0.0
        WRITE(6,*)TIME,H0,H0,HO
10      TIME=TIME+DT
C-------------------------------------------
C Analytic Solution
C-------------------------------------------
C
        HANAL= ( SQRT(H0) - CONS1*TIME )**2
C
C-------------------------------------------
C Forward Euler Solution
C-------------------------------------------
C
        HF=HF - DT *SQRT(CONS0*2.*G*HF)
C
C-------------------------------------------
C Corrected Euler Solution
C-------------------------------------------
C
        HCS=HC - DT *SQRT(CONS0*2.*G*HC)
        HC=HC-0.5*DT*(SQRT(CONS0*2.*G*HC)+SQRT(CONS0*2.*G*HCS))
C
C-------------------------------------------
C
        IF(( SQRT(H0) - CONS1*TIME ).GT.0.0)THEN
        WRITE(6,*)TIME,HANAL,HF,HC
        GO TO 10
        END IF
        STOP
        END
```

D.3.2 Boundary-Layer Similarity Solution

```
C-------------------------------------------
C
C Program Integration of Boundary-Layer Equations
C Numerical Method: Fourth-Order Runge-Kutta
C Written by A. Alexandrou
C
C-------------------------------------------
C Enter Data
C-------------------------------------------
        WRITE(6,*)'ENTER STEP SIZE H=? (Ex. 0.25)'
        READ(5,*)H
        WRITE(6,*)'ENTER MAXIMUM DISTANCE (Ex. 8.)'
        READ(5,*)XMAX
C-------------------------------------------
C Initial Data
```

```
C-------------------------------------------
      X=0.0
      Y=0.0
      Y1=0.0
      WRITE(6,*)'START SHOOTING METHOD BY GUESSING Y"'
      WRITE(6,*)
      WRITE(6,*)'ENTER Y2, (FOR FLAT PLATE =0.33206)'
      READ(5,*)Y2
      WRITE(6,'(F4.1,3(4X,F9.5))')X,Y,Y1,Y2
C-------------------------------------------
C Initiate Integration
C-------------------------------------------
10    RK11=H*FUNC1(Y1)
      RK12=H*FUNC2(Y2)
      RK13=H*FUNC3(Y,Y1,Y2)
C
      F=Y+RK11/2.0
      F1=Y1+RK12/2.0
      F2=Y2+RK13/2.0
C
      RK21=H*FUNC1(F1)
      RK22=H*FUNC2(F2)
      RK23=H*FUNC3(F,F1,F2)
C
      F=Y+RK21/2.0
      F1=Y1+RK22/2.0
      F2=Y2+RK23/2.0
C
      RK31=H*FUNC1(F1)
      RK32=H*FUNC2(F2)
      RK33=H*FUNC3(F,F1,F2)
C
      F=Y+RK31
      F1=Y1+RK32
      F2=Y2+RK33
C
      RK41=H*FUNC1(F1)
      RK42=H*FUNC2(F2)
      RK43=H*FUNC3(F,F1,F2)
C
      Y=Y+(RK11+2.0*RK21+2.0*RK31+RK41)/6.0
      Y1=Y1+(RK12+2.0*RK22+2.0*RK32+RK42)/6.0
      Y2=Y2+(RK13+2.0*RK23+2.0*RK33+RK43)/6.0
C
      X=X+H
      WRITE(6,'(F4.1,3(4X,F9.5))')X,Y,Y1,Y2
      IF(X.LT.XMAX)GO TO 10
      STOP
      END
C
C-------------------------------------------
```

```
C
      FUNCTION FUNC1(R)
      FUNC1=R
      RETURN
      END
C
      FUNCTION FUNC2(R)
      FUNC2=R
      RETURN
      END
C
      FUNCTION FUNC3(R,R1,R2)
      FUNC3=-0.5*R*R2
      RETURN
      END
```

D.4   Numerical Integration of PDEs

D.4.1   Inviscid Flow: Solution of $\nabla^2 \psi = 0$

```
C----------------------------------------
C Inviscid Flow in a 90* corner
C Numerical Method: Finite Difference
C Written by A. Alexandrou
C----------------------------------------
      COMMON/MSH1/NEX1,NEX2,NEX,NEY1,NEY2,NEY,NE
      COMMON/MSH2/NX1,NX2,NX,NY1,NY2,NY
      DIMENSION P(1000000), X(1000000), Y(1000000)
C
C
C NEX1: NUMBER OF CELLS IN SECTION 1
C NEX2: NUMBER OF CELLS IN SECTION 2
C NEY1: NUMBER OF CELLS IN SECTION 1
C NEY2: NUMBER OF CELLS IN SECTION 2
C
C NX1: NUMBER OF NODES IN SECTION 1
C NX2: NUMBER OF NODES IN SECTION 2
C NY1: NUMBER OF NODES IN SECTION 1
C NY2: NUMBER OF NODES IN SECTION 2
C
C NX: TOTAL OF NODES IN X-DIRECTION
C NY: TOTAL OF NODES IN Y-DIRECTION
C
C X: COORDINATE
C Y: COORDINATE
C P: STREAMFUNCTION
C
C----------------------------------------
C Programming Notes on Function NODE:
C
C The domain is divided into two rectangular areas,
```

```
C each having ( NEX1*NEY1 ) and ( NEX2*(NEY1+NEY2))
C cells. Starting from the lower left hand side
C corner of each domain, and by considering the nodes
C to be arranged in a two-dimensional matrix (I,J),
C Function NODE(K,I,J) returns the NUMBER of the node
C at location (I,J) in domain K.
c
C------------------------------------------
C
C Enter Data
C
C------------------------------------------
      WRITE(6,*)'Enter Step size in x-direction'
      READ(5,*)DX
      WRITE(6,*)'Enter Step size in Y-direction'
      READ(5,*)DY
      WRITE(6,*)'Enter NEX1'
      READ(5,^)NEX1
      WRITE(6,*)'Enter NEX2'
      READ(5,*)NEX2
      WRITE(6,*)'Enter NEY1'
      READ(5,*)NEY1
      WRITE(6,*)'Enter NEY2'
      READ(5,*)NEY2
      WRITE(6,*)'Enter PSI1'
      READ(5,*)PSI1
      WRITE(6,*)'Enter PSI2'
      READ(5,*)PSI2
C------------------------------------------
C Create MESH
C------------------------------------------
      HX1=NEX1*DX
      HY1=NEY2*DY
      NEX=NEX1+NEX2
      NX1=NEX1+1
      NX2=NEX2+1
      NX=NEX+1
      NEY=NEY1+NEY2
      NY1=NEY1+1
      NY2=NEY2+1
      NY=NEY+1
C------------------------------------------
C Section 1
C------------------------------------------
      DO I=1,NX1-1
      DO J=1,NY1
      XX=(I-1)*DX
      YY=(J-1)*DY+HY1
      L=NODE(1,I,J)
      X(L)=XX
      Y(L)=YY
```

```
            END DO
            END DO
      C-----------------------------------------------
      C Section 2
      C-----------------------------------------------
            DO I=1,NX2
            DO J=1,NY
            XX=(I-1)*DX+HX1
            YY=(J-1)*DY
            L=NODE(2,I,J)
            X(L)=XX
            Y(L)=YY
            END DO
            END DO
      C-----------------------------------------------
      C Set Boundary Conditions
      C-----------------------------------------------
            NH=NODE(2,NX2,NY)
            DO I=1,NH
            P(I)=0.0
            END DO
      C
      C Inlet
      C
            DO J=1,NY1
            L=NODE(1,1,J)
            P(L)=PSI1+(PSI2-PSI1)/FLOAT(NY1-1)*FLOAT(J-1)
            END DO
      C
      C Exit
      C
            DO I=1,NX2
            L=NODE(2,I,1)
            P(L)=PSI1+(PSI2-PSI1)/FLOAT(NX2-1)*FLOAT(I-1)
            END DO
      C
      C Section 1 Boundary
      C
            DO I=1,NX1
            L=NODE(1,I,1)
            P(L)=PSI1
            L=NODE(1,I,NY1)
            P(L)=PSI2
            END DO
      C
      C Section 2 Boundary
      C
            DO I=1,NX2
            L=NODE(2,I,NY)
            P(L)=PSI2
            END DO
```

```
          DO J=1,NY2
          L=NODE(2,1,J)
          P(L)=PSI1
          END DO
C
          DO J=1,NY
          L=NODE(2,NX2,J)
          P(L)=PSI2
          END DO
C-------------------------------------------
C START ITERATION
C-------------------------------------------
C
          WRITE(6,*)'Enter THETA'
          READ(5,*)TH
          ERRMAX=1.E-5
          ITER=0
   10 ITER=ITER+1
          ERROR=0.0
C
C Section 1
C
          DO I=2,NX1
          DO J=2,NY1-1
          L=NODE(1,I,J)
          LX1=NODE(1,I-1,J)
          LX2=NODE(1,I+1,J)
          LY1=NODE(1,I,J-1)
          LY2=NODE(1,I,J+1)
          Q=(DY*DY*(P(LX1)+P(LX2))+DX*DX*(P(LY1)+P(LY2)))/
             (2.*(DX**2+DY**2))
          P1=TH*Q+(1.-TH)*P(L)
          ERR=ABS(P1- P(L))
          IF(ERR.GT.ERROR)ERROR=ERR
          P(L)=P1
          END DO
          END DO
C
C Section 2
C
          DO I=2,NX2-1
          DO J=2,NY-1
          L=NODE(2,I,J)
          LX1=NODE(2,I-1,J)
          LX2=NODE(2,I+1,J)
          LY1=NODE(2,I,J-1)
          LY2=NODE(2,I,J+1)
          Q=(DY*DY*(P(LX1)+P(LX2))+DX*DX*(P(LY1)+P(LY2)))/
             (2.*(DX**2+DY**2))
          P1=TH*Q+(1.-TH)*P(L)
          ERR=ABS( P1- P(L) )
```

```
      IF(ERR.GT.ERROR)ERROR=ERR
      P(L)=P1
      END DO
      END DO
      IF(ERROR.GT.ERRMAX) GO TO 10
      DO I=1,NH
      WRITE(*,*)I,X(I),Y(I),P(I)
      WRITE(7,*)I,X(I),Y(I),P(I)
      END DO
      WRITE(*,*)'CURRENT ERROR= ',ERROR, '( iter=',ITER,' )'
      STOP
      END
C
C-----------------------------------------------
C
C
      FUNCTION NODE(K,I,J)
      COMMON/MSH1/NEX1,NEX2,NEX,NEY1,NEY2,NEY,NE
      COMMON/MSH2/NX1,NX2,NX,NY1,NY2,NY
      IF(K.EQ.1)THEN
      IF(I.LT.NX1)NODE=(I-1)*NY1+J
      IF(I.EQ.NX1)NODE=(NX1-1)*NY1+J+(NY-NY1)
      IF(I.EQ.(NX1+1))NODE=NY1*(NX1-1)+NY+NY2+J-1
      END IF
      IF(K.EQ.2)NODE=(I-1)*NY+J+NY1*(NX1-1)
      RETURN
      END
```

D.4.2 Boundary-Layer Flow

```
C-----------------------------------------------
C Boundary-Layer Flow Past a Flat Plate
C Numerical Method: Finite Difference
C Written by A. Alexandrou
C-----------------------------------------------------
C
      PARAMETER (NN=1000)
      COMMON/MSH1/NEX,NEY,NE
      COMMON/MSH2/NX,NY
      DIMENSION X(NN), Y(NN)
      DIMENSION UO(NN), VO(NN)
      DIMENSION U(NN), V(NN)
      DIMENSION F(NN), A(NN,NN), B(NN)
C
C
C     NX: TOTAL NUMBER OF NODES IN X-DIRECTION
C     NY: TOTAL NUMBER OF NODES IN Y-DIRECTION
C
C     X: COORDINATE
C     Y: COORDINATE
C
```

```
C-----------------------------------------------
C
C Enter Data
C
C-----------------------------------------------
C
      WRITE(6,*)'Enter Step size in x-direction'
      READ(5,*)DX
      WRITE(6,*)'Enter Step size in Y-direction'
      READ(5,*)DY
      WRITE(6,*)'Enter NX'
      READ(5,*)NX
      WRITE(6,*)'Enter NY'
      READ(5,*)NY
      WRITE(6,*)'Enter RHO'
      READ(5,*)RHO
      WRITE(6,*)'Enter AMU'
      READ(5,*)AMU
      WRITE(6,*)'Enter U FREESTREAM'
      READ(5,*)U8
C
C-----------------------------------------------
C Create  MESH
C-----------------------------------------------
C
      DO J=1,NY
      Y(J)=(J-1)*DY
      END DO
C
C-----------------------------------------------
C Set Boundary Conditions
C-----------------------------------------------
C
      PI=ACOS(-1.)
      NH=2*NY/3
      NP=2*NY
C
      DO J=1,NH
      VO(J)=0.0
      UO(J)=U8*SIN(PI/2.0*(J-1)/(NH-1))
      END DO
      DO J=NH+1,NY
      VO(J)=0.0
      UO(J)=U8
      END DO
C
      DO I=1,NY
      WRITE(6,*)I,Y(I),UO(I),VO(I)
      END DO
C
C-----------------------------------------------
```

```
C Set Initial Guess
C----------------------------------------------
C
      DO J=1,NY
      V(J)=VO(J)
      U(J)=UO(J)
      END DO
C
      ISTEP=0
      ERRMAX=1.E-5
   10 ISTEP=ISTEP+1
C
C-------------------------------------------------------
C Start Iteration Procedure
C-------------------------------------------------------
C
      ITER=0
   20 ITER=ITER+1
C
      DO I=1,NP
      DO J=1,NP
      A(I,J)=0.0
      END DO
      END DO
C
      DO I=1,NY
C
      IF(I.EQ.1)THEN
      F(1)=0.0
      F(2)=0.0
      A(1,1)= 1.0
      A(2,2)= 1.0
      END IF
C
      IF(I.EQ.NY)THEN
      F(NP-1)=(V(I)-V(I-1))/DY
      F(NP)=0.0
C
      A(NP-1,NP-3)= -1.0/DY
      A(NP-1,NP-1)=1./DY
      A(NP,NP)=1.0
C
      END IF
C
      IF((I.LT.NY).AND.(I.GT.1))THEN
      F(2*I-1)=(U(I)-UO(I))/DX+(V(I+1)-V(I-1))/(2.0*DY)
      F(2*I)=U(I)*(U(I)-UO(I))/DX+V(I)*(U(I+1)-U(I-1))/
            (2.0*DY)
      F(2*I)=F(2*I)- AMU/RHO*(U(I+1)-2.0*U(I)+U(I-1))/DY**2
C
      A(2*I-1,2*I-3)= -0.5/DY
```

```
              A(2*I-1,2*I-2)=0.0
              A(2*I-1,2*I-1)=0.0
      C
              A(2*I-1,2*I)=1./DX
              A(2*I-1,2*I+1)=0.5/DY
              A(2*I-1,2*I+2)=0.0
      C
              A(2*I,2*I-3)=0.0
              A(2*I,2*I-2)=-AMU/RHO/DY**2-0.5*V(I)/DY
              A(2*I,2*I-1)=0.5*(U(I+1)-U(I-1))/DY
              A(2*I,2*I)=(U(I)-UO(I))/DX+U(I)/DX+2.*AMU/RHO/DY**2
              A(2*I,2*I+1)=0.0
              A(2*I,2*I+2)=-AMU/RHO/DY**2+0.5*V(I)/DY
      C
              END IF
      C
              END DO
      C
      C-----------------------------------------------------
      C Solution for Increments
      C-----------------------------------------------------
      C
      C
              CALL GAUSS(A,F,NP,NN)
      C
      C-----------------------------------------------------
      C Update variables
      C-----------------------------------------------------
      C
              ERROR=0.0
              DO I=1,NY
              DV=F(2*I-1)
              IF(ABS(DV).GT.ERROR)ERROR=ABS(DV)
              IF(I.EQ.1)DV=0.0
              V(I)=V(I)-DV
              DU=F(2*I)
              IF(ABS(DU).GT.ERROR)ERROR=ABS(DU)
              IF((I.EQ.1).OR.(I.EQ.NY))DU=0.0
              U(I)=U(I)-DU
              END DO
      C
              WRITE(6,*)' ITER=',ITER,'     ERROR= ', ERROR
      C
              IF(ERROR.GT.ERRMAX)GO TO 20
      C
      C-----------------------------------------------------
      C Print solution, advance Solution Downstream
      C-----------------------------------------------------
      C
              DO I=1,NY
              WRITE(6,*),I,Y(I),U(I),V(I)
```

```
              UO(I)=U(I)
              VO(I)=V(I)
              END DO
              WRITE(6,*)' ISTEP= ', ISTEP
              IF(ISTEP.LT.NX)GO TO 10
              STOP
              END
C
              SUBROUTINE GAUSS(A,B,N,NP)
C
C This subroutine was adapted from: Numerical Recipes
C
C
              PARAMETER (NMAX=5000)
              DIMENSION A(NP,NP),B(NP),IPIV(NMAX),INDXR(NMAX),
                  INDXC(NMAX)
C
              DO 11 J=1,N
              IPIV(J)=0
       11 CONTINUE
              DO 22 I=1,N
              BIG=0.
              DO 13 J=1,N
              IF(IPIV(J).NE.1)THEN
              DO 12 K=1,N
              IF (IPIV(K).EQ.0) THEN
              IF (ABS(A(J,K)).GE.BIG)THEN
              BIG=ABS(A(J,K))
              IROW=J
              ICOL=K
              ENDIF
              ELSE IF (IPIV(K).GT.1) THEN
              PAUSE 'Singular matrix'
              ENDIF
       12 CONTINUE
              ENDIF
       13 CONTINUE
              IPIV(ICOL)=IPIV(ICOL)+1
              IF (IROW.NE.ICOL) THEN
              DO 14 L=1,N
              DUM=A(IROW,L)
              A(IROW,L)=A(ICOL,L)
              A(ICOL,L)=DUM
       14 CONTINUE
C
              DUM=B(IROW)
              B(IROW)=B(ICOL)
              B(ICOL)=DUM\
C
              ENDIF
              INDXR(I)=IROW
```

```
      INDXC(I)=ICOL
      IF (A(ICOL,ICOL).EQ.0.) PAUSE 'Singular matrix.'
      PIVINV=1./A(ICOL,ICOL)
      A(ICOL,ICOL)=1.
      DO 16 L=1,N
      A(ICOL,L)=A(ICOL,L)*PIVINV
16    CONTINUE
C
      B(ICOL)=B(ICOL)*PIVINV
C
      DO 21 LL=1,N
      IF(LL.NE.ICOL)THEN
      DUM=A(LL,ICOL)
      A(LL,ICOL)=0.
      DO 18 L=1,N
      A(LL,L)=A(LL,L)-A(ICOL,L)*DUM
18    CONTINUE
C
      B(LL)=B(LL)-B(ICOL)*DUM
C
      ENDIF
21    CONTINUE
22    CONTINUE
      DO 24 L=N,1,-1
      IF(INDXR(L).NE.INDXC(L))THEN
      DO 23 K=1,N
      DUM=A(K,INDXR(L))
      A(K,INDXR(L))=A(K,INDXC(L))
      A(K,INDXC(L))=DUM
23    CONTINUE
      ENDIF
24    CONTINUE
      RETURN
      END
```

Basic Mathematics of Fluid Dynamics

E.1 Scalars

In fluid dynamics, a number of parameters are quantitatively defined by providing a single numerical value. For example, the volume of an amount of liquid is fully described by providing the numerical value of the volume in appropriate units (in the SI system of units, volume is measured in m^3). Other examples of such parameters are the *area, density, mass*, and *energy*. Mathematically, such quantities are known as *scalar* quantities. The numerical value describing the scalar is the *magnitude* of the scalar. Scalar quantities can be manipulated mathematically using the standard arithmetic rules of addition, subtraction, multiplication, and division.

E.2 Vectors and Vector Algebra

In contrast to scalar quantities, some parameters, in addition to their magnitude, are also associated with a certain direction relative to an appropriate coordinate system. Such quantities are known as *vectors*. Typical vector quantities are the position vector, the velocity vector, and acceleration vector. Mathematically, the vector is represented by a straight line whose length is the numerical value of the quantity it represents.

For example, a fluid particle with velocity $10\ m/s$ in the direction shown in Figure E.1 can be represented as a vector using the coordinate system shown in the figure. The direction of the velocity vector can be defined using the angle the vector makes with respect to each coordinate direction as

$$\alpha = \cos^{-1}\frac{u_x}{|\mathbf{u}|} \text{ or } \alpha = \sin^{-1}\frac{u_y}{|\mathbf{u}|} \text{ or } \alpha = \tan^{-1}\frac{u_y}{u_x}.$$

Here $|u|$ represents the magnitude of the vector, which in our example is $10\ m/s$ and u_x, u_y represent the projected lengths along each coordinate direction. Using the geometry of the figure, the magnitude of the vector can be expressed in terms of u_x and u_y as

$$|\mathbf{u}| = \sqrt{u_x^2 + u_y^2}$$

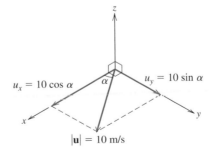

FIGURE E.1 Vector presentation.

Obviously, then, both the direction of the vector and its magnitude can be uniquely defined using the components of the vector in each direction. Therefore, any vector **A** can be defined as

$$\mathbf{A} = a_x\mathbf{i} + a_y\mathbf{j} + a_z\mathbf{k}.$$

A *free vector* is defined as a vector that can be displaced parallel to itself while maintaining the same magnitude and while it is not applied to any point. If a vector is confined to move only in the direction of itself, it is defined as a *sliding vector*. A *bound vector* is a vector that is fixed in space and cannot move.

E.2.1 Coordinate Systems

The components of the vector in Figure E.1 are represented as the projections of the vector on three mutually perpendicular axes. These three axes are sufficient to express all possible three-dimensional vectors. This coordinate system is known as the *Cartesian coordinate system*. As shown in Figure E.2, each axis represents an independent coordinate direction indicated by a *unit vector* for instance, **i**, **j**, **k**.

Vector quantities can also be expressed using the *cylindrical coordinate system* shown in Figure E.3. As shown in the figure, unit vectors in this coordinate system are specified along the radial direction $\mathbf{e}_r$, the azimuthal direction $\mathbf{e}_\theta$ (measured in the counterclockwise direction), and in the direction $\mathbf{e}_z$, which is the direction normal to the plane formed by $\mathbf{e}_r$ and $\mathbf{e}_\theta$. A vector A, then, in cylindrical coordinates is given in terms of its components as

$$\mathbf{A} = A_r\mathbf{e}_r + A_\theta\mathbf{e}_\theta + A_z\mathbf{e}_z.$$

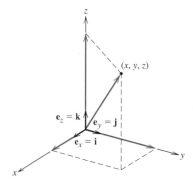

FIGURE E.2 Cartesian coordinate system.

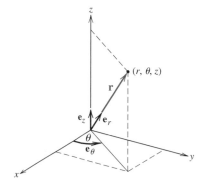

FIGURE E.3 Cylindrical coordinate system.

A vector expressed in Cartesian coordinates can be expressed in terms of cylindrical coordinates using the simple geometric relations:

$$x = r \cos \theta$$

$$y = r \sin \theta$$

$$z = z$$

Also, a vector expressed in cylindrical coordinates can be expressed in terms of Cartesian coordinates using the equivalent relations

$$z = z$$

$$r^2 = x^2 + y^2$$

$$\theta = \tan^{-1} \frac{y}{x}$$

Vectors can also be expressed using the *spherical coordinate systems* shown in Figure E.4. As shown in the figure, unit vectors in this coordinate system are specified along the radial direction $\mathbf{e}_r$ and the angular directions $\mathbf{e}_\theta$ and $\mathbf{e}_\phi$, as shown in the figure. A vector A then in spherical coordinates is given in terms of its components as

$$\mathbf{A} = A_r \mathbf{e}_r + A_\theta \, \mathbf{e}_\theta + A_\phi \, \mathbf{e}_\phi.$$

A vector expressed in Cartesian coordinates can be expressed in terms of spherical coordinates using the simple geometric relations:

$$x = r \sin \phi \cos \theta$$

$$y = r \sin \phi \sin \theta$$

$$z = r \cos \phi.$$

E.2.2   Vector Multiplication by a Scalar

A vector $\mathbf{A}$ multiplied by a scalar m is equivalent to multiplying each component by m:

$$m\mathbf{A} = ma_x \mathbf{i} + ma_y \mathbf{j} + ma_z \mathbf{k}.$$

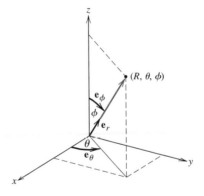

FIGURE E.4   Spherical coordinate system.

Vector Addition and Vector Subtraction

Two vectors **A** and **B** can be added together by simply adding the components of the vectors in each direction as

$$\mathbf{A} + \mathbf{B} = (a_x + b_x)\mathbf{i} + (a_y + b_y)\mathbf{j} + (a_z + b_z)\mathbf{k}.$$

Schematically, this is shown in Figure E.5. The two vectors are subtracted **A** − **B** by multiplying **B** by −1 and then adding the two vectors. Therefore,

$$\mathbf{A} - \mathbf{B} = (a_x - b_x)\mathbf{i} + (a_y - b_y)\mathbf{j} + (a_z - b_z)\mathbf{k}.$$

Schematically, this is shown in Figure E.6. Notice that multiplying a vector by −1 is equivalent to simply reversing the direction of the vector while preserving the magnitude.

E.2.3   Scalar, Vector-Vector Operations

A special operation between two vectors **A**, **B** known as the *dot product* or otherwise as the *scalar product* is defined as

$$\mathbf{A} \cdot \mathbf{B} = a_x b_x + a_y b_y + a_z b_z = |\mathbf{A}||\mathbf{B}|\cos\alpha,$$

where α is the angle between the vectors. The dot product is *commutative* — that is,

$$\mathbf{A} \cdot \mathbf{B} = \mathbf{B} \cdot \mathbf{A}$$

and *distributive*

$$\mathbf{A} \cdot (\mathbf{B} + \mathbf{C}) = \mathbf{A} \cdot \mathbf{B} + \mathbf{A} \cdot \mathbf{C}.$$

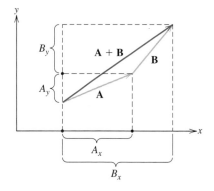

FIGURE E.5   Addition of vectors.

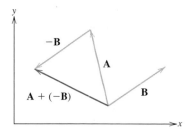

FIGURE E.6   Vector subtraction.

By further exploring the foregoing definitions, we can get some useful relations. First notice that if

$$\mathbf{A} \cdot \mathbf{B} = 0,$$

the two vectors are perpendicular to each other. If $\mathbf{B} = i$, then we have

$$\mathbf{A} \cdot \mathbf{i} = |\mathbf{A}| \cos \alpha = a_x.$$

Similarly,

$$\mathbf{A} \cdot \mathbf{j} = a_y \text{ and } \mathbf{A} \cdot \mathbf{k} = a_z.$$

Thus, the dot product with a unit vector along each direction yields the projection of the vector along the same direction.

E.2.4   Vector, Vector-Vector Operations

Another special operation between two vectors $\mathbf{A}$, $\mathbf{B}$ known as the *cross product* or otherwise as the *vector product*, is defined as

$$\mathbf{A} \times \mathbf{B} = \mathbf{C} = \begin{vmatrix} \mathbf{i} & \mathbf{j} & \mathbf{k} \\ a_x & a_y & a_z \\ b_x & b_y & b_z \end{vmatrix} = (a_y b_z - a_z b_y)\mathbf{i} + (a_z b_x - a_x b_z)\mathbf{j} + (a_x b_y - a_y b_x)\mathbf{k};$$

moreover,

$$|\mathbf{A} \times \mathbf{B}| = |\mathbf{A}||\mathbf{B}| \sin \alpha.$$

Again, α is the angle between the two vectors. The new vector $\mathbf{C}$ is perpendicular to the plane formed by $\mathbf{A}$ and $\mathbf{B}$, its direction is determined by the right-hand rule, and its magnitude is equal to the area of the parallelogram formed by the two vectors.

When

$$\mathbf{A} \times \mathbf{B} = 0,$$

$\mathbf{A}$ and $\mathbf{B}$ are parallel. The vector product is *not commutative* but, in fact, because of the definition of the direction on $\mathbf{C}$,

$$\mathbf{A} \times \mathbf{B} = -\mathbf{B} \times \mathbf{A}.$$

However, the vector product is *distributive* — that is,

$$\mathbf{A} \times (\mathbf{B} + \mathbf{C}) = \mathbf{A} \times \mathbf{B} + \mathbf{A} \times \mathbf{C}.$$

E.3   Elementary Calculus

A variable denoted by a non-numeric character — for example, x — represents a quantity that in a given range can take any numeric value. In most physical phenomena, certain observable quantities are defined by numerical values of the variables that characterize them. For example, the total mass, m, in kgs of any material is directly related to its volume $\mathcal{V}$ — that is,

$$m \propto \mathcal{V} \text{ or } m = \rho \mathcal{V},$$

where ρ is the density. Often, though, relationships of this sort are not so simple, and quantities may depend on a variable in more complicated ways. For example, if the material is a compressible gas, then the relationship can be more complicated, depending on the conditions of the gas. In such a case, we can generally write

$$m = f(\mathcal{V}),$$

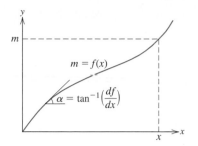

FIGURE E.7 Representation of a function.

which reads as *the mass is a function of the volume*. Therefore, f is a *function* representing the relationship between the independent variable (the volume $\mathcal{V}$) and the dependent variable (the mass m). This is due to the fact that the numerical value of the volume can vary over a given range; thus, it is independent. The mass, on the other hand, cannot arbitrarily assume any value but is uniquely defined by f and is, therefore, dependent.

The above argument can be generalized to consider quantities that are a function of a number of variables. For example, car performance p as measured by fuel consumption may depend on a number of variables (or factors) such as aerodynamic shape, speed of the car, the quality of the fuel type, and the condition of the engine, among other things. In such a case,

$$p = f(x, y, z, \ldots).$$

where $x, y, z, \ldots$ represent all variables that affect the performance.

Figure E.7 shows schematically a single variable function; for a given value of the independent variable x, there is a corresponding value of the dependent variable, m.

Differentiation

The derivative $\dfrac{df}{dx}$ represents the slope the tangent line makes with respect to the x-axis

$$\frac{df}{dx} = \tan \alpha.$$

Since f is function of x, then the derivative of f is also a function of x. The rules of differentiation can be found in elementary calculus textbooks, but it is worth repeating some of them here.

If $A = fg$, where f and g are both functions of x, then

$$\frac{dA}{dx} = f\frac{dg}{dx} + g\frac{df}{dx}.$$

If $A = f/g$, then

$$\frac{dA}{dx} = \frac{f\dfrac{dg}{dx} - g\dfrac{df}{dx}}{g^2}.$$

In the preceding arguments, x could also be a function of another variable as $x = g(t)$, where t is an independent variable. In such a case, if $A = f(x)$, we can then write $A = f(g(t))$. The derivative according to the *chain rule* is, then,

$$\frac{dA}{dt} = \frac{df}{dx}\frac{dx}{dt}.$$

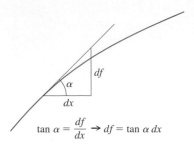

$$\tan \alpha = \frac{df}{dx} \rightarrow df = \tan \alpha\, dx$$

FIGURE E.8   Definition of the local slope.

By considering the geometry of Figure E.8, a small change in the function f, df can be expressed as

$$df = \tan \alpha\, dx = \frac{df}{dx} dx.$$

If $f = f(x_1, x_2, \ldots)$, then

$$df = \frac{\partial f}{\partial x_1} dx_1 + \frac{\partial f}{\partial x_2} dx_2 \ldots$$

The symbol $\partial / \partial x$, known as a *partial derivative*, was used to show that the derivative was taken with respect to a particular variable while the other variables were kept constant.

Example 1

The distance, x, traveled by an object falling under the influence of gravity (neglecting any air resistance) is given as

$$x = \frac{1}{2} g t^2.$$

The object is then falling with velocity u

$$u = \frac{dx}{dt} = gt.$$

The acceleration a is then

$$a = \frac{du}{dt} = g,$$

which is indeed a verification of the fact that the body is falling under the influence of gravity alone.

We can verify these results using the chain rule by rewriting x as

$$x = \frac{1}{2g} u^2$$

(by using $u = gt$) and differentiating with respect to time

$$u = \frac{dx}{dt} = \frac{dx}{du}\frac{du}{dt} = \frac{u}{g} g = u.$$

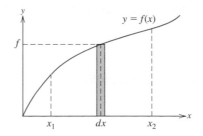

FIGURE E.9 Schematic definition of integration.

Integration

Integration is the opposite of differentiation. For example, given a function

$$f(x) = \frac{dy}{dx},$$

what is y? Otherwise, we can ask, what is function y which, once differentiated, yields $f(x)$?

Example 2

If $f(x) = \dfrac{dy}{dx} = 3x$, then obviously $y = \frac{3}{2}x^2 + C$, where C is a constant. The constant is there because the derivative of a constant is zero. Mathematically, this inverse operation is called *integration* and is expressed as

$$y = \int_{x_1}^{x_2} f(x)dx.$$

x_1 and x_2 are the *limits* of integration, and $f(x)$ is the *argument* of the integral. If the limits are not defined, then the integral is *indefinite*. If the limits are specified, the integral is *definite*.

In such case, by returning to Figure E.7 we see that the argument inside the integral sign represents an infinitesimal area. Therefore, the definite integral form x_1 to x_2 represents the area of the domain as shown in Figure E.9. Thus, as the name reveals, integration is really a summation of infinitesimal quantities over the range defined by the argument of integral. By definition, then, the rules of differentiation are the inverse of those of differentiation.

E.4 Vector Differential Calculus

Quite often in the study of fluid mechanics, we need to evaluate the vector quantity

$$\frac{\partial f}{\partial x}\mathbf{i} + \frac{\partial f}{\partial y}\mathbf{j} + \frac{\partial f}{\partial z}\mathbf{k},$$

where f is a scalar quantity. This vector differentiation (which is very important in fluid mechanics) is usually represented in a mathematically compact notation using the *del* or *nabla* operator ∇. Therefore,

$$\nabla = \frac{\partial}{\partial x}\mathbf{i} + \frac{\partial}{\partial y}\mathbf{j} + \frac{\partial}{\partial z}\mathbf{k}.$$

The ∇ operation applied on a scalar quantity f, as shown above, is known as the *gradient of a scalar*.

The *divergence of a vector*, $\mathbf{u}$, *div* $\mathbf{u}$, is a scalar quantity defined as

$$\nabla \cdot \mathbf{u} = \frac{\partial u_1}{\partial x} + \frac{\partial u_2}{\partial y} + \frac{\partial u_3}{\partial z}.$$

Following the definition, $\nabla \cdot (\nabla f) = \nabla^2 f$. This "second derivative" is the well known *Laplacian operator*.

The *curl of a vector*, $\mathbf{u}$, is a vector quantity defined as

$$\nabla \times \mathbf{u} = \mathbf{C} = \begin{vmatrix} \mathbf{i} & \mathbf{j} & \mathbf{k} \\ \frac{\partial}{\partial x} & \frac{\partial}{\partial y} & \frac{\partial}{\partial z} \\ u_x & u_y & u_z \end{vmatrix} = \left(\frac{\partial u_z}{\partial y} - \frac{\partial u_y}{\partial z} \right) \mathbf{i} + \left(\frac{\partial u_x}{\partial z} - \frac{\partial u_z}{\partial x} \right) \mathbf{j} + \left(\frac{\partial u_y}{\partial x} - \frac{\partial u_x}{\partial y} \right) \mathbf{k}$$

Following are some useful properties of the above operations:

1. $\nabla(fg) = f\nabla g + g\nabla f$
2. $\nabla\left(\frac{f}{g}\right) = \frac{1}{g^2}(g\nabla f - f\nabla g)$
3. $div(f\mathbf{u}) = \nabla \cdot (f\mathbf{u}) = f\nabla \cdot \mathbf{u} + \mathbf{u} \cdot \nabla f$
4. $\nabla \times (f\mathbf{u}) = \nabla f \times \mathbf{u} + f\nabla \times \mathbf{u}$

E.5   Lines, Surfaces, Volumes

The study of fluid mechanics is to a large extent the study and evolution of fluid lines, surfaces, and volumes. It is, therefore, quite appropriate to review here some fundamental mathematical principles related to these concepts.

Lines

An infinite number of mathematical points connected together form a *line*. By definition, then, a line, while having no width, extends in *length* only. Figure E.10 gives an example of a line.

We can then write $y = f(x)$. If $\frac{dy}{dx} = C$ where C is a constant number, then the line is *straight*. The equation of a straight line is then expressed as

$$y = C_0 + C_1 x,$$

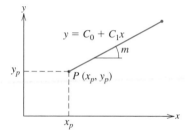

FIGURE E.10   Definition of a line.

where C_0 and C_1 are constants. To uniquely define a straight line, we need two conditions — for example, the slope $\dfrac{dy}{dx} = m$ and a single point on the line $P(x_p, y_p)$. Using m and P, we can readily show that $C_1 = m$ and $C_0 = y_p - mx_p$. Therefore, the equation of the line is given as

$$y = m(x - x_p) + y_p.$$

In general, though, lines may not be straight and $\dfrac{dy}{dx} = m(x) \neq constant$. The equation of the line is obviously given as

$$y = \int g(x)dx.$$

A useful way to describe a line is to introduce the arc-length, s, which is a parameter that follows the length of the line. Using the geometry of the line we can write

$$ds = \sqrt{dx^2 + dy^2} = dx\sqrt{1 + \left(\frac{dy}{dx}\right)^2} = dy\sqrt{1 + \left(\frac{dx}{dy}\right)^2}.$$

Since $y = f(x)$, then $x = g(y)$. Consequently $\dfrac{dy}{dx}$ is a function of x and $\dfrac{dx}{dy}$ is a function of y. Therefore, in principle we can integrate dy and dx to get

$$x = x(s) \quad y = y(s).$$

A *unit tangent vector*, $\mathbf{t}$, is a vector tangent to a line. This unit vector can be constructed at any point on the line by considering the local geometry (i.e., dx and dy) and by dividing by the length $\sqrt{dx^2 + dy^2}$

$$\mathbf{t} = \frac{dx\mathbf{i}}{\sqrt{dx^2 + dy^2}} + \frac{dy\mathbf{j}}{\sqrt{dx^2 + dy^2}} = \frac{dx}{ds}\mathbf{i} + \frac{dy}{ds}\mathbf{j}.$$

The *outward pointing unit normal vector*, $\mathbf{n}$ is the vector perpendicular to $\mathbf{t}$ *pointing in the outward direction*. The vector can be constructed by using the properties $\mathbf{t} \cdot \mathbf{n} = 0$ and $\mathbf{t} \times \mathbf{n} = \mathbf{k}$ as

$$\mathbf{n} = \frac{dy}{ds}\mathbf{i} - \frac{dx}{ds}\mathbf{j}.$$

Surfaces

A surface can be generated by a translation of a line in a direction lateral to its length. Mathematically, a surface is expressed as

$$z = z(x, y)$$

or, more generally, as

$$f(x, y, z) = C,$$

where C is a constant. For example,

$$z = \sqrt{a^2 - x^2 - y^2}$$

or

$$x^2 + y^2 + z^2 - a^2 = 0$$

represents a hemisphere or radius a. Like lines, a surface can be represented in a parametric form as

$$x = x(\xi, \eta) \quad y = y(\xi, \eta) \quad z = z(\xi, \eta),$$

where ξ and η represent appropriately chosen parameters (similar to the arc-length in the description of lines).

As a consequence of the geometry of the surface, at a particular point on the surface there is an infinite number of tangent vectors. Therefore, with surface we can identify a *tangent plane* rather a tangent vector. However, the *outward unit normal vector*, $\mathbf{n}$, is well defined and can be calculated as

$$\mathbf{n} = \frac{\nabla f}{|\nabla f|}.$$

The unit normal vector can also be obtained using the parametric representation in terms of the parameters ξ and η by forming the tangent vectors

$$\mathbf{r}_\xi = \frac{\partial x}{\partial \xi}\mathbf{i} + \frac{\partial y}{\partial \xi}\mathbf{j} + \frac{\partial z}{\partial \xi}\mathbf{k}\bigg|_{\eta = const}$$

$$\mathbf{r}_\eta = \frac{\partial x}{\partial \eta}\mathbf{i} + \frac{\partial y}{\partial \eta}\mathbf{j} + \frac{\partial z}{\partial \eta}\mathbf{k}\bigg|_{\xi = const}$$

Since these two vectors are tangent to the surface, we can form the unit normal vector as

$$\mathbf{n} = \frac{\mathbf{r}_\xi \times \mathbf{r}_\eta}{|\mathbf{r}_\xi \times \mathbf{r}_\eta|}.$$

Volumes

A *volume* quantity can be generated by a lateral translation of a two-dimensional surface. Of course, the concept of volume is pivotal in the study of fluid mechanics for obvious reasons. In practice, we often make use of the divergence theorem given below to switch from volume to surface integrals. In some special cases, we also need to invoke Stokes's theorem, which is presented subsequently.

E.5.1 Gauss Divergence Theorem

Let V be a closed bounded region with a piecewise smooth surface S and a vector function $\mathbf{F}$ that is continuous and has continuous first derivatives in V; then

$$\iiint_V \nabla \cdot \mathbf{F}\, dV = \iint_S \mathbf{F} \cdot \mathbf{n}\, dS,$$

where $\mathbf{n}$ is the outward unit normal vector. If the vector $\mathbf{F}$ is constructed from the gradient of a scalar f as $\mathbf{F} = \nabla f$, then the divergence can be expressed as

$$\iiint_V \nabla^2 f\, dV = \iint_S \nabla f \cdot \mathbf{n}\, dS.$$

E.5.2 Stokes's Theorem

Let S be a piecewise smooth surface in space enclosed by a smooth simple curve C. If $\mathbf{F}$ is a continuous vector with continuous first derivatives in S, then

$$\iint_S (\nabla \times \mathbf{F}) \cdot \mathbf{n}\, dA = \oint_C \mathbf{F} \cdot \mathbf{t}\, ds,$$

where **n** is the outward unit normal vector and **t** is the tangent vector along C moving in the counterclockwise direction.

E.6 Tensors and Tensor Algebra

Tensors are a generalization of vectors in the form

$$\mathcal{A} = \begin{bmatrix} A_{11} & A_{12} & A_{13} \\ A_{21} & A_{22} & A_{23} \\ A_{31} & A_{32} & A_{33} \end{bmatrix}.$$

A component of the tensor $\mathcal{A}$ then is given by A_{ij}, where subscript i represents the row location and subscript j is the column location. The indices i, j range from $1, \ldots .n$, where n is the maximum number of rows and columns or, as otherwise known, the order of the tensor. In the preceding example, $\mathcal{A}$ is a third-order tensor.

A tensor $\mathcal{A}$ multiplied by a scalar m is equivalent to multiplying each component by the same scalar — that is,

$$m\mathcal{A} = mA_{ij}.$$

Two tensors $\mathcal{A}$ and $\mathcal{B}$ are added or subtracted by adding/subtracting their corresponding components — that is,

$$\mathcal{A} \pm \mathcal{B} = A_{ij} \pm B_{ij}.$$

In fluid mechanics, typical tensors are the stress tensor σ expressed as

$$\sigma = \begin{bmatrix} \sigma_{11} & \sigma_{12} & \sigma_{13} \\ \sigma_{12} & \sigma_{22} & \sigma_{23} \\ \sigma_{13} & \sigma_{23} & \sigma_{33} \end{bmatrix} = \begin{bmatrix} -P + \tau_{11} & \tau_{12} & \tau_{13} \\ \tau_{12} & -P + \tau_{22} & \tau_{23} \\ \tau_{13} & \tau_{23} & -P + \tau_{33} \end{bmatrix},$$

where indices $1, 2, 3$ correspond to the Cartesian directions x, y, z. P is the hydrostatic pressure and τ the viscous stress tensor.

Tensors often appear with vectors in various operations. Depending on the operation, the result of such operations could be a vector or a tensor. The most common vector-tensor operation found in fluid mechanics is the product

$$\mathbf{T}_r = \mathbf{n} \cdot \sigma,$$

where $\mathbf{T}_r$ is the traction vector (force per unit area) on a surface with outward pointing unit normal vector **n** because of the local stress state denoted by σ. In index form, the product is obtained by $T_{ri} = \sigma_{ij} n_j$. In component form, the traction vector is given as

$$\mathbf{T}_r = (\sigma_{11} n_1 + \sigma_{12} n_2 + \sigma_{13} n_3)\,\mathbf{i} + (\sigma_{21} n_1 + \sigma_{22} n_2 + \sigma_{23} n_3)\,\mathbf{j} + (\sigma_{31} n_1 + \sigma_{32} n_2 + \sigma_{33} n_3)\,\mathbf{k}.$$

REFERENCES

E. KREYSZIG, *Advanced Engineering Mathematics*, 6th ed., New York, Wiley, 1988.

F. HILDEBRAND, *Advanced Calculus for Applications*, 2nd ed., Englewood Cliffs, NJ, Prentice Hall, 1976.

W.A. SMITH, "Elementary Numerical Analysis", Harper and Row, New York, 1979.

A.I. BORISENKO, and I.E. TAPAROV, *Vector and Tensor Analysis with Applications*, Translated from Russian by A. Silverman, New York, Dover, 1979.

R. ARIS, *Vectors, Tensors and the Basic Equations of Fluid Mechanics*, New York, Dover, 1989.

C.A.J. FLETCHER, *Computational Techniques for Fluid Mechanics*, vol. I, *Fundamental and General Techniques*, 2nd ed., Berlin, Germany, Springer Verlag, 1990.

Index